MW00910196

Papermaking Science and Technology

a series of 19 books
covering the latest
technology and
future trends

Book 6B

Chemical Pulping

Series editors
Johan Gullichsen, Helsinki University of Technology
Hannu Paulapuro, Helsinki University of Technology

Book editors
Johan Gullichsen, Helsinki University of Technology
Carl-Johan Fogelholm, Helsinki University of Technology

Series reviewer
Brian Attwood, St. Anne's Paper and Paperboard Developments, Ltd.

Book reviewers
Martin MacLeod, PAPRICAN
Desmond Smith, Acrowood
Bill Fuller, Weyerhaeuser
Hongi Tran, Pulp & Paper Centre, University of Toronto
Norman Duke, Avenor

Published in cooperation with the Finnish Paper Engineers' Association and TAPPI

ISBN 952-5216-00-4 (the series)
ISBN 952-5216-06-3 (book 6)

Published by Fapet Oy
(Fapet Oy, PO BOX 146, FIN-00171 HELSINKI, FINLAND)

Printed by Gummerus Printing, Jyväskylä, Finland 2000

Printed on LumiMatt 100 g/m^2, Stora Enso Fine Paper, Imatra Mill

Certain figures in this publication have been reprinted by permission of TAPPI.

Table of Contents

BOOK A

BOOK B

Table of Contents

Table of Contents

CHAPTER 11

Chemical recovery

CHAPTER 11

Esa Vakkilainen

Chemical recovery

1 Overview of kraft recovery

Washing separates spent cooking chemicals and dissolved organics from pulp. The initial procedure was to discard this black, alkaline liquor. Chemical recovery systems were available, but their use did not become widespread until the 1930s and 1940s with the modern regeneration of spent liquor. Development of new equipment and an increase in mill size made it more economical to process black liquor than to buy new chemicals.

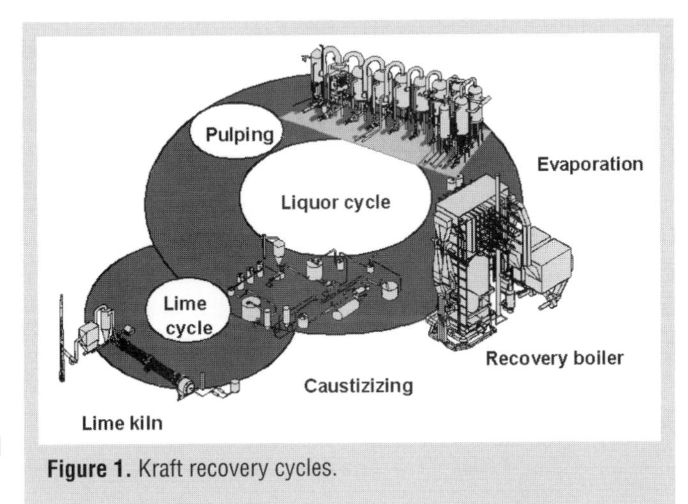

Figure 1. Kraft recovery cycles.

Recovery of black liquor has several advantages. Incineration of concentrated black liquor releases energy to generate steam and electricity. Regeneration of energy in contemporary mills exceeds internal needs. Mills therefore produce excess energy from their own waste.

1.1 Kraft recovery unit operations

Figure 1 shows the principal unit operations of the kraft recovery process:

- Evaporation of black liquor
- Combustion of black liquor in a recovery furnace to form sodium sulfide and sodium carbonate
- Causticizing sodium carbonate to sodium hydroxide
- Regeneration of lime mud in a lime kiln.

Other minor operations ensure the continuous operation of the recovery cycle. Removal of soap in the black liquor produces tall oil. Adding makeup chemicals such as sodium sulfate or sodium carbonate to a mixing tank or removing recovery boiler fly ash controls the balance of sodium to sulfur. Disposal of dregs and grits prevents accumulation of nonprocess compounds. Combustion processes odorous gases. Most modern or closed cycle mills use chlorine and potassium removal processes. With additional closure, new internal chemical manufacturing methods are sometimes necessary.

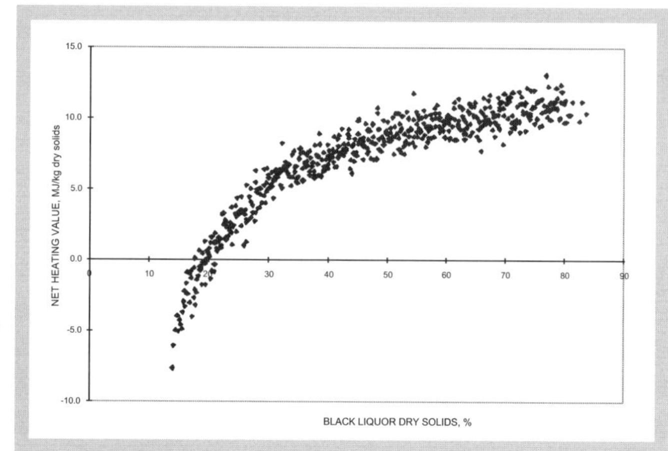

Figure 2. Net heating value of typical kraft liquor at various dry solids content.

1.1.1 Evaporation

The aim of evaporation is to produce black liquor of sufficiently high concentration with minimum chemical losses. Washing separates pulp and black liquor. The resulting weak black liquor contains 12%–20% organic and inorganic solids. Figures 2 and 3 show that burning this weak black liquor would require more heat than it would produce. The black liquor must therefore undergo concentration for efficient energy recovery.

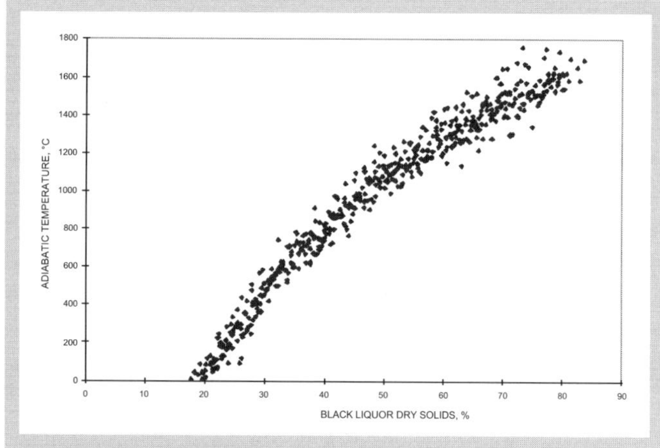

Figure 3. Adiabatic combustion temperatures of typical kraft liquors vs. dry solids content.

The evaporation of black liquor in Fig. 4 has three principal unit operations:

- Separation of water from black liquor to generate concentrated black liquor and condensate
- Processing of condensate to segregate clean and fouled condensate fractions
- Separation of soap from black liquor.

A liquor processing stage such as a liquor heat treatment (LHT) unit can also be present to decrease black liquor viscosity. Another possible component could be a black liquor oxidation stage to decrease total reduced sulfur (TRS) emissions. In modern high solids evaporators, the mixing of recovery boiler electrostatic precipitator (ESP) ash occurs with 30%–45% dry solids black liquor. Noncondensable gases from evaporation require collection for processing. When using a direct contact evaporator stage, efficient oxidation of black liquor is necessary to suppress release of odorous gases into the flue gas stream.

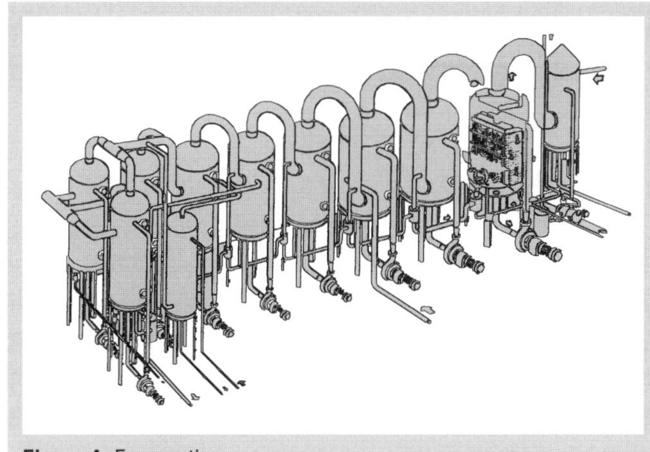

Figure 4. Evaporation.

Evaporation of black liquor uses direct or indirect heating and flashing of black liquor. Most industrial evaporators are the multiple effect, steam heated type. Vapor recompression evaporation is often a component in the first stage of evaporation of weak black liquor as a capacity booster.

Small mills use direct contact evaporation especially when processing nonwood black liquors. Hot flue gas from the recovery boiler heats a film or spray of black liquor. This technique can only evaporate to a maximum 65% dry solids content due to the steeply increasing liquor viscosity at higher dry solids contents. Unoxidized black liquor releases organic sulfur compounds on contact with flue gases. Oxidation of weak black liquor can partly avoid this. Economics favor the installation of indirect heating as unit size increases. Then all flue gas heat generates steam and electricity.

1.1.2 Recovery boiler

Concentrated black liquor contains organic dissolved wood residues and inorganic cooking chemicals. Combustion of the organic portion of liquor produces heat. In the recovery boiler, heat produces high pressure steam that generates electricity and low pressure steam for process use.

Combustion in the recovery furnace needs careful control. High concentration of sulfur requires optimum process conditions to avoid production of sulfur dioxide and reduced sulfur gas emissions. Besides

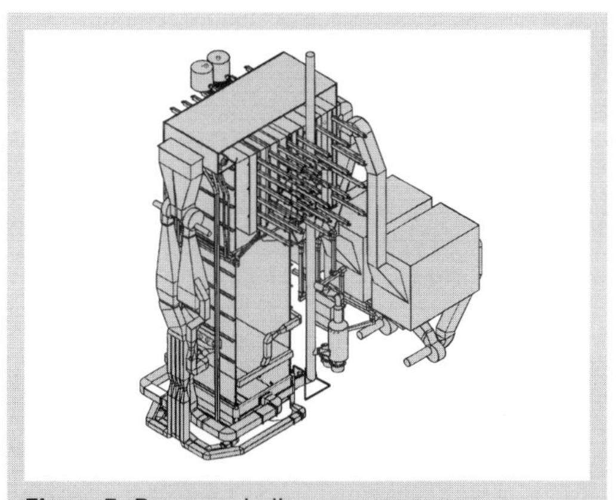

Figure 5. Recovery boiler.

environmentally clean combustion, efficient reduction of inorganic sulfur must occur in the char bed.

The process of the recovery boiler in Fig. 5 includes several unit processes:

- Combustion of organic material in black liquor to generate steam
- Reduction of inorganic sulfur compounds to sodium sulfide
- Production of molten inorganic flow consisting primarily of sodium carbonate and sodium sulfide
- Recovery of inorganic dust from flue gas
- Production of a sodium fume to capture combustion residues of released organic sulfur compounds.

1.1.3 Lime kiln

A lime kiln calcines lime mud to reactive lime (CaO) by drying and subsequent heating as Fig. 6 shows. The calcining process can use a rotary furnace or a fluidized bed reactor.

The main unit processes of the lime kiln are the following:

- Drying of lime mud
- Calcining of calcium carbonate

Some additional operations can also be present. The lime kiln combusts small amounts of odorous noncondensable gases. The lime kiln process produces dust that requires capture. For larger amounts of oxidized sulfur gases, flue gas scrubbers are necessary.

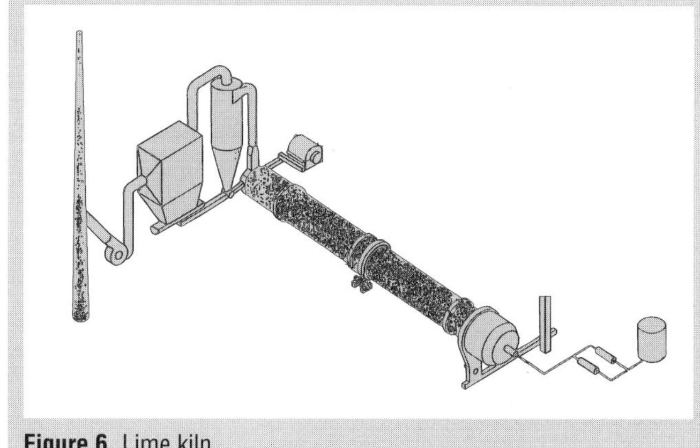

Figure 6. Lime kiln.

The drying of lime mud and the calcining of calcium carbonate to calcium oxide require heat. This heat comes from burning oil or natural gas in the lime kiln. The lime kiln can also use other fuels such as gasified biomass.

1.1.4 Causticizing

The causticizing process in Fig. 7 converts sodium carbonate in green liquor to caustic soda.

The unit operations in causticizing include:

- Dissolving of molten smelt to weak white liquor to produce green liquor

- Green liquor clarification or filtration
- Mixing green liquor and lime in a slaker to form sodium hydroxide and lime mud with subsequent completion of the causticizing reaction in reaction tanks
- White liquor clarification and filtration for lime mud separation
- Lime mud washing.

Molten smelt from the recovery boiler contains small amounts of unreacted carbon and nonprocess elements. The small undissolved particles in green liquor require separation for disposal. Separation can use settling or filtration. Washing the dregs minimizes chemical losses.

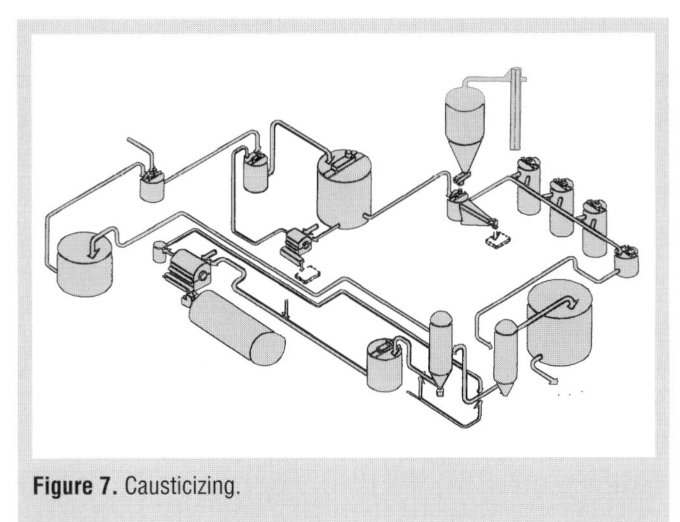

Figure 7. Causticizing.

1.2 Gasification

Gasification provides a way to convert solid fuel to a combustible gas. Use of combined cycle power generation where a gas turbine and a steam process produce electricity increases energy efficiency. Higher electricity conversion efficiency and high unit cost of the recovery boiler are the primary driving forces for development of new processes to replace recovery boilers.

Development of more efficient processes would offset the high unit cost of gasification. The recovery boiler has several weaknesses from an energy efficiency standpoint. The steam temperature and pressure are low, the energy of the smelt is not recovered, and the combustion temperature is low. Higher steam temperature and pressures could be possible if the product gas had proper cleaning. Recovery of energy of the smelt is possible if dissolving occurs under pressure. Higher combustion temperatures are possible when using oxygen instead of air in gasification.

From a process viewpoint, the recovery boiler also has several significant weaknesses. Recovery boiler downtime and accidents are expensive, and they limit pulp production. In reductive and oxidative processes, the possibility of smelt water explosions and the corrosive process media cause high unit costs and require extra safety features. Although the modern recovery boiler is very environmentally friendly, new processes are under development to reduce emissions of NOx, SOx, and particles.

All the leading recovery boiler manufacturers are doing research on gasification. Several obstacles are hindering their progress. Energy conversion always entails extra losses. Cleaning gases is expensive. Reduction requires extra process equipment. Table 1 shows that all commercial processes have less energy efficiency than conventional recovery boilers.

Table 1. Effectiveness of converting black liquor with high heating value (HHV) into fuel value in net product gas[1].

Original HHV, MJ/kgbl	14.7 HHV in net gas % of original	13.3 HHV in net gas % of original
Commercial processes		
Chemrec (Kvaerner)	55.9	50.7
StoneChem (MTCI)	49.0	42.6
Piloted processes		
ABB	70.4	66.6
Tampella	60.7	55.9
Conventional RB	61.1	58.6

A totally practical gasification process is still a long way from becoming commercially feasible.

1.3 Direct alkali recovery

Other processes can replace the conventional evaporator recovery boiler in the causticizing process of the lime kiln cycle. Because the number of new process stages is high and heavy investment is necessary, only few processes have evolved to the mill scale.

Australian Paper has processed liquor from soda pulping in a fluidized bed with ferric oxide[2]. Water dissolves the resulting sodium ferrite to form sodium hydroxide and ferric oxides for recovery and reuse of the ferrous salt. The mill has accumulated several years of operational experience with this process.

The Fredericia mill used straw to make pulp for various applications. Despite the partially successful recovery operation, the mill closed because of other economic reasons.

The development of the Direct Alkali Recovery System (DARS) process is slow because using the fluidized bed with subsequent pelletizing and leaching operations are unfamiliar to typical pulp mill personnel.

2 Properties of kraft recovery process streams

Process calculations require property estimations of kraft recovery process streams. Individual operating practices, equipment differences, streams from bleaching chemical preparation, and fresh chemical makeup cause variations in process stream composition. When designing equipment or considering process changes, one should make every effort to measure all relevant property data. Solid fuel properties are given in Chapter 15.

2.1 Black liquor properties

Black liquor properties depend on the raw materials used for pulping, the pulping conditions, the equipment used for pulping, and the treatment of the liquor after pulping. The largest supply of raw material for pulp production is definitely wood with a combination of softwood and hardwood finding frequent use. Interest in the use of other fibrous raw materials has revived. Mills in India, southeast Asia, and South America use bagasse, straw, reed, and bamboo.

The main variables in the conditions are the concentrations of the different chemicals in the cooking liquor, the chemical charge per weight of wood, the liquor-to-wood ratio, the cooking temperature, and the length of the cook.

Black liquor properties are not constant. They change as liquor flows from the digesters, pulp washing, evaporation, soap skimming, and storage. In optional processes, oxidation and liquor heat treatment have the greatest impact on liquor properties. Black liquor viscosity, boiling point rise, and heating value will especially vary.

2.1.1 Composition of black liquor

Black liquor contains water, organic residue from pulping, and inorganic cooking chemicals. The primary organic compounds are lignin, polysaccharides, carboxylic acids, and extractives. Table 2 gives a typical analysis of kraft black liquor[3].

Table 2. Composition of black liquor dry solids from kraft pulping of birch[3].

Organics, % by weight	**78**
Degraded lignin, including Na and S, %	37.5
Isosaccharinic acids, including Na, %	22.6
Aliphatic acids, including Na, %	14.4
Resin and fatty acids, including Na, %	0.5
Polysaccharides, %	3.0
Inorganics, % by weight	**22**
NaOH, %	2.4
NaHS, %	3.6
Na_2CO_3 and K_2CO_3, %	9.2
Na_2SO_4, %	4.8
$Na_2S_2O_3$, Na_2SO_3 and Na_2S_x, %	0.5
NaCl, %	0.5
Nonprocess elements (Si, Ca, Fe, Mn, Mg, etc.), %	0.2

The proximate analysis of black liquor determines the main components necessary for combustion calculations. It is used in typical mill recovery boiler calculations. Evaporator evaluations need boiling point rise and viscosity.

Proximate analysis usually consists of HHV, dry solids content of black liquor, nonreactive material in black liquor dry solids, the inert material in organics vs. inorganics, and the ratio of reactive to nonreactive portions in organics vs. inorganics.

The advantage of proximate analysis is that it is simple to perform. No complicated equipment is necessary. Proximate analysis can estimate black liquor ultimate analysis and combustion properties.

2.1.2 Elementary analysis of black liquor dry solids

An elementary analysis provides the main chemical components in the black liquor. Another name for the elementary analysis is the ultimate analysis. For black liquor recovery boiler material and energy balance calculations, the elementary analysis of the dry solids in the black liquor should include the weight fraction or percentage of the following components:

- C, carbon
- H, hydrogen
- O, oxygen
- Na, sodium
- K, potassium
- S, sulfur
- Cl, chlorine
- Inert.

The oxygen content is often determined as the difference between the total and the analyzed components. Inert includes minor solid components such as the following materials:

- N, nitrogen
- Ca, calcium
- F, fluorine
- Fe, iron
- Al, aluminum
- Mg, magnesium
- P, phosphorus
- V, vanadium
- Si, silica.

Some minor compounds are important. Nitrogen influences the recovery boiler NOx emissions. Calcium, aluminum, and silica form scales in evaporators. Vanadium influences the corrosion rate at high temperatures. Fluorine is a hazardous chemical. Tables 3–6 show typical elementary compositions for different raw materials.

Table 3. Typical composition of virgin black liquor from Scandinavian wood.

	Softwood (pine)		Hardwood (birch)	
	Typical	Range	Typical	Range
Carbon, %	35.0	32–37	32.5	31–35
Hydrogen, %	3.6	3.2–3.7	3.3	3.2–3.5
Nitrogen, %	0.1	0.06–0.12	0.2	0.14–0.2
Oxygen, %	33.9	33–36	35.5	33–37
Sodium, %	19.0	18–22	19.8	18–22
Potassium, %	2.2	1.5–2.5	2.0	1.5–2.5
Sulfur, %	5.5	4–7	6.0	4–7
Chlorine, %	0.5	0.1–0.8	0.5	0.1–0.8
Inert, %	0.2	0.1–0.3	0.2	0.1–0.3
Total, %	100.0		100.0	

Table 4. Typical composition of virgin black liquor from North American wood.

	Softwood (pine)		Hardwood	
	Typical	Range	Typical	Range
Carbon, %	35.0	32–37.5	34.0	31–36.5
Hydrogen, %	3.5	3.4 –4.3	3.4	2.9–3.8
Nitrogen, %	0.1	0.06–0.12	0.2	0.14–0.2
Oxygen, %	35.4	32 –38	35.0	33–39
Sodium, %	19.4	17.3–22.4	20.0	18–23
Potassium, %	1.6	0.3–3.7	2.0	1–4.7
Sulfur, %	4.2	2.9–5.2	4.3	3.2–5.2
Chlorine, %	0.6	0.1–3.3	0.6	0.1–3.3
Inert, %	0.2	0.1–2.0	0.5	0.1–2.0
Total, %	100.0		100.0	

Table 5. Typical composition of virgin black liquor from tropical wood.

	Hardwood (eucalyptus)		Mixed tropical wood	
	Typical	Range	Typical	Range
Carbon, %	34.8	33–37	35.2	34–37
Hydrogen, %	3.3	2.7–3.9	3.6	3.1–4.2
Nitrogen, %	0.2	0.1–0.6	0.3	0.1–0.9
Oxygen, %	35.5	33–39	35.5	33–39
Sodium, %	19.1	16.2–22.2	18.8	16.5–22.5
Potassium, %	1.8	0.4–9.2	2.3	0.5–6.3
Sulfur, %	4.1	2.4–7.0	3.0	2.4–5.0
Chlorine, %	0.7	0.1–3.3	0.8	0.5–2.4
Inert, %	0.5	0.2 –3.0	0.5	0.2–3.3
Total, %	100.0		100.0	

Table 6. Typical composition of virgin black liquor from miscellaneous pulping materials.

	Bagasse Typical	Bamboo Typical	Straw Typical
Carbon, %	36.9	34.5	36.5
Hydrogen, %	3.9	3.3	3.9
Nitrogen, %	0.3	0.4	0.7
Oxygen, %	36.3	34.1	33.9
Sodium, %	18.6	18.3	17.5
Potassium, %	0.6	4.0	2.1
Sulfur, %	2.5	3.3	2.8
Chlorine, %	0.4	1.6	2.1
Inert, %	0.5	0.5	0.5
Total, %	100.0	100.0	100.0

2.1.3 Heating value

The heating value of fuel signifies the amount of heat that fuel can produce upon incineration. *HHV* reflects the maximum chemical energy released. Measuring the *HHV* of a fuel sample involves oxidization in a special reaction vessel called a calorimeter. All water vapor produced in chemical reactions condenses. Products are cooled to reference temperature and pressure. The chemicals in ash are fully oxidized species (Na_2CO_3 and Na_2SO_4) in the solid state at the test temperature. The gas contains only noncondensable combustibles such as CO_2 and SO_2 with some water vapor. This heating value is the calorimetric or gross heating value. *HHV* is the normal term for the heating value of the black liquor dry solids.

Table 7. Typical *HHV*.

	Typical		Range	
	MJ/kg	Btu/lb	MJ/kg	Btu/lb
Nordic softwood	14.2	6 100	13.3–14.8	5700–6350
Nordic hardwood	13.5	5 800	13.0–14.3	5550–6150
North American softwood	14.2	6 100	13.3–15.0	5700–6450
North American hardwood	13.9	5 975	13.0–14.8	5550–6350
Tropical hardwood	14.1	6 050	13.4–14.8	5750–6350
Bagasse	14.8	6 350		
Bamboo	14.1	6 050		
Straw	14.7	6 325		

Table 7 shows typical heating values for different raw materials. The *HHV* depends strongly on the carbon content of the dry solids[4]. The following formula came from analysis of approximately 500 different black liquors. It can estimate the heating value with a known carbon content:

$$HHV = 29.35(C_C) + 3.959 \pm 0.42 \quad (1)$$

where *HHV* is the higher heating value of kraft black liquor, MJ/kg dry solids
C_C the carbon content of the dry solids, kg C/kg dry solids.

Figure 8 shows this relationship. The equation is accurate for northern softwoods and hardwoods within ± 0.4 MJ/kg dry solids.

Earlier equations developed for black liquors with high carbon content[5] give 39.27 and 0.2230 for respective coefficients of the *HHV* equation The equations give similar estimates for the *HHV* of black liquor when carbon content is 30%–40%.

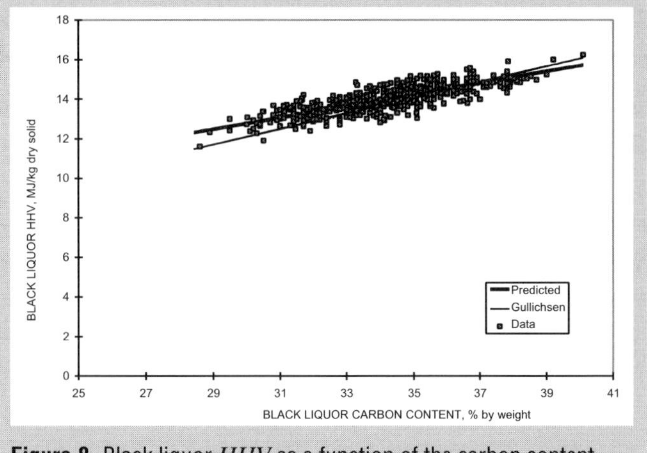

Figure 8. Black liquor *HHV* as a function of the carbon content.

If the levels of sodium, sulfur, and hydrogen contents are available, the equation becomes:

$$HHV = 25.04(C_C) + 0.1769(C_S) - 2.582(C_{Na}) + 48.92(C_H) + 42.31 \pm 0.41 \qquad (2)$$

where C_S is the sulfur content of the dry solids, kg S/kg dry solids
C_{Na} the sodium content of the dry solids, kg Na/kg dry solids
C_H the hydrogen content of the dry solids, kg H/kg dry solids.

Note that inclusion of S, Na, or H does not significantly increase the accuracy of *HHV* predictions. The ratio of carbon to hydrogen is almost constant. Most sulfur is inorganic and does not react during *HHV* determination. Poor correlation with sodium indicates that very little inorganic reactions occur.

Recovering heat from combustion of fuel indicated by its *HHV* requires condensation of all water vapor produced. For typical combustion applications, this is not possible. To indicate the effectiveness of typical boiler applications, the use of lower heating value *(LHV)* is common for oil, coal, and biofuel boilers. Calculation of *LHV* results by subtracting the heat of evaporation of water generated by combustion of hydrogen and water in the fuel from the *HHV*:

$$LHV = HHV - l_{25}[(M_{H_2O}/M_{H_2})C_H + (1 - X)/X] \qquad (3)$$

where LHV is the lower heating value, MJ/kg dry solids
HHV the higher heating value, MJ/kg dry solids
l_{25} the heat of evaporation of water at 25°C, MJ/kg (= 2.443)
C_H the hydrogen content of the dry solids, kg H/kg dry solids
X the dry solids concentration, kg dry solids/kg fuel
M_{H_2O} the molar mass of water, kg/kmol (= 18.015)
M_{H_2} the molar mass of hydrogen, kg/kmol (= 2.016).

In kraft recovery boilers, significant heat-requiring reduction reactions occur in ash. The heat release potential expression uses net heating value *(NHV)*. It reflects more accurately the heat available for steam generation. Calculation of the net heating value involves subtracting the heats of evaporation and the heat required for the reduction reaction from the *HHV*:

$$NHV = HHV - I_{25}[(M_{H_2O}/M_{H_2})C_H + (1-X)/X] - (78/32)Dh_R C_S \eta_{red} \quad (4)$$

where NHV is the net heating value of black liquor, MJ/kg dry solids
Dh_R the heat of reduction, MJ/kg (= 13.1 MJ/kg Na_2S)
C_S the sulfur content of the dry solids, kg S/kg dry solids
η_{red} the reduction efficiency in the smelt.

In mills where the black liquor heating value is unknown, one must assume it or calculate it using wood species, cooking conditions, etc. Useful estimation methods are available in Adams *et al.*[6] as Table 8 shows.

Table 8. Black liquor component heating values[6].

Component	MJ/kg	Btu/lb m
Softwood lignin	26.9	11.57
Hardwood lignin	25.11	10.8
Carbohydrates	13.555	5.83
Resins, fatty acids	37.71	16.22
Sodium sulfide	12.9	5.55
Sodium thiosulfate	5.79	2.49

2.1.4 Density

At low solids concentration, the density of the black liquor is near that of water. At higher solids concentrations, the density depends on the inorganic and organic materials that constitute the solids. A good approximation for the density to 50% dry solids is the following[7]:

$$\rho_{25} = 997 + 649X \quad (5)$$

where ρ_{25} is the black liquor density at 25°C, kg/m^3
X the dry solids concentration, kg dry solids/kg.
Increasing the temperature will decrease the black liquor density

$$\rho_T/\rho_{25} = 1.008 - 0.237t/1000 - 1.94(t/1000)^2 \quad (6)$$

where t is the black liquor temperature, °C.

An equation for the black liquor density based on experimental data is the following:

$$1/\rho_{90} = (1-X)/\rho_{H_2O,90} + 0.476X + 0.0911X(X\rho_{90})^{0.5} \quad (7)$$

where ρ_{90} is the black liquor density at 90°C, kg/dm^3
$\rho_{H_2O,90}$ the water density at 90°C, kg/dm^3
X the dry solids concentration, kg dry solids/kg

This equation applies to solids concentrations of 12%–68% and assumes that the liquor temperature is 90°C.

Figure 9 shows the liquor density as a function of dry solids concentration

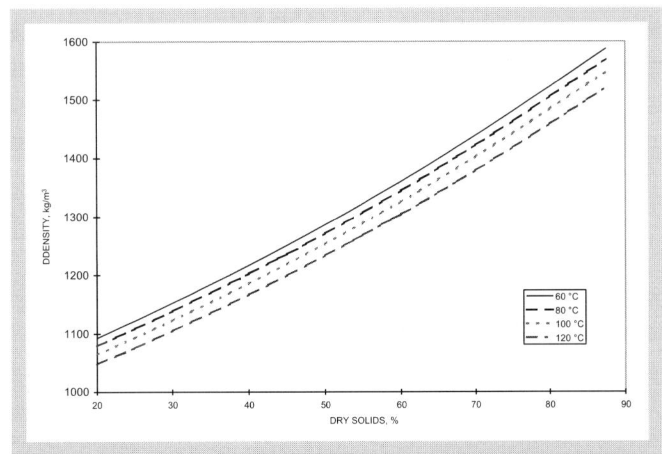

Figure 9. Black liquor density vs. dry solids concentration.

2.1.5 Specific heat capacity

The specific heat capacity of black liquor depends on the heat capacities of the constituents. Any increase in dry solids will decrease specific heat capacity, and an increase in temperature will decrease it.

An empirical formula to estimate the effect of dry solids on the specific heat capacity is the following:

$$c_{p,X}/c_{p,ref} = 1-(X-X_{ref})/(2.14-X_{ref}) \quad (8)$$

where $c_{p,X}$ is the black liquor specific heat, kJ/kg • °C
$c_{p,ref}$ the reference black liquor specific heat, kJ/kg • °C
X the dry solids concentration, kg dry solids/kg
X_{ref} the reference dry solids concentration, kg dry solids/kg.

Increasing the temperature of the black liquor will decrease the black liquor specific heat. An empirical formula to estimate the effect of temperature to the specific heat capacity is the following:

$$c_{p,T}/c_{p,ref} = 1-(t-t_{ref})/(377-t_{ref}) \quad (9)$$

where $c_{p,T}$ is the black liquor specific heat, kJ/kg • °C
$c_{p,ref}$ the black liquor specific heat at reference temperature, kJ/kg • °C
t the temperature of the black liquor, °C
t_{ref} the reference temperature of the black liquor, °C.

The following equation by Masse *et al.* estimates the black liquor heat capacity[8]:

$$c_p = 4.216(1-X)+(1.675+(3.31t)/1000)X+(4.87-20t/1000)(1-X)X^3 \quad (9)$$

where c_p is the black liquor specific heat, kJ/kg • °C
X the dry solids concentration, kg dry solids/kg
t the temperature of the black liquor, °C.

Figure 10 shows a typical black liquor enthalpy as a function of dry solids.

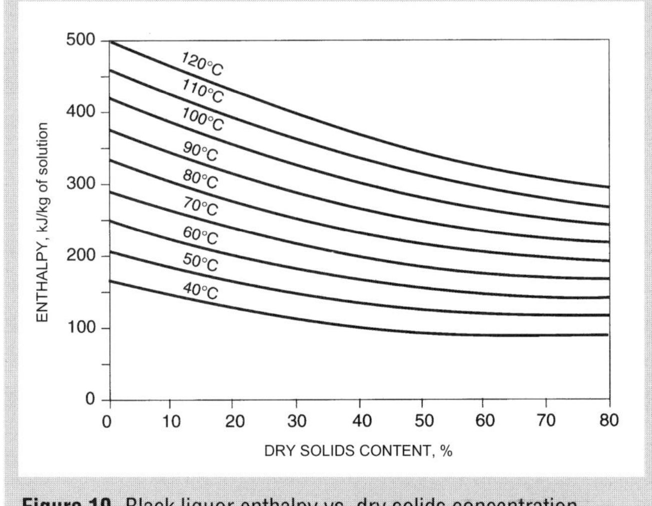

Figure 10. Black liquor enthalpy vs. dry solids concentration.

2.1.6 Boiling point rise

The temperature where black liquor evaporation occurs is higher than the evaporation temperature of water at the same pressure. As black liquor dry solids increase, this difference between black liquor and water boiling temperatures increases.

The following equation estimates the black liquor boiling point rise at atmospheric pressure:

$$\Delta T = 6.173X - 7.48X(X)^{0.5} + 32.747X^2 \quad (10)$$

where ΔT is the boiling point rise, K
X the dry solids concentration, kg dry solids/kg liquid

Figure 11 shows the boiling point rise as a function of dry solids concentration.

For other pressures the following correction is necessary:

$$\Delta T_p/\Delta T = 1 + 0.6(T_p - 3.7316)/100 \quad (11)$$

where ΔT_p is the boiling point rise, K
ΔT the boiling point rise at atmospheric pressure, K
T_p the boiling temperature of water at specified pressure, K

2.1.7 Thermal conductivity

Thermal conductivity of the black liquor depends on the temperature and dry solids content. A temperature increase will increase the thermal conductivity, and an increase in dry solids content will decrease the thermal conductivity.

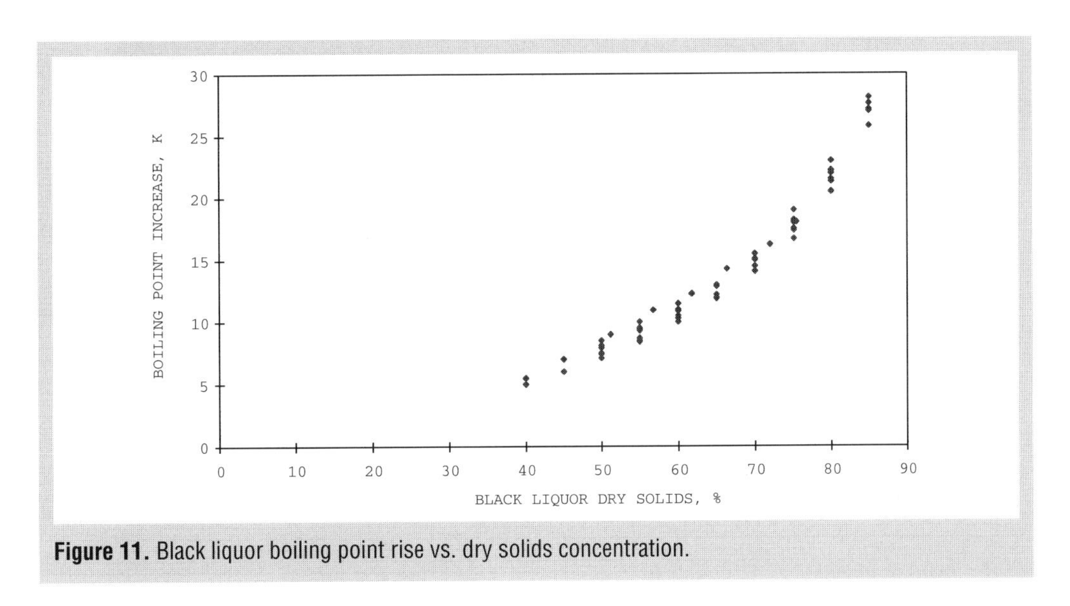

Figure 11. Black liquor boiling point rise vs. dry solids concentration.

The following equation provides an estimate of black liquor thermal conductivity:

$$\lambda = \lambda_{H_2O}(1 - X) + aX + bX^2 \tag{12}$$

where λ is the thermal conductivity, W/m °C
λ_{H_2O} the thermal conductivity of water, W/m • °C
X the dry solids concentration, kg dry solids/kg
T the temperature of black liquor, °C

$a = a_1 + a_2T$
$b = b_1 + b_2T$
$a_1 = 0.3176$
$a_2 = 0.002268$
$b_1 = -0.01394$
$b_2 = -0.003069$

Figure 12 shows thermal conductivity as a function of dry solids concentration[9].

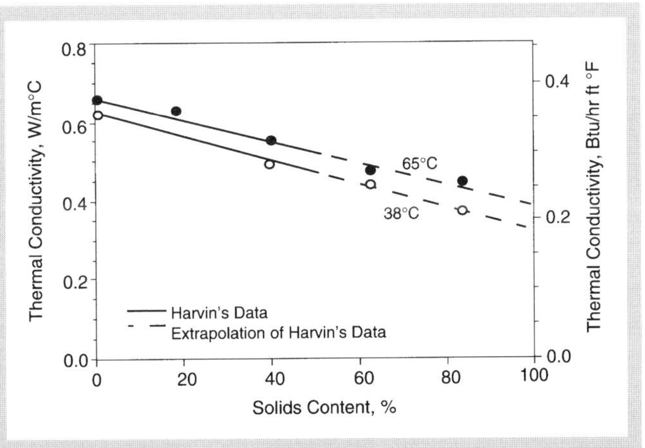

Figure 12. Black liquor thermal conductivity vs. dry solids concentration.

2.1.8 Viscosity

The common definition for viscosity of black liquor is the ratio of shear stress to shear rate. The concentrated black liquor is typically not a Newtonian fluid. The measured viscosity of black liquor depends on the shear rate. One must exercise great care to ensure measuring the viscosity values as near the actual operating conditions as possible. The error is typically smaller than the error associated with other operating conditions.

Two common ways to indicate viscosity are dynamic and kinematic viscosity. They are related. Dynamic viscosity calculation relates to kinematic viscosity as follows:

$$\eta = \mu\rho \tag{13}$$

where η is the dynamic viscosity, Pa · s (= N · s/m^2 = kg/m · s)
μ the kinematic viscosity, m^2/s
ρ the density, kg/m^3.

Sometimes the dynamic viscosity uses older units where 1 mPa · s = 1 cP.
The following equations provide an estimate of the kinematic viscosity:

$$\begin{aligned} ln\mu &= A + B/T^3 \\ A &= A_{H_2O} + a_1X + a_2X^2 + a_3X^3 \\ B &= B_{H_2O} + b_1X + b_2X^2 + b_3X^3 \end{aligned} \tag{14}$$

where μ is the kinematic viscosity, cSt = mm^2/s
T the temperature, K
X the dry solids concentration, kg dry solids/kg
A_{H_2O}= –2.4273
B_{H_2O} = 6.1347 x 10^7
a_i = value in Table 9
b_i = value in Table 9.

Table 9. Coefficients for viscosity equations.

	Softwood	Hardwood	Tropical
a_1	9.1578	3.3532	10.482
a_2	-56.723	3.7654	-54.046
a_3	72.666	-2.4907	61.933
b_1	-42.178 • 10^7	-5.442 • 10^7	-40.165 • 10^7
b_2	335.12 • 10^7	21.915 • 10^7	300.55 • 10^7
b_3	-349.23• 10^7	17.042 • 10^7	-266.47 • 10^7

When black liquor temperature increases, its viscosity decreases as Fig. 13 shows. An increase in dry solids content will increase the liquor viscosity. The dynamic viscosity of black liquor varies considerably. Figure 14 shows selected black liquor viscosities at 115°C as a function of dry solids concentration. The range of viscosities at any dry solids value differs more than an order of magnitude. Note that the slopes of viscosity curves differ from each other and are functions of dry solids.

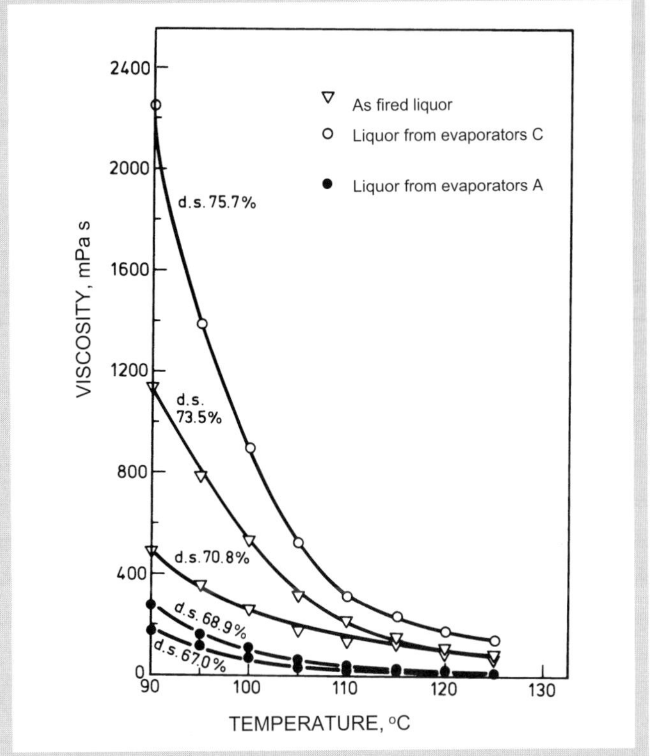

Figure 13. Effects of temperature and dry solids on black liquor viscosity[2].

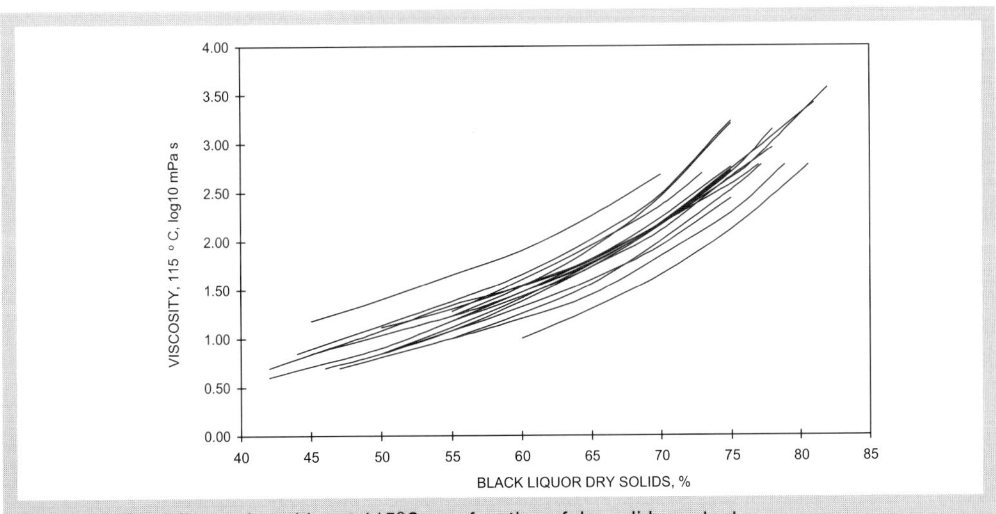

Figure 14. Black liquor viscosities at 115°C as a function of dry solids content.

2.1.9 Surface tension

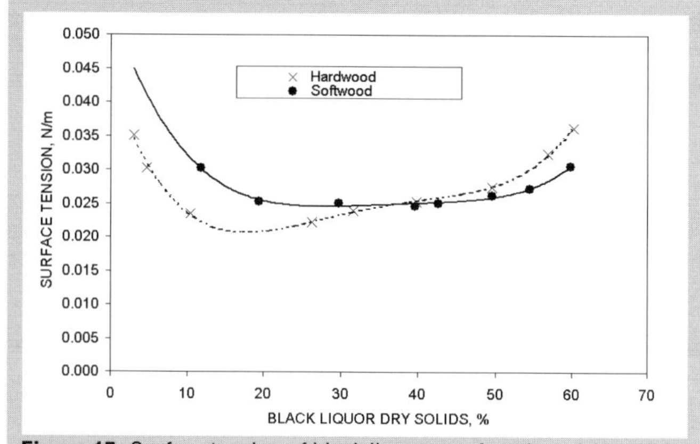

Figure 15. Surface tension of black liquor as a function of dry solids content.

Figure 15 shows surface tension of black liquor vs. dry solids content. Black liquor viscosity cannot be extrapolated to very high dry solids. Thermal degradation and tar formation can radically change the viscosity.

2.2 Smelt properties

Smelt is the molten ash produced in recovery boiler furnace. Important properties of the smelt components in the recovery process operation are heat capacity, melting heat, and heats of formation[10].

2.2.1 Enthalpy

The following equation expresses the enthalpy of different smelt components:

$$h_c = \Delta h_m + c_p(T_{smelt} - T_{ref}) \tag{15}$$

where h_c is the specific enthalpy of a component, J/kg
Δh_m the specific enthalpy of melting for a component, J/kg
c_p the specific heat capacity of a component, J/kg · °C
T_{smelt} the smelt temperature, °C
T_{ref} the reference temperature, °C.

The enthalpy for a smelt with a given composition is the following:

$$H_{SMELT} = \sum_{i-1}^{n} m_i h_{c,i} \tag{16}$$

where H_{smelt} is the tenthalpy of smelt, J
m_i the mass of component i, kg
$h_{c,i}$ the specific enthalpy of component i, J/kg.

Table 10 shows specific heats of melting and specific heat capacities for typical smelt compounds.

Table 10. Specific heats of melting and specific heat capacities for typical smelt compounds.

Compound	h_m(25 °C) kJ/mol	c_p kJ/mol °C	h_c(850 °C) kJ/mol
Na_2CO_3	29.7	0.157	163.4
Na_2S	19.2	0.096	100.7
Na_2SO_4	23.8	0.188	183.4
NaCl	28.3	0.063	81.6
$Na_2S_2O_3$	29.7	0.133	142.9
K_2CO_3	27.9	0.159	163.3
K_2S	16.2	0.098	99.2
K_2SO_4	34.4	0.187	193.5

For other smelt compounds, the values h_m = 200 kJ/kg, c_p = 0.94 kJ/mol °C and h_c = 1350 kJ/kg can be used.

2.2.2 Heat of formation

The heat of formation is the energy required for the reaction of components in the base state to form that compound as Table 11 shows. Heats of formation are important in the proper calculation of energy balances.

Table 11. Specific heats of formation for typical compounds in a recovery boiler.

Compound	Mole weight	$-Dh_F$ (25 °C)	
	kg/kmol	MJ/kg	kJ/mol
Na_2S	78.04	4.691	366.1
Na_2SO_3	126.05	8.687	1095.0
Na_2SO_4	142.04	9.768	1387.4
$Na_2S_2O_3$	158.11	7.107	1123.7
Na_2CO_3	105.99	10.669	1130.8
K_2S	110.26	3.416	376.6
K_2SO_4	174.25	8.251	1437.7
K_2CO_3	138.2	8.329	1151.0
NaCl	58.443	7.036	411.2
SO_2	64.06	4.633	296.8
CO_2	44.01	8.948	393.8

2.2.3 Heat of reduction

A primary function of the recovery boiler is to reduce oxidized sodium and sulfur compounds to useful cooking chemicals. Reduction reactions require heat to proceed.

For each individual compound, the reduction heat is the energy required for reaction of components in the base state to form that compound as Table 12 shows.

Table 12. Specific reduction heats for typical smelt compounds.

Compound	Dh_R, kJ/kg	Btu/lb	Dh_{Rm}, kJ/mol	Btu/lbm
Na_2S	13099	5631	10217	439252
NaCl	0	0	0	0
Na_2SO_4	0	0	0	0
Na_2CO_3	0	0	0	0
Na_2SO_3	2325	1000	292.9	125924
$Na_2S_2O_3$	5787	2488	915.0	393380
K_2S	9629	4140	1061.1	456191
K_2SO_4	0	0	0	0
K_2CO_3	0	0	0	0
SO_2	5531	2378	354.0	152193

2.3 Flue gas and air properties

Data on properties for individual gases are available in many reference books[11–16]. To estimate the properties of a mixture, the usual procedure is to calculate first the property values for each gaseous component with a mole fraction, x_i. for a given temperature, T, and pressure, p. The property value, g, of the gaseous mixture of m gases is a function of the calculated component properties:

$$g = g(f(T,p,x_1),f(T,p,x_2),\ldots,f(T,p,x_m)) \tag{17}$$

For quick determination of physical properties, all gaseous properties are reduced to polynomials of temperature, T, and pressure, p. Usually, one can express the properties as polynomials of T with a pressure correction for flue gas temperatures at 250 K–1500 K and pressures of 0.1–6.0 MPa. This results in the following equation:

$$g = \sum_{i-1}^{n} a_i T^{i-1} + e(p) \tag{18}$$

In the previous equation, e(p) represents a pressure correction function.

2.3.1 Density

The density, ρ, of a pure fluid at low reduced pressure and high reduced temperature is a function of temperature, T, and pressure of that gas, p_i:

$$\rho_i = \frac{p_i}{R_i T} \tag{19}$$

where ρ is the density of the pure fluid, kg/m^3
p the partial pressure of the pure fluid, Pa (bar)
T the temperature of the pure fluid, K
R the gas constant of the pure fluid, J/kg • K.

The usual procedure to calculate the density for a mixture of gases is to use the weighted average of molar composition:

$$\rho = \sum_{i=1}^{m} \rho_i \tag{20}$$

For a mixture of gases, we can then use the ideal gas law and substitute to obtain the following for given T and p:

$$\rho = \frac{1}{T}\sum_{i=1}^{m}\left(\frac{p_i}{R_i}\right) = \frac{p}{TR} \tag{21}$$

where ρ is the density of the mixture, kg/m^3
p the pressure of the mixture, Pa (bar)
T the temperature of the mixture, K
R the gas constant of the mixture, J/kg • K.

Table 13 shows constants for typical gases.

Table 13. Constants for gas components.

Component	Mole weight, kg/kmol	Density, 0°C, 1.01325 bar kg/m^{3}n	Specific volume, m^{3}n/kmol	Gas constant, J/kgK
CO_2	44.01	1.977	22.263	188.9
H_2O	18.0154	0.804	22.408	461.5
SO_2	64.06	2.927	21.887	129.8
N_2 (pure)	28.0134	1.250	22.403	296.8
O_2	31.9988	1.429	22.394	259.8
CO	28.01	1.250	22.404	296.8
H_2	2.01594	0.090	22.432	4124.
N_2 (with Ar)	28.1539	1.257	22.403	295.3
Air	28.96	1.293	22.4	287.1

2.3.2 Enthalpy

The specific enthalpy of a pure fluid at low reduced pressure and high reduced temperature is a polynomial function of temperature, T:

$$h_i = \sum_{j-1}^{n} a_{ij} T^{j-1} \tag{22}$$

where h is the specific enthalpy of the pure fluid, kJ/kmol
T the temperature of the pure fluid, K.

The typical procedure to calculate the enthalpy value of a gaseous mixture is to sum the individual enthalpies as weighted averages:

$$h = \sum_{i-1}^{m} x_i h_i \tag{23}$$

where x is the mole fraction of the pure fluid.

This becomes

$$h = \sum_{i-1}^{m} x_i \sum_{j-1}^{n} a_{ij} T^{j-1} \tag{24}$$

Table 14 shows constants for typical gases.

Table 14. Enthalpy equation constants for gas components.

	a_1	a_2	a_3	a_4
CO_2	8.9725	27.64	20.32	-4.600
H_2O	8.5205	29.71	5.366	0.2340
SO_2	9.8315	31.22	18.75	4.672
N_2	7.4640	26.40	3.451	0.2365
O_2	7.3969	25.27	7.028	1.481
CO	7.3908	25.89	4.423	0.5551
H_2	7.9803	29.35	0.6991	0.7498

2.3.3 Specific heat capacity

The specific heat capacity of a pure fluid, i, at low reduced pressure and high reduced temperature is the following:

$$c_{pi} = \frac{\partial}{\partial T} h_i(T) \tag{25}$$

We can substitute the specific enthalpy function to obtain a similar polynomial function of temperature, T, as the one for specific enthalpy:

$$c_{pi} = \sum_{j-1}^{n} b_{ij} T^{j-1} \tag{26}$$

The usual procedure to calculate the specific heat capacity of a gaseous mixture is to sum individual heat capacities as weighted averages:

$$c_p = \sum_{i-1}^{m} x_i c_{pi} \tag{27}$$

This is also the following:

$$c_p = \sum_{i-1}^{m} x_i \sum_{j-1}^{n} b_{ij} T^{j-1} \tag{28}$$

Table 15 shows constants for typical gases.

Table 15. Specific heat capacity equation constants for gas components.

	b_1 (=a_2)	b_2 (=$2a_3$)	b_3 (=$3a_4$)	b_4
CO_2	27.64	40.64	-13.800	0
H_2O	29.71	10.732	0.7020	0
SO_2	31.22	37.50	-14.016	0
N_2	26.40	6.902	-0.7095	0
O_2	25.27	14.056	-4.443	0
CO	25.89	8.846	-1.1653	0
H_2	29.35	-1.3982	2.249	0

2.3.4 Conductivity

The conductivity of a pure fluid, i, at low reduced pressure and high reduced temperature is a polynomial function of temperature, T, with a pressure correction:

$$\lambda_1 = \sum_{j-1}^{n} c_{ij} T^{j-1} + d_{ij} P \tag{29}$$

where λ is the conductivity of the pure fluid, W/mK
T the temperature of the pure fluid, K.

The usual procedure to calculate the conductivity of a gaseous mixture is to sum individual conductivities as weighted averages of molar weights raised to the power of one-third. Since the pressure in the recovery boiler calculations is approximately atmospheric, no pressure correction is necessary, i.e., $d_{ij} = 0$.

$$\lambda = \frac{\sum_{i=1}^{m} X_i \lambda_i M_i^{1/3}}{\sum_{i=1}^{m} m_i M_i^{1/3}} \tag{30}$$

This becomes

$$\lambda = \sum_{i=1}^{m} X_i M_i^{1/3} \frac{\left(\sum_{j=1}^{n} c_{ij} T^{j-1}\right)}{\sum_{i=1}^{m} X_i M_i^{1/3}} \tag{31}$$

Table 16 shows constants for typical gases.

Table 16. Conductivity equation constants for gas components.

	c_1	c_2	$c_3 10^4$	$c_4 10^7$
CO_2	7.75822	0.080927	0.034848	0.080138
H_2O	7.32096	0.0098527	1.098610	0.297037
SO_2	8.04021	0.0684547	0.323878	0.183626
N_2	1.74403	0.0946376	0.506615	0.154557
O_2	0.58820	0.0950439	0.348834	0.101551
CO	0.98633	0.1018060	0.594083	0.276143
H_2	57.69140	0.4343330	0.703374	0.108389

2.3.5 Viscosity

The dynamic viscosity of a pure fluid, i, at low reduced pressure and high reduced temperature is a polynomial function of temperature, T, with pressure correction as follows:

$$\eta_i = \sum_{j=1}^{n} e_{ij} T^{j-1} + f_{ij} P \tag{32}$$

where η is the viscosity of the pure fluid, kg/ms
T the temperature of the pure fluid, K.

The usual procedure to calculate the viscosity of a gaseous mixture is to sum individual viscosities as weighted averages of square roots of the molar weights. Since the pressure in boilers is nearly atmospheric, no correction is necessary, i.e., $f_{ij} = 0$.

$$\eta = \frac{\sum_{i=1}^{m} X_i \eta_i M_i}{\sum_{i=1}^{m} x_i M_i^{1/2}} \tag{33}$$

This is also the following:

$$\eta = \sum_{i=1}^{m} X_i M_i^{1/2} \frac{\left(\sum_{j=1}^{n} e_{ij} T^{j-1}\right)}{\sum_{i=1}^{m} X_i M_i^{1/2}} \tag{34}$$

Table 17 shows constants for typical gases.

Table 17. Viscosity equation constants for gas components.

	e_1	e_2	e_3	e_4
CO_2	1.99687	0.0669782	0.389360	1.244840
H_2O	3.37994	0.0108578	0.420030	1.835200
SO_2	0.11323	0.0438088	0.008726	0.523698
N_2	3.21160	0.0541257	0.194993	0.268087
O_2	3.64206	0.0640418	0.264129	0.659854
CO	1.17308	0.0788968	0.587696	2.173350
H_2	1.88838	0.0272931	0.137478	0.457806

2.4 Air

Air contains hundreds of different gases. Most of these occur in very small quantities. Making calculations possible requires assuming that air contains only a few components. The standard dry air (Bureau of Mines Standard Air Composition) contains 79.02% nitrogen, 20.95% oxygen, and 0.03% carbon dioxide on a dry mole basis. This assumes that all other air compounds (mainly argon) can be expressed as nitrogen. The standard dry air molar weight is 28.85.

Simplifying the air composition to speed calculations is sometimes useful. One possibility is that air contains 79.0% nitrogen and 21.0% oxygen on a dry volume basis[17]. This corresponds to 76.8% nitrogen and 23.2% oxygen on a dry weight basis.

The air is never dry but contains some water vapor. Specific humidity expresses the amount of water vapor or moisture in the air. This is grams of water in one kilogram dry air, g/kg dry air. Another expression is percent relative humidity. The water content is then a percentage of the maximum humidity in air. If a specific amount of moisture in the air is not available, a value of 13.5 g of moisture/kg of dry air will suffice.

2.5 Properties of green liquor

Green liquor results when smelt from the dissolving tank dissolves into weak white liquor, weak wash, or water. Note that all streams contain impurities and alkali.

Green liquor is an aqueous solution of sodium sulfides, sulfates, and carbonates. It contains some unburned char. Table 18 provides an analysis of green liquor.

Table 18. Typical green liquor analysis.

Location		US	US	US	Finland	Finland
Dry solids,	%	18.9	18.5	21.8		
Suspended solids,	mg/L	580	809	534	500	800
NaOH,	g/L	15.5	17.8	18.8	11.7	14.1
Na_2S,	g/L	30.8	32.5	42.7	46.8	49.2
Na_2CO_3,	g/L	143.2	133.3	134.9	95.4	105.3
Na_2SO_3,	g/L	2.19	2.09	1.41	0.2	0.9
$Na_2S_2O_3$,	g/L	4.57	5.10	7.08	2.4	3.8
Na_2SO_4,	g/L	3.28	4.50	8.79	6.5	3.1
Total alkali,	g Na_2O/L	124.6	122.6	134.7	106.	114.9
Active alkali,	g Na_2O/L	36.5	39.6	48.5	46.3	50.0
Effective alkali,	g Na_2O/L	24.3	26.7	31.5	27.7	30.5
Reduction, $Na_2S/(Na_2S+Na_2SO_4)$		94.5	92.9	89.8	92.9	96.7
Reduction, $Na_2S/(Na_2S+Na_2SO_3+Na_2S_2O_3+Na_2SO_4)$		85.1	83.8	82.3	90.6	92.2

Table 19 shows the metal content after the green liquor clarifier

Table 19. Metal content after green liquor clarifier.

Component	Amount, mg/L
Ca	10–30
Na	100 000–120 000
Al	5–10
Si	100–250
Mg	2–4
Mn	0.5–2
Fe	0.5–1

2.6 Properties of miscellaneous chemical compounds

Table 20 shows the values for several additional chemical compounds in the recovery process.

Table 20. Properties of miscellaneous chemical compounds.

Component	Melting, °C	Boiling, °C	Density, kg/m^3_{solid}	Molecular wt., kg/kmol
Calcium carbonate, $CaCO_3$	Dec. 825	-	2930	100.09
Calcium oxide, CaO	2570	2850	3320	56.08
Chromium, Cr	1615	2200	7100	52.01
Copper, Cu	1083	2300	8920	63.54
Ferric oxide, Fe_2O_3	Dec. 1560	-	5120	159.70
Magnesium, Mg	651	1110	1740	24.32
Magnesium oxide, MgO	2800	3600	3650	40.32
Potassium, K	62	760	860	39.102
Potassium carbonate, K_2CO_3	891	dec.	2290	138.20
Potassium cyanide, KCN	635	-	1520	65.11
Potassium chloride, KCl	790	1500	1988	74.56
Potassium hydroxide, KOH	380	1320	2044	56.10
Potassium sulfate, K_2SO_4	588	-	2662	174.25
Potassium sulfide, K_2S	60	-	-	110.27
Silicon dioxide, SiO_2	1670	2230	2230	60.09
Sodium, Na	98	880	970	22.99
Sodium carbonate, Na_2CO_3	851	dec.	2533	105.99
Sodium cyanide, NaCN	564	1496	-	49.01
Sodium chloride, NaCl	800	1413	2163	58.443
Sodium hydroxide, NaOH	318	1390	2130	40.00
Sodium sulfate, Na_2SO_4	884	-	2698	142.04
Sodium sulfide, Na_2S	-	-	1856	78.04
Sulfur, S	120	445	2070	32.06

Dec.=decomposes before this stage

References

1. Grace, T. M. and Timmer, W. A., TAPPI 195 International Chemical Recovery Conference Proceedings, TAPPI PRESS, Atlanta, p. B269.
2. Scott-Young, R. E. and Cukier, M., TAPPI 1995 International Chemical Recovery Conference Proceedings, TAPPI PRESS, Atlantn, p. B263.
3. Söderhjelm, L., Black liquor properties. 30 Years Recovery Boiler Co-operation in Finland, Finnish Recovery Boiler Users Association, Helsinki, 1994, p. 23.
4. Gullichsen, J., Proceedings of the Symposium on Recovery of Pulping Chemicals, The Finnish Pulp and Paper Research Institute, Helsinki, 1968, p. 211.
5. Hultin, S. O., Proceedings of the Symposium on Recovery of Pulping Chemicals, Helsinki, The Finnish Pulp and Paper Research Institute, Helsinki, 1968, p. 165.
6. Adams, T. N., Frederick, W. J., Grace, T. M., et al., Kraft recovery boilers, TAPPI PRESS, Atlanta, 1997, pp. 59–99.
7. Fricke, A. L., Physical properties of kraft black liquors: Interim report © Phase II,. report No. DOE/CE 40606©T5 (DE88002991), U.S. Dept. of Energy, Washington, 1987, p. 66.
8. Masse, M. A., Kiran, E., and Fricke, A. L., Polymer 27:619(1986).
9. Ramamurthy, P., et. al., TAPPI Journal, 76(1993)117, pp 175–179.
10. Knacke, O., Kubaschewski, O., and Hesselmann, K., Thermochemical properties of inorganic substances, II, Springer-Verlag, Berlin, 1991.
11. Maddox, R. N., in Heat Exchanger Design Handbook, Part 5. Physical properties (Shlünder, E. U. et.al., Ed.),. Hemisphere Publishing Corp., New York,1983, pp. 5.2–1–5.2–11.
12. Schunk, M., in Heat Exchanger Design Handbook, Part 5. Physical properties, Hemisphere Publishing Corp., New York, 1983, pp. 5.2–1–5.2–11.
13. Stelzer, F. J., Physical property algorithms, Karl Thiemig Ag, Munich, 1984.
14. Sychev, V. V., et. al., Thermodynamic Properties of Nitrogen, Hemisphere Publishing Corp., New York, 1987.
15. Sychev, V. V., et. al., Thermodynamic Properties of Methane, Hemisphere Publishing Corp., New York, 1987.
16. Sychev, V. V., et. al., Thermodynamic Properties of Oxygen, Hemisphere Publishing Corp., New York, 1987.
17. Anon., Performance test for recovery units, TAPPI PRESS, Atlanta, 19

CHAPTER 12

Evaporation of black liquor

CHAPTER 12

CHAPTER 12

Karl Holmlund and Kari Parviainen

Evaporation of black liquor

1 General

When processing wood chips into pulp in the cooking plant, organic materials such as lignins, hemicellulose, and a minor part of the cellulose dissolve in the cooking liquor. These organic materials with the inorganic components of the cooking liquor form the spent cooking liquor, commonly referred to as black liquor.

The organic material in the black liquor contains chemical energy recovered as heat in the recovery boiler. Inorganic materials are recovered simultaneously and regenerated in the recovery boiler and recausticizing plant to a form that allows their reuse in the pulp cooking process.

The black liquor is separated from the pulp during the pulp washing as a watery solution with a dry solids content of 14%–18% depending on the raw material and the efficiency of the washing plant. This weak black liquor contains too much water for direct use as fuel in the recovery boiler. The energy from burning the organic material is less than the energy necessary to evaporate the water in the liquor.

The main purpose of the evaporation plant is to increase the dry solids content of the black liquor by evaporating water until reaching a concentration that allows burning in the recovery boiler. This concentration is normally 65%–75% dry solids. Many modern installations operate at a level above 80% dry solids.

2 Black liquor properties

2.1 Composition of black liquor dry solids

Black liquor contains water and dry solids. The main components of the dry solids are the following:

- Organic substances dissolved from wood during cooking
- Lignins
- Hemicellulose
- Cellulose
- Inorganic substances
- Sodium compounds

- Sulfur compounds (or sodium-sulfur compounds)
- Inert material.

The organic material dissolved from wood is approximately 60% of the total black liquor dry solids.

2.2 Physical properties of black liquors

An evaporation plant consists of heat transfer units connected in various configurations. To design the plant properly, several physical properties of the black liquor require consideration.

Viscosity

The viscosity of black liquor is a function of concentration and temperature and is specific for each liquor. It depends on wood species, cooking method, further thermal treatment, etc. The viscosity may vary considerably from mill to mill[1]. Available formulas for calculating viscosity primarily use statistical data. Large deviations are possible.

Viscosity increases with increased dry solids content. In some cases, this is very steep after a certain point. An increase in temperature lowers the viscosity as chapeter 11.2.1.8 mentions. The practical limit for handling the liquor is the pumping limit of 300–500 cP. The viscosity must always be below this level, and is usually much lower in the evaporation plant. If the liquor at final concentration is stored at atmospheric pressure, the limit of the final concentration is 70%–75% dry solids at the maximum temperature of 115°C. If the final product liquor is stored in a pressurized storage tank, the final concentration can be 75%– 85% dry solids and even higher at a storage temperature of 125°C–150°C. These extremely high concentrations might require using medium pressure steam in the evaporation plant. This is a disadvantage when considering energy economy and investment cost.

A heat treatment process (HTP) can also reduce the viscosity. This method heats liquor to approximately 180°C for 30 min. This breaks the long organic molecules into shorter forms and results in an irreversible reduction of the liquor viscosity. Section 6.4 provides a more detailed description.

Boiling point rise (BPR)

A liquid mixture that contains dissolved organic substances, inorganic substances, or both, will boil at a higher temperature than water at the same pressure. This temperature difference is the boiling point rise (BPR).

BPR is specific for each black liquor and depends on the amount and composition of the dissolved substances. BPR increases with dry solids concentration. As an example, in one liquor BPR is less than 2°C at the feed concentration of 20% dry solids but 18°C at a final product liquor concentration of 70% dry solids. Chapter 11.2.1.6 provides more details.

An example can illustrate the practical consequence of BPR. A liquor with concentration of 64% dry solids and BPR of 15°C evaporated in an evaporation stage.

Steam of 0.3 MPa absolute is available for the evaporation. This means that the liquor theoretically can be heated to a temperature of 133.5°C, which is the saturation temperature of the steam at this pressure. The vapor generated on the liquor side that is the heating medium for the following stage will have a saturation temperature of 133.5–BPR = 118.5°C. This corresponds to a steam saturation pressure of 0.19 MPa absolute. The BPR temperature and pressure difference is therefore lost in the evaporation stage.

This is a simplified example because the actual pressure and temperature loss over the stage must also include the necessary ΔT over the heat transfer surface and the losses in piping and ducts. The generated vapor will also be slightly superheated which does not markedly influence the heat transfer. Over the entire evaporation plant, the sum of the BPR values in the various stages will consume a large part of the total available ΔT.

Density

Black liquor density increases with increased concentration as chapter 11.2.1.4 indicates. A liquor with a concentration of 16% dry solids would have a density of 1.05 t/m^3. At 70% dry solids, it has a density of 1.43 t/m^3. Both values for a reference temperature of 90°C. The actual density will also depend on the temperature of the liquor. It does require consideration in determining the equipment characteristics.

Surface tension

The practical implication of surface tension in an evaporation plant is that a low surface tension will increase the tendency to foam. Surface tension increases with an increase in dry solids concentration. It decreases with an increase in temperature. Foaming is a problem primarily in the evaporator elements operating at low concentration. To overcome this, the feed liquor concentration is often increased by recirculating a portion of the concentrated liquor, ("sweetening" the feed liquor). Crude tall oil and soap will decrease the surface tension. They therefore need to be separated from the liquor if the contents are high, especially when softwood is the raw material for the pulp.

Specific heat

Specific heat decreases with increasing black liquor concentration. Equipment design must consider this. Chapter 11.2.1.5 gives formulas for the calculation.

Solublity of substances in black liquor

The inorganic substances in the black liquor are mostly bound to the dissolved organic compounds. A minor portion of the inorganic substances such as NaOH, Na_2S, Na_2CO_3, Na_2SO_4, etc., dissolve as salts in the weak black liquor. During concentration of the liquor, a point occurs when Na_2CO_3, Na_2SO_4, and/or their double salt $2\ Na_2SO_4 \cdot Na_2CO_3$ (burkeite) begin to precipitate. This normally occurs at 45%–60% dry solids, which is the solubility limit, or the critical concentration of the liquor. The salt precipitation begins at a lower concentration with a higher total effective Na content in black liquor as Section 9.3.4 discusses. The crystallized salts foul the heating surfaces very rapidly but are easily washable. Keeping the liquor in circulation or applying crystallization evaporation techniques can minimize the fouling.

Normally, weak black liquor contains 5 to 10 g/L of residual NaOH and Na_2S giving a pH value of approximately 12 in the black liquor. If this residual alkaline material (NaOH and Na_2S) decreases below 7 to 9 g NaOH/L and the pH is below 11, lignin compounds start to precipitate and foul the evaporator heating surfaces.

3 Principles of evaporation

Evaporation is the separation of water from a solution containing nonvolatile solutes. Supplying heat to the solution causes vaporization. An evaporation plant usually consists of several heat transfer units connected in series with steam or vapor as the heating media and black liquor as the heated media. The following are some basic definitions and terminology:

- **Multiple stage evaporation** consists of a number of heat exchangers connected in series.
- **Evaporation stage or effect** is one or possibly several heat exchangers operating at the same steam pressure level. A modern evaporation plant normally consists of 5–7 stages in series. The effects are numbered in the direction of the steam flow with the first effect operating at the highest steam pressure.
- **Evaporator body** is the heat exchanger unit. One effect might have several bodies in parallel.
- **Live steam, primary steam,** or fresh steam is the clean steam from the mill´s steam distribution system. It usually has a pressure of 0.3–0.4 MPa (abs). Live steam is mostly used only in the first effect. The steam must be at saturation temperature or only slightly superheated.
- **Primary condensate** is the clean condensate from the live steam.
- **Secondary steam or vapor** is the steam evaporated from the black liquor. Vapor from the first effect is used as a heating medium in the second effect. Vapor from the second effect is then used in the third effect etc. The vapor always contains some organic materials evaporated from the liquor or even liquor droplets entrained in the steam flow.
- **Secondary condensate** is condensate derived from the vapor. It contains various levels of organic contaminants. It is therefore not as clean as the primary condensate.
- **Foul condensate** is the most contaminated secondary condensate and shall be purified by stripping.
- **Noncondensible gases** (NCG) are gaseous compounds that are liberated from the black liquor during evaporation. They are odorous, poisonous, and inflammable. They require extraction from the evaporation plant and treatment such as incineration in suitable process equipment.

- **Concentrator** refers to the first effect where the liquor is evaporated to its final concentration. The name has a historic background. Earlier, the final concentrators were clearly separated from the multiple-effect evaporation train and had a different design. In a modern plant, this difference does not exist. The terminology is still in use, although it is not entirely correct.
- **Surface condenser** is a water-cooled heat exchanger that condenses the vapor from the last evaporation effect.

3.1 Multiple effect evaporation

Figure 1 shows the simplified diagrams of one-stage and three-stage evaporation plants to illustrate the principles of multiple-effect evaporation.

In one-stage evaporation, the live steam consumption is the same or theoretically slightly more than the amount of vapor generated on the liquor side. This assumes that black liquor enters the evaporator at the boiling point. If the liquor feed temperature is lower, energy is necessary for liquor preheating. Then even less secondary vapor is generated.

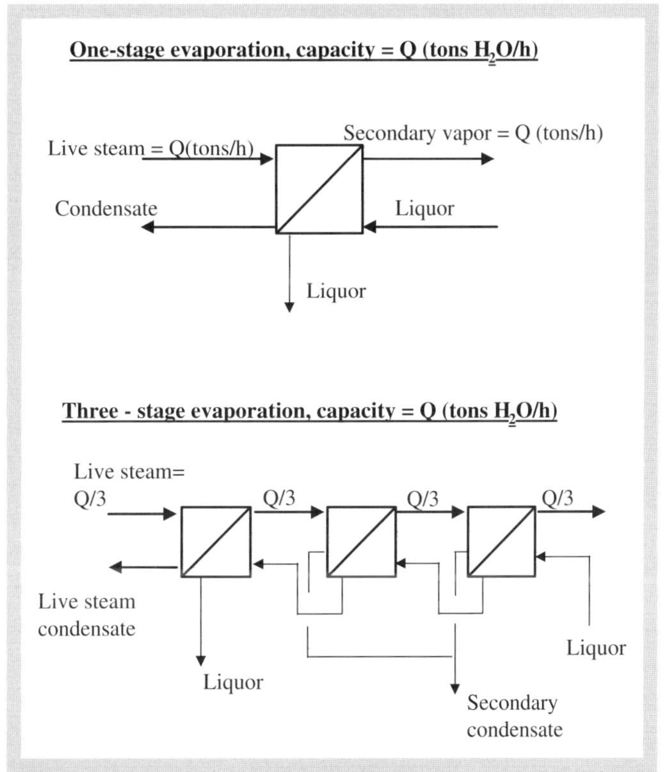

Figure 1. Principle of multiple evaporation.

For the three-stage evaporating plant, the vapor generated in the first stage is the heating medium in the second stage. The vapor generated in the second stage is the heating medium for the third stage. As a result, the live steam consumption will be approximately one-third of the steam consumption of the one-stage plant for the same amount of evaporated water.

In other words, the steam economy, defined as tons evaporated water per ton fresh steam (ton H_2O/t steam), would be one for the one-stage evaporator and three for the three-stage plant. In reality, the real steam economy is smaller due to the ΔT losses in the multiple evaporation. Table 1 lists measured steam economies for a practical operation.

Another definition used more often in Scandinavia is the specific heat consumption defined as the energy in fresh steam minus the energy in the retrieved condensate divided by water evaporated (MJ/t H_2O) as Table 1 shows.

Table 1. Steam economy and specific heat consumption in multiple evaporation including an integrated stripper column.

Number of ostages	Steam economy, ton H_2O/t steam	Specific heat consumption, MJ/t H_2O
4	3.7–3.6	630–650
5	4.3–4.1	550–570
6	5.1–4.9	460–480
7	6.2–5.9	390–400

3.2 Vapor compression evaporation

Figure 2 shows the principle of vapor compression evaporation (VCE) or also called mechanical vapor recompression (MVR). A mechanical compressor or a turboblower increases the pressure and consequently the saturation temperature of the secondary vapor for reuse on the primary side of the heating element. The heat energy is regained at the cost of the mechanical energy required for the compression of the vapor. This energy depends on the apparent ΔT over the effect, i.e., the BPR + the piping losses + the effective ΔT (driving force) over the heat transfer surface. Since the BPR represents most of the total required ΔT, the VCE is used primarily for pre-evaporation of liquor with a concentration less than 25% dry solids,which has a moderate BPR. The normal power consumption at a vapor compression pre-evaporator is 11–14 kWh/t H_2O.

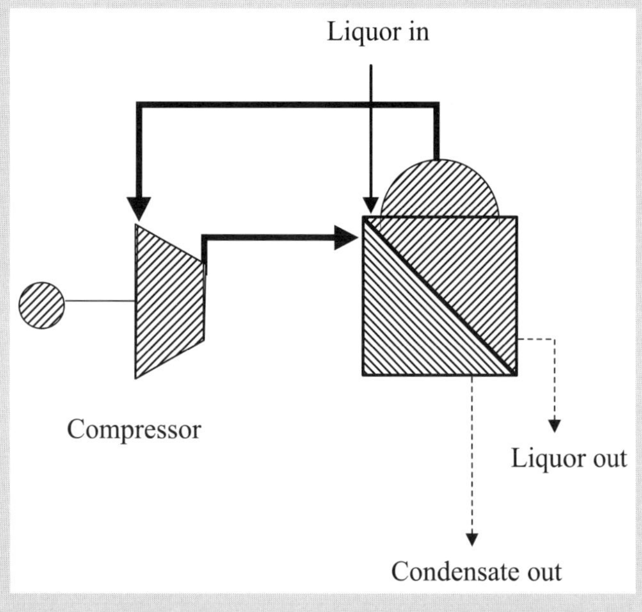

Figure 2. Principle of vapor compression evaporator (VCE).

The economic feasibility of the VCE depends on the cost of power being sufficiently low compared with the cost of available heat energy. Advantages of the VCE are the compact design and the fact that no external steam supply is necessary. Installing such a unit to boost evaporation capacity can be favorable despite a higher operating cost.

3.3 Flash steam evaporation

This type of evaporation uses the vapor generated when hot weak black liquor is flashed to a lower pressure. The following applications are typical in the pulp mill:

- Blow heat evaporator using heat from the hot water accumulator of a batch digester system
- Flash steam evaporator using flash steam from the flash cyclone of a continuous digester system
- Multiflash evaporator

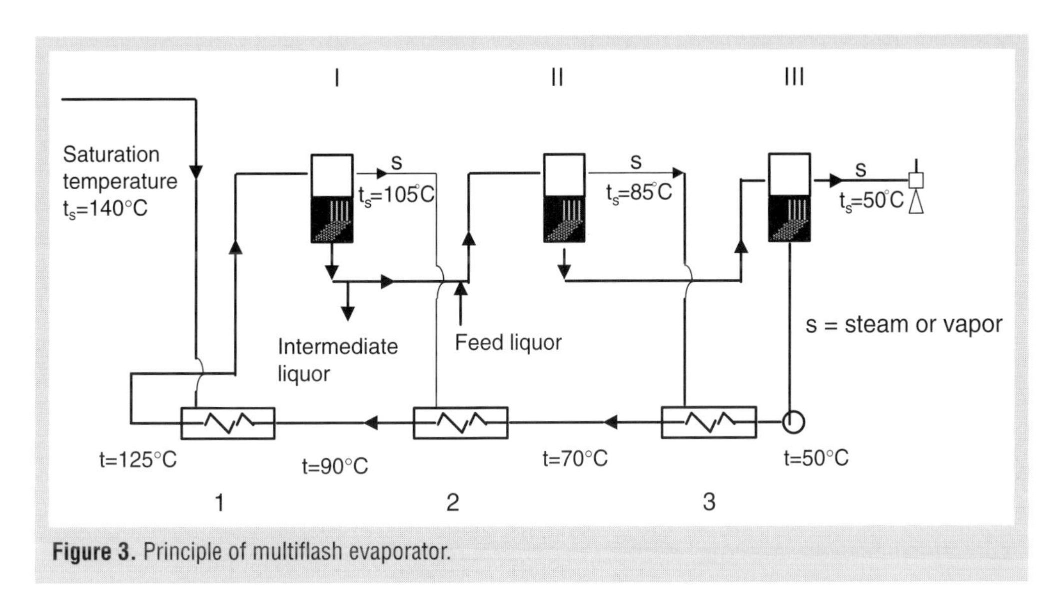

Figure 3. Principle of multiflash evaporator.

Figure 3 shows the principle of the multiflash evaporator. Liquor travels through several heating elements that receive flash steam. The circulation is sufficiently large that all vapor from the flashing stages is consumed for circulating liquor preheating.

3.4 Evaporator plant coupling

The steam flow sequence in a multiple effect evaporation plant is almost always a straight flow from effect 1 downwards. In certain conditions, a parallel feed of live steam to various effects or a partial bypass of one effect might be necessary for capacity reasons.

The black liquor sequence usually (falling film evaporators) has counter flow toward the first effect. The feed liquor temperature will determine the optimum feeding arrangement. The last effect will operate at a temperature of 55°C–60°C. Since the feed liquor temperature is above this, the heat energy in the liquor has best use in the evaporation by flashing the liquor in external vessels or by feeding the liquor into the stage that corresponds to the same temperature. In the second case, the liquor flow in the last effects will have concurrent flow with the steam. In addition, other considerations such as hot water heating for process, soap skimming, separation and treatment of noncondensable gases, integration of a stripper column, fouling considerations, etc., might determine the sequence.

Despite the apparent simplicity of the process, almost all evaporation plants have a design tailored to the individual demands of the mill in question.

Figures 4 and 5 show some typical liquor sequences.

Counter-flow Liquor Sequence

Typical configuration of a counter flow liquor sequence for a 7-effect plant for hardwood without soap skimming with liquor feed → 7 → 6 → 5 → 4 → 3 → 2 → 1A → 1B → 1C is shown in Fig. 4.

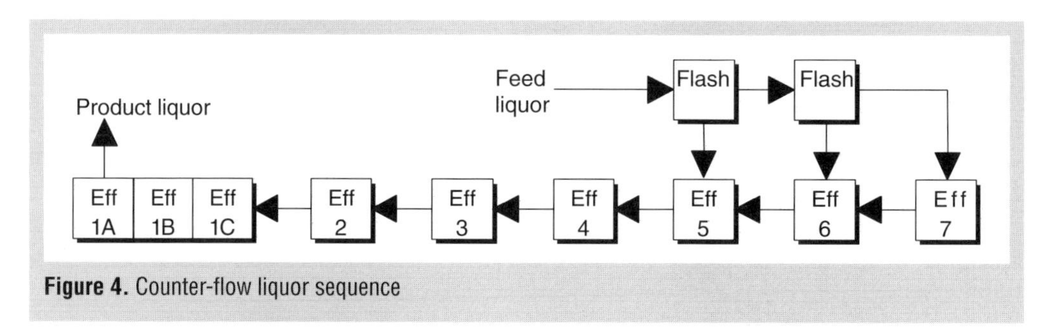

Figure 4. Counter-flow liquor sequence

Parallel-flow liquor sequence

For a parallel flow liquor sequence, like the liquor feed in Figure 3 is → 1 → 2 → 3. This configuration is used in flash steam evaporators as described previously and in evaporators for black liquor that polymerizes easily at a higher temperature (such as ammonium base spent liquor, for example).

Mixed flow liquor sequence

A typical configuration for a 5-effect plant for softwood with soap skimming is shown in Fig. 5.

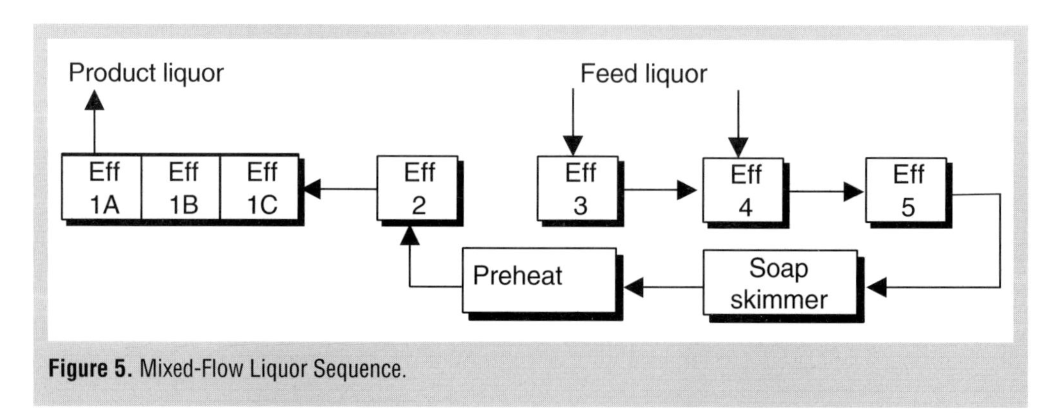

Figure 5. Mixed-Flow Liquor Sequence.

The mixed liquor flow pattern gives the following advantages:

- Soap separation is optimum due to the concentration of 27%–30% dry solids and the low temperature of approximately 60°C.
- Liquor feed to effects 3 and 4 in parallel allows balancing the feed according to the feed liquor temperature and operating temperature in the effects.
- Release of noncondensible gases that enter with the feed liquor is evened between the effects. The gases vent from effects 4 and 5.

3.5 Concentrator coupling

The concentrator is the first effect that evaporates the liquor to the final concentration. It is normally split into several bodies operating in parallel on the steam side and in series on the liquor side. Fouling of the first effect cannot be avoided totally. The arrangement aims to wash the bodies during operation with a minimum loss of capacity. In many cases, switching the sequence of three bodies so that each in turn receives the liquor with the lowest concentration is sufficient to minimize fouling. Another arrangement is to provide a fourth body which is used as a spare, and can be separated for washing on a sequential switching basis.

The modern concept is to mix the recovery boiler precipitator dust and ashes in the mix tank into the liquor stream before the final concentration unit because the ashes can reduce fouling. One arrangement is to make a very small final concentrator unit that requires washing at infrequent intervals. During the washing, the evaporation capacity decreases slightly. This capacity loss is compensated with the buffer storage tanks.

With the very high final concentrations used today, the pressure of the low pressure steam might be insufficient. Some end concentrators require feeding with steam of a higher pressure taken from the mill´s medium pressure steam system.

3.6 Stripper column integration

The stripper column cleans the most contaminated secondary condensate from the evaporation plant and from the digester by separating organic compounds (mostly methanol) using a distillation process.The stripper is a considerable steam consumer and requires a steam flow of approximately 20% of the condensate to be stripped. Most energy in this steam can be recovered and used in the evaporation plant. The stripper can be integrated between effects I and II. It can also be fed from the live steam system and bypass the first effect depending on the relative capacities and available pressure levels. Section 4.5 will provide more details.

The energy from the stripper can also be recovered for other useful purposes such as hot water production, boiler makeup water preheating, etc. Integration into the evaporation plant is the most effective technique.

4 Evaporator design features

4.1 Rising film (RF) evaporator

The rising film evaporators [also called long-tube-vertical (LTV) or Kestner evaporator] were widely used for black liquor evaporation in the pulp industry until mid 1980's. In new installations, the falling film type predominates.

Figure 6 shows the principle of the RF evaporator. The heating element is a single pass, shell and tube heat exchanger mounted vertically with the liquor flow inside the tubes. The tube length is typically 8.5 m with nominal diameter of 50 mm (2 in.) and a wall thickness of 1.5 mm. The tubes are expanded or rolled into the holes of the tube sheets. Liquor is fed into the bottom liquor box where it is distributed to the tubes. Steam or vapor is fed into the shell side of the heating element, and the heat is transferred through the tube walls into the liquor.

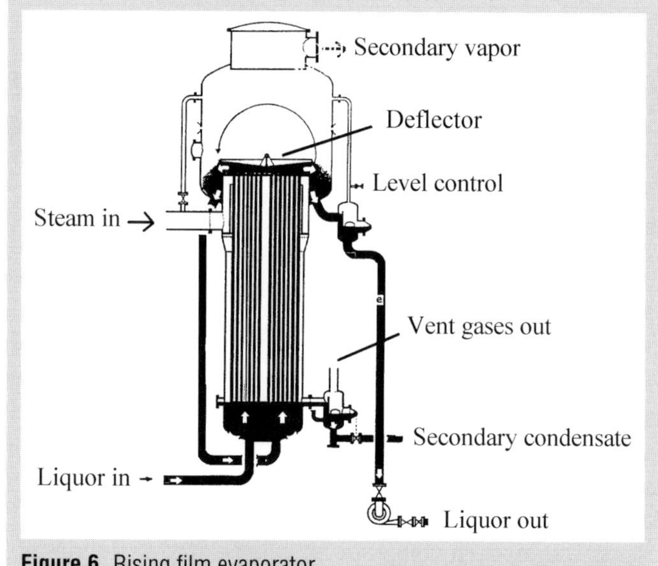

Figure 6. Rising film evaporator.

The liquor rises inside the tubes where it first undergoes preheating and then gradually begins to boil. The vapor released at the boiling has a high specific volume that increases the velocity of the mixture of liquor and vapor, and the heat transfer rate in the heat exchanger. At low evaporation rates, the boiling is unstable. The average heat transfer rate is low, and scaling and foaming problems are common because of local hot spots. Above a partial load of 50%, the flow conditions normally stabilize since the vapor velocity inside the tubes is sufficiently high. This behavior means that RF cannot be used at a low partial load.

Vapor and liquor are separated first by a deflector plate placed above the upper tube sheet. The vapor flows through the vapor head also called vapor body or separator. It is cleaned by an impingement-type drop separator before exhausting to the succeeding evaporator effect heating element. The RF unit especially in the last effects often has a baffling plate in the shell side. Its purpose is to separate the condensate into two fractions so that cleaner vapor condenses before the baffle. More contaminated vapor containing volatile organic sulfur compounds and methanol (MeOH) condenses behind the baffling plate in the after-condensing section. The term for this is condensate segregation.

A disadvantage of the RF type is that a plugged tube cannot be cleaned by washing. It requires manual cleaning by mechanical means or by hydroblasting.

4.2 Falling film (FF) evaporators

In this evaporator type, the liquor is fed to the bottom of the evaporator body where a fixed level is maintained. It is then raised to the top of the heating element by a circulating pump, and flows downward on the heating surface by gravity. The concentration within the effect is practically constant at the concentration of the out-feed liquor. The circulation rate is also constant. These features make the FF evaporator insensitive to variations in the evaporation load. FF evaporators can therefore operate in the load range 30%–100% of the rated capacity, depending mostly on the precision of the control elements. These advantages have resulted in the selection of FF evaporators in most new installations in recent years.

The following designs have common use:

- Plate (lamella)-type FF with liquor flowing on the outside of the vertical plates
- Tubular-type FF with two alternatives:
  - Liquor flows inside vertical tubes
  - Liquor flows outside vertical tubes.

Plate (lamella)-type FF

Black liquor is fed to the bottom of the unit. A circulation pump lifts the liquor to the top part. A liquor distribution system (usually a box with a perforated plate bottom) spreads the liquor evenly on the external surfaces of the heating elements. The liquor flows down and begins to boil. The vapor is separated immediately after generation from the black liquor and flows to the surrounding space. A drop separator at the top of the evaporator ensures the purity of secondary vapor. The heating medium (steam or vapor) is inside the lamella heat transfer surfaces. In effects where segregation of the condensate is desirable, the vapor is fed to the bottom of the heating element. In other

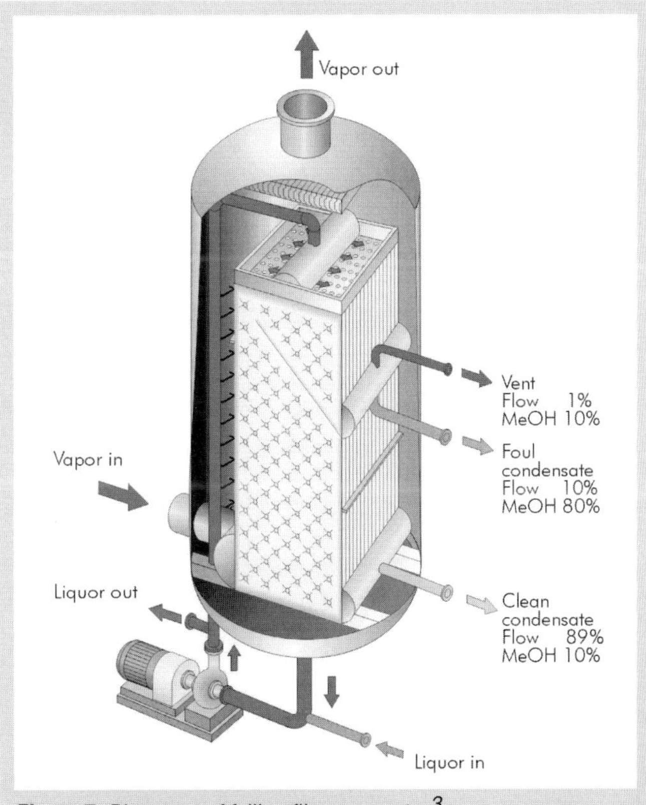

Figure 7. Plate type of falling film evaporator[3].

cases, a top feed is preferable. Figure 7 shows a plate type of falling film evaporator.

Condensate segregation occurs by dividing the heating surface into pre- and after-condensing sections as Fig. 8 shows. Cleaner vapor condenses in the pre-condensation section and volatile, odorous organic sulfur compounds and methanol are carried into the after-condensing section with a small part of the vapor.The vapor feed to the bottom of the heating element increases the efficiency of the segregation through an internal counter flow stripping of the cleaner condensate fraction as it flows down the heating surfaces.

Table 2 shows the efficiency of the condensate segregation when using an internal stripping system.

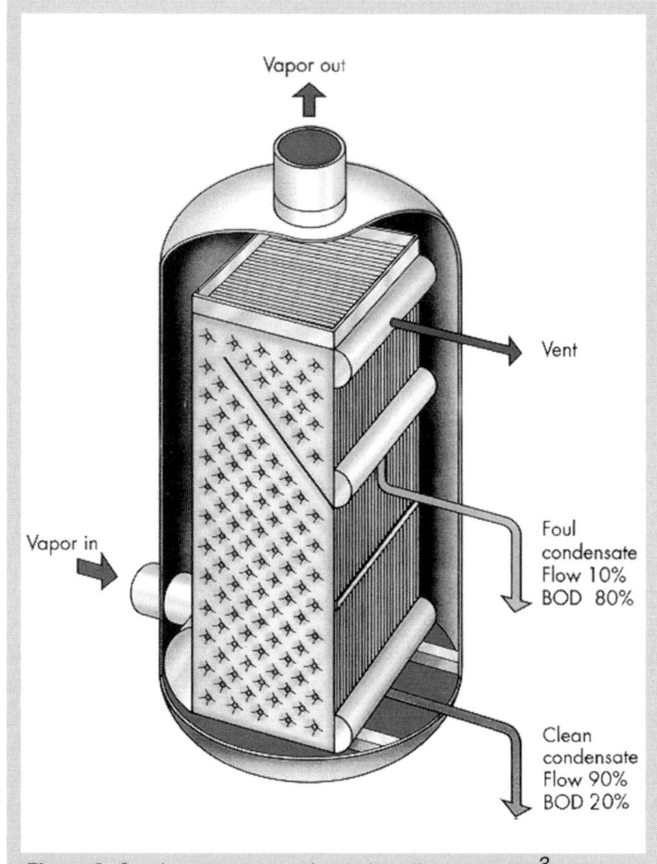

Figure 8. Condensate segregation in lamella evaporator[3].

Table 2. BOD-distribution in evaporation condensates.

	Mass distribution	BOD distribution
Clean condensate fraction	80%–90%	20%
Foul condensate fraction	10%–20%	80%

Tube-type FF

Figure 9 shows a tubular FF with liquor flow inside tubes. The evaporator has a heating element and a vapor body. The heating element is very similar to that of an LTV evaporator consisting of a vertically mounted shell and tube heat exchanger. Liquor is pumped to the liquor distributor that is a tray with a perforated bottom or a spray nozzle type. The liquor flows down inside the tubes mainly by gravity but also assisted by the vapor flow.

The liquor and vapor mixture leaving the tubes passes through the vapor body in the bottom of the evaporator where vapor separates from the liquor. A drop separator cleans the vapor before it exits.

Condensate segregation into clean and foul condensate is achieved by dividing the shell side into pre- and after-condensing sections using the baffle plate of Fig. 10.

For cleaning plugged tubes, the same comments apply as for the RF. Washing cannot clean them.

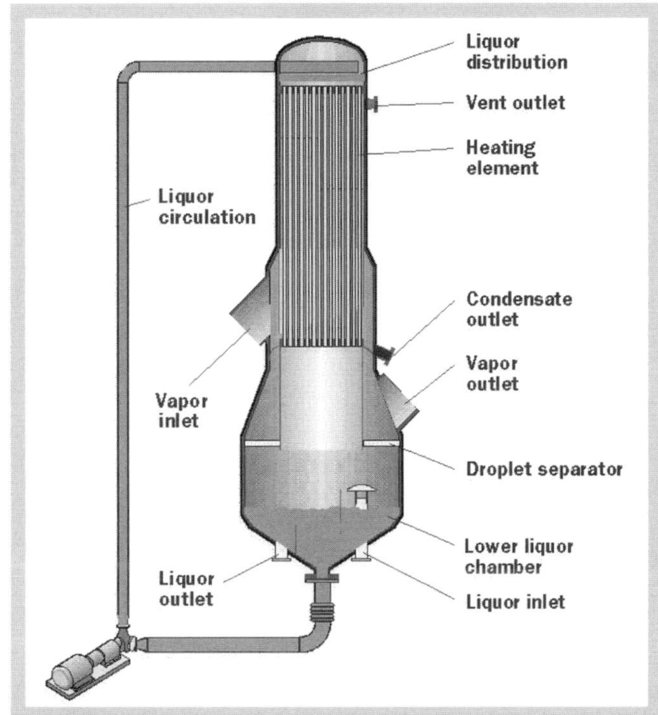

Figure 9. Tube type of falling film evaporator where liquor flows inside the tubes[2].

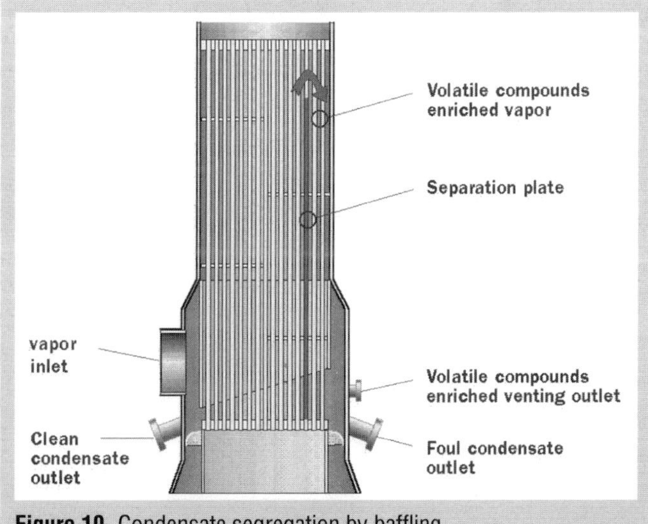

Figure 10. Condensate segregation by baffling.

Figure 11 shows liquor flow outside the tubes. The evaporator construction is more similar to a lamella-type evaporator than to the tubular evaporator with liquor flowing inside the tubes. The tubes are connected to headers that allow an even liquor distribution for the outside of the vertical tubes and the vapor inside the tubes. With the present designs, a disadvantage is that the design limits the vapor side cross-section area. This type cannot be used in the last effects due to insufficient room for the vapor to pass through. Both lamella and tube falling film designs with liquor outside the tubes are available today for medium pressure steam applications.

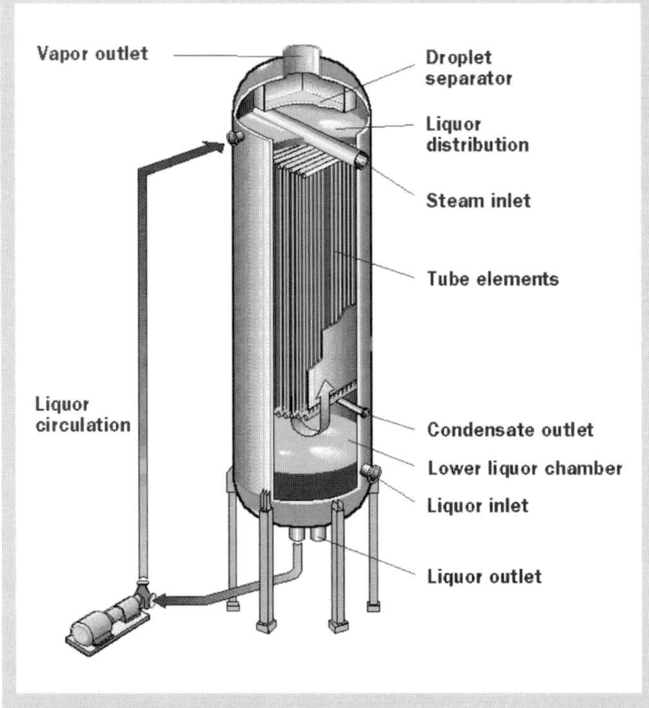

Figure 11. Tube type falling film evaporator where liquor flows outside the tubes[2].

4.3 Concentrators (first effect evaporators)

Evaporation of liquor up to the final concentration of 70%–85% dry solids usually occurs today with the falling film evaporators (now called concentrators) presented above. The applications with high pressure lamella and the tubular falling film with liquor flow outside tubes are the most recent concentrator developments. The main reason for this is the necessity to use a higher steam pressure with the higher final concentrations to overcome the high boiling point rise and operate at higher temperatures due to the increasing viscosity. The practical experience is that both designs are easily washable and the evaporator can run very high concentrations with minor scaling tendency.

The present practice of mixing the sulfate ash from the recovery boiler to the liquor before the final concentration has considerably reduced fouling of the heat transfer surfaces. One reason is that the sulfate crystals act as nuclei for the precipitation of burkeite that form crystals in suspension in the liquor and do not adhere to the heat transfer surfaces.

Concentrators with forced circulation are also used for black liquor final concentration. Figure 12 shows a typical configuration using horizontal heaters. The circulating liquor is heated in the horizontal heat exchangers and then flashed in the vapor head.

Boiling is suppressed in the heater, and the vapor head has a sufficiently large retention time to form crystals from super-saturated liquor. The crystal sludge is recirculated with the incoming black liquor into the heater, and the "mother crystals" effectively prevent fouling of the heating surfaces as explained earlier. The circulation rate and heater tubes have a size such that the flow is turbulent in the heater. This minimizes scaling of the heater tubes. The disadvantage of the design is that the power consumption is high.

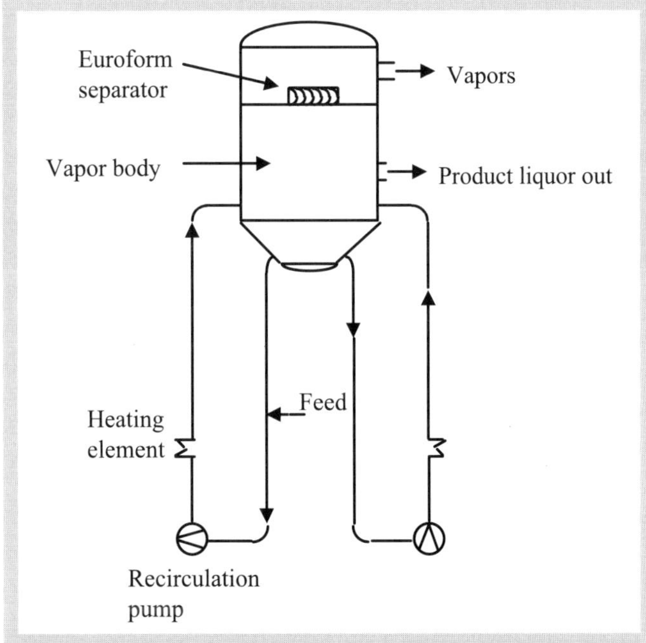

Figure 12. Forced circulation evaporator or crystallizer evaporator.

4.4 Evaporator plant auxiliary equipment

The definition of auxiliary equipment is a matter of preference. In this context, the evaporator effects are the primary equipment. The other smaller equipment is auxiliary. The evaporator plant also includes piping, valves, instrumentation, electrical equipment, etc. They are standard components and so, are not described in detail.

Surface condenser

A surface condenser condenses the secondary vapor from the last evaporator effect. The pressure in the surface condenser and the live steam pressure determines the pressure span and the available total temperature difference over the evaporation plant. The surface condenser pressure depends on the selected heating surface and the temperature of the cooling water. It could theoretically be very low in regions where the cooling water is cold. The cooling water outlet temperature is normally at 45°C–50°C. This is suitable for the warm water demand. Another reason for not using an excessively low surface condenser pressure is that the vapor ducts would become impractically large.

To maintain vacuum and the heat transfer rate, the noncondensible gases that enter the system with the liquor require evacuation. A vacuum device discussed in detail later accomplishes this.

The most common designs of a surface condenser are the following:

- Vapor-water condenser using the falling film principle

- Vapor-water tubular heat exchangers.

The surface condenser with falling film operation is very similar to the FF evaporator presented in Section 4.2. Vapor from the last effect condenses first in the pre- and then the after-condensing section from which the vent gases are drawn into a vacuum device. Cooling water goes to the outside of heating elements, and the water side works at atmospheric pressure preparing warm water at 45°C–50°C. A special application of the FF lamella type surface condenser is to combine it with a cooling tower function. Air is blown with a fan through the lamella battery to dissipate the heat to the ambient air. This arrangement has certain advantages through reduction of the total investment cost and space use. The disadvantage is that it is not possible to produce warm water for process demands.

For tubular-type heat exchangers, the surface condenser has separate pre- and after-condensers. Vapor from the last evaporator effect condenses in the shell side under vacuum. The water side remains under pressure and delivers warm water of 45°C–50°C.

Vacuum device

The vacuum device evacuates noncondensible vent gases from the evaporators and surface condenser. A steam-driven ejector-set or a vacuum pump (normally of the liquid-ring type) are the most commonly used vacuum devices.

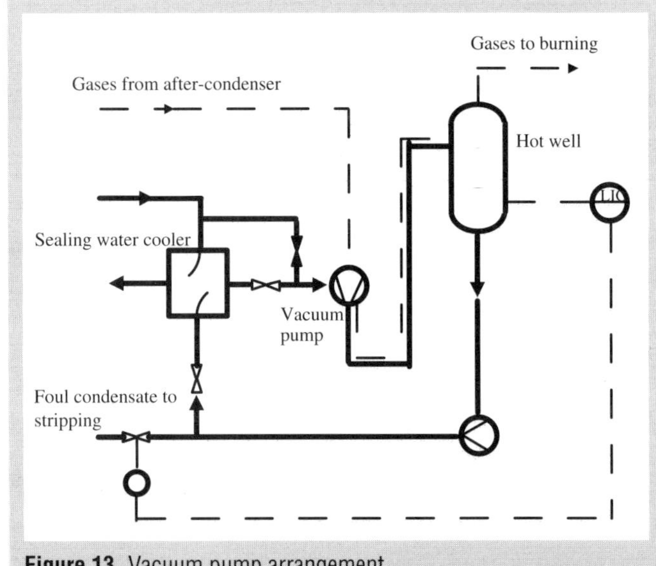

Figure 13. Vacuum pump arrangement.

The selection between a vacuum pump and an ejector depends on the costs assigned to the steam and to the electric power. The installation cost of the ejector is small, but the operation cost often makes it a less economical alternative. Technically, both solutions are acceptable.

To reach the required vacuum of 0.01–0.015 MPa(absolute) in the condenser, a two-stage ejector system is necessary. To reduce the flow through the vacuum device, a system of auxiliary units condenses the vapor in the evacuated gas mixture. These are installed before and after the vacuum device. For the case of an ejector system, installation is also between the stages.

The noncondensible gases are sent into the gas collection and incineration system from the vacuum system. Figure 13 shows a typical vacuum system using a vacuum pump. The vacuum pump has its own closed sealing water circulation system.

Flash and pumping tanks

Flash tanks are used for flashing condensate or liquor between evaporator stages. They are small vessels having inlet and outlet for condensate or liquor with vapor outlet and internals to separate the flash vapor from the liquid in question effectively.

The main functions of the pumping tank for liquor or condensate are to provide a location for a level control measurement and to even small variations in the liquid flow. The tank has a vent discharge on its top.

Pumps

Pumps are necessary for transporting liquor and condensate to, from, and within the evaporation plant. They are usually conventional one-stage centrifugal pumps. In some cases such as circulation pumps for the evaporation effects, special attention is necessary in installation to compensate for thermal expansion of the piping, either by using metallic expansion joints, or by supporting the pumps on a spring-mounted baseplate to allow the entire pump assembly to follow the piping expansion.

Some applications require special pumps. Earlier discussion mentioned the vacuum pump. Soap handling requires positive displacement pumps. For explosion hazard fluids such as methanol or methanol-rich condensate, zero-leakage pumps are necessary.

Some installations have used positive-displacement pumps for the final concentrated liquor. The general experience is that conventional centrifugal pumps are adequate for this purpose. The additional investment and maintenance costs for the positive displacement pumps are not justifiable.

Due to the nature of the fluids handled in the evaporation, all pumps with the possible exception of the primary condensate pumps have mechanical seals. These reduce leaking and avoid dilution of the liquor through leakage of sealing water into the system.

4.5 Stripping column and methanol liquification

Stripping column

The stripping column purifies the foul or contaminated condensate from the evaporation and the cooking plant. Stripping is a mass transfer process where volatile gas components such as methanol and reduced organic sulfur compounds transfer from the liquid phase into the gas phase. Figure 14 shows a typical stripping column. The stripping column is a tall cylindrical vessel where the liquid for stripping flows down by gravity. A mixture of steam and noncondensible gases rises upward in counter flow. The mass transfer process is enhanced by intermediate bottoms in the column that divide the heating and degasification of the liquor in stages. The bottoms have valves or bubble caps that disperse the vapor and gas mixture in a liquid layer.

The bottom on which the feed enters is the feed bottom. Its location is in an intermediate stage of the column. All bottoms above the feed bottom constitute the rectifying section, and all bottoms below the feed including the feed bottom itself constitute the stripping section.

The vapor (approximately 20% of the flow for stripping) can come from an evaporation effect or be generated by a reboiler. The reboiler is a steam- or vapor-heated heat exchanger that generates vapor. The reboiler can be the natural circulation type or have a circulation pump working with the falling film principle. If the additional cost of losing the fresh steam condensate is acceptable, direct live steam use instead of a reboiler is also possible.

The stripped condensate is withdrawn from the stripper column bottom and used to preheat the foul condensate feed. The vapor and volatile compounds rising through the rectifying section partly condense in an overhead condenser that can be a section of an evaporator dedicated for this purpose or a separate black liquor pre-heater.

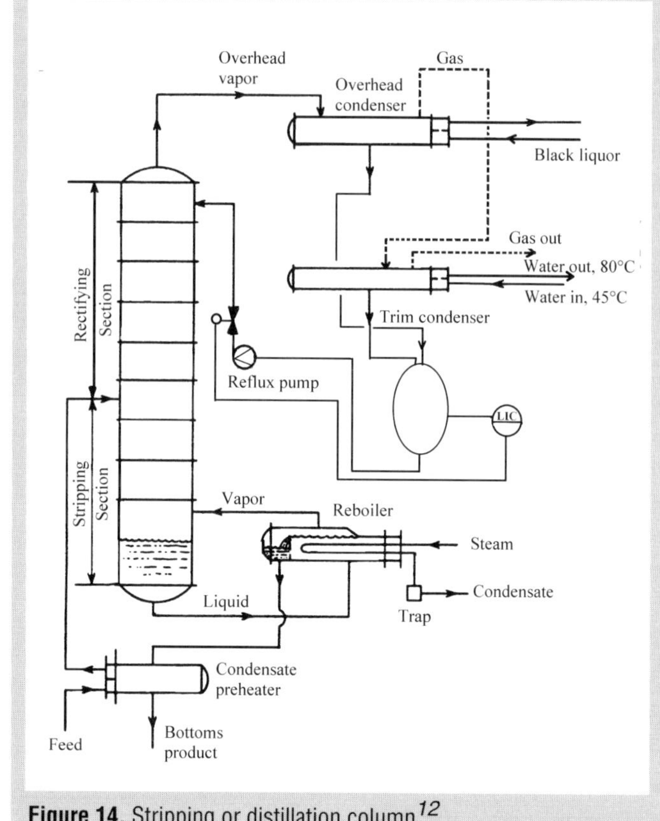

Figure 14. Stripping or distillation column[12].

Vapor and gases proceed into the trim condenser where the methanol content of the outlet gas is adjusted to approximately 35%–40% by volume. Warm water is used as coolant in the trim condenser. If heated to 70°C–80°C, it is suitable for use as hot water in the process.

The exit gas from the stripping column top contains vapor, methanol, and organic sulfur compounds. The purpose of the rectifying section of the stripping column where the vapor rises in counter flow to the recirculated (refluxed) condensate from the condensers is to concentrate the contaminants in the exit gas.

The purification rate of foul condensate is normally 90%–95% for methanol and 98%–99% for organic sulfur compounds.

Methanol liquefaction

The methanol liquefaction uses the same type of distillation column as the foul condensate stripping in Fig. 14. The methanol-containing vapor from the stripping at a concentration of 35%–45% methanol by volume is introduced into the bottom part of the distillation column. Below the vapor and methanol feeding point, some fresh steam or vapor is added.

In the methanol distillation, an end product concentration of 80% methanol in the liquid phase is reached in the condenser. The methanol is pumped to an intermediate storage tank and then to the incinerator. The bottom product from the column is often handled in a turpentine decanter before returning it to the foul condensate tank of the stripper system.

4.6 Direct contact (DC) evaporator

The DC evaporator is an integral part of the recovery boiler. The description here is short only to complete the discussion of evaporation. Note that this type is not acceptable in areas with strict regulations for environmental protection due to the difficulties of handling the emissions caused by the DC evaporator.Some installations of this type still exist, although the tendency is to modify the boilers to the "low-odor" type, i.e., the normal concept for the modern recovery boiler.

In the DC evaporator, black liquor is concentrated to the firing liquor concentration by allowing hot flue gas to enter in direct contact with the liquor. The flue gases vaporize water from the black liquor and simultaneously cool from 350°C–400°C to 160°C–180°C. The liquor is concentrated from 50%–55% dry solids to 60%–67% dry solids. To achieve its function, the contact area between the gases and the liquor must be large, and the liquor must remain in motion. The most common configurations of the DC evaporator are the following:

Cascade type DC evaporator
This type has a rotary element made of tubes that pick the liquor from a vat and pass it through the flue gas stream where the evaporation occurs. (Fig. 15)

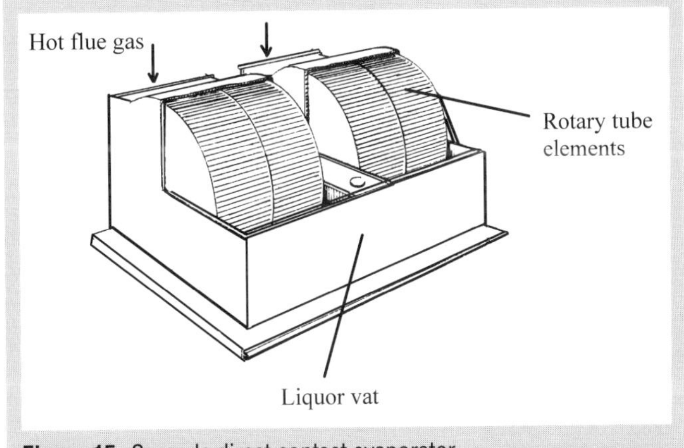

Figure 15. Cascade direct contact evaporator.

Cyclone DC evaporator
In the cyclone evaporator, the liquor for evaporation sprays into the flue gas flow, and the concentrated liquor droplets are thrown by centrifugal force to the walls of the cylindrical cyclone. The concentrated liquor is collected from the cyclone bottom and pumped to the recovery boiler mix tank. (Fig. 16.)

The DC evaporator releases large amounts of hydrogen sulfide and other odorous sulfur compounds into the atmosphere, due to the reaction between sodium sulfide in the liquor, water and carbon dioxide in the flue gas.

To reduce the hydrogen sulfide generation in the DC evaporator, the black liquor is often oxidized to convert sodium sulfide to sodium thiosulfate and sodium sulfate. The oxidation uses air or molecular oxygen. To avoid odorous emissions, the entire pulping process sometimes operates at very low sulfidity of 7%–10% compared with a normal value of 25%–35%. This influences the economy of the mill operation negatively in several ways.

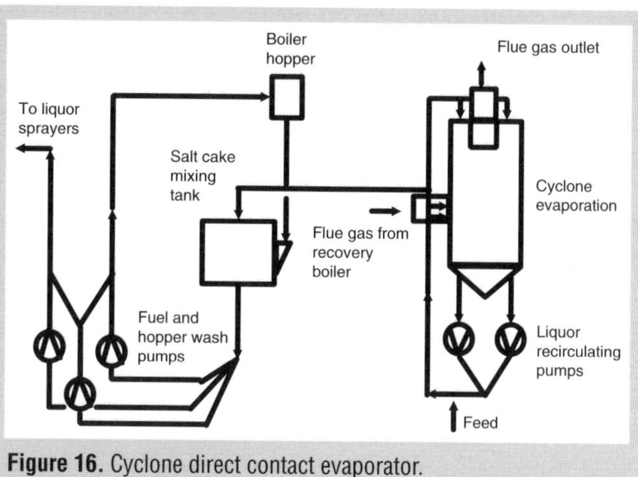

Figure 16. Cyclone direct contact evaporator.

5 Heat transfer

5.1 Heat transfer of evaporators

The general heat transfer equation below is valid for the heat transfer through a heating element of an evaporator:

$$Q = U \times A \times \Delta t \tag{1}$$

where Q is rate of heat transfer, W
U heat transfer coefficient, W/m^2°C
A heat transfer area, m^2
Δt effective temperature difference or thermal driving force (= steam or vapor saturated temperature minus liquor temperature, °C.) (Note that this is not the logarithmic Δt normally used in heat transfer calculations.)

The heat transfer coefficient, U, is the inverse of the sum of the heat transfer resistances:

$$1/U = R_C + R_A + R_S + R_L \tag{2}$$

where R_C is resistance of the condensate film (steam/condensate side)
R_A resistance of the heating surface material
R_S resistance due to scaling
R_L resistance of the liquor side.

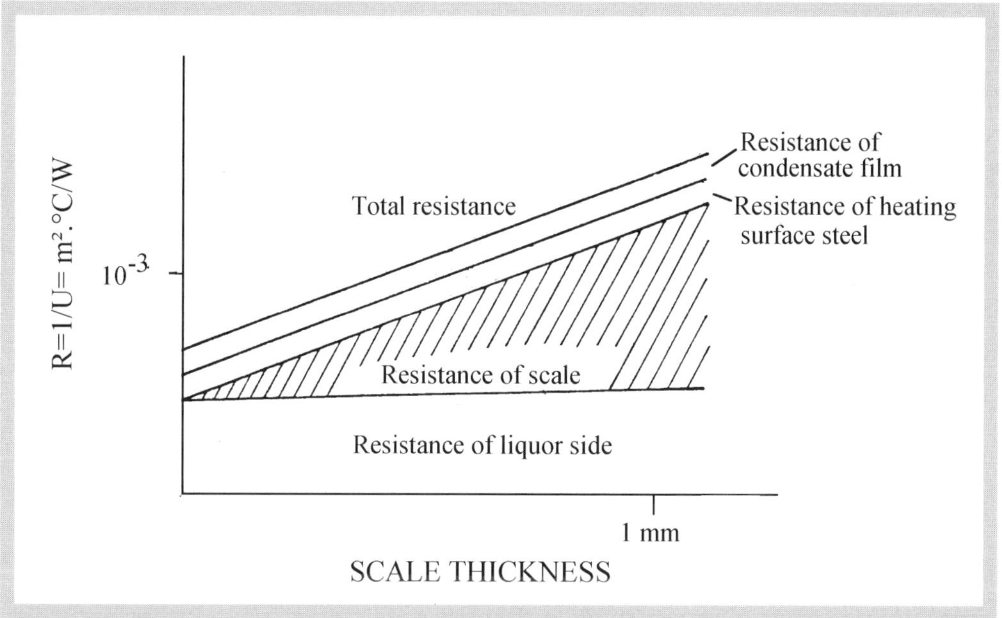

Figure 17. Different heat transfer resistances and how scaling influences total resistance in a forced circulation evaporator.

Figure 17 shows the magnitudes of the heat transfer resistances of an evaporator. The resistance of the scale is predominant when the heating surface becomes fouled.

If the venting of noncondensible gases from the heating element is not done properly, the resistance on the condensate side increases considerably.

The value of U depends on the evaporator type and the physical properties of liquor. It is determined empirically by the evaporator supplier. The values of U are very similar for the tubular and lamella falling film evaporators. They are almost independent of the evaporation rate. The falling film evaporator can operate down to 30% of the rated capacity.

The values of U for the rising film evaporator depend on the evaporation rate. They are low when the specific evaporation rate is low. The tubular rising film evaporator needs a sufficiently high vapor velocity inside the tubes so a stable film can form on the tube walls. In practice, the rising film evaporator cannot operate under 50%–60% of the plant design capacity. Table 3 gives the typical values for U for the falling and rising film evaporators for use in evaporator design.

In Fig. 18, the value of U is a function of the evaporation rate (the heat flux) of the rising film evaporator[7].

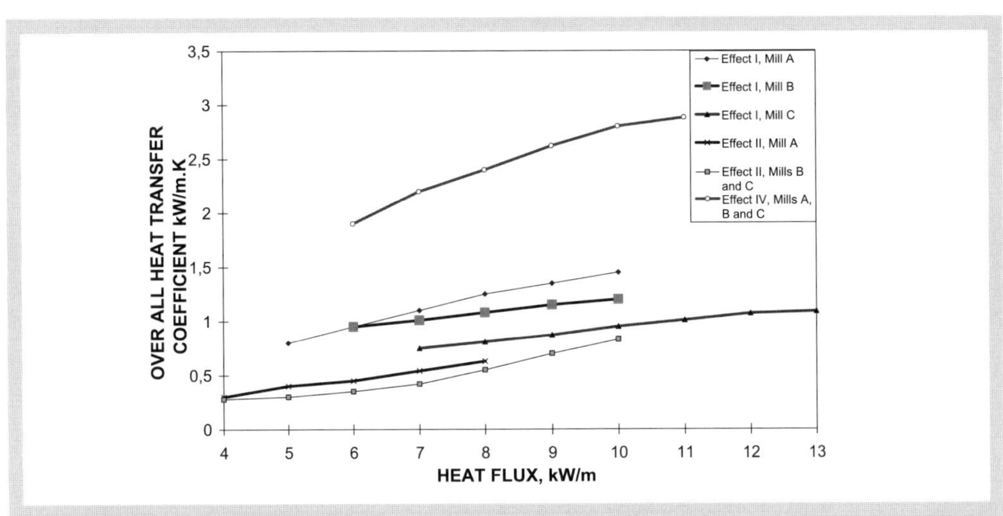

Figure 18. Total heat transfer coefficient vs. heat flux in rising film effects I, II, and IV. Liquor flow pattern: effects IV – V – VI – III – I – II[7].

Table 3. Heat transfer coefficient of the falling and rising film evaporators for heating surface design.

Effect	Falling film, W/°C • m²	Effect	Rising film, W/°C • m²
IA			
IB	Average for IA/IB/IC	I	400–600
IC	500–1 000		
II	1 300–1 550	II	700–1 000
III	1 600–1 700	III	1 200–1 500
IV	1 900–2 000	IV	1 700–1 900
V	1 900–2 000	V	1 100–1 400
VI	1 800–2 000	VI	600–800
VII	1 600–2 000		

Liquor flow pattern:	Liquor flow pattern:
Effects VII => VI => V => IV => III => II => IA => IB => IC	Effects VI => V => VI => III =>II => I

Product liquor concentration	Product liquor concentration
73% dry solids	63% dry solids

5.2 Heat transfer theory and calculation in multiple effect evaporation

The heat transfer calculation for a multiple effect evaporation plant is special for the following reasons:

- The liquor concentration and also the steam conditions change through the plant. This influences the characteristics of the liquor so that each effect must be calculated with different parameters.
- The BPR has a considerable influence on the heat transfer, since it defines the available effective temperature difference as explained in various contexts before.
- In a multieffect plant, the total temperature difference is split between the effects according to various parameters. A method for calculating this follows below.
- The calculation is iterative by its nature. It is extremely elaborate to make manually especially when trying for high precision. A spreadsheet PC program provides the easiest calculation method.

Calculation process sequence

The following gives the typical sequence for the calculations in an evaporation plant.

Given parameters:

- Dry solids flow
- Liquor concentration into the evaporation plant
- Liquor temperature into the evaporation plant
- Liquor concentration from the evaporation plant
- Steam pressure and temperature into the evaporation plant
- Primary condensate temperature (if specified)
- Secondary condenser cooling water outlet temperature (warm water production)
- Number of stages in the plant.

Define the concentrations:

- Inlet and outlet concentrations are given. This defines the total evaporation in the plant.
- Assume that the evaporation in each stage is the same (as a first approximation). This gives the evaporation in each stage and consequently the concentrations.

Define the liquor characteristics in all the stages:

- BPR
- Specific heat
- Heat transfer coefficients

Define the effective temperature difference in all the stages:

- The live steam saturation temperature and the cooling water outlet temperature define the total temperature difference.
- Pressure and temperature differences due to piping losses are defined. Most important are to define the pressure loss of live steam before the first effect and the temperature difference in the surface condenser.
- When the effective temperature difference is defined, this is distributed to the different stages with the method described in detail below. The approximate total effective temperature difference of the evaporation plant is in Fig. 19.

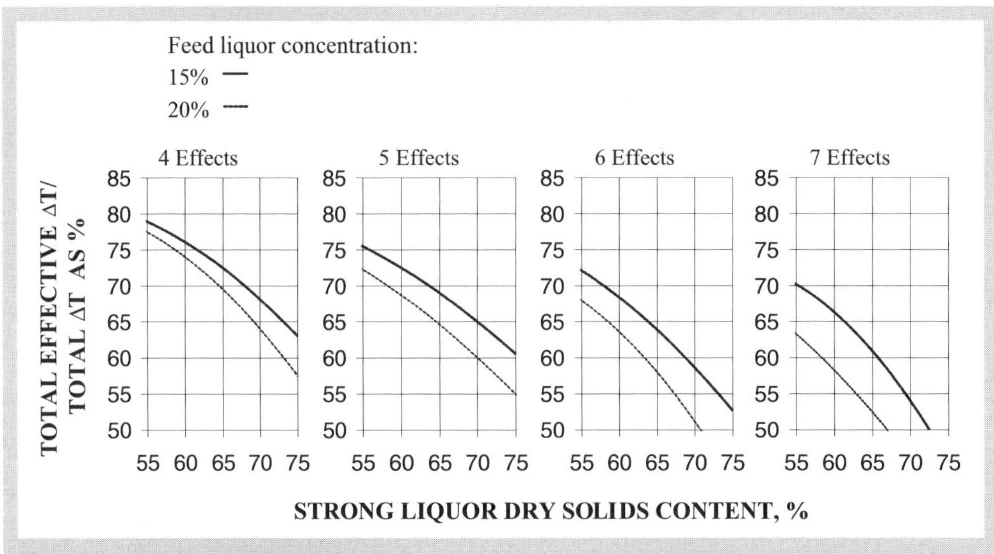

Figure 19. How the total effective Δt of the evaporation plant depends on the number of stages and the feed liquor and strong liquor concentrations.

Prepare the heat balance for each stage:

The main heat flow is in the steam or vapor transferred from one effect to another.

Besides this, smaller energy flows require consideration such as

- liquor preheating

- flashing of condensate
- flashing of feed liquor to one or several stages

- The necessary heat transfer surface is defined.

Distributing the effective Δt between the effects

The concepts used in the evaporation calculations have the following definitions:

Total temperature difference

= Total Δt

= Live steam saturation temperature of the heating element of the first effect minus condensing temperature in the surface condenser.

Total effective temperature difference

= Total effective Δt = ΣΔt

= Total thermal driving force

= Total Δt minus sum of boiling point rises minus sum of temperature drops due to pressure losses (vapor ducts, heating elements).

The total effective Δt for the multiple evaporation is the sum of the effective Δt of the effects:

$$\Sigma\Delta t = \Delta t_1 + \Delta t_2 \ldots + \Delta t_n \tag{3}$$

where Δt_1 is thermal driving force of effect 1
Δt_2 thermal driving force of effect 2
Δt_n thermal driving force of effect n.

For each effect, the general heat transfer formula is valid:

$$Q = U \times A \times \Delta t \tag{4}$$

Rewriting this gives:

$$\Delta t = \frac{Q}{A \times U} \tag{5}$$

For the evaporator train, an equation group can be constructed:

$$\Delta t_1 = \frac{Q_1}{A_1 \times U_1}, \Delta t_2 = \frac{Q_2}{A_2 \times U_2}, \Delta t_3 = \frac{Q_3}{A_3 \times U_3} \ldots\ldots \Delta t_n = \frac{Q_n}{A_n \times U_n} \tag{6}$$

To solve this equation group, the relations between the effects are formed by dividing the equations so that all are expressed as a function of Δt_1:

$$\Delta t_2 = \Delta t_1 \times \frac{Q_2 \times A_1 \times U_1}{Q_1 \times A_2 \times U_2},\ \Delta t_3 = \Delta t_1 \frac{Q_3 \times A_1 \times U_1}{Q_1 \times A_3 \times U_3} \ldots\ldots \Delta t_n = \Delta t_1 \times \frac{Q_n \times A_1 \times U_1}{Q_1 \times A_n \times U_n} \quad (7)$$

The effective temperature difference for effect 1 can then be calculated:

$$\Delta t_1 = \frac{\Sigma \Delta t}{\left(1 + \frac{Q_2 \times A_1 \times U_1}{Q_1 \times A_2 \times U_2} + \frac{Q_3 \times A_1 \times U_1}{Q_1 \times A_3 \times U_3} + \ldots\ldots \frac{Q_n \times A_1 \times U_1}{Q_1 \times A_n \times U_n}\right)} \quad (8)$$

When Δt_1 has been determined, the temperature difference for the other effects can be calculated with the formulas (8).

5.3 Heat balance

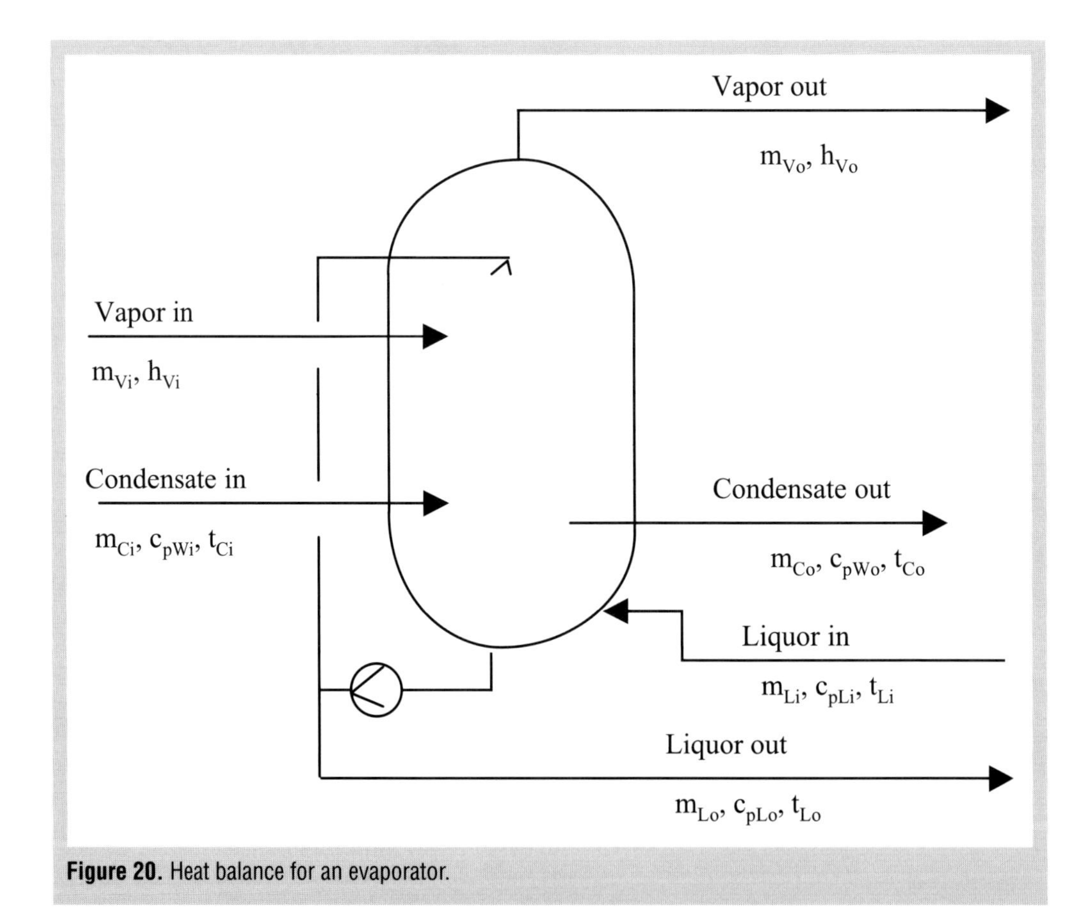

Figure 20. Heat balance for an evaporator.

Based on the designations of Fig. 20, Table 4 shows the heat balance of an evaporator.

Table 4. Heat balance of an evaporator

	Flow	Enthalpy or specific heat	Temperature
Heat flow in			
Vapor (or live steam)	m_{Vi}	h_{Vi}	
Condensate	m_{Ci}	c_{pwi}	t_{Ci}
Liquor	m_{Li}	c_{pLi}	t_{Li}
Heat flow out			
Vapor	m_{Vo}	h_{Vo}	
Condensate	m_{Co}	c_{pwo}	t_{Co}
Liquor	m_{Lo}	c_{pLo}	t_{Lo}

When heat in = heat out, the following is true:

$$m_{Vi}{}^*h_{Vi} + m_{Ci}{}^*c_{pwi}{}^*t_{Ci} + m_{Li}{}^*c_{pLi}{}^*t_{Li} = m_{Vo}{}^*h_{Vo} + m_{Co}{}^*c_{pwo}{}^*t_{Co} + m_{Lo}{}^*c_{pLo}{}^*t_{Lo} \tag{9}$$

Liquor preheat or flashing

If the black liquor entering into an effect has a temperature lower than the boiling point, it will consume steam for preheating. If the temperature is higher than the boiling point temperature in the effect, it will contribute to the vapor flow by flashing off steam.

The following equation that gives a positive value for flashing and a negative value for preheating can calculate the vapor flow in both cases:

$$\text{Vapor flow} = \frac{m_{Li}{}^*c_{pLi}{}^*t_{Li} - m_{Lo}{}^*c_{pLo}{}^*t_{Lo}}{r_2} \tag{10}$$

where r_2 is latent heat of vapor outlet.

Condensate flashing

Flashing the vapor condensate from one effect into the following effect is usually advantageous. The amount of vapor gained from the flashing is the energy difference in the condensate in and out of the flash tank.

Condensate flashing = energy in condensate in – energy in condensate out

$$\text{Flashing} = \frac{m_{Ci}{}^*c_{pWi}{}^*t_{Ci} - m_{Co}{}^*c_{pWo}{}^*t_{Co}}{r_2} \tag{11}$$

5.4 Evaporator calculation

Example 1

An Excel spreadsheet shows the calculation for a three-stage evaporation. The basic design data are in lines 4–11. The calculation starts by assuming the evaporation in the first stage. This then determines the heat energy available in the following effects. The characteristics of the liquor and vapor are calculated for each stage. A flashing of the condensate from effects 2 and 3 has been considered. The energy demand for the effects are calculated in lines 76–79. After the heat transfer areas are given in lines 91–94, the heat transfer capacity is calculated in lines 95–98. This should be larger than the heat demand. Figure 21 shows the temperature profile of the plant.

This is a simplified calculation. It does not consider internal pressure losses between the effects or the losses of noncondensible gases that vent from the evaporation. However, the impact of these factors on the heat balance is small.

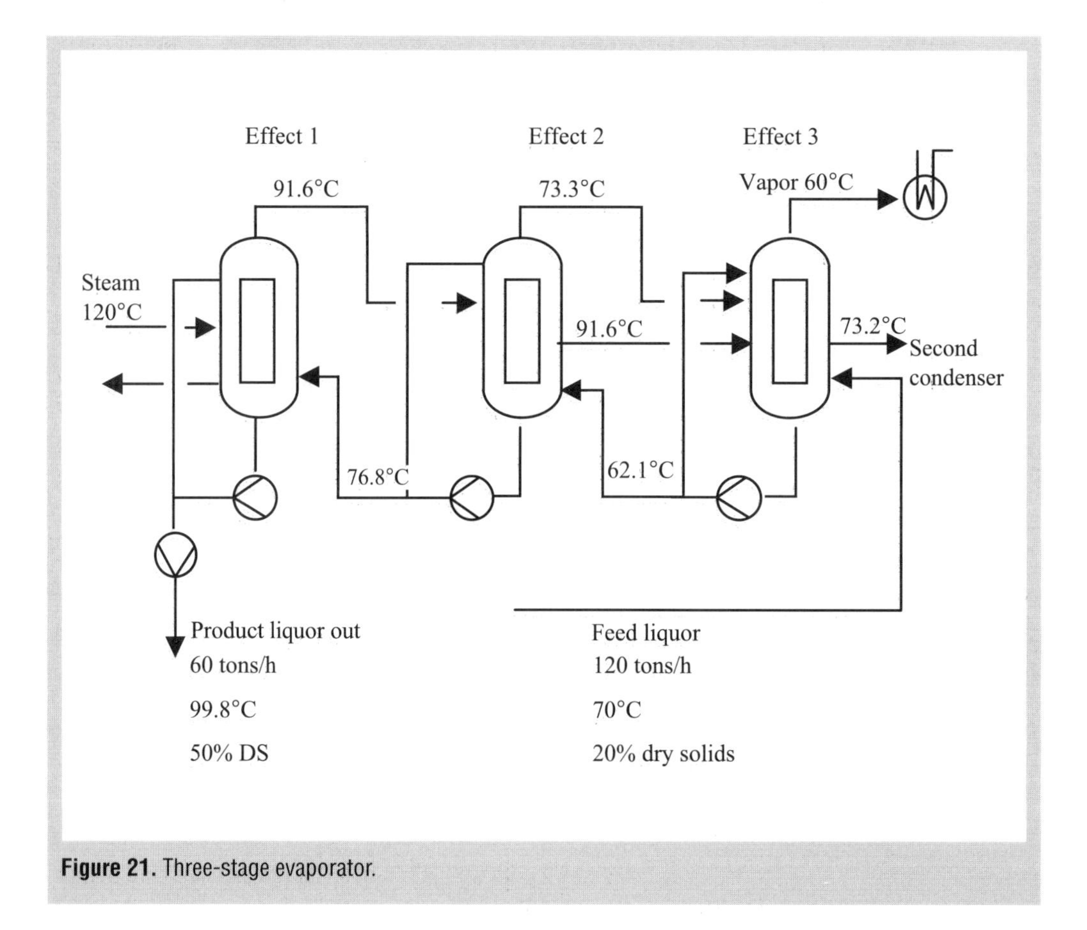

Figure 21. Three-stage evaporator.

A	B	C	D	
2	SIMPLIFIED EVAPORATION CALCULATION			
3	**Basic design data**			
4	Number of effects	–	3	Input value
5	Liquor flow pattern	–	**Effect 3 > effect 2 > effect 1**	
6	Black liquor inlet concentration	–	**0.20**	Input value
7	Black liquor outlet concentration	–	**0.50**	Input value
8	Black liquor inlet temp. effect 3	°C	**70**	Input value
9	Evaporation capacity	kg/s	**30**	Input value
10	Live steam sat. temp. to effect 1	°C	**120**	Input value
11	Sat. vapor temp. out of effect 3	°C	**60**	Input value
12	**Dry solids calculation**			
13	Dry solids flow	kg/s	10.00	= D9/(1/D6–1/D7)
14	Evaporation			
15	Effect 1	kg/s	**10.0**	Input value
16	Effect 2	kg/s	9.5	= D91
17	Effect 3	kg/s	10.4	= D92
18	Total evaporation	kg/s	29.9	= SUM(D15:D17)
19	Liquor flows			
20	Effect 1 out	kg/s	20.0	= D13/D7
21	Effect 2 out	kg/s	30.0	= D20+D15
22	Effect 3 out	kg/s	39,5	= D21+D16
23	Effect 3 in	kg/s	49.,9	= D22+D17
24	**Liquor concentrations**			
25	Effect 1 out	–	0.50	= D13/D20
26	Effect 2 out	–	0.33	= D13/D21
27	Effect 3 out	–	0.25	= D13/D22
28	Effect 3 in	–	0.20	= D13/D23
29	**Boiling point rise**			
30	Effect 1	°C	8.2	BPR formula, see 11.2.1.6
31	Effect 2	°C	3.6	BPR formula, see 11.2.1.6
32	Effect 3	°C	2.1	BPR formula, see 11.2.1.6
33	**Liquor temperatures**			
34	Effect 1 out	°C	99.8	= D10–D46
35	Effect 2 out	°C	76.8	= D52–D47
36	Effect 3 out	°C	62.1	= D53–D48
37	Effect 3 in		70.0	= D8
38	**Heat transfer coefficient**			

A	B	C	D	
2	SIMPLIFIED EVAPORATION CALCULATION			
39	Effect 1	$kW/m^2°C$	**1.2**	Input value
40	Effect 2	$kW/m^2°C$	**1.6**	Input value
41	Effect 3	$kW/m^2°C$	**2.0**	Input value
42	**Effective Δt calculation**			
43	Total available Δt	°C	60.0	= D10–D11
44	Sum of BPR	°C	13.8	= D30+D31+D32
45	Total effective Δt	°C	46.2	= D43-D44
46	Effective Δt, effect 1	°C	20.2	= D45/(1+(D82*D96*D39)/(D81*D97*D40)+ (D83*D96*D39)/(D81*D98*D41))
47	Effective Δt, effect 2	°C	14.7	= D46*(D82*D96*D39)/(D81*D97*D40)
48	Effective Δt, effect 3	°C	11.2	= D46*(D83*D96*D39)/(D81*D98*D41)
49	Check sum		46.2	= SUM(D46:D48)
51	**Vapor saturation temperatures**			
52	Live steam	°C	120.0	= D10
53	Effect 1	°C	91.6	= D34–D30
54	Effect 2	°C	73.3	= D35–D31
55	Effect 3	°C	60.0	= D36–D32
56	**Vapor enthalpy**			
57	Live steam	kJ/kg	2704	Steam table or enthalpy formula
58	Effect 1	kJ/kg	2661	Steam table or enthalpy formula
59	Effect 2	kJ/kg	2631	Steam table or enthalpy formula
60	Effect 3	kJ/kg	2609	Steam table or enthalpy formula
61	**Condensate enthalpy**			
62	Live steam	kJ/kg	503	Steam table
63	Effect 1	kJ/kg	384	Steam table
64	Effect 2	kJ/kg	307	Steam table
65	Effect 3	kJ/kg	251	Steam table
66	**Liquor specific heat**			
67	Effect 1	kJ/kg, °C	3.0	Formula for liquor specific heat 11.2.1.5
68	Effect 2	kJ/kg, °C	3.4	Formula for liquor specific heat 11.2.1.5
69	Effect 3	kJ/kg, °C	3.6	Formula for liquor specific heat 11.2.1.5
70	Inlet liquor	kJ/kg, °C	3.7	Formula for liquor specific heat 11.2.1.5
71	**Energy in liquor**			
72	Effect 1 out	kJ/s	6063	= D20*D34*D67
73	Effect 2 out	kJ/s	7856	= D21*D35*D68
74	Effect 3 out	kJ/s	8783	= D22*D36*D69
75	Inlet liquor	kJ/s	12964	= D23*D37*D70

A	B	C	D	
2	SIMPLIFIED EVAPORATION CALCULATION			
76	**Energy demand, preheat+evaporation**			
77	Effect 1	kJ/s	24819	= D15*D58+D72-D73
78	Effect 2	kJ/s	24116	= (D15+D89)*(D58-D63)
79	Effect 3	kJ/s	22934	= (D82+D90)*(D59-D64)
80	**Evaporation**			
81	Effect 1	kg/s	10.0	= D15
82	Effect 2	kg/s	9.5	= (D78+D74-D73)/D59
83	Effect 3	kg/s	10.4	= (D79+D75-D74)/D60
84	**Vapor production/demand**			
85	Effect 1	kg/s	11.3	= D77/(D57-D62)
86	Effect 1 to effect 2	kg/s	10.6	= D81+D89
87	Effect 2 to effect 3	kg/s	9.9	= D82+D90
88	**Condensate flash**			
89	Effect 1 to effect 2	kg/s	0.6	= D85*(D62-D63)/(D58-D63)
90	Effect 2 to effect 3	kg/s	0.3	= D86*(D63-D64)/(D59-D64)
91	**Heat transfer area**			
92	Effect 1	m^2	**1040**	Input value
93	Effect 2	m^2	**1040**	Input value
94	Effect 3	m^2	**1040**	Input value
95	**Heat transfer capacity**			
96	Effect 1	kJ/s	25240	= D39*D46*D92
97	Effect 2	kJ/s	24525	= D40*D47*D93
98	Effect 3	kJ/s	23323	= D41*D48*D94

5.5 Multiple effect evaporation

The following summarizes some practical applications of evaporation technology.

The multiple effect evaporation has use in almost all pulp mills operating with the sulfate process. The number of effects is usually 5–7. The optimum number of effects depends on the steam balance of the mill. Saving steam is not always economical.

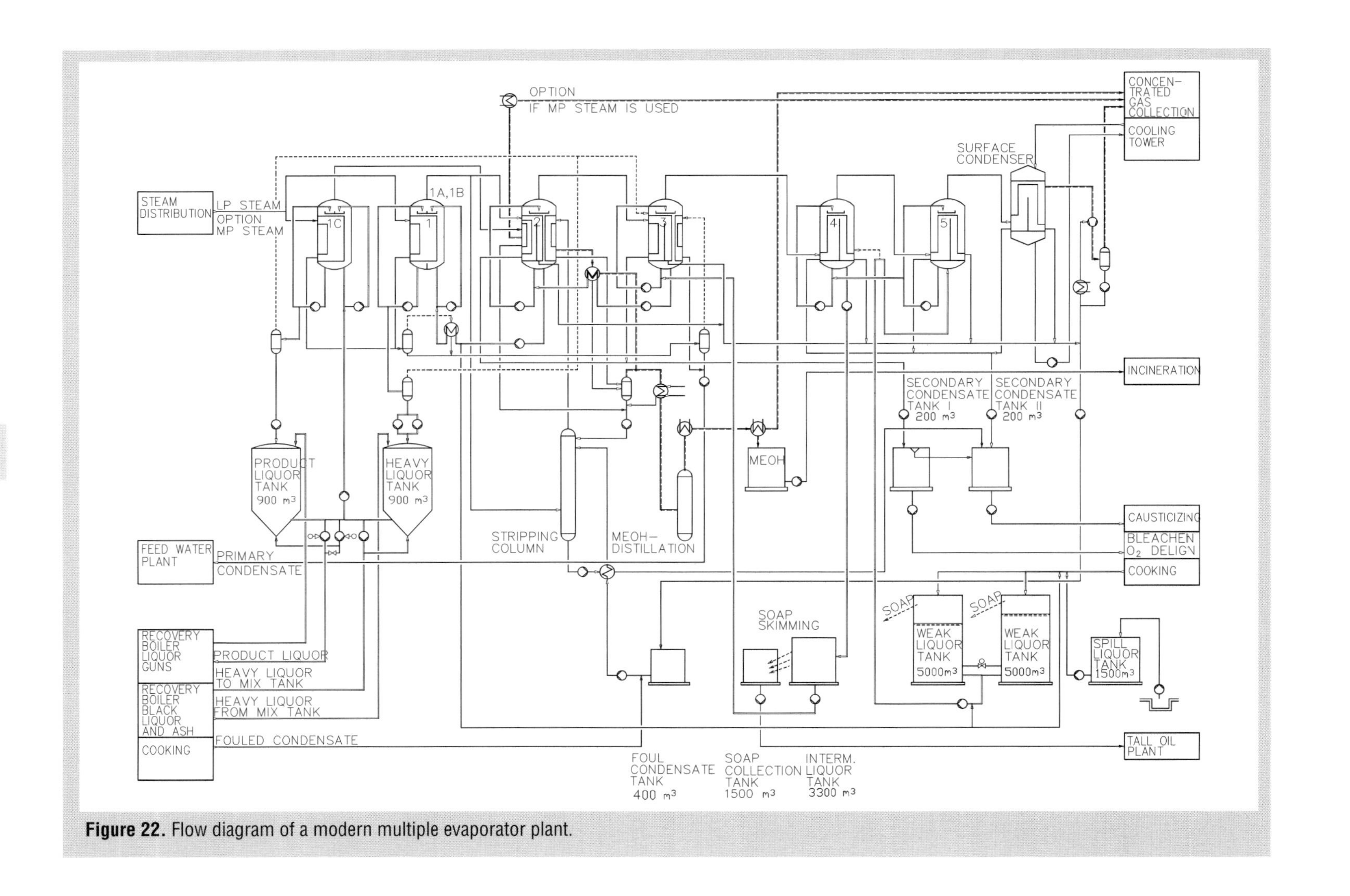

Figure 22. Flow diagram of a modern multiple evaporator plant.

The application in Fig. 22 is typical for a northern pulp mill using softwood as raw material.

Steam sequence

The steam sequence is straight downstream. This is almost always true. Live steam comes from the mill´s low pressure steam distribution system at a pressure of 0.35–0.45 MPa (absolute). This corresponds to a saturation temperature of 139°C–148°C. The live steam is fed into the heating element of the first effect. Vapor generated in the liquor side of effect 1 is led into the heating element at effect 2 and from there into effect 3 and so on. Finally, vapor from the last effect at 57°C–60°C condenses in the surface condenser.

Liquor flow sequence

Several possibilities exist to arrange the liquor flow sequence in the evaporation plant. The following shows a sequence suitable for evaporation of softwood liquor to a final concentration of 75%–80% dry solids:

Weak liquor tank => flashing to effect 4 => effect 5 => effect 4; => intermediate liquor tank (for soap separation) => evaporation in effect 3 => effect 2 => effect 1A => effect 1B => intermediate heavy liquor tank => evaporation in effect 1C; => product liquor tank (pressurized) => recovery boiler furnace

Feed liquor concentration

The weak black liquor from the washing normally has a concentration of 14%–18% dry solids. Occasionally, it may be lower due to washing or spills. In some cases, this is too low because of foaming, soap separation, etc, Then the liquor is "sweetened." This means increasing the concentration by recirculating an heavier intermediate liquor into the feed liquor or into the weak liquor tank.

If the liquor contains soap, the feed liquor is usually sweetened to a concentration of 18%–22% dry solids to facilitate the separation of easily washable soap already in the feed liquor tanks. The rest of the soap is then skimmed at a concentration of 25%–30% dry solids in the intermediate liquor tank = soap skimmer tank.

For hardwood liquor with a low soap content, the sweetening can be smaller or not made at all.

Recovery boiler ash handling

Until recently, the normal technique was to pump the product liquor from the evaporation plant at a concentration of approximately 65% dry solids to the recovery boiler mixing tank or tanks. Here the ash and the makeup salt cake (Na_2SO_4) were mixed into the liquor. Then the mixture went through a preheater directly to the liquor sprayers.

The modern concept is to bring the ash and the salt cake makeup from the recovery boiler mixing tank with an intermediate liquor before the final concentration and feed the final product liquor directly to the liquor sprayers with or without minor additional preheating.

Normally a liquor of 60%–72% dry solids is used for ash transporting. Some suppliers prefer to use a liquor at a lower concentration. An arrangement including an intermediate concentrated liquor storage is possible, but the trend is to reduce the number of storage tanks. The liquor is then pumped from one evaporator to another through the mixing tank without any intermediate storage.

The mixing of salt crystals into the liquor reduces fouling of the heat transfer surfaces. Experience from some recent installations has shown that concentrator units up to 80% dry solids can operate without fouling for several weeks using this technique without requiring separate washing or liquor in-feed switching. Since the behavior of liquors can vary considerably from one plant to another, this is not conclusive evidence for all plants.

Concentrator arrangement

The final evaporator or the "concentrator" units normally work as the first effect. Arrangements are possible where a concentrator operates in parallel with the first or second effect. These are usually the result of upgrading the capacity of an already existing installation.

Three or four concentrator bodies indicated as 1ABC or 1ABCD usually can be isolated for individual washing with weak liquor or condensate. The praxis of switching the sequence of the first effect bodies at intervals of 4–8 h so they receive the weakest liquor in turn is still common but is perhaps not so suitable for installations operating with very high final liquor concentrations.

Pressurized storage

Product liquor at 75%–85% dry solids can normally not be stored at atmospheric pressure since the corresponding temperature would result in an excessively high viscosity for handling. A solution is to store the liquor in a pressurized tank before pumping to the recovery boiler furnace for burning. Liquor flashing at the evaporation plant controls the product liquor temperature at 125°C–150°C. The liquor viscosity characteristics and the atomization properties of liquor at the liquor guns of the recovery boiler determine the optimum temperature. The required over-pressure is moderate. A value of 1 bar gauge is normally sufficient. In a large tank, this results in considerable additional cost. The tank volume is often smaller than what is usual for product liquor. The minimum is a volume that is sufficient for washing the final evaporation effect.

Condensate handling

Primary condensate

Live steam condensate from effect I returns through a small pump tank to the power plant. This primary condensate is sometimes flashed to a lower temperature for handling convenience. Although this will marginally improve the apparent steam economy of the plant, no real advantage results. The flashing may waste clean condensate that must be replaced with demineralized water.

Secondary condensate

The vapor condensate that has some degree of contamination has four fractions determined by the degree of contamination indicated in Table 5. Often, fractions 2 and 3 are combined. Normally, vapor condensate fraction 1 and some fraction 2 are used in pulp washing. Fraction 3 is used in the causticizing plant. Fraction 4 is treated in the stripper. It is then equal to fraction 2.

Table 5. Vapor condensate fractions from a modern 7-stage multiple falling film evaporation.

Vapor condensate fraction	Vapor condensate drawn	Flow, m3/adt	MeOH content, mg/L	MeOH amount, g/adt	Remarks
Fraction 1	From effects 2, 3, and 4	3.4	30	100	
Fraction 2	From effects 5 and 6	2.5	300*	700	
Fraction 3	From effect 7 and surface condensate	2.0	600*	1 200	Slightly odorous
Fraction 4	Foul condensate from evaporator	1.0	6 000*	6 000	Malodorous
Total		8.9		8 000	
Stripped Condensate	From evaporation and cooking foul condensate stripping	1.6	300	500	Stripper MeOH purification efficiency = 95%

* *Provided with internal stripping*

Fouled condensate stripping

The stripping column for the foul condensate purification is usually integrated into the evaporation plant to minimize the heat consumption of the stripping. Vapor to the stripper is taken from effect I or via a reboiler that produces vapor for the stripper with live steam. Stripper overhead vapor goes into a dedicated section of effect 2, the black liquor preheater or preheaters, or both. The stripper gases then enter the trim condenser where the gas temperature and pres-

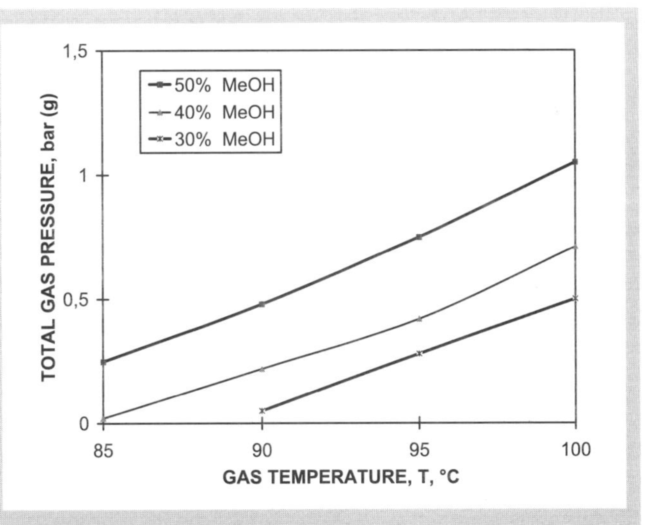

Figure 23. MeOH concentration dependency on temperature and total pressure in a stripper gas.

sure are controlled to keep the MeOH concentration in the outlet gas constant. The trim condenser uses warm water of 45°C as coolant. Using cooling water at 20°C–25°C can result in cold spots on the heating surface. This may cause methanol to totally condense instead of exiting in gaseous form. If only cold water is available, one should use a circulation pump to maintain a water inlet temperature of 40°C–50°C. The target concentration in the outlet gas is approximately 35%–40% methanol by volume. Figure 23 shows the concentration dependency on temperature and total pressure for MeOH.

The stripper gas travels by its own pressure directly to the incineration or into a methanol liquefaction unit for concentration in a distillation column and then condensation in the final condenser. The rest gases from the final condenser go to the gas collection system. The liquid end product containing approximately 80% methanol is stored in a tank and then burned in the incinerator. The advantage here is that the liquefied methanol can be stored and used in a controlled manner as an auxiliary fuel in the lime kiln or in the gas incinerator. This also reduces the amount of odorous gases handled. The methanol liquefaction plant does cause additional investment costs, and it does not fully solve the problem of odorous gas destruction.

Evaporation plant tanks

Storage tanks for liquor at various concentrations and condensate of various quality form part of the evaporation plant. The tank sizes must be sufficiently large to absorb the variations in liquor flow to and from the plant. Table 8 of Section 7.2 gives the retention times for storage tanks considered normal in Scandinavia. Tank storage time depends primarily on the operational philosophy of the mill. Large storage volumes used intelligently can benefit the operational availability of the mill, since the buffer capacity prevents disturbances in one department from immediately influencing the operation of other departments. A large storage capacity increases investments costs. The end result is often a compromise.

6 Evaporation systems and applications

6.1 Application of vapor compression evaporation (VCE)

Despite being an elegant solution, the cost of power and heat energy has not been favorable to VCE, and few references exist.

The VCE installation has primarily been for use as a pre-evaporator to increase the evaporation capacity in the pulp mill. The available size of suitable compressor units has limited the maximum size of VCE. The evaporation capacity is 20–40 kg H_2O/s depending on the boiling temperature.

The limitation is due to the present market situation. If other conditions such as customer requirements and cost of electric energy favor the installation of larger units, the necessary equipment would certainly be developed.

Table 6. VCE design data for one pine liquor application.

Evaporation with one compressor	kg H_2O/s	20
Installed power	kW	1 250
Noise level	dB	80–90
Pine black liquor concentration		
From washing	% dry solids	16
Feed liquor	% dry solids	19
Liquor in circulation	% dry solids	25
Liquor out	% dry solids	25
Secondary condensates		
Semi-clean	kg/s	14
Foul	kg/s	6

6.2 Application of flash steam evaporation

This evaporation technique has use for pre-concentration of weak liquor at the cooking plant:

Blow heat evaporator

Figure 24 shows a blow heat pre-evaporator connected to the batch digester. In this evaporator, hot condensate from the hot water accumulator goes into a flash tank from which vapor is used for the two- or three-effect evaporation.

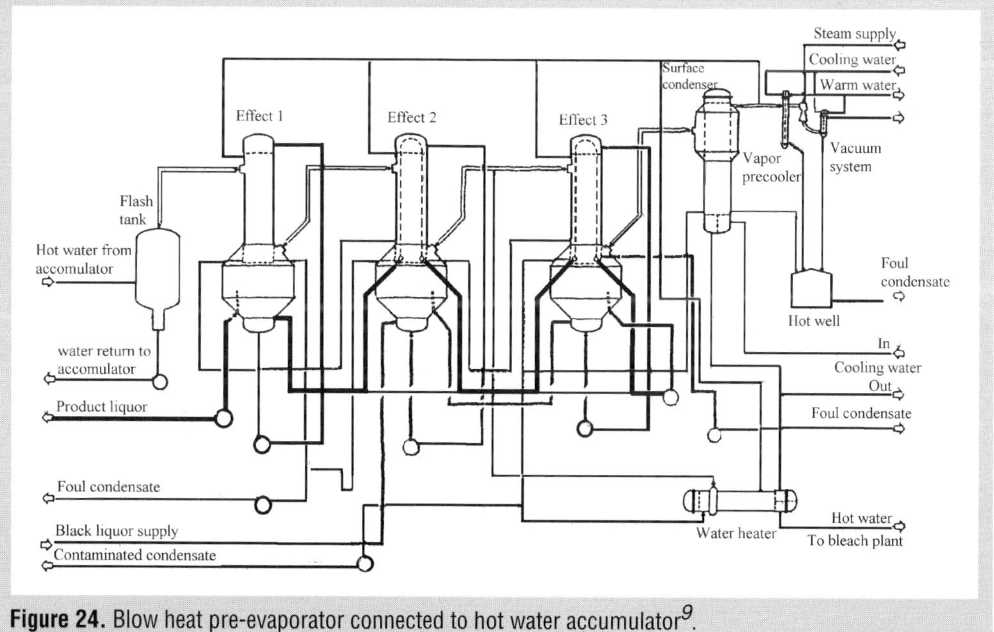

Figure 24. Blow heat pre-evaporator connected to hot water accumulator[9].

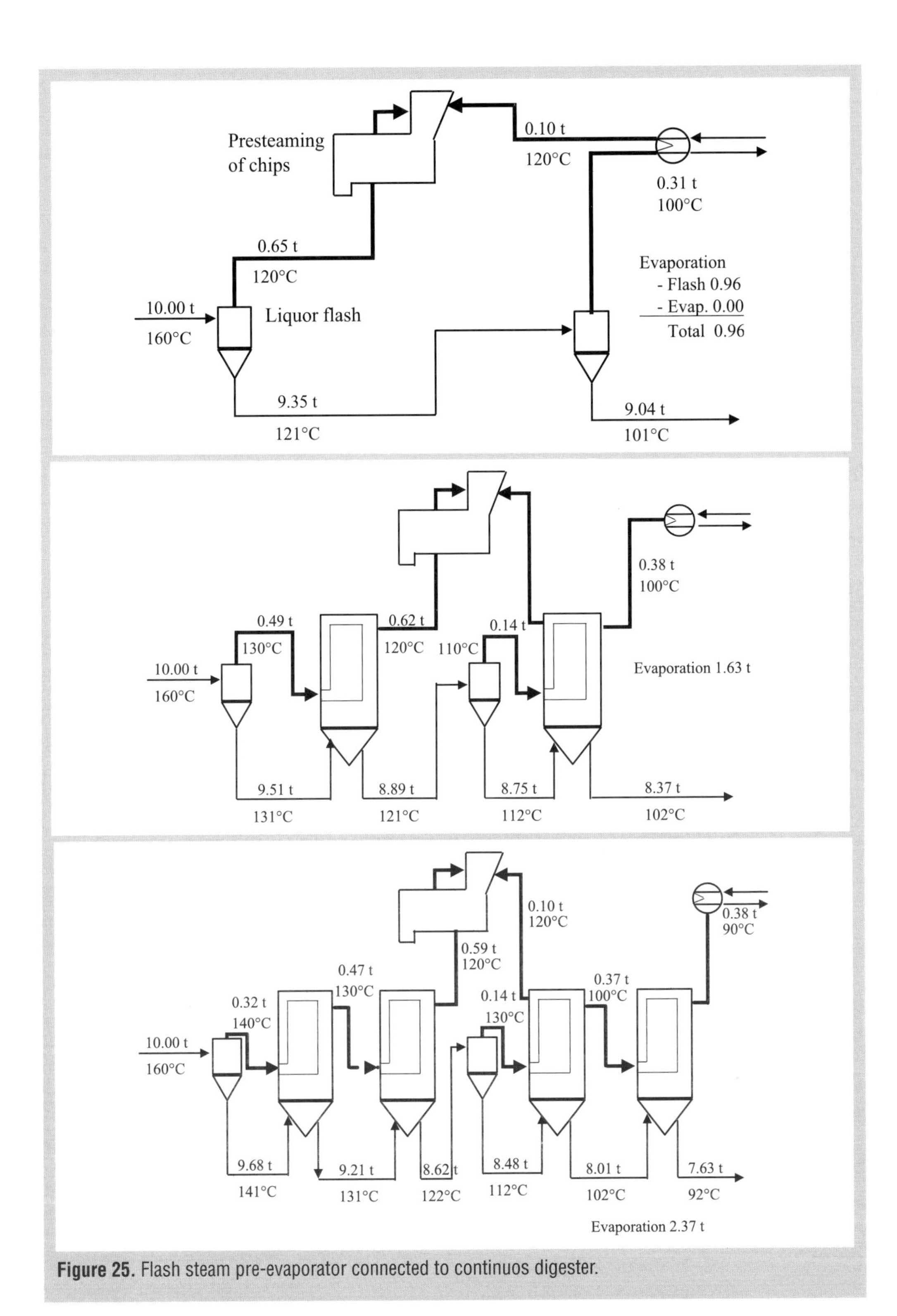

Figure 25. Flash steam pre-evaporator connected to continuos digester.

Flash steam evaporator

Figure 25 shows flash steam evaporators connected to the continuous digester indicating the connection possibilities with process flows and temperatures[8].

Hot black liquor extracted from the digester at 160°C and 15%–17% dry solids goes into the pre-evaporators of the flash evaporator type normally having 2–5 evaporation effects. Weak liquor goes from the flash cyclone into the first effect. Pre-concentrated liquor at 18%–25% dry solids is removed from the last effect that has a slight vacuum. The surface condenser prepares hot water for use in the pulp mill. The capacity of the blow heat evaporation can increase by admitting fresh steam into the first effect. Normally, that is not feasible since using the live steam in the main multiple evaporation is more economical.

A stripper for foul condensate could be integrated with the blow heat evaporation, if the condensate amount is limited to the condensate produced in the cooking plant. If the amount of foul condensate is large, the stripper column should be integrated with the multiple effect evaporators for better heat economy.

Foaming at the blow heat pre-evaporator is a problem. Increasing the dry solids content in the evaporator through recirculation or sweetening the inlet liquor with heavier liquor from the heavy liquor tank can reduce this. In some mills in northern regions, liquor with high soap content requires concentration to 20%–23% dry solids before entering the first evaporator.

Multi flash evaporator

The Lockman multi flash evaporator has been used for kraft liquor pre-evaporation up to 20% dry solids. In this evaporator, liquor circulates through a series of heat exchangers and then through multiple flash chambers in series. These components are installed in a common vertical pressure vessel. The pressure and temperature successively decrease from the column top fed by fresh steam to the vacuum side temperature of 55°C.

The Lockman evaporator can be connected to continuous and batch cooking plants. Few references for this technology are available, and none in recent years.

6.3 High solids concentrators

The current trend is to design and operate the evaporation plant at increasingly higher final dry solids concentrations. This increased concentration has resulted in some advantages in operating the recovery boiler:

- Better stability of the boiler furnace conditions
- Higher combustion temperatures in the lower furnace
- Lower flue gas temperatures in the upper furnace because of better heat transfer in the furnace
- Reduced emissions as a result of the above factors
- Increased boiler capacity due to smaller specific flue gas volume

- Less plugging of the boiler
- Smaller soot blowing steam consumption
- Higher smelt reduction rate.

The operation of the boiler usually improves with no real disadvantages. The combustion of liquor with high dry solids will increase the dust content of the flue gases due to increased fume generation. However, when this is considered in the dimensioning of the electrostatic precipitators, it has normally been possible to maintain the emissions within acceptable limits. The problems and limitations to high solids firing occur in the evaporation plant.

The following factors define and limit the final product liquor concentration:

- Temperature limitations due to the available steam pressure
    - When using low pressure steam of 0.35–0.45 MPa absolute, the final concentration is limited to 75%–80% dry solids depending on the liquor quality.
    - If a partial use of medium pressure steam is acceptable, concentrations well above 80% dry solids are possible.
- Added value, i.e., consideration of cost vs. benefits
    - The advantage of increasing the concentration from 65% dry solids to 75% dry solids has proven worthwhile. The advantage of increasing to even higher concentrations needs to be evaluated, since the enviromenta benefits are already available at a concentration of 75% dry solids. Further improvements are marginal.
- New installation or retrofit of existing plant
    - A high end concentration can be selected for a new plant with only a marginal additional cost.
    - Upgrading an old installation by installing a concentrator in one form or another requires a major investment. If this results in an increase of the mill production, even a large investment may be acceptable.

Dry solids concentrations up to 80%–85% have been available for all concentrator designs.

6.4 Liquor heat treatment (LHT) method

The black liquor viscosity can be decreased by holding liquor at a high temperature for a period of time. The lowered viscosity makes pumping and storing liquor at 75%–80% dry solids in an atmospheric tank possible.

The liquor heat treatment system (LHT) takes advantage of depolymerization of large lignin and hemicellulose molecules of the black liquor at high temperatures. LHT-treated liquor viscosity can be comparable with the viscosity of the normal liquor with 10% lower dry solids content (Fig. 26).

Figure 27 shows Ahlstrom liquor heat treatment. Liquor from the second effect at 45% dry solids is heated to 180°C–185°C and maintained at this temperature for

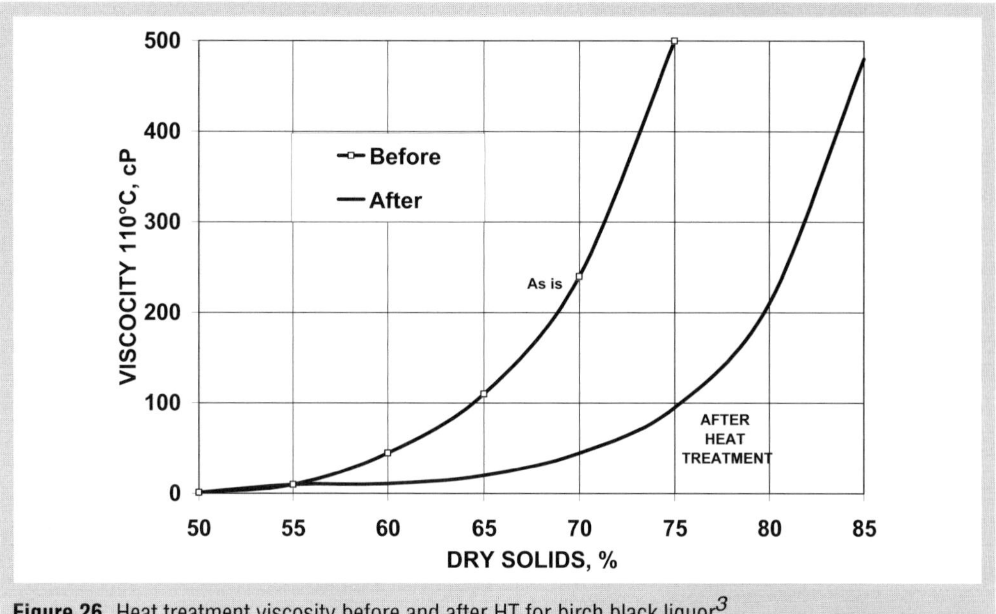

Figure 26. Heat treatment viscosity before and after HT for birch black liquor[3].

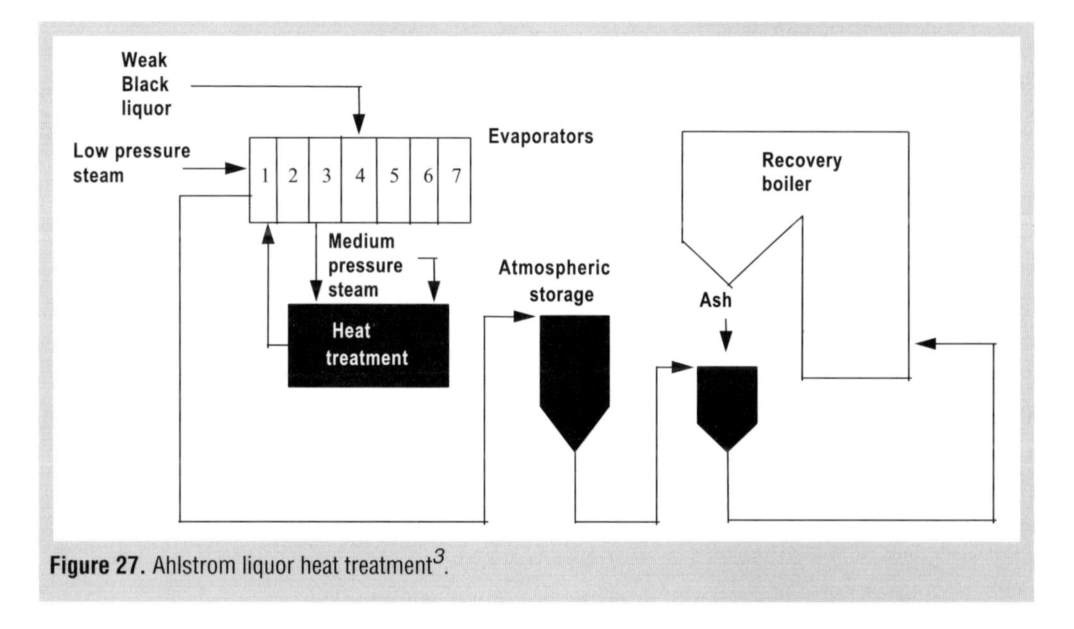

Figure 27. Ahlstrom liquor heat treatment[3].

approximately 30 min. The incoming liquor is pre-heated in several stages by vapor from flashing the treated liquor. Medium pressure steam (nominally at 11 to 13 bar) is necessary only for the final heating. The treated liquor then goes to the first effect at 130°C for evaporation to the final concentration of 75%–80% dry solids. Storage of this liquor at atmospheric pressure is possible.

Heat treatment will release a considerable amount of organic sulfur compounds in gaseous form corresponding to 2 to 4 kg organic sulfur per ton of produced pulp. These gases are collected by an NCG (non condensible gas) system and are incineraed in lime kilns, separate incinerators, recovery boilers, or power boilers.

6.5 Pressurized tank method

At high dry solids concentrations, the viscosity increases. Pumping the liquor requires heating to a sufficiently high temperature. At a certain concentration, this temperature will reach the boiling point of the liquor at atmospheric pressure. If a higher concentration is necessary, the concentrated product liquor requires storage in a pressurized tank to maintain the temperature and viscosity at acceptable levels. The limit for storage at atmospheric pressure is approximately 73%–75% dry solids.

The normal design pressure of a pressurized tank is 0.1 MPa gauge. The storage temperature is selected so, that the liquor has a viscosity that is suitable for proper sprays in the recovery boiler, without additional liquor pre-heating. A direct steam heater is usually available for startup and emergency situations. The pressurized liquor storage tank should have agitation to keep the liquor in motion. A pressure controller admitting steam to the tank or blowing out tank vapor controls the relationship of pressure, temperature, and viscosity. Figure 28 shows high solids liquor storing in the pressurized tank.

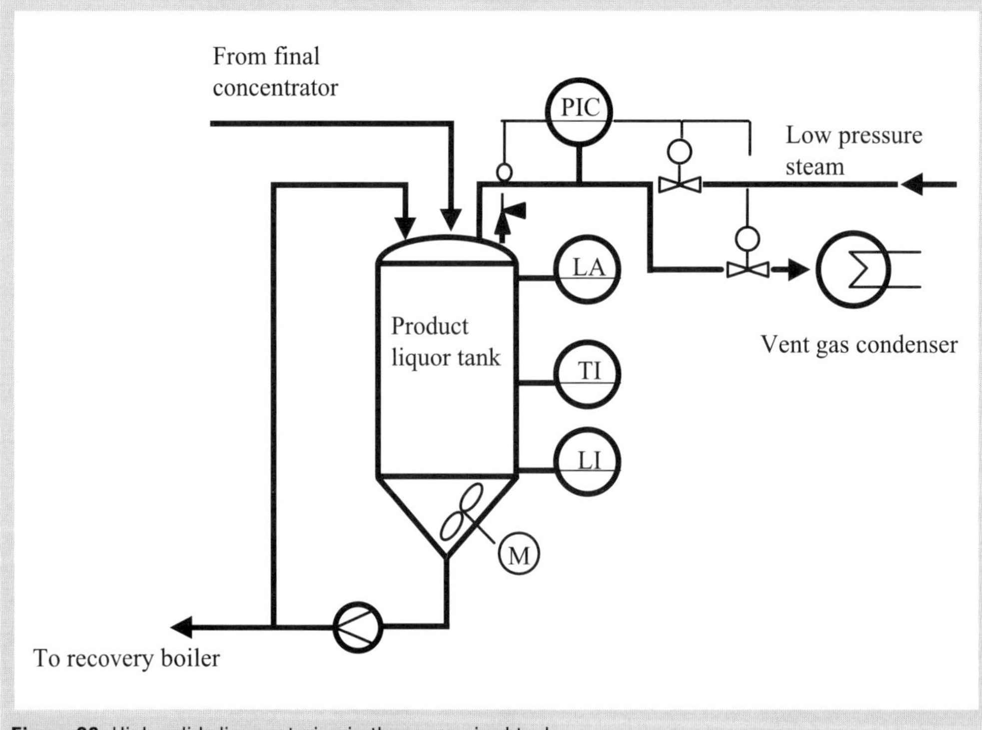

Figure 28. High solids liquor storing in the pressurized tank.

6.6 Typical liquor specification for black liquor evaporation

An evaporation plant design assumes certain liquor characteristics. Table 7 shows typical guarantee conditions such as liquor quality for a plant with a target product liquor concentration of 75%–80% dry solids.

Table 7. Black liquor specifications.

Feed black liquor		
$Na_2CO_3 + Na_2SO_4$	kg/t dry solids	Max. 100
Residual alkali	g NaOH/L	Min. 8
Fiber content with 80 mesh filter	mg/L	Max. 40
Soap content	% of dry solids	Max. 1.2
Liquor boiling point rise at 50% of dry solids	°C	Max. 8.5
Liquor viscosity at 80% dry solids and at 130°C	cP	Max. 350

When possible, the plant supplier receives a sample of the liquor in question for analysis or conducting pilot plant tests.

7 Design

7.1 Evaporator plant design

The evaporators should have the necessary process measurement connections, manholes, and sight glasses. The evaporators should also have connections and piping necessary for pressure testing and the necessary measurement connections for a performance test.

The minimum material qualities according to Scandinavian practices are the following:

- Heat transfer surface — AISI 304
- (high concentration units) — AISI 316
- Internal steel structures — AISI 304
- Drop separators — AISI 304
- Shell
- Front effects 1–3 — AISI 304
- Effect 4 and higher
    - Bottom part of liquor space — AISI 304
    - Upper part of liquor space — carbon steel (or AISI 304)
    - Corrosion allowance — 5 mm
    - Vapor ducts — AISI 304

- Piping
  - Live steam and live steam condensate — carbon steel
  - NCG — AISI 316
  - MeOH from methanol liquefaction — AISI 316
  - Piping for high concentration units — AISI 316
  - Other pipes — AISI 304.

The following design aspects are important considerations for the evaporator design:

- Heating surface specific loads and heat fluxes
- Heating surface areas and heat transfer coefficients
- Liquor circulation rates
- Type and efficiency of drop separators, liquor entrainment losses
- Amount of gases vented
- Segregation of foul condensate by baffling or use of a dedicated section of the heating element
- TRS and MeOH contents in vapor condensate
- Capacity reservation for washing as design contingency or as a spare evaporation body
- Boil-out time and frequency
- Heat economy
- Power consumption
- Sufficient dimensioning of vacuum equipment
- Pump capacity (taking into account pump tank level variations and occasional peak demands).

The most important design parameters are the amount of evaporated water per heating surface (g/s,m^2) and the heat flux (kW/m^2) defined as the heat transfer rate divided by the heating surface area. Figures 29 and 31 show typical variations of these parameters as a function of liquor concentration from each effect or concentrator body.

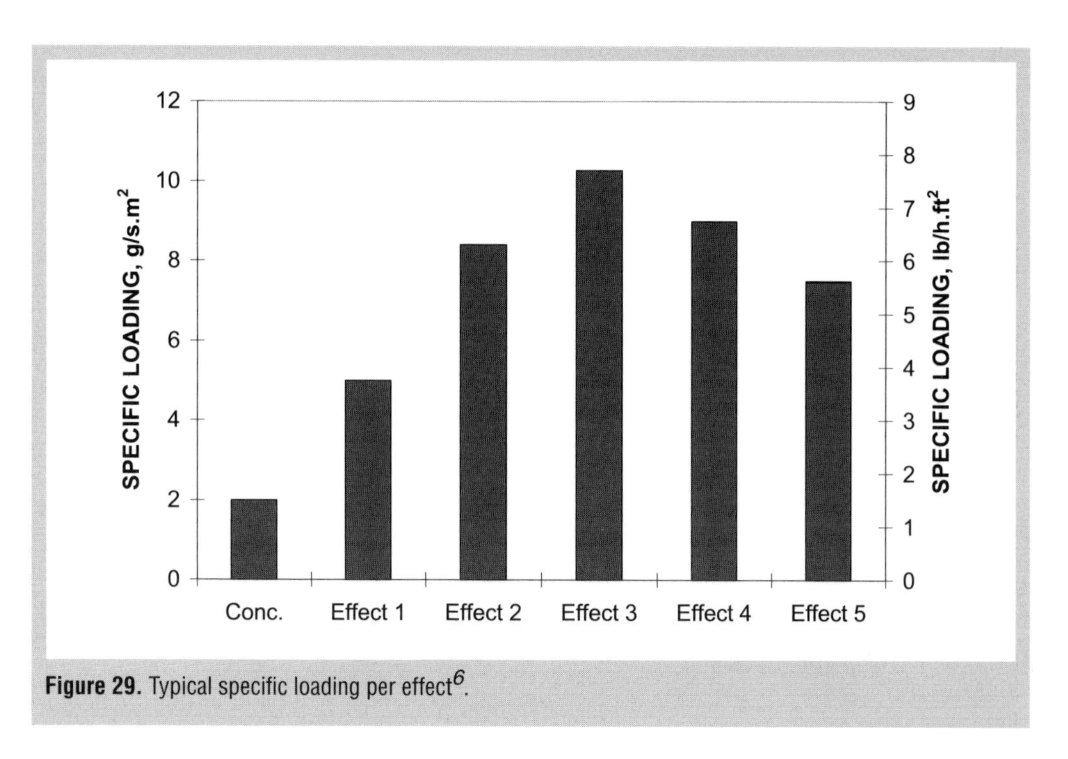

Figure 29. Typical specific loading per effect[6].

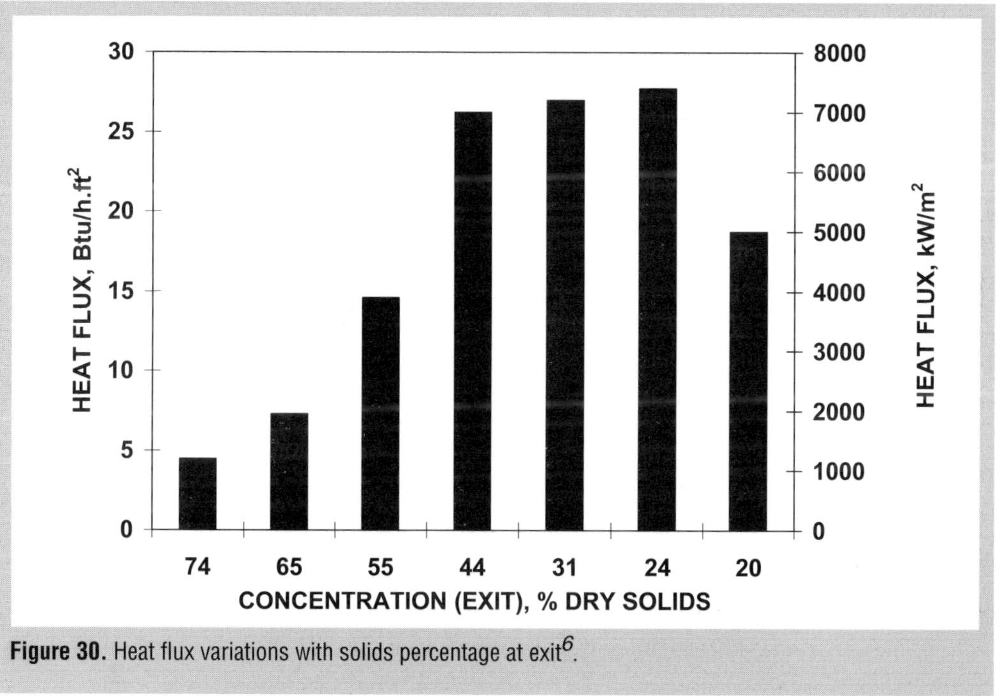

Figure 30. Heat flux variations with solids percentage at exit[6].

Condensate stripping design parameters requiring consideration are:

- Number of theoretical and real bottoms required to achieve the desired purification rate.
- Bottom type.
- Definitions of the stripping and rectifying portions in the column.
- Heat recovery efficiency of the overhead vapor.

7.2 Storage tank design

Table 8 shows the typical storage tank retention times and material selection for a new evaporation plant,

Table 8. Storage tank design data for Scandinavian projects.

Storage tank	Retention time, h	Material	Concentration range,%
Weak black liquor tanks with soap spout	12 (normally in two tanks)	Carbon steel	15–18
Intermediate liquor tank with soap spout	7	Carbon steel	25–30
Soap collection tank	Volume sufficient for three soap batches from weak liquor tank.	Carbon steel	
Spill liquor tank with soap spout	1 m^3/air dry ton of pulp per day	AISI 304	
Intermediate heavy liquor tank - conical bottom with agitator - design pressure = atmospheric	7	Carbon steel or AISI 304 lining	over 65
Heavy product liquor tank - conical bottom with agitator - construction pressure = 0.1 Mpa	8	Carbon steel or AISI 304 lining	75–83
Foul condensate tank	6	AISI 304	
Secondary condensate tanks	2 x 0.3	AISI 304	
MeOH storage tank	40	AISI 304	
Warm water tank	0.3	AISI 304	Can be located in cooking plant

8 Soap skimming

High levels of soap in the black liquor will cause excessive scaling and foaming. Soap compounds therefore require separation from the liquor.

8.1 Soap separation theory

Soap characteristics

The alkaline pulping in the kraft process converts the rosin acids and fatty acids in wood to their sodium salts. These sodium compounds with the neutral or so-called unsaponifiable organic materials form raw soap that separates from black liquor upon concentration. Table 9. shows the composition of raw soap, and Table 10 gives the distribution of tall oil extractives.

Table 9. Composition of raw soap.

	Range,%	Recommended after soap storage,%
Tall oil content	45–55	Min. 50
Black liquor	41–53	
Lignin	0.2–2.0%	Max. 1.0
Fiber	0.1–1.5	Max. 0.5
Na	3.5–5.0	
Ca as $CaSO_4$	0.05–0.3	Max. 0.25
Organic acids	0.5–2	

Table 10. Distribution of tall oil extractives.

	Rosin acids,%	Fatty acids,%	Neutrals,% kg/adt	Tail oil yield
Softwood				
North America	42	47	11	
Scandinavian pine	30–35	50–55	5–10	40–80
Scandinavian spruce	20–30	35–55	18–25	20–40
Hardwood				
North America		76	24	
Scandinavian, birch		55–90	5–35	15–30

Soap separation

The main factors affecting soap separation from black liquor are the following:

- Soap solubility
- Soap skimming tank configuration
- Black liquor viscosity
- Black liquor concentration

Several process conditions define the soap solubility:

- Soap is more soluble at a higher temperature when the dry solids content is below 34% (Fig. 31).

- Soap solubility is minimum at a certain residual effective alkali. This minimum changes according to the ratio of fatty acid to rosin acid (Fig. 32)
- For mixed liquors, soap solubility increases when the share of hardwood liquor is over 60% (Fig. 33)

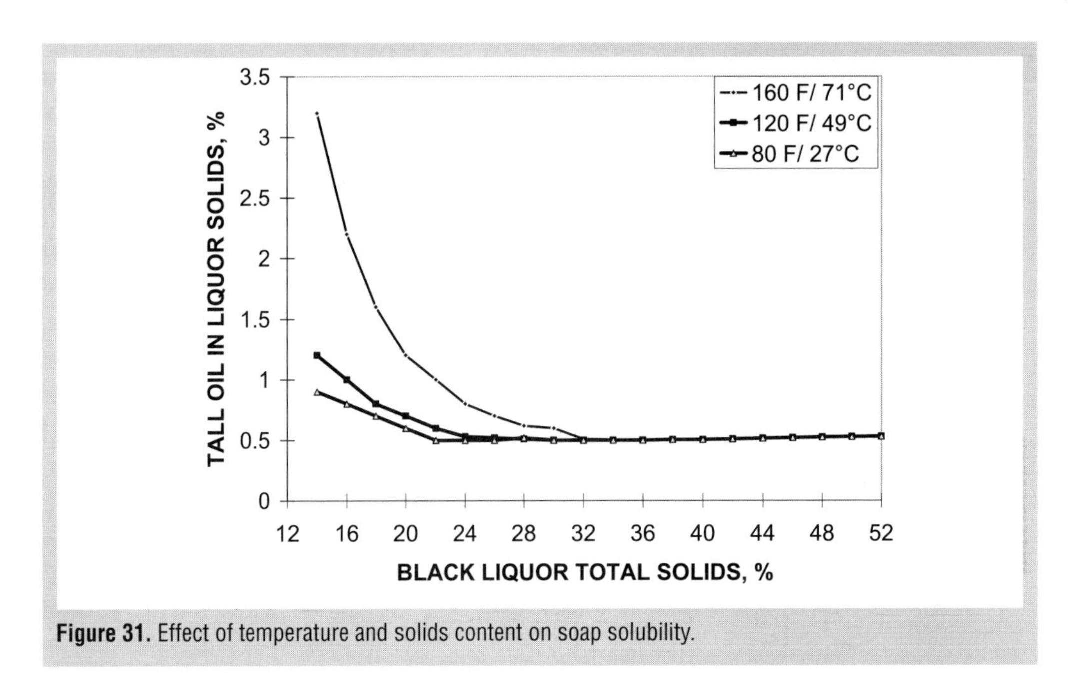

Figure 31. Effect of temperature and solids content on soap solubility.

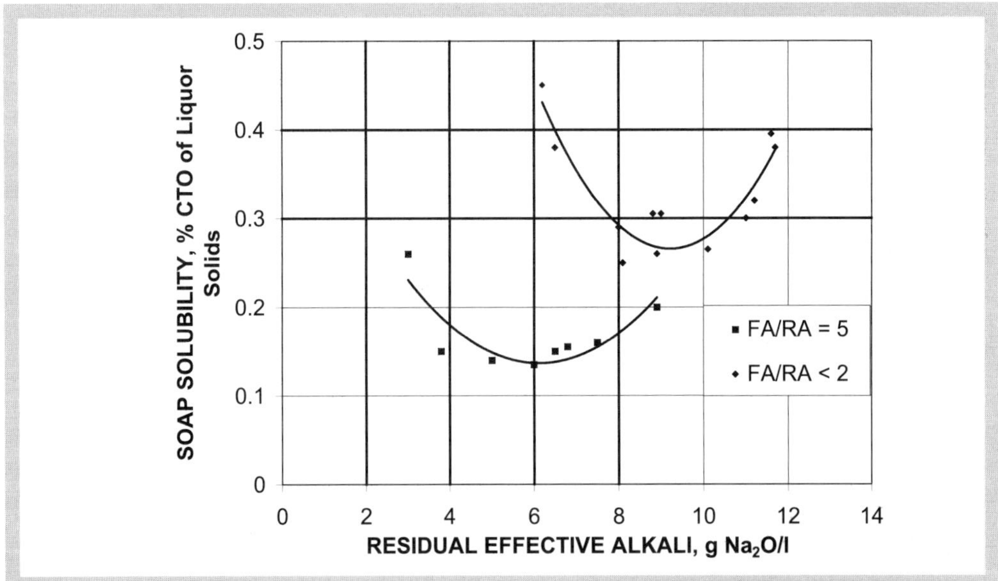

Figure 32. Effect of effective alkali content and fatty acid to rosin acid ratio. *FA* = fatty acids, *RA* = rosin acid.

The soap separation from black liquor uses the difference of the velocity vectors of soap and liquor. Soap is lighter and rises with a velocity 1.2–7.6 m/h. Larger soap particles rise faster at higher temperature. The soap particles must rise against a liquor downdraft or superficial velocity that the liquor flow, m^3/h, divided by skimmer area, m^2, defines. The superficial velocity should be 2.0–2.5 m/h to reach a soap skimming efficiency of 80%–90% when defining efficiency as follows:

$$\text{Soap skimmer efficiency} = \frac{\text{Inlet soap} - \text{outlet soap}}{\text{Inlet soap} - \text{soap solubility}}$$

Differences in soap skimmer design in North America and Scandinavia are topics in the next section.

The rising velocity of soap particles is maximum at dry solids concentration of 18%–22%, and then decreases as concentration increases. When the black liquor viscosity is excessively high, soap does not separate. Almost no soap forms above the liquor surface in a storage tank for concentrations over 45% dry solids.

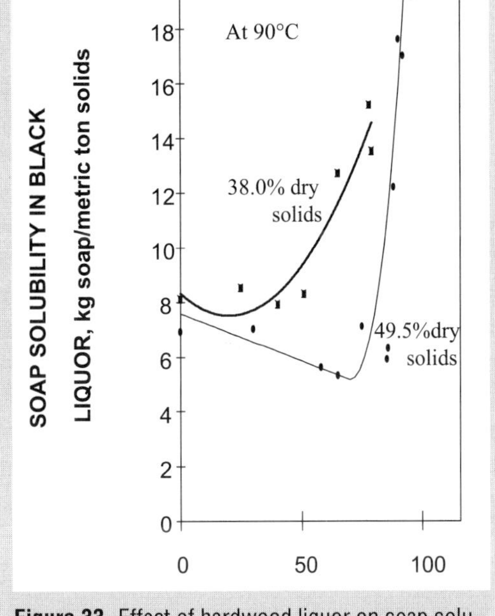

Figure 33. Effect of hardwood liquor on soap solubility at 90°C.

8.2 Soap skimming procedures

North America practice[4]

Weak liquor tank soap recovery

Both manual and automatic floating soap skimmers are used. Figure 34 shows the general equipment configuration for soap skimming and handling in North America.

Soap removal from soap skimmer

A skimmer tank with a retention time of at least 2.5 h is used. Baffling prevents short-circuiting. Level control is by an overflow standpipe and adjustable weir. The recommended skimmer operating practice is the following:

- Skimmer efficiency > 85%
- Superficial liquor velocity <1.5 m/h
- Inlet wall baffled from the outlet
- Liquor throughput = feed rate divided by channel area >0.7 m/h*
- Residence time > 2 h
- Soap bed depth < 0.6 m
- Outlet residual < 0.6% of black liquor solids.

** A lower velocity may result in excessive solids settling in the skimmer.*

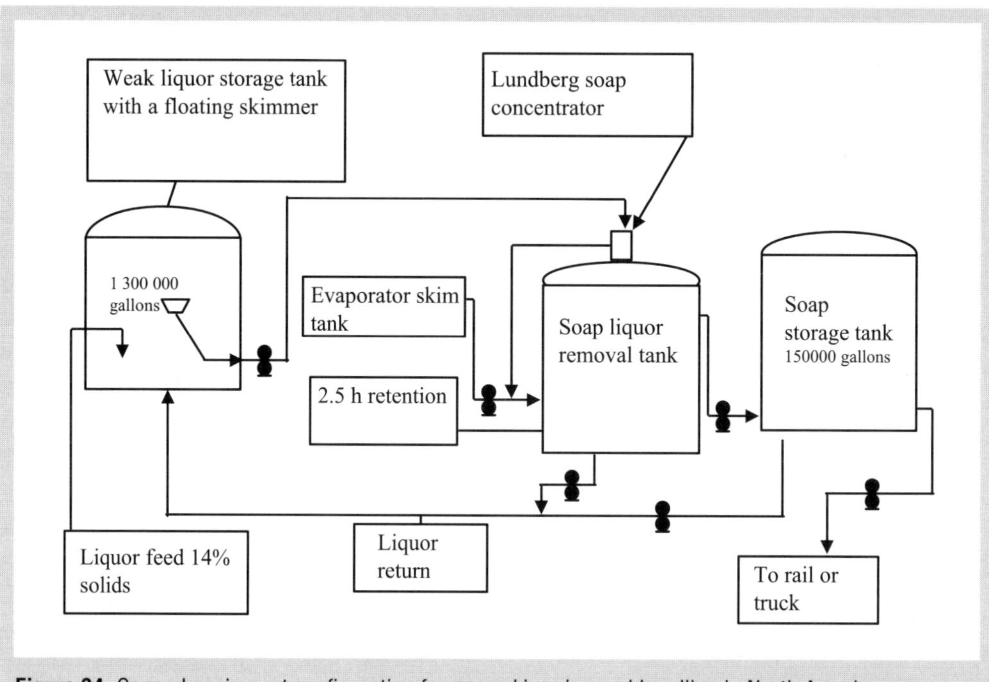

Figure 34. General equipment configuration for soap skimming and handling in North America.

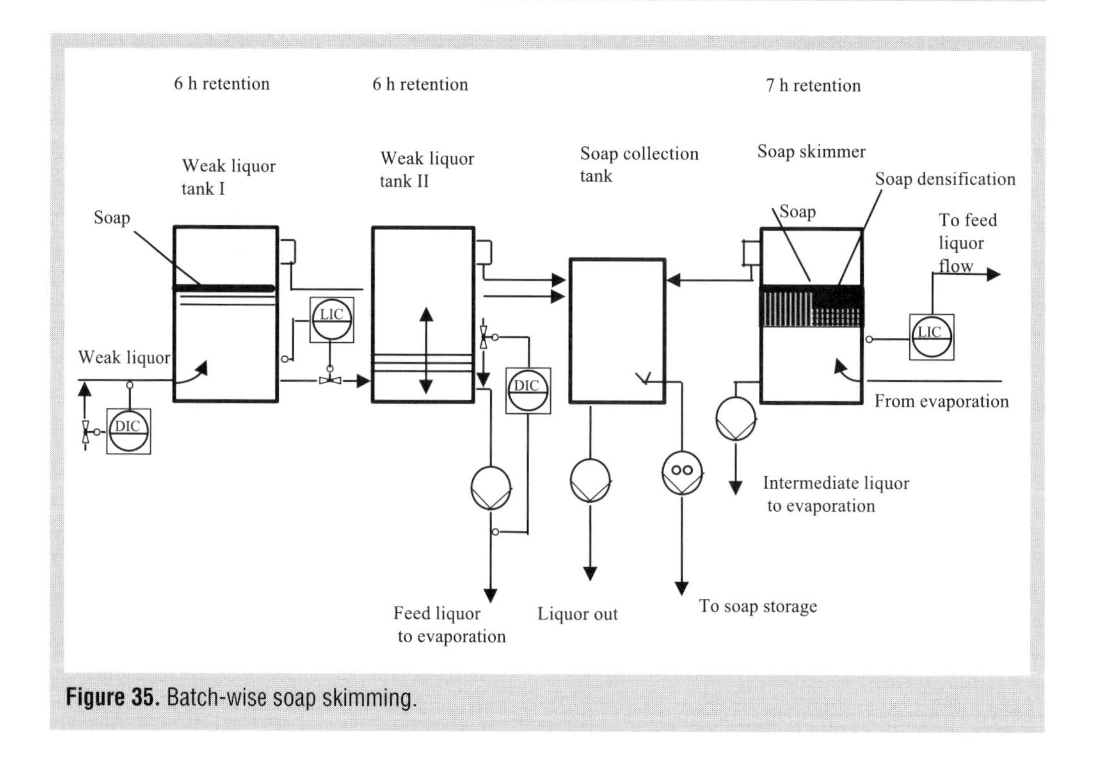

Figure 35. Batch-wise soap skimming.

Scandinavian practice

Weak liquor tank soap separation

A double weak liquor tank arrangement is commonly used. Weak liquor enters tank I and flows by gravity to tank II. Tank I is full of black liquor creating the optimum conditions for soap separation and for minimizing the effect of of liquor density variation. Soap skimming is carried out batch-wise. (Fig. 35).

Soap removal from soap skimmer tank

A skimmer tank or intermediate liquor tank with a retention time of 6–7 h is used. The height of the tank is much higher than used in North America. The tank operation is as follows:

- Liquor inlet is directed upwards to give the correct direction to the soap velocity vector at an elevation 3–4 m from the bottom.
- The tank elevation of 7–14 m works as the soap separation section. The route of liquor is maximized to provide time for soap separation.
- The soap compaction section is above the soap separation section. The layer height is 3–5 m. (Compaction means that the entrained air gradually separates from the soap, and the specific density increases.)
- Well compacted soap collects on the liquor surface. When the soap layer is approximately 1-m thick, the soap is skimmed batch-wise once-a-week or as necessary.

Increasing the feed liquor flow to the evaporation increases the liquor level. A temperature indicator located at the upper edge of the soap spout indicates when all compacted soap is skimmed. Then the level returns to the soap separation level of approximately 90%–95%.

The skimmer efficiency is 85%–90% maximum. The efficiency decreases to 70%–75% when a high black liquor fiber content with a very low residual alkali occurs.

Typical features of soap skimming in Scandinavian practice are the following:

- Skimmer efficiency > 85%
- Superfical liquor velocity < 2.5 m/h
- No baffling (The liquor inlet is on the opposite side to the liquor outlet. In a bigger tank, the liquor inlets, normally 3 inlet funnels or boxes, are symmetrically placed near the tank wall and the liquor outlet at the center of the tank.)
- The liquor route inside the tank is designed to provide maximal time for soap separation.
- Retention time > 5 h
- Soap bed depth 1–1.5 m
- Batch-wise soap skimming (This allows sufficient time for soap compaction above the liquor surface.)

Soap flows into the soap collection tank. The separated black liquor is transported from the bottom to the liquor tank by a centrifugal pump. Soap is pumped in recirculation with a positive displacement pump.

9 Scaling of the heat transfer surfaces

9.1 Scaling and heat transfer

When scale forms on the heating surface, the heat transfer coefficient decreases considerably. In the multiple evaporation unit where the aim is to keep the evaporation rate and live steam flow into the first effect constant, a higher steam temperature (or pressure) can compensate for the effect of the fouling. This results in an increased Δt over the heat transfer surface.

9.2 Reasons for scaling

Evaporator scaling can be an indicator that something is wrong somewhere else in the pulp mill. When black liquor is concentrated, the scaling or fouling of heating surfaces will appear at the concentration typical for each type of scale deposit.

When scaling starts to plague an evaporator, the following malfunctions of the pulp mill departments are possible:

- Recovery boiler
  - Low reduction efficiency -> high Na_2SO_4 content of black liquor
- Causticizing plant
  - Low causticizing efficiency -> high Na_2CO_3 content
  - Poor white liquor settling or filtering -> high $CaCO_3$
  - Poor dregs separation from green liquor -> high content of Si, Al and other non-process elements
- Debarking plant
  - Poor debarking -> high bark content of chips -> high Al, Si and Ca contents of black liquor
  - Logs are contaminated with soil when trees are harvested -> high Si, Al, or both in the black liquor. (Some trees assimilate Si, Al, and Ca from the soil when growing.)
- Cooking and washing
  - Wash filter broken -> high fiber content in black liquor
  - Black liquor filter out of operation or by-passed -> high fiber content
  - Excessively low residual alkali in black liquor -> lignin precipitation
- Evaporation plant
  - Poor soap separation -> high soap content
  - Unstable operation -> over concentrating

- Excessive amount of spent acid from the ClO_2 plant or tall oil plant is introduced to evaporators -> lignin precipitation. (Lignin phase from tall oil plant gives Ca scaling.)

Black liquor from pulping of bagasse, bamboo, and various agricultural straws contains considerable silica and aluminum that contribute to hard, glossy scaling. This scaling is very difficult to remove. Normally, a heavy liquor concentration of 50% dry solids maximum can is reached in the multiple evaporation, and the final concentration is tried to make in a cascade or cyclone type DC evaporator.

9.3 Types of scales

9.3.1 General

The kraft black liquor evaporator can be subject to fouling or scaling by the following:

- Calcium carbonate scaling
- Sodium carbonate or sulfate scaling (Burkeite)
- Soap or fiber scaling
- Aluminum or silicate scaling

The deposits formed on an heating surface may sometimes consist of some or all the above scaling products. Normally, only the liquor side surfaces are fouled. Practically no vapor-side organic deposits or corrosion products have occurred since the introduction of stainless steel materials for heating elements.

When heating surface fouling occurs in an evaporator, one should analyze the scaling products in a laboratory. Studying the solubility of scale in various chemical solvents can indicate the proper scaling abatement control to apply.

9.3.2 Calcium carbonate scaling

Calcium scaling is very sensitive to temperature. Calcium binds to organic compounds such as lignin complexes, oxalate, and soap. When the temperature reaches 90°C–130°C, the calcium ion becomes free. $CaCO_3$ forms and precipitates on heating surfaces. When the temperature increases further, calcium scaling accelerates rapidly, due to the lowered solubility of calcium compounds at higher temperatures.

Calcium scale deposits primarily form on the heating surfaces at the first effect or its preheater. Second effect surfaces have occasionally encountered $CaCO_3$ precipitation with some residues of Na_2CO_3, Na_2SO_4, soap, and lignin.

9.3.3 Calcium carbonate scale control

Section 9.2 describes the general preventive control measures. In-plant scaling abatement measures:

- Low steam temperature with maximum of 115°C–125°C
- Sufficiently high residual alkali with a minimum of 5 g NaOH/L weak liquor. (The calcium-scaled evaporator can be cleaned by water boil-outs with ther-

mal shocks or acid wash. The evaporator body must be disconnected for cleaning with acid circulation from the boil-out tank. The possible acid agents are sulfaminacid with inhibitor, HNO_3 and formic.)

- Hydroblasting if needed
- Using thermal deactivation. (In this method, black liquor is heated to 120°C–150°C and held for 10–15 min in a reaction vessel. This may reduce $CaCO_3$ scaling considerably.)

9.3.4 Sodium carbonate and sulfate scaling

Sodium carbonate and sulfate salts precipitate when their concentration in black liquor exceeds the solubility limit or critical concentration. The composition of the precipitation is mainly burkeite ($2Na_2SO_4$–Na_2CO_3). Some hardwood pulp mills especially when using weak liquor use oxidation so sodium oxalate, $Na_2C_2O_4$, occurs in the scale. Table 11 shows the critical black liquor concentrations at which burkeite salts begin to precipitate. A higher effective sodium content lowers this concentration (Fig. 36)[10].

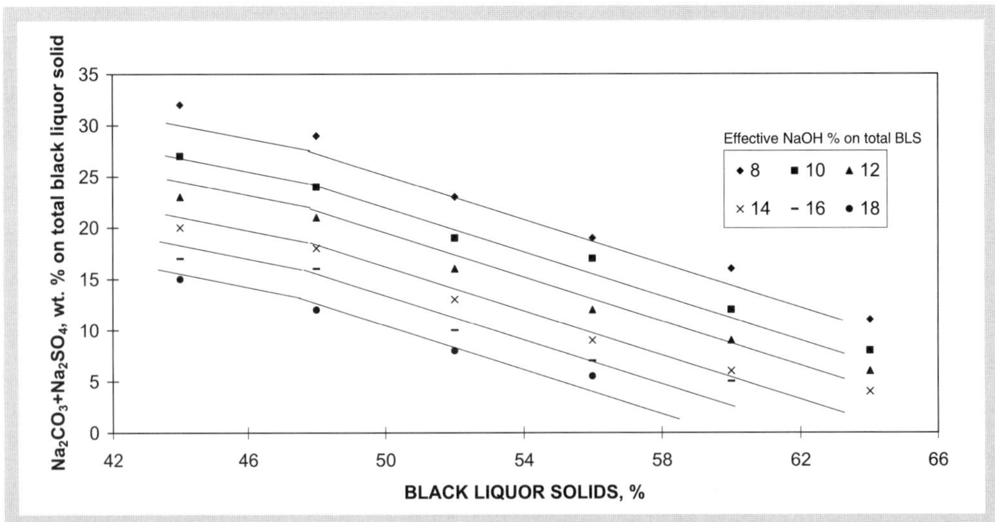

Figure 36. Observed vs. predicted solubility of burkeite in black liquor at Na_2CO_3/Na_2SO_4 ratio of 80/20 at six levels of effective alkali.

Table 11. The solubility limits or critical concentrations of some black liquor samples.

Case	A	B	C	D
Na_2CO_3,% of dry solids	7.1	8.1	12.3	6.9
Na_2SO_4,% of dry solids	2.3	4.5	3.2	8.3
Total Na,% of dry solids	17.3	18.2	20.5	19.8
Solubility limit,% dry solids	56.3	53.1	49.1	48.4

The solubility of sodium carbonate and sodium sulfate decreases slightly with an increase in liquor temperature above 40°C. The precipitated products are readily washable by weak liquor or vapor condensate.

9.3.5 Sodium carbonate and sulfate scale control

In-plant measures for scale abatement include the following:

- Recirculating after surpassing the critical solubility limit
- Flushing heating surfaces with a lower concentration liquor (This is the concentration below the critical solubility limit. It is done by changing the liquor infeed sequence regularly in parallel evaporators.)
- Introducing Na_2SO_4 "seeding" to the liquor in-feed of an evaporator

  (The "seeding" crystals in the liquor promote crystal growth in the slurry instead of depositing them at the heating surface.)
- Modifying the rising film unit to the falling film unit. (Normally, the evaporation effects I and II are the last effect upgraded. This is obviously an extreme measure.)

9.3.6 Fouling by fiber and soap

Fiber can plug the distribution boxes of the evaporators if its content in weak liquor exceeds 50–100 mg/L. This type of plugging occurs especially in the distribution boxes of the effects 2, 3, and 4. Increasing the hole diameter in the distribution boxes can solve the problem.

High fiber content hinders soap separation, since the soap adheres to the fiber surfaces. Both substances are common in the fouling of heating surfaces of the medium consentration effects 2-3-4. To control the fiber content, a filter is necessary in the black liquor flow from the pulp washing plant.

Soap fouling is avoided by keeping the soap content after soap skimming below 0.8%–1.0% crude tall oil (CTO) of the liquor dry solids.

9.3.7 Aluminum silicate scaling

Sodium aluminum silicate scales are hard, glassy deposits. They form a persistent layer that grows very slowly on the heating surfaces. Even a thin layer reduces the heat transfer markedly. The scale consists primarily of alumunosilicate with a small amount of Na_2CO_3 and $NaSO_4$. The deposits normally form on heating surfaces of the first effect and the final concentrators. The other effects can also become fouled. The aluminum concentration determines the scaling rate. An aluminum content of 0.02% dry solids is the limit above which one can expect problems with Al and Si scaling.

In North American and Scandinavian mills, silicate scaling is not a common problem because its input to the recovery cycle is small. Silicate scaling does plague evaporators for bamboo, bagasse, and various straw liquors and sometimes mixed tropical hardwood liquors. In these mills, silicate may be controlled by purging lime mud or lime kiln ash.

Silicate removal systems for black liquor operate in soda pulping mills. Silicate removal from green liquor in a soda-based pulp mill may be achieved by lowering of the pH of liquor with CO_2-containing flue gases from the lime kiln or other sources. No commercial silicate removal method for kraft mill liquors has been available.

9.3.8 Aluminum silicate scale control

Section 9.2 suggests the general preventive control measures. In-plant measures for scaling abatement include a silicate removal system. This is applicable for liquor from soda based pulping only. Other possibilities are thermal shocks or cleaning. The second option uses an acid wash or a chemical treatment as Table 12 shows. The acid wash uses an inhibitor that reduces the corrosion on the metal surfaces of the evaporator.

Table 12. Results of laboratory trials for treating 3 $NaAlSiO_4 \cdot Na_2CO_3$[5].

Agent	Treatment °C	Time needed to temperature become soluble min
5% NH_2SO_3H (sulfaminacid)	80	40
10% HCl	25	5
10% HNO_3	25	10
10% H_2SO_4	25	10–15
5% H_2SO_4	25	40
5% H_2SO_4	50	10
2% H_2SO_4	25	80
2% H_2SO_4	50	35
NaOH, 80 g/L	80	No effect
White liquor	80	No effect

References

1. *Söderhjelm, L., Communications from the Finnish Pulp and Paper Research Institute, Finnish Pulp and Paper Research Institute, Helsinki, 1988, .*
2. *Anon., Several Technology Marketing Guidelines, Kvaerner Pulping Tech., 1995-1997*
3. *Anon., Several Marketing Guidelines, Ahlström, , 1994-1997*
4. *Grace, T.M., Black liquor evaporation, Pulp and Paper Manufacture, Alkaline Pulping, TAPPI, 1989, p. 477-530*
5. *SPCI´s sulfatinkrustkommitte, Inkruster I sulfatfabriker, SPCI, 1981, p. 1-52*
6. *McCANN, D., Design review of black liquor evaporators,Pulp and Paper Canada, 79(4):T138(1995)*

Suggested Reading

7. *G. Gudmundson, G., Alsholm,H., Hedström, B., Heat transfer in industrial black liquor evaporator plants Svensk Papperstidning 75 (1972) 4, p 778*
8. *Beckström, B.W., The evaluation and optimization of blow heat pre-evaporation systems, Pulp and Paper Canada 88 (1987) 7, p 48*
9. *Sandell, A., Pre-evaporation and batch digester PPI (1983)*
10. *Rosies, M., Model to predict the precipitation of burkeite in the multiple-effect evaporator and techniques for controlling scaling, Tappi 80 (1997) 4, p 205*
11. *B. Jungerstam – P. Kock, Puumassan valmistus, 1983, p. 1164, Jäteliemen haihdutus (in Finnish)*
12. *McCabe, W. and Smith, J Unit Operations of Chemical Engineering Second Edition, p 549*

CHAPTER 13

Recovery boiler

Esa Vakkilainen

Recovery boiler

The recovery boiler serves three main purposes. It burns the organic material in the black liquor to generate high pressure steam. The second purpose is to recycle and regenerate used chemicals in black liquor. Finally, it minimizes discharges from several waste streams in an environmentally friendly way. In a recovery boiler, concentrated black liquor burns in the furnace with the simultaneous emergence of reduced inorganic chemicals in molten form.

Significant changes in kraft pulping have occurred in recent years[1,2,3]. Combined use of extended cooking, hot alkali extraction, and oxygen delignification make it possible to reach kappa numbers as low as 10 before bleaching. This has increased the degree of organic residue recovery. Table 1 shows that black liquor properties have reflected these changes.

Table 1. Development of black liquor properties.

Property	1982	1992	2002
Liquor dry solids, kg dry solids/t pulp	1 700	1 680	1 760
Sulfidity, $Na_2S/(Na_2S+NaOH)$	42	45	41
Black liquor HHV, MJ/kg dry solids	15.0	13.9	13.0
Liquor dry solids, %	64	72	80
Elemental analysis, % weight			
C	36.4	34	31.6
H	3.75	3.5	3.4
N	0.1	0.1	0.1
Na	18	18.4	19.8
S	5.4	5.9	6
Cl	0.2	0.4	0.8
K	0.75	1.0	1.8
Cl/(Na+K), mol-%	0.70	1.37	2.49
K/(Na+K), mol-%	2.39	3.10	5.07
Net heat to furnace, kW/kg dry solids	13 600	12 250	11 200
Combustion air* required, m^3n/kg dry solids	4.1	3.7	3.4
Flue gas* produced, m^3n/kg dry solids	4.9	4.3	3.9

* *At air ratio 1.2*

1 Evolution of recovery boiler design

Changes in investment costs, increase in scale, demands placed on energy efficiency, and environmental requirements are the primary factors directing development of the recovery boiler[4].

Recovery boiler sizes have been increasing at a steady pace. The chemical recovery system and its equipment must be able to handle more material than before. The nominal capacity of new recovery boilers at the beginning of the 1980s was 1700 t of dry solids per day. This was regarded as the absolute maximum then. At the end of the 1990s, more than ten recovery boilers rated 2 500–3 500 t of dry solids per day are operating. The maximum design capacity has increased because black liquor has less water, liquor spraying is now more uniform, and new computer controls allow better stability and control.

Steam generation increases with increasing dry solids content. For a rise in dry solids content from 65% to 80%, the main steam flow increases by about 7%. The increase is more than 2% for each 5% increase in dry solids. Steam generation efficiency improves slightly more than steam generation itself. This is primarily because the drier black liquor requires less preheating.

Recovery boilers that burn liquor with solids concentration higher than 80% are available. Unreliable liquor handling, the need for pressurized storage, and high pressure steam demand in the concentrator have frequently prevented sustained operation at very high solids levels. The main reason for the handling problems is the high viscosity associated with high solids contents. Black liquor heat treatment (LHT) can reduce viscosity at high solids.

The significance of electricity generation from the recovery boiler has been secondary. The most important factor in the recovery boiler has been high availability. Electricity generation in the recovery boiler process and the steam cycle can be increased by elevated main steam pressure and temperature or by higher black liquor dry solids.

Increasing the main steam outlet temperature increases the available enthalpy drop in the turbine. The normal recovery boiler main steam temperature of 480°C is lower than the typical main steam temperature of 540°C for coal and oil fired boilers. The main reason for choosing a lower steam temperature is to control corrosion. Requirements for high availability, avoidance of superheater corrosion, and avoidance of more expensive materials are the important reasons.

1.1 Two drum recovery boiler

Most recovery boilers operating today have the two drum design of Fig.1. Their main steam pressure is typically about 85 bar(absolute) with temperature of 480°C. The maximum design solids handling capacity of the two drum boiler is about 1 700 t dry solids/day. The units use three level air intakes and stationary firing. The two drum boiler represents one successful stage in a long evolutionary path and signifies a design to minimize sulfur emissions successfully. A main steam temperature of 480°C is still the accepted design value for most new recovery boilers. Lower values are necessary if high potassium and chlorine levels exist in the black liquor.

These recovery boilers have water screens to protect the superheaters from direct furnace radiation, lower flue gas temperatures, and decrease combustible material carryover to superheaters. The two drum boiler has vertical flow economizers that replaced horizontal economizers because of their improved resistance to fouling.

1.2 Modern recovery boiler

The modern recovery boiler has a single drum design with vertical steam generating bank and widely spaced superheaters. The most marked change in the past ten years has been the adoption of single drum construction. The construction of the vertical steam generating bank is similar to the vertical economizer. Experience shows this is easy to keep clean. The spacing between superheater panels has increased. Wide spacing helps minimize fouling. This arrangement with sweet water addition ensures maximum protection against corrosion. Numerous improvements in recovery boiler materials to limit corrosion have been developed[5–8].

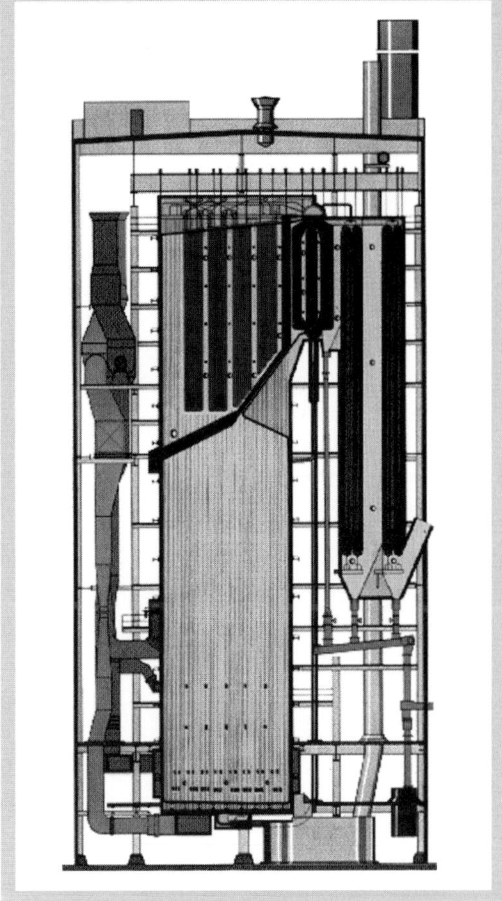

Figure 1. Two drum recovery boiler.

The effect of increasing dry solids concentration has significantly influenced the main operating variables. Steam flow increases with increasing black liquor dry solids content. Increasing closure of the pulp mill means that less heat per unit of black liquor dry solids will be available in the furnace. The flue gas heat loss will decrease as the flue gas flow diminishes. Increasing black liquor dry solids is especially helpful since the flue gas flow often limits the recovery boiler capacity.

The modern recovery boiler in Fig. 2 consists of heat transfer surfaces made of steel tubing. The components are furnace (1), superheaters (2), boiler generating bank (3), and economizers (4). The steam drum (5) design is the single drum type. The air and black liquor enter through primary and secondary air ports (6), liquor guns (7), and tertiary air (8). The combustion residue and smelt exit through smelt spouts (9) to the dissolving tank (10).

Nominal furnace loading has increased during the last ten years and will continue to increase[9]. Changes in air distribution design have increased furnace temperatures[10–13]. This has enabled an increase in hearth solids loading (HSL) with only a modest design increase in hearth heat release rate (HHRR). The net heat input per unit area of furnace floor has increased noticeably. The average flue gas flow decreases since less water vapor is present. The vertical flue gas velocities can therefore be lower even with increasing temperatures in the lower furnace.

The most marked change has been the adoption of the single drum construction. This change is possible partly due to more reliable quality control of water. The advantages of a single drum boiler compared to a two drum design are improved safety and availability. Single drum boilers can withstand higher pressures and have bigger capacities. Savings are available through decreased erection time. Less tube joints in the single drum type allow construction of drums with improved startup curves.

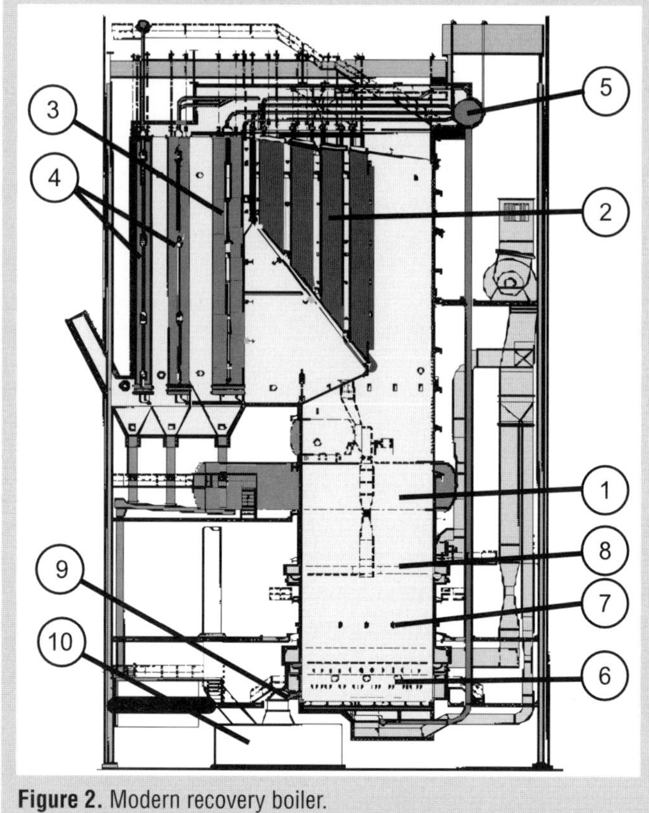

Figure 2. Modern recovery boiler.

The construction of the vertical steam generating bank is similar to the vertical economizer. Experience shows this is very easy to keep clean[14]. A vertical flue gas flow path improves the cleaning with high dust loading[15]. To minimize the risk for plugging and maximize the efficiency of cleaning, the generating bank and the economizers have generous side spacing. Two drum boiler bank plugging is often due to excessively tight spacing between tubes.

The spacing between superheater panels has increased. All superheaters are now widely spaced to minimize fouling. This arrangement with sweet water addition ensures maximum safety from the potential for corrosion. With an increased number of superheaters, the difference in the heat transfer between a clean and fouled surface is less. This facilitates the control of superheater outlet steam temperature especially during startups. Plugging of the superheaters is unlikely, the cleaning is easier, and the soot blowing steam consumption is low.

The lower loops of the hottest superheaters can use austenitic material with better corrosion resistance. The steam velocity in the hottest superheater tubes is high. This decreases the tube surface temperature. Low tube surface temperatures are essential to superheater corrosion resistance. A high steam side pressure loss over the hot superheaters ensures uniform steam flow in tube elements.

1.3 Future recovery boilers

Recovery boiler main steam pressures will increase in the future. If main steam pressure increases to 104 bar(absolute) and temperature increases to 520°C, then the electricity generation from recovery boiler plant will increase about 7%. For a design dry solids load of 4 000 t/day, this produces an additional 7 MW of electricity.

The future superheater arrangement will have optimum heat transfer with extra protection from furnace radiation. Higher heat fluxes and higher chloride and potassium contents will require more attention to design. Almost all superheaters are behind the bullnose to minimize direct radiative heat transfer from the furnace. Increasing superheating demand has eliminated the need for screen arrangement. A straight flow path in the furnace ensures a flat temperature profile (Fig. 3).

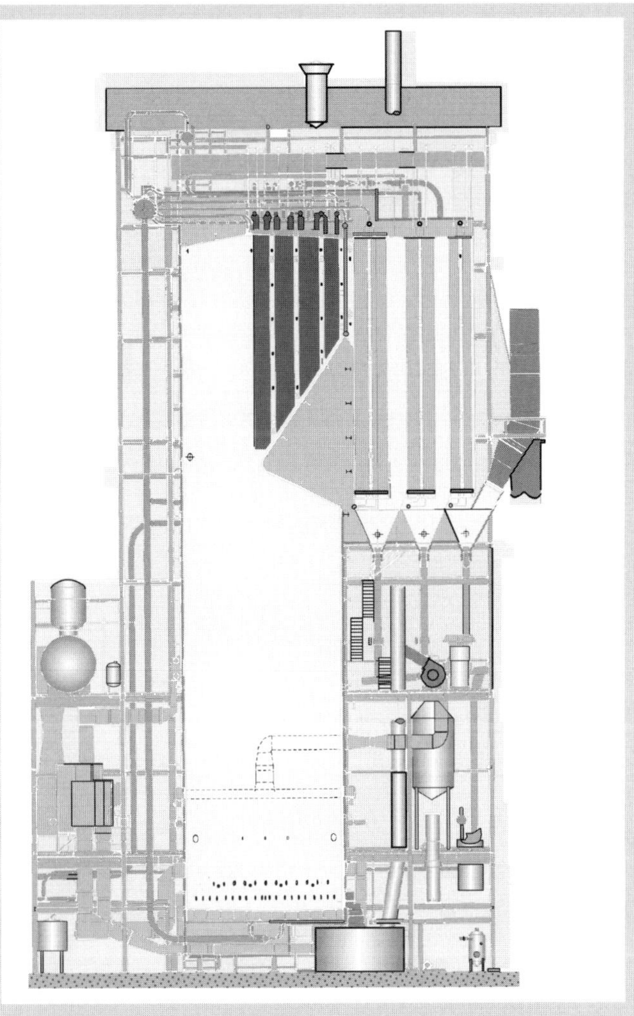

Figure 3. Future recovery boiler.

A higher main steam outlet temperature requires more heat in the superheating section. Either the furnace outlet temperature must increase or boiler bank inlet temperature must decrease. If boiler bank inlet temperature decreases, the average temperature difference between flue gas and steam also decreases. This reduces heat transfer, and substantially more superheating surface is necessary. With increasing dry solids content, the furnace exit temperature can safely increase.

Increasing recovery boiler main steam temperature influences the corrosion of the superheaters. Designing for higher recovery boiler main steam pressure increases the design pressure for all boiler parts. The recovery boiler lower furnace wall temperatures increase with higher operating pressure. The air flow per unit of black liquor burned in the recovery boiler furnace decreases. The number of air ports will therefore decrease.

2 Material and energy balance

Calculation of material and energy balances is essential for determining dimensions of recovery boilers. Proper calculation is important for the mill energy and mass balances necessary to determine economics.

2.1 Material balance

Control of combustion requires air flow that matches fuel flow. The air required to burn a mass unit of black liquor depends primarily on water content and the heating value of the fuel. Figure 4 shows that increasing black liquor higher heating value increases black liquor air demand. Similar figures can estimate the required air and the possible range of air flows that a recovery boiler should be able to handle. For design and performance estimating purposes, calculation of air demand using fuel chemical composition is necessary.

Typical fuels contain carbon, hydrogen, nitrogen, sulfur, and oxygen. Fuels pertinent to kraft recovery also have sodium, potassium, and chlorine. Defining the individual chemical compounds that form the fuel is often impossible. Calculation of the stoichiometric air demand assumes that the fuel has three fractions:

- Organic portion that combusts fully
- Reactive inorganic portion that reacts to predefined end products
- Inactive portion that passes through the combustion system unchanged.

In calculating recovery boiler inorganic reactions, one can assume that all chloride reacts to form sodium chloride and all potassium forms potassium sulfide. For air

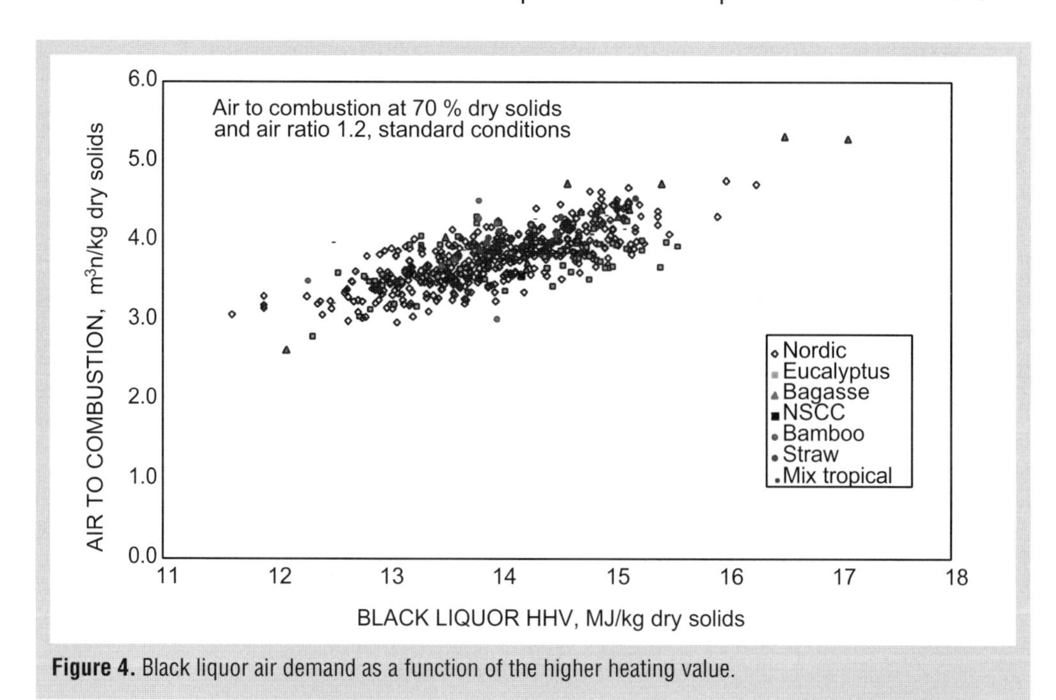

Figure 4. Black liquor air demand as a function of the higher heating value.

calculations, one assumes that no different potassium and chloride compounds exist. Another assumption is that all sulfur that is not escaping with the flue gas is present as sodium sulfide or sodium sulfate.

The basis for material calculations is typically one mass unit of black liquor. If values for one mass unit are known, multiplying these values for the desired load is easy.

Example

Calculate required air flow for eucalyptus black liquor. Liquor dry solids analysis as mass percentage is as follows:

Carbon	33.7
Hydrogen	3.43
Nitrogen	0.16
Sodium	21.1
Sulfur	4.5
Potassium	1.9
Chloride	0.53
Oxygen	33.6
Other inorganic	0.1
Total	100.0

In addition, assume that the reduction degree is 95% expressed as the ratio of sulfide to sum of sulfate and sulfide in the smelt. Black liquor dry solids is 80%. Air ratio is 1.2 (20% excess air). SO_2 emissions are 0 ppm in dry gas, and dust emission is 0.5 g/kg dry solids as Na_2SO_4. Soot blowing steam consumption is 150 g/kg dry solids.

Input flow per 1 kg of black liquor dry solids:

	Mass, g/kg dry solids	Mass, mol/kg dry solids	End product
Carbon	337	337/12.011 = 28.058	CO_2, Na_2CO_3
Hydrogen	343	343/2.016 = 17.014	H_2O
Nitrogen	1.6	1.6/28.015 = 0.057	N_2
Oxygen	345.8	345.8/31.999 = 10.807	CO_2, Na_2CO_3, SO_2, Na_2SO_4, H_2O
Chloride	5.3	5.3/34.53 = 0.149	NaCl
Potassium	19	26/78.204 = 0.243	K_2S
Sulfur	45	45/32.060 = 1.404	SO_2, K_2S, Na_2S, Na_2SO_4
Sodium	211	211/45.980 = 4.589	Na_2S, Na_2SO_4, Na_2CO_3
Water	1 000(1/0.80-1) = 250	250/18.015 = 13.877	H_2O

Sulfur balance:

	Mass, mol/kg dry solids	Mass, g/kg dry solids	End product available
	1.404	45	
	-0	-0	SO_2
	-0.5/142.040 = -0.003	-0.003 x 32.060 = -0.1	Dust as Na_2SO_4
	-0.243	-0.243 x 32.060 = -7.8	K_2S in smelt
Sum	1.158	37.1	Na_2S, Na_2SO_4
	-0.05 x 1.158 = -0.058	-0.058 x 32.060 = -1.9	Na_2SO_4 in smelt
	-0.95 x 1.158 = -1.100	-1.100 x 32.060 = -35.2	Na_2S in smelt

Sodium balance:

	Mass, mol/kg dry solids	Mass, g/kg dry solids	End product available
	4.589	211	
	-0.003	-0.003 x 45.980 = -0.1	Dust as Na_2SO_4
	-0.058	-0.058 x 45.980 = -2.7	Na_2SO_4 in smelt
	-1.100	-1.100 x 45.980 = -50.6	Na_2S in smelt
	-0.149/2 = -0.075	-0.075 x 45.980 = -3.4	NaCl in smelt
Sum	3.353	154.2	Na_2CO_3 in smelt

Other inorganic: 1 g/kg dry solids is assumed to pass through to smelt unreacted.

Carbon balance:

	Mass, mol/kg dry solids	Mass, g/kg dry solids	End product available
	28.058	337	
	-3.353	-3.353 x 12.011 = -40.3	Na_2CO_3 in smelt
Sum	24705	296.7	CO_2

Oxygen balance:

	Mass, mol/kg dry solids	Mass, g/kg dry solids	End product available
	10.807	345.8	
	-24705	-24705 x 31.999 = -790.5	CO_2
	-3.353 x 3/2 = -5.030	-5.030 x 31.999 = -160.9	Na_2CO_3 in smelt
	-0	-0	SO_2
	-0.058 x 2=-0.116	-0.116 x 31.999 = -3.7	Na_2SO_4 in smelt
	-17.014/2 = -8.507	-8.507 x 31.999 = -272.2	H_2O
Sum	-27.551	-881.5	

The humid air demand is then 1.2 x 0.8816/0.2281 = 4.638 kg/kg dry solids.

Air required could have been estimated: Applying the carbon content vs. higher heating value equation gives 13.8 MJ/kg dry solids for HHV. From Fig. 4, the air demand is 3.2…4 m^3n/kg dry solids or 3.8…4.8 kg/kg dry solids.

The smelt flow calculation sums all mass flows to the smelt.

Example
Calculate smelt flow for the previous example.

Smelt balance:

	Mass, mol/kg dry solids	Mass, g/kg dry solids	End product available
	0.243	0.243 x 110.26 = 26.8	K_2S in smelt
	-1.100	1.100 x 78.04 = 85.8	Na_2S in smelt
	-0.058	0.058 x 142.04 = 8.2	Na_2SO_4 in smelt
	-0.149	0.149 x 58.443 = 8.7	NaCl in smelt
	3.353	3.353 x 105.99 = 351.5	Na_2CO_3 in smelt
	-	1	Other inorganics
Sum		482	

The smelt flow is then 0.482 kg/kg dry solids.

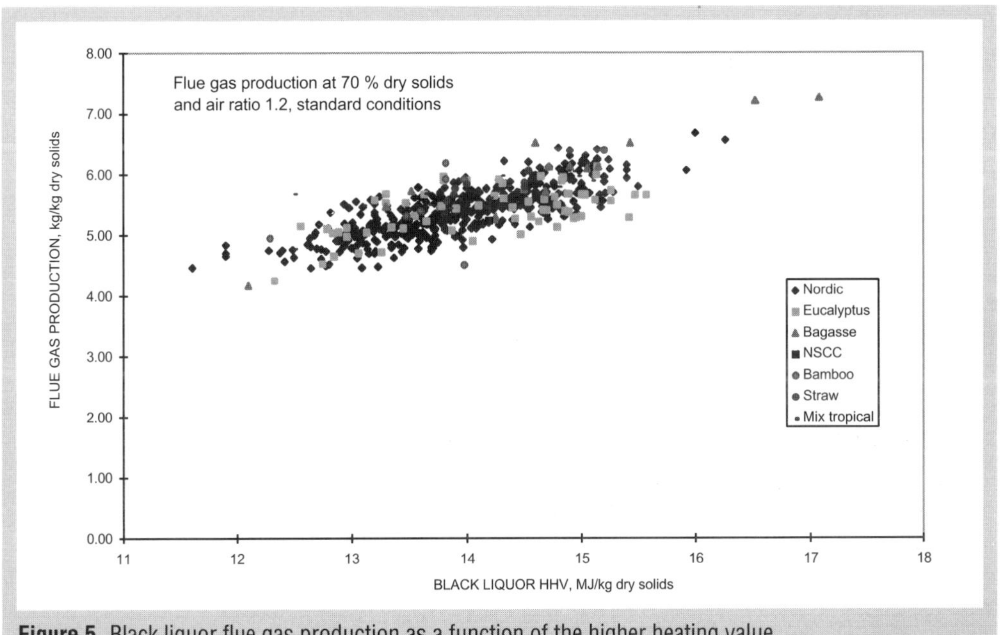

Figure 5. Black liquor flue gas production as a function of the higher heating value.

Figure 5 shows that black liquor flue gas production depends on the black liquor heating value. For a simple analysis, estimation uses the predetermined mass ratio of flue gas to air. An accurate flue gas flow calculation results from simple mass balance with known air flow, black liquor flow, and smelt flow.

Example

Calculate flue gas flow for the previous example.

Mass balance:

	Mass, g/kg dry solids	Product
	1000	Dry black liquor
	250	Water with black liquor
	4638	Air
	150	Soot blowing steam
	-482	Smelt
Sum	5556	Flue gas

The flue gas flow is then 5.556 kg/kg dry solids. For environmental calculations, knowing the dry flue gas flow is useful. This is 5.556–0.250–0.150–0.0135 x 4.638–17.014 x 0.01802 = 4.786 kg/kg dry solids.

2.2 Energy balances

To calculate energy balances for a recovery boiler, one must draw an imaginary boundary around the recovery boiler and then calculate all energy flows in and out of the boiler. This is easy after calculation of all mass flows to calculate the energy balance.

The basis for energy balance is the so-called heat loss method. This calculates the sum of input energy flows. Losses are subtracted from the input energy. The result is the net heat.

Example

Calculate the main steam flow and feed water flow for the previous example. The main steam values are 8.0 MPa (absolute) and 480°C. Feed water process values are 9.0 MPa (absolute) and 120°C. Flue gas flow exits at 180°C. Air enters at 30°C and is preheated to 120°C. The blow down is 0.03 kg/kg dry solids at drum pressure 8.8 MPa (absolute). Black liquor HHV is 13.84 MJ/kg dry solids, and it enters at 125°C. Radiation and convection losses are 0.3% of total heat input. Unaccountable losses are 200 kJ/kg dry solids.

Input	Mass flow, kg/kg dry solids	Enthalpy, kJ/kg	Enthalpy, kJ/kg dry solids
Heating value of black liquor	1	13840	13840.0
Black liquor sensible heat	1	125 x 2.5	312.5
Air	4.64	30 x 1.006	140.0
Air preheat	4.64	(120–30) x1.006	420.1
Soot blowing	0.125	3000	375.0
Total heat input			15087.6
Losses			
Wet flue gas	5.57	180 x 1.2	1202.7
Water in black liquor	0.25	2440	610.0
Water in H_2	0.307	2440	748.1
Smelt	0.482	1350	650.7
Reduction to Na_2S	0.0858	13099	1123.9
Reduction to K_2S	0.0027	9629	258.0
Radiation, convection heat losses	0.003	15088	45.3
Unaccounted losses	1	200	200.0
Total losses			4838.7
Net heat available			10248.9

Enthalpy of steam at 8.0 MPa (absolute) and 480°C is 3349.6 kJ/kg.
Enthalpy of water at 9.0 MPa (absolute) and 120°C is 509.9 kJ/kg.
Enthalpy of saturated water at 8.8 MPa (absolute) is 1354.6 kJ/kg.
Enthalpy of soot blowing steam at 180°C is 2840.3 kJ/kg.
Steam mass flow can then be calculated from the simple balance:
X x 3349.6+0.030 x 13546–(X+0.030) x 509.9+0.150 x 2840.3 = 10248.9
Steam mass flow, X, is then 3.450 kg/kg dry solids.
Feedwater mass flow is 3.450+0.030+0.150 = 3.630 kg/kg dry solids.

2.3 High dry solids

A major trend in recent years has been the increase of dry solids from evaporators. The data presented here is from a study of a 3 000 tons dry solids per day recovery boiler. Although this size recovery boiler is only an example, the results should apply to most current recovery boilers. The more conventional cases are complimented with a case where black liquor undergoes heat treatment (LHT)[11,16,17]. Then the process will affect the black liquor elemental analysis and the amount of black liquor to the furnace.

Table 2 shows the design parameters for the studied recovery boiler. Table 3 presents the material balance data from a low dry solids value of 60% to a high value of 85%. In these calculations, the assumption is that black liquor elemental analysis is constant. The typical change in existing recovery boilers has been an increase of liquor solids content from 65% to nearly 80% dry solids. Table 4 presents the corresponding energy balance values and steam flows.

Table 2. 3 000 tons dry solids/day recovery boiler main parameters.

Maximum continuous firing rate	3000	ton solids/day
Steam pressure	90.0	bar
Steam temperature	490	°C
Feedwater pressure	120	bar
Feedwater temperature	120	°C
Primary air percentage	35.0	%
Primary air temperature	120	°C
Secondary air percentage	50.0	%
Secondary air temperature	120	°C
Tertiary air percentage	15.0	%
Tertiary air temperature	50	°C
Flue gas temperature after economizer	150	°C
Black liquor analysis		
C	37.6	% weight
Na	19.9	% weight
S	4.8	% weight
O_2	32.9	% weight
H_2	3.5	% weight
K	1.0	% weight
Others	0.2	% weight
Higher heating value of dry solids	15.0	MJ/kg dry solids
Dry solids content before mixing tank	60...85	% weight
Liquor temperature before mixing tank	115.0	°C
Liquor temperature after mixing tank	119.0	°C
Chemical loss to stack	0.5	g/kgds
Reduction	97	%
Soot blowing steam flow	1.5	%
Balance reference temperature	0	°C

Table 3. Material balance for the boiler in the example.

Liquor dry solids, %	60	65	70	75	80	85	LHT
Liquor flow, kg/s	57.9	53.4	49.6	46.3	43.4	40.8	42.1
Air flow, kg/s	177.4	177.4	177.4	177.4	177.4	177.4	170.0
Smelt flow, kg/s	15.3	15.3	15.3	15.3	15.3	15.3	14.8
Flue gas flow, kg/s	221.0	216.9	212.8	209.5	206.7	204.2	197.2

In the material balance, the combustion air and black liquor dry solids flows remain constant except for the LHT case. The flue gas flow decreases as less water enters the furnace with increasing black liquor dry solids content. The flow of smelt remains constant since reduction degree has been assumed to stay at 95%.

Table 4. Energy balance for the boiler in the example.

Liquor dry solids, %	60	65	70	75	80	85	LHT
Heat in black liquor, kJ/kg dry solids	15 000	15 000	15 000	15 000	15 000	15 000	14 815
Sensible heat in BL, kJ/kg dry solids	526	485	451	420	394	371	379
Liquor steam preheat, kJ/kg dry solids	18	17	16	15	14	13	13
Heat in cold air, kJ/kg dry solids	153	153	153	153	153	153	153
Heat for air preheating, kJ/kg dry solids	413	413	413	413	413	413	412
Reduction and heat in smelt, kJ/kg dry solids	2 047	2 047	2 047	2 047	2 047	2 047	1 722
Heat available in furnace, kJ/kg dry solids	14 063	14 022	13 986	13 955	13 927	13 903	1 405
Heat in H_2O, kJ/kg dry solids	2 456	2 135	1 861	1 622	1 414	1 230	1 297
Heat in dry flue gases, kJ/kg dry solids	1 112	1 076	1 045	1 018	995	974	976
Sootblowing and misc. losses, kJ/kg dry solids	416	416	416	416	416	416	416
Heat available for steam, kJ/kg dry solids	10 080	10394	10 664	10 898	11 102	11 283	11 443
Steam flow, kg/s	122.8	126.6	129.9	132.8	135.3	137.5	139.4
Efficiency of steam generation, %	62.6	64.7	66.5	68.1	69.5	70.7	71.8

Although the steam flow increases with increasing black liquor dry solids content, the soot blowing can remain constant. The blow down has increased with increasing steam flow since it depends primarily on the quality of the incoming feed water. The total heat input falls slightly with increasing dry solids, since the heat in the black liquor preheat will decrease. Because of this decrease, Fig. 6 shows that less heat will be available in the furnace when the reduction and smelt losses remain constant.

Flue gas heat loss will decrease as flue gas flow diminishes. The H_2O loss in flue gas will decrease since less water evaporation is necessary. The heat to steam will increase as the losses decrease more than the total heat input.

Steam generation increases with increasing dry solids. For an increase in dry solids content from 65% to 80%, the main steam flow increases by 7%. The increase is more than 2% for each 5% increase in dry solids. The superheater pressure loss

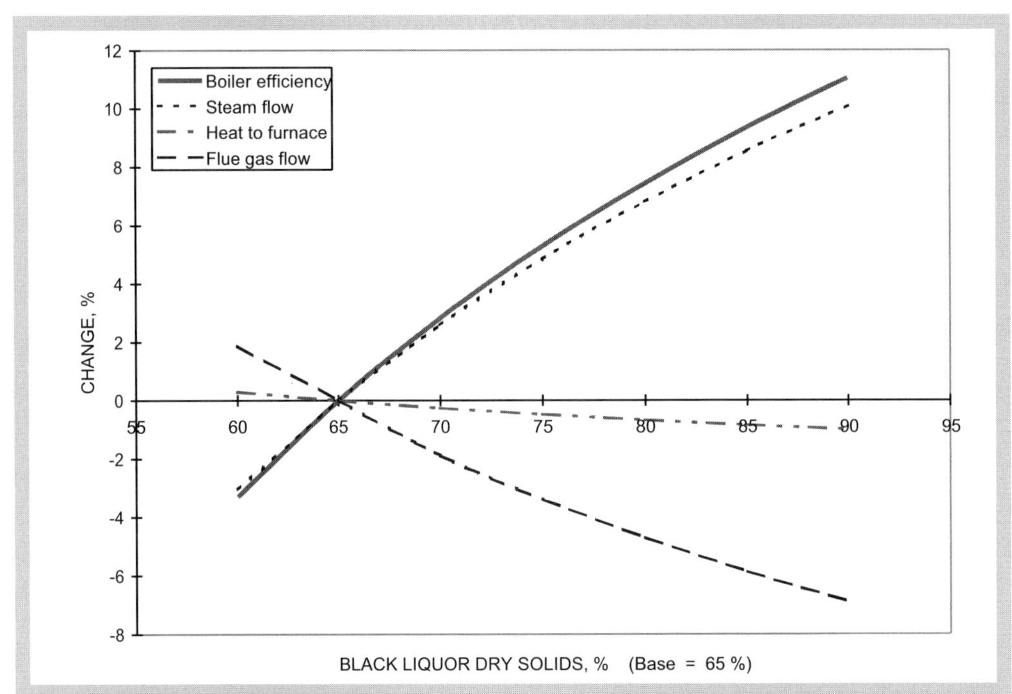

Figure 6. Effect of black liquor dry solids on amount of flue gas, boiler efficiency, steam flow, and heat input to furnace.

increases as the main steam generation increases. From an electricity generation standpoint, pressure loss in superheaters should remain low by using larger tubes or more parallel tubes.

Steam generation efficiency improves more than steam generation itself. This is because preheating the black liquor requires less heat. For the same change in dry solids, the steam generating efficiency improves from about 65% to almost 70%. The decrease in flue gas flow causes the greatest increase in steam generating efficiency. For an increase of dry solids content from 65% to 80%, the amount of flue gas generated falls by 7%. The flue gas passages can be smaller as the flue gas flow decreases. At the same time, less flue gas fan capacity will be necessary.

Liquor heat treatment separates some combustible material as noncondensable gases. The main effect is a decrease in black liquor sulfur content. Liquor heat treatment influences recovery boiler performance in the same way as the increase in black liquor dry solids. The efficiency of steam generation increases, and the amount of flue gas falls. The lower sulfur load to the furnace causes a sharp decrease in the reduction heat.

3 Combustion of black liquor

The black liquor contains many organic and inorganic compounds. The amount and the composition of the black liquor depend on the wood species and the cooking method. The organics burn in the recovery boiler furnace. The inorganic portion is recovered as smelt. Black liquor has the lowest heating value for an industrial fuel because of its large inorganic content. High water content, low heating value, and huge ash content make combustion of black liquor difficult.

Black liquor combustion occurs as a droplet sprayed to the furnace from a liquor gun or in the char bed at the bottom of the recovery boiler furnace[18]. Black liquor is sprayed into the furnace through many liquor guns. In many combustion applications, the aim is to produce very small droplets to maximize combustion rates and temperature. The black liquor is not finely atomized as it enters the furnace. Black liquor is sprayed as very coarse droplets. The average droplet diameter is about 2…3 mm. The aim is to produce sufficiently large droplets so that unburned material can reach the char bed. Table 5 shows the four typical stages for black liquor droplet combustion.

Table 5. Stages in black liquor combustion.

Stage	Characterized by	Time scale in furnace for a 2 mm droplet
Drying	Water evaporation Constant diameter after initial swelling	0.1–0.2 s
Devolatilization	Appearance of flame, ignition Swelling of the droplet Release of volatiles	0.2 –0.3 s
Char burning	Disappearance of flame Decreasing diameter Reduction reactions	0.5–1 s
Smelt	Constant or increasing diameter Reoxidation	Long

Figure 6 shows that black liquor swells during combustion[19]. Very few industrial fuels swell as much as black liquor during combustion. The swelling behavior is due to high volatile yields and suitable surface properties[20–23]. Figure 7 shows the different combustion steps as vertical lines, although all the combustion stages overlap. A reason for this is that different combustion speeds exist at different droplet regions. While black liquor can be dry and undergoing volatile release at one point, drying is not complete at some other point.

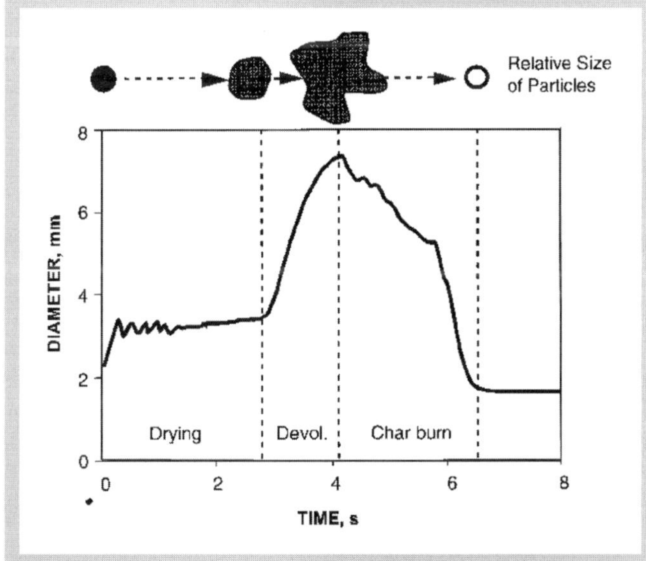

Figure 7. Characteristic swelling behavior of black liquor during combustion.

3.1 Drying

Drying is the evaporation of water from the black liquor droplet with the absence of combustion (visible flame). Evaporation of water requires heat. The drying of black liquor proceeds as fast as the heat transfers to the droplet. Even in the furnace, the heat flux to the droplet limits drying.

The black liquor diameter increases to 1.3...1.6 times the original diameter during the first couple of milliseconds after insertion into the furnace. As water evaporates, the density decreases, but the diameter stays constant. Swelling begins again with the onset of volatile release. The black liquor droplet is not completely dry at the beginning of volatile release. Typically, approximately 5% moisture remains. Drying rate for pine, birch, and sodium sulfite liquors is constant for black liquor droplets with various dry solids contents at 700°C and 800°C.

3.2 Devolatilization

As black liquor dries and temperature increases, reactions with the lowest activation energies start to occur. For organic fuels, release of low molecular weight component gases such as methane, carbon dioxide, hydrogen, and hydrogen sulfide begins. Devolatilization is typically the onset of the increase of black liquor droplet volume, release of volatile gases from the black liquor droplet, and appearance of visible flame. The last condition is the most typical criteria for determining the length of devolatilization time in experimental droplet combustion studies.

During devolatilization, the black liquor droplets swell considerably. The swelling is continuous from the onset of ignition until the devolatilization is complete. Pyrolysis or gasification conditions often study black liquor devolatilization. The release of the volatile fraction occurs whether oxygen is present or not.

Two conditions must be met for the char particle to swell. Gas generation must occur, and the droplet must have plastic surface properties. Swollen, devolatilized particles have a porous, foam-like structure. Swelling for softwood and hardwood liquors is proportional to the ratio of lignin to aliphatic acids. A longer cooking time typically increases swelling. The maximum swelling can change from one type of liquor to another. The maximum swollen volume for industrial black liquors can change from less than 10 to 50 cm^3/g. The well swollen droplets exhibit extensive macroporosity and often have a hollow central core. Adding sodium sulfate or sodium carbonate decreases swelling.

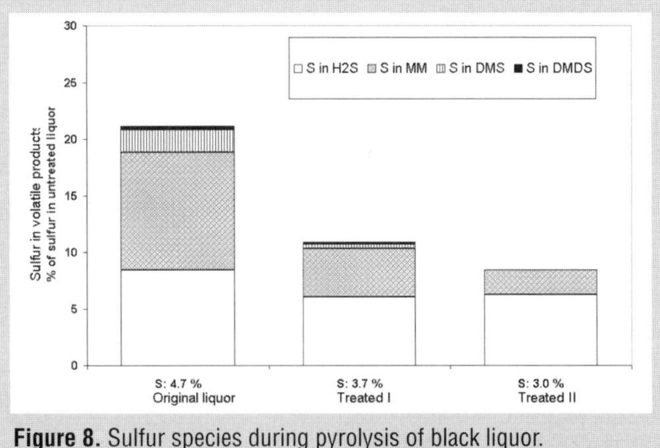

Figure 8. Sulfur species during pyrolysis of black liquor.

During devolatilization, many reactions occur. The main form of sulfur release is dimethyl sulfide and methyl mercaptan as Fig. 8 shows. Hydrogen sulfide, H_2S, forms rapidly during decomposition reactions after gases release from the droplet.

Water evolution ends at the beginning of char oxidation. This means that substantially all hydrogen reacts during pyrolysis.

3.3 Char combustion

Char combustion starts as the volatile release finishes. In laboratory tests, this means the visible flame goes out. In practice, char combustion and devolatilization overlap considerably. The term organic combustion time that is the sum of devolatilization and char combustion times often describes combustion. The combustion residue after the devolatilization is large, but the structure is porous.

Almost all the inorganic matter remains with the carbon char. Half the carbon typically remains in the char. For most purposes, the char consists of carbon with three inorganic salts: sodium carbonate, sodium sulfate, and sodium sulfide. Table 6 shows a representative composition of char.

Table 6. Composition of kraft char.

		Moles/Mole Na_2	Weight, %
Sodium sulfide	Na_2S	1/6	9.0
Sodium sulfate	Na_2SO_4	1/6	16.4
Sodium carbonate	Na_2CO_3	2/3	49.0
Carbon	C	3	24.9
Hydrogen	H	1	0.7

Approximately two-thirds of carbon and less than one-third of hydrogen is present, but no organic oxygen remains. The reduction rate is about 50%. Char continues to burn with the droplet temperature increasing from outside to inside. The inorganic residue remains in the droplet as liquid smelt.

During the char combustion, reduction reactions occur[24-27]. Carbon has a major role in the reduction reaction. Sodium sulfate, Na_2SO_4, reacts with carbon to form sodium sulfide, Na_2S. While the carbon in the char bed burns, it causes the reduction of sodium.

$Na_2S + 2O_2 \rightarrow Na_2SO_4$

$Na_2SO_4 + 2C \rightarrow Na_2S + 2CO_2$

$Na_2SO_4 + 4C \rightarrow Na_2S + 2CO$

The rate of the reduction reactions depends on the char carbon content. This rate is the following:

$$\frac{\partial[SO_4]}{\partial t} = K_{red}\frac{[SO_4]}{B + [SO_4]}[C]e^{-\frac{E_a}{RT}} \tag{1}$$

The constants measured for kraft char were as follows:

K_{red}= 1.31±0.41*10^3 1/s
B = 0.022±0.008 kmol/m^3
E_a = 122 kJ/kmol.

The reduction rate depends on temperature. In modern kraft recovery boilers, high reduction efficiencies are typical. From thermodynamic equilibrium, we note very little sodium oxides and thiosulfite. The final reduction process is slow and requires significant residence time to complete to a high degree.

If sufficient oxygen reaches the char surface, the carbon in the char reacts with the oxygen. If oxygen has a deficiency, the char gasifies with the carbon dioxide, CO_2, and water vapor, H_2O. Carbon dioxide and water vapor react with char to form carbon monoxide, CO.

$C_{char}+CO_2$ –> 2CO

$C_{char}+H_2O$ –> $CO+H_2$

The CO further oxidizes to CO_2 when it contacts oxygen.

The rate of combustion increases with higher swelling as Fig. 9 shows[19]. This is due to the larger available surface area for highly swelled black liquors.

3.4 Smelt reactions

At the end of char combustion, the inorganic residue remains. The black liquor droplet first enlarges and then shrinks to a liquid droplet. If oxygen contacts the smelt, the smelt particles reoxidize to sodium carbonate, Na_2CO_3, and sodium sulfate, Na_2SO_4, In the recovery boiler, having sufficient reacting material on top of the smelt is necessary to avoid smelt reoxidation.

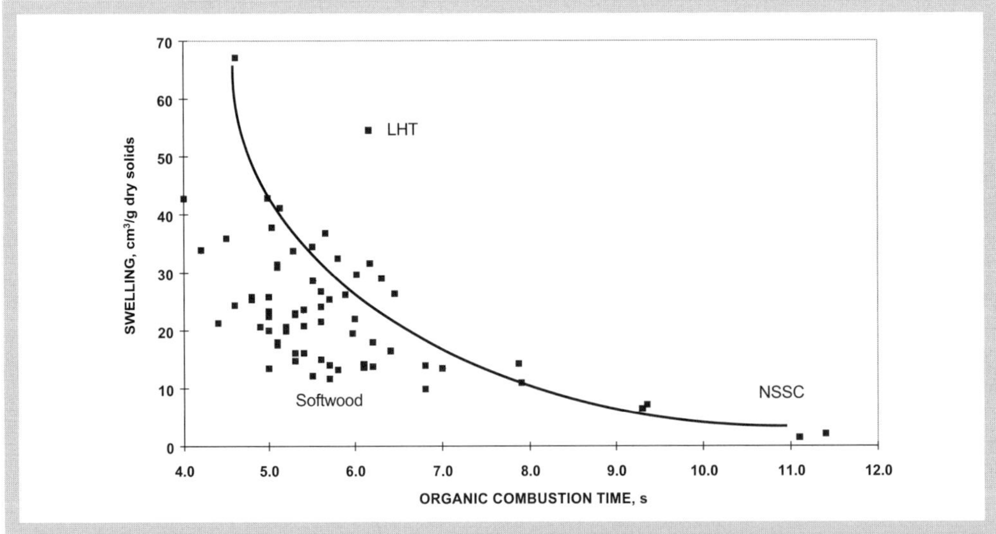

Figure 9. Swelling vs. organic combustion time for single droplet studies at 800°C (= sum of devolatilization and char burning times).

3.5 Combustion of black liquor droplet in the furnace

Figure 10 shows a typical lower part of the recovery furnace. It has three air levels: primary, secondary, and tertiary. A char bed covers the furnace bottom. The black liquor sprays from black liquor guns.

Combustion of black liquor in the furnace can be computer simulated[28–37]. Figure 11 shows the results of one such simulation[30]. Here two black liquor droplets of 1.5-mm diameter but with different combustion properties are sprayed into the furnace. Each droplet is drawn in 0.1-s intervals.

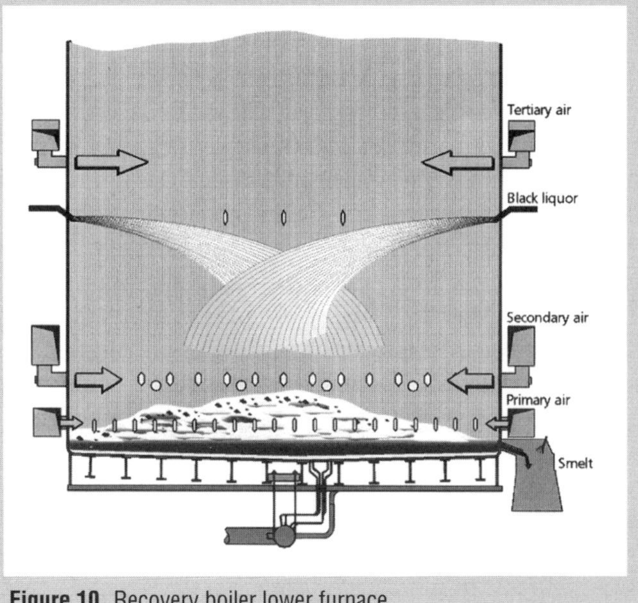

Figure 10. Recovery boiler lower furnace.

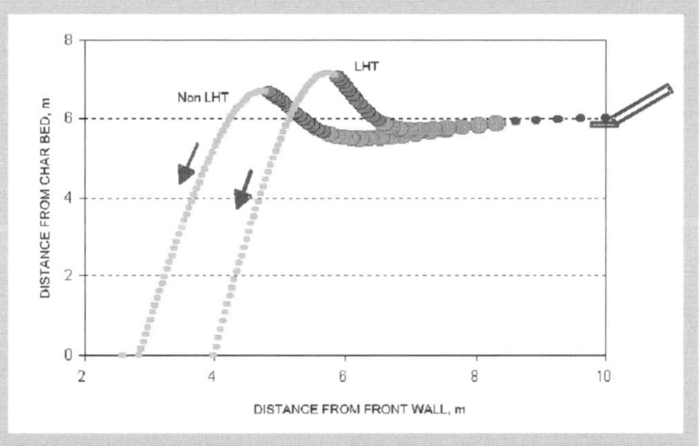

Figure 11. Combustion of 1.5 mm black liquor droplets fired to furnace.

Black liquor drying occurs near the liquor gun at the right wall. Droplet velocity is high, about 10 m/s. As volatiles start to release, the droplet slows. Swelling of black liquor causes the slowing. The droplet flight path curves upward as its density decreases because of drag from the flue gas flow. During char burning, the horizontal velocity is low. As char burning completes, the droplet density increases. It starts falling again until it hits the char bed.

Figure 11 also shows the effect of different combustion properties. Because heat treated liquors swell more than untreated liquors, they burn faster. Since LHT liquor swells more, its speed decreases faster and droplets do not fly as far into the furnace.

4 Dimensioning of boiler heat transfer surfaces

Recovery boiler dimensioning depends on the properties of the design fuel or black liquor. Examination of the differences in construction of past and present recovery boilers illustrates the evolution of recovery boiler design. Predicted black liquor properties and current trends will form the basis of a discussion on requirements for future design.

Chapter 15 provides a detailed discussion of general boiler heat transfer and pressure loss calculations. Exact design rules and especially margins differ. Recovery boiler design has improved over time. One must exercise caution in using the following set of guidelines too rigidly.

4.1 Main design criteria

The main design criteria for recovery boilers are furnace exit temperature, furnace hearth heat release rate (HHRR), hearth solids loading (HSL), boiler bank inlet temperature, and specific heat transfer surface areas.

Recovery boiler furnace temperature depends strongly on black liquor dry solids and heating value. The adiabatic combustion temperature represents the maximum temperature achievable by black liquor combustion. The minimum practical operating dry solids is 55%. In a recovery furnace, the combustion black liquor radiates heat to furnace walls. Maximum temperatures are therefore about one-third lower than adiabatic temperatures. Char bed operation requires temperatures more than 900°C. Typical design furnace exit temperatures are 950°C...980°C. In practice, they can exceed this by several hundred degrees.

The main parameters for the recovery boiler furnace dimensioning are the following:

- Furnace floor area = width x depth
- Hearth solids loading = dry solids firing rate/furnace floor area
- Heat input = dry solids firing rate x black liquor HHV
- Hearth heat release rate = heat input/furnace floor area
- Effective projected surface area = projected furnace wall area excluding furnace floor.

A major dimensioning criteria for recovery boilers is the furnace floor area. Figure 12 shows that the hearth solids loading is typically near 0.2 kg dry solids/s m^2. Hearth solids loading is misleading because it does not consider the variation in liquor properties. A better indicator is the hearth heat release rate. In new boilers with low higher heating value and high dry solids, the hearth heat release rate is 2 600...2 800 kW/m^2.

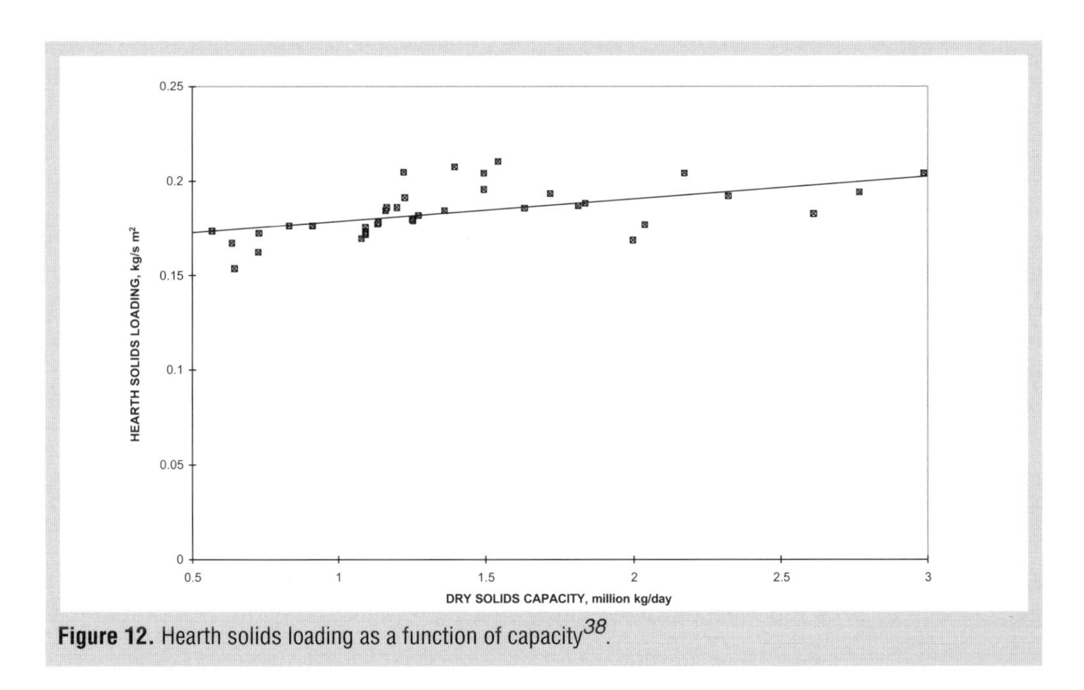

Figure 12. Hearth solids loading as a function of capacity[38].

Furnace volume does not limit processes in the recovery boiler. After selecting the furnace bottom area, the furnace height is determined to correspond to furnace exit temperature of 950°C...980°C. Calculating furnace height requires determining where heat release occurs. Since this depends on the air splits and liquor firing strategy, results differ from one boiler design to another. Typically, furnace heights correspond to average heat flux of 60...100 kW/m^2.

To show the effect of changing recovery boiler design, Table 7 shows the different design bases for the three recovery boilers discussed in Section 13.1.

Table 7. Design basis for three recovery boilers (See Table 1 for design liquors.).

	Current	Modern	Future
Design dry solids capacity, ton dry solids/day	1 700	2 400	4 000
Design black liquor HHV, MJ/kg dry solid	15.0	13.9	13.0
Design dry solids, %	64	74	80
Design steam temperature, °C	480	480	520
Design steam pressure, bar(absolute)	85	85	104
Steam production, kg steam/kg dry solids	3.584	3.357	3.004
Specific steam production*, kg steam/MJ	0.238	0.242	0.231
Furnace design HSL, ton dry solids/m^2 day	15.6	17.1	20.1
Furnace design HHRR, kW/m^2	2715	2760	3 031
Superheater design area, m^2 day/ton dry solids	6.0	5.2	6.8
Boiler bank design area, m^2 day/ton dry solids	3.5	3.0	3.5
Economizers design area, m^2 day/ton dry solids	7.1	10.1	9.1

** steam production per unit of black liquor divided by the HHV of unit of black liquor*

Nominal furnace loading has increased and will continue to increase. The increase of black liquor dry solids from 65% to 80% has increased furnace temperatures. This has allowed an increase in HSL with only a modest design increase in HHRR. Recovery boiler manufacturers have introduced improved air designs. A major factor helping these innovations has been the use of new three-dimensional computational methods for furnace calculation.

The net heat input per unit area of furnace floor will also increase. The average flue gas flow decreases as less water vapor is present. This decrease is not significant but is less than 0.5 m/s. The main impact from evolving designs is the increase in required heat transfer surface after the furnace. This is due to changing liquor properties that shift more heat to transfer after the furnace bullnose.

The processes and reactions occurring on the surface and immediately above the char bed have the greatest impact on boiler operation and performance. About 90% of the total combustion air (equal to all theoretical air required) enters immediately above the char bed to maintain a high temperature in the lower furnace. This high temperature leads to low emissions and high reduction efficiency. Earlier the main problem was to keep the char bed hot to achieve low sulfur emissions and high reduction. With increased dry solids the main problem now is keeping sufficient char bed for proper operation.

Example

Estimate the required furnace dimensions for the 3 000 tons dry solids/day recovery boiler in Section 13.1. The black liquor dry solids is 70%.

Dry solids firing rate is 3 000/86.4 = 34.72 kg dry solids/s.

If we choose the HSL of 0.2 kg dry solids/sm^2, then the furnace floor area is 34.72 kg dry solids/s/0.2 kg dry solids/s m^2 = 173.6 m^2. This corresponds to 13.2 m x13.2 m square.

The black liquor higher heating value is 15 MJ/kg dry solids. Heat input is then 15 x 34.72 = 520.8 MW. (Note that in addition to HHV, heat enters with air preheat, etc.)

The HHRR is then 520 800/173.6 = 3 000 kW/m^2. The area chosen assures that the boiler is of average HSL load. The HHRR is higher than typical for new boilers.

Heat available in the furnace is (13 986–1 861–290) kJ/kg dry solids x 34.72 = 410.9 MW.

If we choose the furnace exit temperature as 936°C, then with flue gas flow of 209.6 kg/s and c_p of 1.355 the heat transferred in the furnace is 410 900–209.6 x 1.355 x 936 = 145.1 MW.

If the furnace height corresponds to heat flux of 80 kW/m^2, then the effective projected area is 1 800 m^2. The furnace height is then 1 800/(13.2+13.2+13.2+13.2) = 34.3 m.

4.2 Effect of high dry solids on recovery boiler dimensioning

The lower furnace plays a key role for lowering emissions and raising capacity[39, 40]. The essential item is to control the reactions occurring in the lowest part of the furnace, on the bed surface, and immediately above it. Increasing black liquor dry solids will increase lower furnace temperatures[16, 28]. Temperature will have a significant impact on emissions, reduction efficiency, and the fouling and corrosion of the convective heat surfaces.

The air system introduces most combustion air into the lower furnace immediately above the furnace floor. As the dry solids increase, the penetration of air jets increases. With increasing dry solids and improved design, emissions have decreased significantly. The nominal furnace loading changes only slightly with an increase in black liquor dry solids content. The hearth heat release rate (HHRR) or the firing capacity usually determine the furnace loading. Because the higher black liquor heating value and the dry solids rate remain constant, these indicators show constant loading. The hearth heat release rate in the furnace remains constant.

As dry solids increase from 65% to 80%, the total heat input per plant area decreases. This is due to a reduction in the black liquor sensible heat input. The net heat input to the furnace increases noticeably, since less water requires evaporation. The average flue gas flow decreases since less water vapor is present. If the furnace temperature remained constant, this would reduce the flue gas flow rates, although reduction would be less than 0.5 m/s.

Table 8 lists the loading values for the example of 3 000 tons dry solids per day recovery boiler in Section 13.1 for a furnace of similar size. A case with 6.9% more than 65% dry solids content liquor firing is added to the previous cases. The steam generation with 80% dry solids equals the steam generation at 6.9% more than 65% dry solids black liquor.

Table 8. Furnace operating conditions for the boiler in the example with constant HHRR.

Liquor dry solids, %	60	65	70	75	80	85	LHT
Sum of heat inputs, MW	559.4	557.9	556.7	555.6	554.7	553.8	525.8
Heat in furnace, MW	393.0	402.6	410.9	418.1	424.4	430.0	423.9
C_p gas, kJ/kg °C	1.391	1.372	1.355	1.340	1.324	1.314	1.321
Liquor flow, kg/s	57.9	53.4	49.6	46.3	43.4	40.9	41.7
Flue gas flow, kg/s	221.0	216.9	212.8	209.5	206.7	204.2	197.2
Furnace outlet temperature, °C	913	926	936	943	949	953	937
Firing capacity, kg dry solids/m^2s	0.2	0.2	0.2	0.2	0.2	0.2	0.192
HHRR, MW/m^2	3.0	3.0	3.0	3.0	3.0	3.0	2.8
Heat release rate, MW/m^2	2.26	2.32	2.37	2.41	2.44	2.48	2.44

With the same steam flow, the furnace loading is higher with 65% dry solids content black liquor than with 80% dry solids black liquor. All the nominal furnace loading indicators are higher including the hearth heat release rate, solids loading, and total and net heat inputs. This is because the flue gas losses are greater for the 65% liquor.

Liquor heat treatment reduces the furnace loading. The net heat input remains almost constant. The liquor dry solids flow decreases as does the black liquor volumetric flow. The average flue gas flow falls more than 10%. The hearth heat release rate falls by 11%, and the nominal firing capacity decreases by 8%. Because the higher black liquor heating value and the dry solids firing rate decrease, the net heat release rate remains constant. Even using a conservative estimate, the furnace bottom area could decrease 4%...8%.

Furnace temperature profiles are available in the literature[32, 41]. A heat release profile can be found by fitting to a known temperature profile. Making correct furnace outlet temperature predictions requires fine tuning to available measurements.

In typical furnace temperature profiles, the temperature starts dropping from a maximum in the lower furnace. The lower furnace temperature increases 100°C...120°C with an increase in dry solids from 65% to 80%. This is significantly less than the 200°C change in the adiabatic flame temperature. The radiative heat flux to the walls will increase by 30%...40% with increasing dry solids. The average heat flux to the walls will increase by about half this amount. The exit heat flux increase is only slight. Increased lower furnace temperatures can speed local corrosion in recovery boilers.

The furnace outlet temperatures increase as the dry solids content rises from 65% to 80%. The measured increase is 20°C...35°C. Note that the measurement error of furnace outlet temperatures is at least 20°C. The emissivity of the flue gas decreases with increasing dry solids content. The actual emissivity is almost constant because of the large furnace volume. The increase in furnace outlet temperatures is due to changes in the mass flow and the net heat to the furnace.

Table 8 shows the furnace outlet temperature variations for a fixed-size furnace. Understandably, liquor heat treatment lowers the furnace exit temperature. The change in furnace exit temperature will be evident with reduced magnitude in the flue gas temperature entering the boiler bank.

The basic dimensioning of recovery boiler superheaters will not change much in response to the change in black liquor dry solids content. Since these surfaces are radiative surfaces, their heat transfer does not depend on flue gas speed.

The radiative heat transfer at the superheaters increases by 10%–20% with the increase in black liquor dry solids. The predicted heat flow increase will not require major changes in the design of the superheater. More important than the change in radiative heat transfer is the change in fouling of superheater surfaces.

The heat available for superheating decreases for a similarly sized boiler as the black liquor dry solids content increases. The flue gas flow is slower with higher dry solids liquor. The heat flux required to superheat steam increases, but the heat transfer coefficient remains almost constant since the heat transfer is primarily radiative. The temperature drop required for the flue gas will increase.

To maintain constant superheating, the furnace outlet temperature must increase or the boiler bank inlet temperature must decrease. If the boiler bank inlet temperature is lower, the average temperature difference in the superheater section also decreases. This reduces heat transfer and requires substantially more superheating surface.

With increasing dry solids content, no obstacle to increasing the furnace exit temperatures should exist. For new recovery boilers, less than 1 000°C is typically specified although nose temperatures of up to 1 100°C have been recorded. With higher dry solids and liquor heat treatment, the superheater corrosion rates are much lower.

Mixed results have occurred from the effect of high dry solids on superheating. Jones has recorded a significant superheating temperature drop with the Arkansas kraft boiler[42]. The Metsä-Sellu Äänekoski boiler does not show such behavior[16]. One possible reason for this is the smaller amount of fouling of heat transfer surfaces.

Table 9 shows that heat treatment changes the heat available for superheating. The flue gas flow decreases to reduce the heat available for superheating further. This behavior is similar to the decrease in superheating with falling boiler load.

Table 9. Heat available for superheating.

Liquor dry solids, %	60	65	70	75	80	85	LHT
Flue gas flow, kg/s	221.0	216.9	212.8	209.5	206.7	204.2	197.2
Furnace exit temperature, °C	913	926	936	943	949	953	937
C_p gas, kJ/kg °C	1.391	1.372	1.355	1.340	1.324	1.314	1.321
Steam flow, kg/s	122.8	126.6	129.9	132.8	135.3	137.5	132.8
Heat for superheat, MW	109.9	110.4	109.7	108.8	107.5	106.4	98.6
Heat for superheat, kW/kg	894.6	871.3	844.0	819.2	794.3	773.6	742.7
If furnace outlet temperature fixed = 926 °C							
Heat for superheat, MW	113.9	110.4	106.8	103.9	101.2	99.1	95.9
Heat for superheat, kW/kg	927.5	871.3	821.7	782.5	748.1	721.2	721.9

To produce the same amount of steam, the recovery boiler must be designed for 6.9% higher solids flow with 65% dry solids liquor than with 80% dry solids liquor. The steam generating efficiency is substantially lower with 65% dry solids compared with 80% dry solids.

For a similar furnace, the firing of 80% dry solids liquor always produces lower loading than the 65% dry solids liquor if firing uses a constant steam flow. The furnace outlet temperatures show that the capacity increase resulting from the higher firing rate produces higher loading than a capacity increase resulting from a dry solids increase. The economizers, boiler banks, and furnaces can all be smaller with an increase in the black liquor dry solids content. An important feature of high dry solids combustion is the increased availability in an overloaded boiler. Firing 80% dry solids instead of 65% dry solids increases firing capacity by 5%. The steam flow increase is much higher at 12%.

Increasing black liquor dry solids will slow the black liquor droplets faster, shorten their active flight path, and burn dry solids black liquor faster. The combustion zone will therefore try to move upward. The amount of water and pyrolysis products landing in the char bed will decrease and increase the char bed temperature. Positioning the liquor guns at a steeper, downward angle would further reduce carryover. Increasing dry solids increases temperatures in the lower furnace. This high temperature leads to low emissions and high reduction efficiency. With increased dry solids, the main problem is keeping enough char bed for proper operation.

The main problem with high dry solids is the decrease in the heat available for superheating. For constant furnace exit temperatures, the required superheater size will increase dramatically. Because of the smaller flue gas flow, less heat is present in the flue gases for superheating.

Example

During operation with 65% dry solids and at 90 bar steam drum pressure, total superheating equals 230°C of which 50°C is for attemperating. Assuming that furnace exit temperature remains constant at 950°C and boiler bank inlet temperature is now 580°C, how would superheating change when black liquor dry solids is 80%?

Solution can use the nondimensional calculation and assume that heat transfer coefficient do not change.

Total temperature difference, θ, is about 950–310 = 640°C.

Temperature difference of flue gas is 950–580 = 370°C.

Temperature difference of steam is 480–310 = 170°C.

Nondimensional parameters are the following:

$$R = 170/370 = 0.459$$

$$e = 370/640 = 0.578$$

then

$$R\,e = 0.578 \text{ x } 0.459 = 0.266$$

And we can solve z

$$z = ln((1-R\,e)/(1-e))/(1-R) = 1.025$$

The flue gas flow will now decrease 5% (Fig. 6), and the steam flow will increase 7%. The value of z will decrease 5% to 0.974, and R will decrease to 0.406.

We can then solve the new e by assuming that z remains constant

$$e = 1-(1-\mathrm{R})/\,(e^{-z(1-R)}-R) = 0.569$$

then

$$Re = 0.406 \text{ x } 0.569 = 0.231.$$

Heat to superheat therefore decreases 1–1.05 x 0.231/0.266 = 9% or 20°C.

Example

How would an increase in furnace exit temperature of 23°C influence the boiler bank inlet temperature and heat to superheat?

Solution uses the nondimensional calculation and assumes that heat transfer coefficient does not change.

Total temperature difference, θ, is $973-310 = 663$°C.

All dimensionless parameters, R, e, and z remain constant. From the previous example $R = 0.406$, $e = 0.569$, and $z = 1.025$

Temperature difference of flue gas is $\theta = 0.569 \times 663 = 377$ °C.

Boiler bank inlet temperature = 973–377 = 596°C. The heat capacity ratio, R, is the same so heat to steam increases by 377/370–1 = 2%.

Table 9 shows that the available heat for that boiler to superheat decreased 107.5/110.4 = 2.6%, steam flow increased 135.3/126.6 = 6.9%, and superheating decreased 20°C.

5 Boiler processes

The processes occurring in boilers are divided according to the process fluid to various subcategories. Typically the main categories are air system, flue gas system and water, and steam system.

5.1 Air system

Fuels consist of combustible organic material and inorganic material that react in the furnace. Air is necessary to sustain combustion. The main requirements for the boiler air system are maximum mixing and proper air ratio profile.

The air is typically delivered at several horizontal elevations to ensure complete combustion and minimize emissions. Figure 13 shows a boiler air system. It has an inlet duct with silencer (1), venturi for air flow measurement (2), air blower (3), air heater (4), distribution ducts (5), and air ducts(6 and 7). Pressure and temperature measuring devices are also necessary at proper locations.

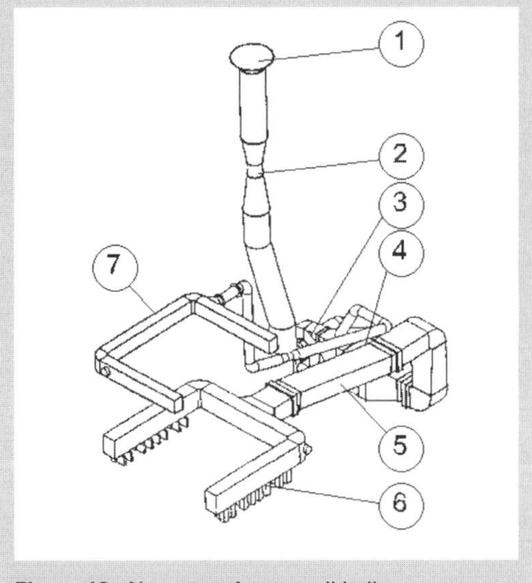

Figure 13. Air system for a small boiler.

Dampers in the openings and air blowers control air flow. A damper in the duct changing the rotating speed of the blower, or inlet vane dampers control the air flow through blowers. The air flow control with dampers in the openings produces the desired duct air pressures. High duct pressure increases blower power consumption but ensures uniform flow though each opening.

Air intake is typically high inside the boiler house. This ensures an even vertical temperature profile and use of heat losses through boiler walls and openings. Upstream location of the air blower from the air preheaters decreases air blower power consumption. After the air heater, the air splits into separate ducts for even air distribution to different sides of the boiler. Using low design velocities is important, since very small form variations in parallel air ducts cause large flow differences.

Air heaters of several types control air temperature. When flue gas is clean, it can preheat air. Bark boilers use several pass steel tube cross flow air heaters. Dirty environments find use of recovery boilers and steam and feed water air heaters. The role of an air heater is to increase the furnace temperature. This decreases the necessary heat transfer surfaces and increases heat to steam.

5.2 Flue gas system

The flue gas system transfers combusted material from the furnace through heat transfer surfaces and emission control devices safely to the atmosphere. The flue gas duct starts from the last heat transfer surface in the pressure part. It has flue gas ducts, dampers, flue gas fan, electrostatic precipitator, scrubber, and stack.

Flue gases from the furnace pass through heat transfer surfaces before entering the flue gas duct. Flue gas fans control the flue gas flow. The flue gas fan also controls the furnace draft. The furnace pressure must be below boiler house pressure.

5.3 Water and steam

Boiler water and steam circulation starts with low pressure feed water and ends with steam at high pressure and temperature. The water and steam circulation components transport, pressurize, preheat, vaporize, and superheat the water into steam as Fig. 14 shows.

The starting point for boiler water and steam circulation is the feed water system. For reliable operation of the boiler, the feed water must be low in oxygen and minerals. The feed water system consists of the feed water tank (1), deaerator, boiler feed water pump (2), control valve, and feed water piping (3). The feed water then enters the first heat transfer surface or economizer. In other than recovery boilers, feed water is usually preheated with turbine extraction steam in high pressure preheaters.

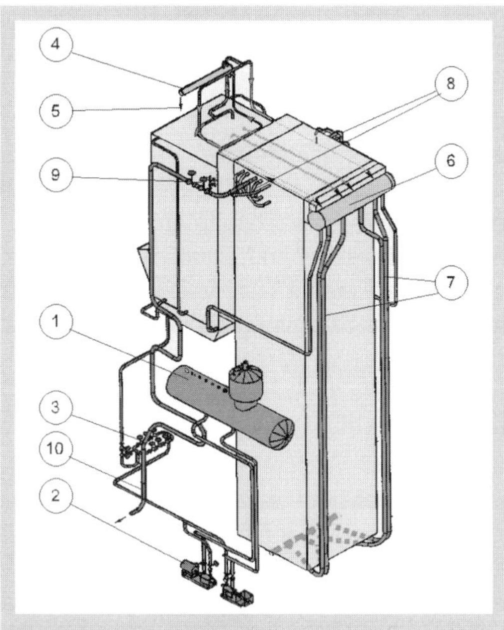

Figure 14. Water circulation system.

The economizer recovers heat available in the coolest flue gases. This heat preheats the water almost to the boiling point. If the flue gas is clean, horizontal flow tubes in staggered arrangement are possible. For dirtier surfaces, vertical flow is common. Counter flow arrangement is preferable.

The water from the economizer generates clean attemperating water (5) from steam in a sweet water condenser (4). Then it flows to the steam drum (6). In the steam drum, steam separates from water. Separation occurs by gravity, screens, and cyclone separators.

Downcomers (7) feed saturated water to evaporative surfaces. Risers to the steam drum collect partially evaporated water. Furnace walls do most evaporation. Depending on boiler operating pressure, 10%...25% evaporation occurs in the boiler bank. The two predominant boiler bank arrangements are vertical and horizontal flow. Both are suitable for dirty and clean gases.

Saturated steam exits from the drum and flows through superheaters. The role of superheaters is to heat the steam well above saturation (8). Because turbines can only operate down to about 95% steam content, proper choice of the main steam pressure and temperature is necessary. The main steam valves control steam flow (9). The steam from superheaters flows through the main steam line (10) to the turbine.

Attemperating controls the steam temperature. Condensed steam or feed water is sprayed into the steam line between superheaters. Typically, only two-stage desuperheating is used. Available feed water purity determines the source of the spray water.

The recovery boiler has a risk of smelt and water explosion due to the presence of the porous char bed. An emergency, rapid drain system minimizes possible damage. The rapid drain system consists of fast valves and pipes to major pressure parts. When the rapid drain opens, pressure inside the boiler drives water from the pressurized part. The rapid drain system has separate lines from the furnace walls, the boiler generating bank, and the economizers. Water remains at the lowest part of the furnace to provide cooling for the floor tubes. If all water drains from the floor tubes, the heat from the char bed damages the tubes.

Water circulation is the result of the density difference between steam filled heating surfaces and downcomers. Friction losses in surfaces, headers, risers, and downcomers limit the circulation. Increasing boiler operating pressure decreases the driving force or density difference. High pressure operation requires pump assisted circulation or once-through operation.

6 Recovery boiler operation

Many factors require proper control during recovery boiler operation. The most problematic factors are proper timing of soot blowing and minimization of potassium and chloride enrichment to dust.

6.1 Soot blowing control

Spent pulping liquors contain inorganic salts. During combustion and reduction reactions, temperatures increase to levels such that significant amounts of chemicals vaporize. Small char fragments and liquor particles may entrain to flue gas flow. This ash collects on all heat transfer surfaces and fouls them. Fouling and fouling related phenomena have long been a concern in recovery boiler design and operation. Several studies of recovery boiler plugging and fouling mechanisms are available[43-49].

All recovery boilers have fouling as a limiting factor for load increase to some degree. With the introduction of modern computer based control, recognizing the changes in the fouling of individual heat transfer surfaces is now possible. Soot blowing can be directed to those heat transfer surfaces where it is most necessary[48]. Soot blowers are usually in pairs on the left and right side of the boiler. The soot blowing sequence typically uses continuous operation of a single soot blower. This type of soot blowing keeps the furnace pressure stable.

A large decrease in soot blowing steam consumption has not materialized even with computerized control allowing various methods for soot blowing. The reason is the lack of understanding of best possible control strategy.

The two main methods to predict the fouling of the heat transfer surfaces are visual observation of the heat transfer surfaces by experienced operators and prediction of fouling from operating parameters such as temperature and pressure measurements[49]. Prediction of fouling through estimation from operating parameters is the most common method in modern soot blowing control systems. The main fouling indicators are pressure loss measurements, flue gas temperature measurements, and steam and water temperature measurements.

The heat flow in a heat transfer surface is a function of heat transfer area and temperature difference between flue gas and steam and water. As the heat transfer surface becomes fouled, this heat flow decreases. A reduction in performance is due to increased heat transfer resistance of deposits.

As the heat transfer surface becomes fouled, the accumulation of deposits decreases available gas flow area. With decreased area, flue gas flow velocity increases. The pressure loss in a heat transfer surface increases as flue gas velocity increases.

The pressure loss is a more important parameter than the decreasing heat flow. The flue gas side pressure loss usually limits the recovery boiler maximum capacity. Fouling can increase until the flue gas fan cannot supply sufficient pressure and flow. Then the recovery boiler load must decrease, or the boiler water must be washed.

Positioning a gas side pressure measurement element before the heat transfer surface and another after the heat transfer surface provides pressure difference information to predict fouling. As Fig. 15 shows, the pressure decreases of the economizers and the boiler bank are significant. The pressure decreases of the superheater section are always lower than 10 Pa.

Pressure losses increase during the first hours after startup because of load increase. This rise decreases after a few days.

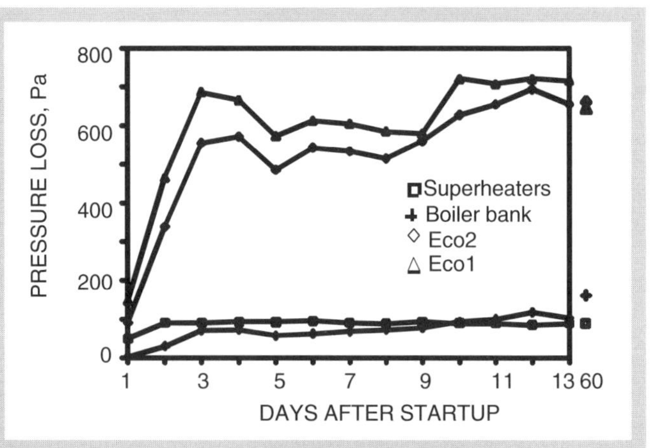

Figure 15. Pressure losses in heat transfer surfaces during 14 days after startup (60 = situation after 60 days).

Flue gas flow also influences measured pressure loss. The flue gas flow can be assumed directly proportional to the boiler load. The most reliable boiler load measurement is main steam flow. To compensate for the changes in boiler load, the measured pressure differences can be scaled with the square of the steam flow as follows:

$$dp^* = dp(q_{nom}/q_{steam})^2$$

where dp^* is the scaled pressure difference, Pa
dp the measured pressure difference, Pa
q_{steam} the measured steam flow, kg/s
q_{nom} the nominal steam capacity, kg/s.

After the startup, a level of stable pressure difference occured in a few days. The pressure difference at the superheater area and economizers did not increase or change after that. These levels were still the same after several months of operation. The boiler bank fouling seems to be a slower and more continuous process. Two weeks after the startup, the boiler bank pressure difference increased periodically. This increase was very slow. With modern instrumentation, pressure loss measurements can predict the fouling of boiler banks and economizers.

Measurement of flue gas temperatures occurs at several locations. These typically include char bed temperatures, furnace temperatures from different heights, boiler bank inlet temperature, and flue gas exit temperature. The furnace temperature measurements are mainly for operating purposes and are not normally used to predict fouling. Temperature measurements at superheaters, boiler bank, and economizers can predict fouling. An increase in operating temperatures is due to increased fouling at the preceding heat transfer surfaces. The main difficulty in predicting fouling with flue gas side temperature measurements is the limited number of measuring points. Accurate prediction of fouling at individual superheaters needs far more gas side temperature measurements than are usually available.

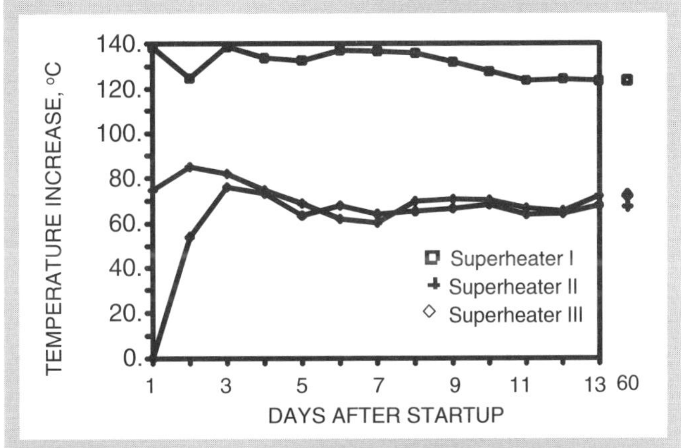

Figure 16. Temperature increases in superheaters 14 days after startup (60 = situation after 60 days).

Using the steam side temperature increases of the superheaters can predict fouling. These temperatures are reliable and easily measured. At constant steam flow, the temperature increase in a superheater is directly proportional to the heat flow to that surface. When the superheating decreases, the fouling in the superheater has increased.

Figure 16 presents the hourly averages of superheating immediately after the startup of a recovery boiler. The figure shows that each superheater achieves its equilibrium fouling degree in two weeks. The superheating in all superheaters is then constant.

The attemperating flow is another good index of total superheater fouling. Since the heat flow to superheaters decreases, less attemperating is necessary. The attemperating water flow describes how much the superheated steam requires cooling to keep the main steam flow temperature constant. A clean heat transfer surface conducts heat better, and the superheating is more effective. Direct measurements of attemperating flows are usually not done. They can be measured as the temperature decrease between superheaters.

Figure 17 presents the attemperating flows between the superheaters as hourly averages after startup. The lowest curve, attemperating flow I, is between primary and secondary superheaters. The middle curve, attemperating flow II, is between secondary and tertiary superheaters. The highest curve is the sum of these two curves.

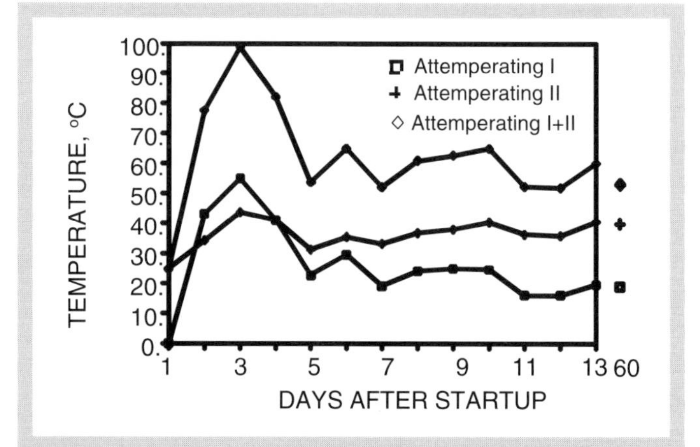

Figure 17. Attemperating flow as the sum of temperature drops between superheaters II – III and IB – II.

Immediately after startup when the superheaters are clean, the attemperating flow is at its maximum. The steep decrease ends after 30 h of operating. The changes in attemperating flow are small after 2 weeks of operating. The boiler operating mode influences the superheating and attemperating flow. The same attemperating flow as observed during the stable boiler operating period was achieved two weeks after startup.

The best fouling indicators are degradation of thermal performance and visual observations. Degradation of thermal performance is best indicated by decreased steam side temperature increases as a function of time and load, flue gas side temperatures as a function of time and load, and attemperating mass flows as a function of time and load.

6.2 Enrichment of potassium and chlorides to dust

Potassium and chloride are important for recovery boiler fouling. Even small amounts of potassium and chloride can influence the sticky temperature of dust. If their concentration in the dust is high, they can cause serious plugging problems at a certain temperature range.

The amounts of chloride and potassium are typically expressed as a mole ratio of substance to alkali metals. Examining the enrichment with molar ratios of potassium and chloride to alkali metals is easy. These ratios have the following definitions:

Cl/(Na+K) mole/mole

K/(Na+K) mole/mole

The enrichment is usually a ratio between the molar ratios of chloride and potassium, respectively, in black liquor and in dust:

$Cl/(Na+K)_{in\ dust}/Cl/(Na+K)_{in\ black\ liquor}$

$K/(Na+K)_{in\ dust}/K/(Na+K)_{in\ black\ liquor}$

Chloride levels at Scandinavian and North American mills are typically 0.2%...0.8% of mass. Potassium levels at Scandinavian and North American mills can vary widely: 1%...4% of mass.

The recovery boiler dust contains primarily sodium sulfate and sodium carbonate with small amounts of chlorides and potassium. It is a mixture of deposited fume and carryover particles. The coarser carryover particles contain a high percentage of sodium carbonate. The small fume particles consist mainly of sodium sulfate and chloride and potassium salts.

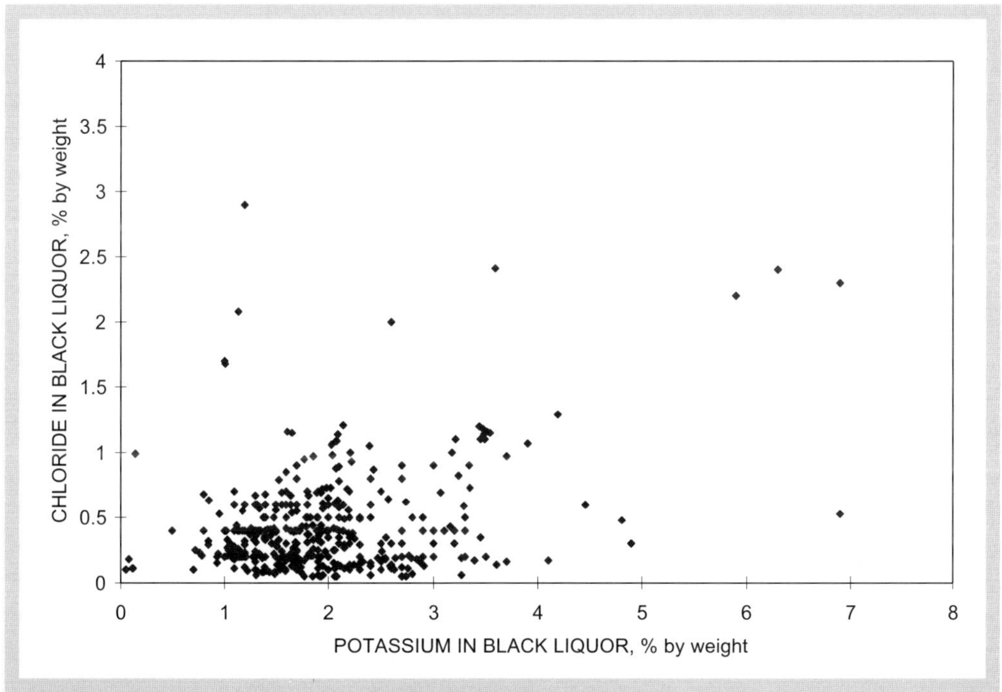

Figure 18. Chloride and potassium in black liquors.

Figure 19 shows the average chloride content in dust at various sampling points vs. virgin black liquor chloride content. The enrichment of different dust samples seems similar. The amount of chlorides is highest in the electrostatic precipitator dust and lower in the hottest surfaces at the boiler bank and economizer II. This is in accordance with the theory that NaCl in fume forms by reaction of HCl vapor with sodium salts. The amount of chloride in electrostatic precipitator dust seems insensitive to changes in chloride level in black liquor.

The amount of potassium in black liquor does not influence the amount of potassium in dust as Fig. 20 shows. A clear difference exists between the dust from hot surfaces and cool surfaces. The economizer I and electrostatic precipitator dust samples contain significantly more potassium than the economizer II and boiler bank dust at Scandinavian and North American mills.

A clear difference in enrichment between hot surface dusts and cool surface dusts exists. The ratio of carryover to fume in dust decreases when taking the sample further in the flue gas passage.

Increasing chloride content in the electrostatic precipitator dust with higher black liquor dry solids has occurred in many boilers. The sulfur content in black liquor influences the chloride enrichment. High sulfidity in liquors will decrease the chloride level in electrostatic precipitator dust.

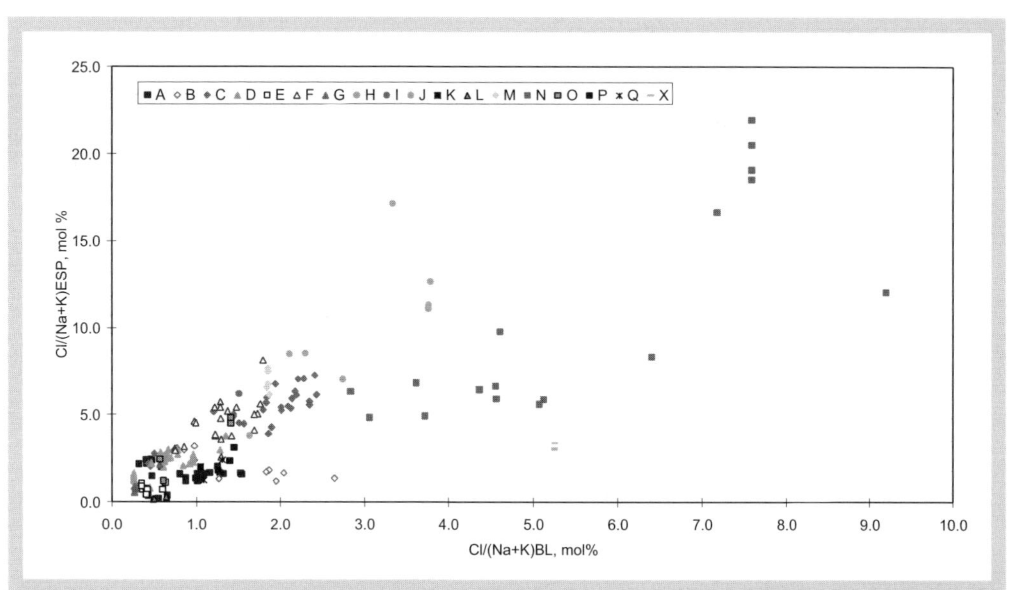

Figure 19. Chlorine content of electrostatic precipitator ash vs. chlorine in black liquor with data from 15 boilers.

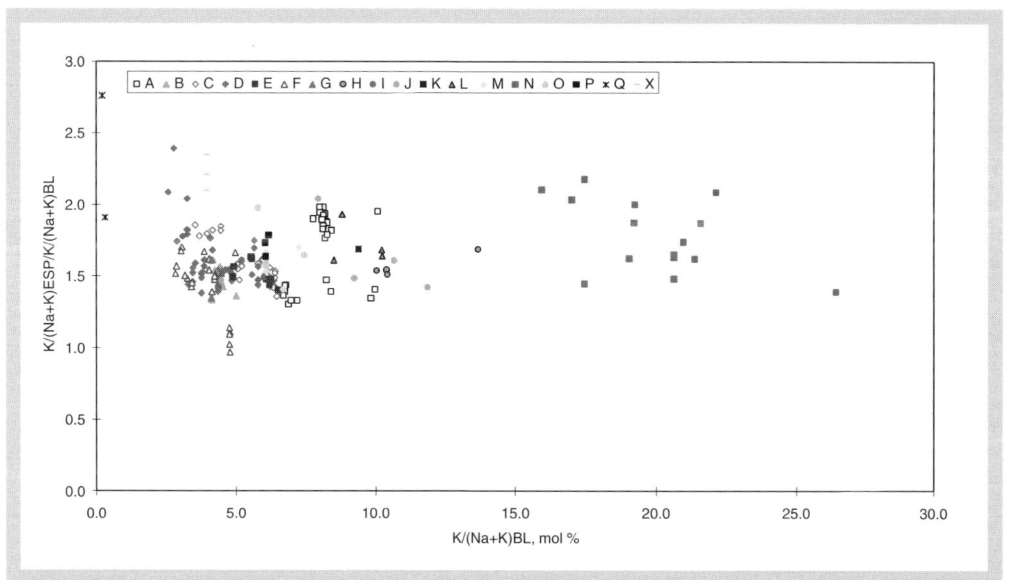

Figure 20. Potassium enrichment of electrostatic precipitator ash vs. potassium in black liquor with data from 15 boilers.

References

1. *Lindberg, H., Ryham, R., CPPA 80th Annual meeting proceedings, CPPA, Montreal p. A73.*
2. *Vakkilainen, E., 30 Years Recovery Boiler Cooperation in Finland, Finnish Recovery Boiler Users Association, Helsinki, p. 132*
3. *Ryham, R., TAPPI 1992 International Chemical Recovery Conference Proceedings, TAPPI PRESS, Atlanta,. p. 581.*
4. *Kiiskilä, E., Lääveri, A., Nikkanen, S., et al., Bioresource Technology 46(2):129(1993).*
5. *Ahlers, P. E., Swedish Corrosion Institute, Stockholm, 1983.*
6. *Hänninen, H., 30 Years Recovery Boiler Cooperation in Finland, Finnish Recovery Boiler Users Association, Helsinki, pp. 121–132.*
7. *Klarin, A., Tappi J., 76(12):183(1993).*
8. *Nikkanen, S., Tervo, O., Lounasvuori, R.,et al., TAPPI 1989 International Chemical Recovery Conference Proceedings, TAPPI PRESS, Atlanta, p. 39.*
9. *McCann, C., CPPA 77th Annual meeting proceedings, CPPA, Montreal, p. A49.*
10. *Adams, T. N., 30 Years Recovery Boiler Cooperation in Finland, Finnish Recovery Boiler Users Association, Helsinki, pp. 61–78.*
11. *Lankinen, M., Paldy, I. V., Ryham, R., et al., CPPA 77th Annual Meeting Proceedings, CPPA, Montreal, p. A373.*
12. *MacCallum, C., TAPPI 1992 International Chemical Recovery Conference Proceedings, TAPPI PRESS, Atlanta, p. 45.*
13. *MacCallum, C. and Blackwell, B. R., TAPPI 1985 International Chemical Recovery Conference Proceedings, TAPPI PRESS, Atlanta, p. 33.*
14. *Tran, H., TAPPI 1988 Kraft Recovery Operations Seminar Proceedings, TAPPI PRESS, Atlanta, p. 175.*
15. *Vakkilainen, E. and Niemitalo, H., TAPPI 1994 Engineering Conference Proceedings, TAPPI PRESS, Atlanta.*
16. *Ryham, R. and Nikkanen, S., TAPPI 1992 Kraft Recovery Operations Short Course Proceedings, TAPPI PRESS, Atlanta, p. 233.*
17. *Ryham, R., TAPPI 1990 Engineering Conference Proceedings, TAPPI PRESS, Atlanta, p. 677.*

18. Hupa, M. and Solin, P., TAPPI 1985 International Chemical Recovery Conference Proceedings, TAPPI PRESS, Atlanta, p. 335.

19. Hupa, M., Backman, R., Frederick, W. J., 30 Years Recovery Boiler Cooperation in Finland, Finnish Recovery Boiler Users Association, Helsinki, p. 37.

20. Noopila, T., Alén, R., Hupa, M., TAPPI 1989 International Chemical Recovery Conference Proceedings, TAPPI PRESS, Atlanta, p. 75.

21. Noopila, T, Alén, R., Hupa, M., Journal of Pulp and Paper Science, 17(4): J105(1991).

22. Milanova, E., Journal of Pulp and Paper Science, 14(4):J95(1988).

23. Miller, P. T. and Clay, D. T., Lonsky, W. F. W., TAPPI 1986 Engineering Conference Proceedings, TAPPI PRESS, Atlanta, pp. 225–234.

24. Grace, T. M., Kraft Chemical Recovery Notes,1990.

25. Grace, T. M., Cameron, J. H., Clay, D. T., TAPPI 1985 International Chemical Recovery Conference Proceedings, TAPPI PRESS, Atlanta, p. 371.

26. Grace, T. M., Cameron, J. H., Clay, D. T., TAPPI 1988 Kraft Recovery Operations Seminar Proceedings, TAPPI PRESS, Atlanta, p. 157.

27. Grace, T. M., Cameron, J. H., Clay, D. T., TAPPI 1989 Kraft Recovery Operations Seminar Proceedings, TAPPI PRESS, Atlanta, p. 159.

28. Haynes, J. B., Adams, T. N., Edwards, L. L., TAPPI 1988 Engineering Conference Proceedings, TAPPI PRESS, Atlanta 355.

29. Horton, R. R. and Vakkilainen, E. K., TAPPI 1993 Engineering Conference Proceedings, TAPPI PRESS, Atlanta, pp. 20–23.

30. Horton, R. R., Black liquor chemical recovery research, DOE, 1991.

31. Horton, R. R., DOE Annual Report on Recovery Furnace Modeling, 1991.

32. Jones, A. K., Ph.D. Thesis, Appleton, Wisconsin, The Institute of Paper Chemistry, 1989.

33. Shiang, N. T., Ph.D. Thesis, University of Idaho, Moscow, 1986.

34. Shiang, N. T. and Edwards, L. L., AIChE 1986 Annual Meeting, AIChE, New York, p. 85.

35. Shiang, N. T., Edwards, L. L., AIChE 1988 Annual Meeting, AIChE, New York, p. 105.

36. Sumnicht, D. W., Ph. D. Thesis, The Institute of Paper Chemistry, Atlanta, Georgia, 1989.

37. Walsh, A. R. and Grace, T. M., Journal of Pulp and Paper Science; 15(3):J84(1989).

38. McCann, C., CPPA 77th Annual meeting proceedings, CPPA, Montreal, p. A49.

39. Borg, A., Teder, A. and Warnquist, B., Tappi, 57(1):126(1973).

40. Borg, A., Nilsson, C., Teder, A., Warnquist, B., STFI Meddelande, serie B nr. 203 (MA B:40). (in Swedish).

41. *Jutila, E., Pantsar, O., Uronen, P., Pulp and Paper Canada, 79(4):61(1978).*

42. *Jones, A. K., and Anderson, M. J., CPPA 78th Annual meeting proceedings, CPPA, Montreal, p. A39.*

43. *Hupa, M., TAPPI 1989 Kraft Recovery Operations Seminar Proceedings, TAPPI PRESS, Atlanta, p. 153.*

44. *Skrifvars, B-J., The midnight sun colloquium on recovery research Proceedings, STFi, Stockholm, 1989, p. 8.*

45. *Tran, H., Reeve, D., and Barham, D., Pulp and Paper Canada 84(1):36(1983).*

46. *Vakkilainen, E., Frederick, W. J., Reis, V. V., et al., TAPPI 1995 International Chemical Recovery Conference Proceedings, TAPPI PRESS, Atlanta,. p. B63.*

47. *Bunton, M. A. and Moskal, T. E., TAPPI 1995 Engineering Conference Proceedings, TAPPI PRESS, Atlanta, p. 706.*

48. *Jameel, M. I., Schwade, H. and Easterwood, M. W., TAPPI 1995 Engineering Conference Proceedings, TAPPI PRESS, Atlanta, p. 695.*

49. *Vakkilainen, E. and Vihavainen E., TAPPI 1992 International Chemical Recovery Conference Proceedings, TAPPI PRESS, Atlanta, p. 146.*

CHAPTER 14

White liquor preparation

CHAPTER 14

CHAPTER 14

Olli Arpalahti, Holger Engdahl, Jouni Jäntti, Erkki Kiiskilä, Osmo Liiri, Jukka Pekkinen, Raili Puumalainen, Hannu Sankala and Juhani Vehmaan-Kreula

White liquor preparation

1 General

An efficient and closed chemical recovery is a great benefit of the kraft process that makes efficient recirculation of cooking chemicals inside the process possible while using only marginal amounts of makeup chemicals. Although the equipment used for this purpose has changed considerably and improved during the existence of the kraft process, the basic process and chemistry are still much the same as in the very beginning. In recent years, the major driving forces behind developments in the process have been energy efficiency, environmental issues, and fast and large scale-up of equipment.

The recausticizing process has two targets. One is to produce clean, hot white liquor containing a minimum amount of unreactive chemicals for the cooking process. The other is to prepare clean and dry lime mud to burn in the lime kiln for reuse as lime with minimum energy usage. The amount of white liquor needed depends on the effective alkali charge in cooking. This is typically 3.5–4.0 m^3/adt. The production capacity of a recausticizing plant can be 8 000–10 000 m^3 of white liquor per day.

Fig. 1 shows the main process steps in a recausticizing plant converting sodium carbonate (Na_2CO_3) in the smelt to sodium hydroxide (NaOH) in the white liquor. Smelt contains primarily Na_2CO_3 and sodium sulfide (Na_2S). Dissolving smelt in water or weak wash produces white liquor that is an aqueous solution of NaOH and Na_2S and simultaneously precipitates lime mud ($CaCO_3$). Lime mud is filtered and washed with secondary condensates or water. The resulting filtrate is weak wash (or weak white liquor) that is used to dissolve smelt in the dissolving tank.

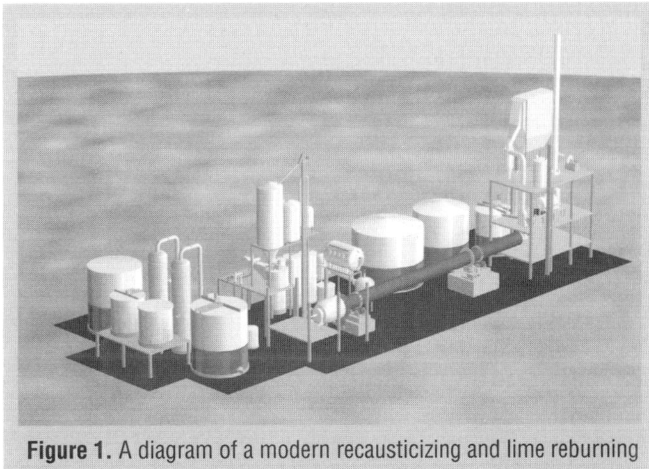

Figure 1. A diagram of a modern recausticizing and lime reburning plant.

There are seven separate unit operations necessary to perform the sodium hydroxide recovery:

- Dregs removal from green liquor
- Slaking of lime
- Recausticizing reaction
- White liquor separation from lime mud
- Lime mud washing
- Lime mud drying
- Lime reburning.

During this treatment, sodium carbonate reacts with slaked lime, and solid calcium carbonate forms. This lime mud is then separated from the white liquor, dried, calcinated, and recycled. The lime reburning system is also a closed recovery system where fresh chemicals are used only to replace process losses and the purge of lime that is necessary to correct the contamination of chemical cycles.

Alkali that is needed in oxygen delignification is also causticized before sulfide oxidation. One suggestion is to prepare all the alkali used in a closed bleach plant in the recausticizing plant.

1.1 Characterization of the liquors

The major compounds in green and white liquors are sodium hydroxide, sodium sulfide, sodium carbonate, and sodium sulfate. TAPPI and SCAN standard methods are commonly used to characterize the process liquors.

The main properties of the green liquor made by dissolving smelt from the recovery boiler into weak white liquor are:

- Density, kg/L
- Total titratable alkali, TTA, g NaOH/L
- Temperature, °C
- Dregs content, mg/L
- Reduction efficiency (degree), $Na_2S/(Na_2S+Na_2SO_4)$,% of equivalents.

The green liquor density can be used as a measure of TTA that describes the chemical content or total alkali content of the liquor.

The dregs content gives the amount of solid impurities in the liquor. These impurities consist primarily of carbon and lime mud particles but also contain nonprocess impurities such as metal hydroxides and sulfides. These impurities are separated from the green liquor by sedimentation or filtration before the recausticizing process. If not removed, they could disturb the filtration properties of lime mud and become enriched in the recirculated lime.

The important properties of the dregs separated from the green liquor are the alkali content (g NaOH/kg) and moisture. The normal expression for moisture is percent dry solids content. The alkali content of dregs depends strongly on the moisture content and describes the alkali loss caused by purging dregs from the process.

When considering the reduction degree of the green liquor and other parameters, it is important to note the effect of the weak wash composition. The reduction degree when measured from smelt is often a few percentage points higher than in green liquor. If direct information from recovery boiler operation and performance is necessary, the analysis of reduction degree using smelt samples is required.

White liquor properties include[1]:

- NaOH, g NaOH/L
- Na_2S, g NaOH/L
- Na_2CO_3, g NaOH/L
- Causticizing efficiency (degree), NaOH/(NaOH+Na_2CO_3),% based on molar equivalents
- Sulfidity, Na_2S/(Na_2S+NaOH),% based on molar equivalents
- Total suspended solids, mg/L
- Temperature, °C
- Density, kg/L.

Table 1 is an example of white liquor composition. Potassium has not been considered in the total mass concentration. Table 2 gives characteristics of white liquor.

Note that sulfidity is also influenced by causticizing efficiency according to the definition. If causticizing efficiency drops for some reason, the sulfidity rises even if the S/Na ratio in the system remains constant. The sulfidity is therefore not a direct measure of the balance between sodium and sulfur in a mill.

The total suspended solids in white liquor is mostly lime mud particles that have passed the white liquor filtration or sedimentation. The purity of the white liquor therefore does not relate to green liquor purity.

The above mentioned analysis allows calculation of the following parameters. Calculation of the chemical charge in cooking usually uses the terms effective and active alkali. By definition, the compounds are treated in terms of NaOH equivalents[1]:

- Effective alkali, EA, (NaOH+1/2 Na_2S), g/L
- Active alkali, AA, (NaOH+Na_2S), g/L
- Total titratable alkali, TTA, (NaOH+Na_2S+Na_2CO_3), g/L.

The equivalent concentrations can also be expressed as g Na_2O/L instead of g NaOH/L.

Green liquor and white liquor contain minor amounts of other sodium compounds such as Na_2SO_3, $Na_2S_2O_3$, Na_2S_2, and NaCl. These compounds require consideration when calculating the total alkali amount using sodium analysis in the liquor. The liquors also contain a variable amount of potassium that requires consideration in cation vs. anion comparisons.

The properties usually measured from lime mud for process control purposes are consistency in various process stages (% dry solids), final dryness (% dry solids), and residual water soluble alkali content after lime mud washing (% Na_2O on dry mud basis). These properties influence lime reburning and lime kiln operation.

For design purposes, knowing the porosity or the free surface of the lime mud and its filterability are also useful.

The residual carbonate content (% $CaCO_3$) and lime availability (% active CaO) usually describe the quality of the burned lime. The residual carbonate tells how well the calcination has succeeded. Available lime is a measure of the alkaline content of reburned lime that reacts with acid as described in the standards. Availability is affected by the contamination of lime. Low availability usually means that phosphorous, silicon, magnesium, aluminum, or iron have enriched the lime cycle requiring replacement of some lime with fresh makeup lime.

Instead of lime availability, the causticizing power of lime is often measured. The causticizing power is a measure of how efficient the lime is; it represents the percentage of CaO and $Ca(OH)_2$ in lime that reacts with Na_2CO_3. The causticizing power also correlates with the reactivity of lime. It depends on its porosity, free surface area, and purity.

Table 1. An example of white liquor composition[1].

Compound	Concentration, g NaOH/L	Concentration, g Na_2O/L	As compound, g/L
NaOH	95	73.6	95
Na_2S	40	31.8	39
Na_2CO_3	23	17.8	30.5
Na_2SO_4	4	3.1	7.1
$Na_2S_2O_3$	2	1.6	4.0
Na_2SO_3	0.5	0.4	0.8
Other compounds	-	-	3.0
Effective alkali	135	105.4	-
Active alkali	115	89.5	-
Total alkali	164.5	128.3	179.4

Table 2. Characteristics of white liquor.

Sulfidity,%	29.6
S/Na equivalent ratio	0.58
Causticizing degree,%	80.5
Reduction degree,% based on total S	82.2
Reduction degree,% based on SO_4	90.9

1.2 Recausticizing chemistry

Two consecutive reactions occur during recausticizing: slaking and causticizing. Both occur in the solid phase of the heterogeneous mixture of lime and green liquor.

When green liquor is mixed with calcium oxide, CaO, it slakes with water and forms calcium hydroxide, $Ca(OH)_2$. This continues to react with sodium carbonate, Na_2CO_3, in the green liquor to produce sodium hydroxide, NaOH, and calcium carbonate, $CaCO_3$.

The slaking of lime is a strongly exothermic reaction[2]:

$$CaO + H_2O \rightarrow Ca(OH)_2 + 65\ \text{kJ/mol} \quad (1)$$

Slaking occurs rapidly at elevated temperatures of green liquor and lime. The subsequent causticizing reaction begins simultaneously with slaking. The slaking reaction takes about 10–30 min to complete depending on the quality of lime. The temperature of the solution still has a major influence on the reaction rate. Below 70°C, the reaction rate is significantly slower than at the normal operating temperatures close to 100°C. In practice, all the lime that is capable of slaking will react with water.

The causticizing reaction is an equilibrium reaction:

$$Ca(OH)_2\,(s) + Na_2CO_3\,(aq) \leftrightarrows 2\,NaOH\,(aq) + CaCO_3(s) \quad (2)$$

The reaction does not have any significant heat effect. Calcium hydroxide and calcium carbonate are insoluble and participate in the reaction as solids.

The equilibrium in the causticizing reaction is reached with some reactants still present in the mixture. The total liquor strength, TTA, will determine the equilibrium composition. An increase in the concentration of the liquor will shift the equilibrium to the reactant side. In dilute liquors, the equilibrium will shift to the product side resulting in higher causticizing efficiency.

The equilibrium causticizing efficiency is a function of total titratable alkali at different sulfidity levels. Goodwin's curve shown in Fig. 2 is a representation of the data graphically. Figure 3 shows another representation of the data as the strength of the liquor vs. total concentration: AA or EA vs. TTA.

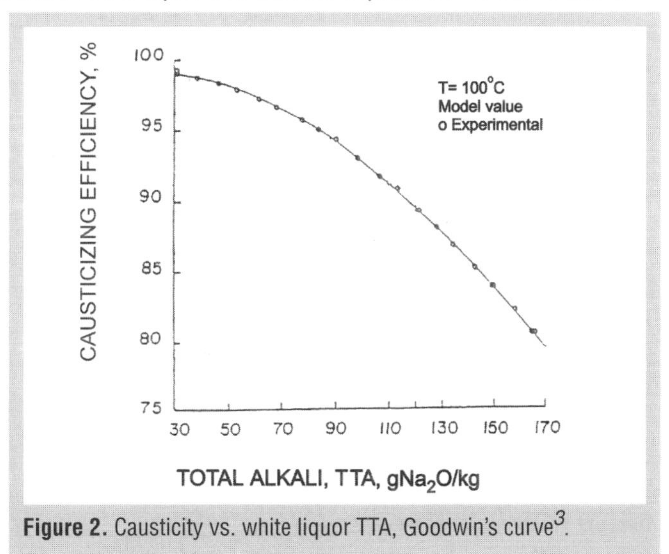

Figure 2. Causticity vs. white liquor TTA, Goodwin's curve[3].

The causticizing reaction occurs between solid and dissolved species. It depends on the rate of diffusion of the species. This is a slow reaction that takes about 1–1.5 h to complete to an adequate extent for white liquor separation. A continuous reaction that occurs in mixing tanks in series will require approximately 2–2.5 h average residence time. If the residence time is shorter, excessive unreacted calcium hydroxide will remain in the lime mud and will deteriorate the filterability or dewatering properties and blind the filter media with precipitated calcium carbonate. Figure 4 shows the dependence of causticizing efficiency on lime quality, and Fig. 5 shows the relationship of causticizing efficiency to temperature.

Chemical equilibrium determines the ratio of sodium hydroxide and sodium carbonate expressed as causticizing efficiency (causticity degree). The causticity and therefore also sodium hydroxide content in white liquor can only vary in the recausticizing plant within the limits set by the composition of green liquor. The sodium to sulfur balance in the mill liquor inventory and the reduction in the recovery boiler determine the

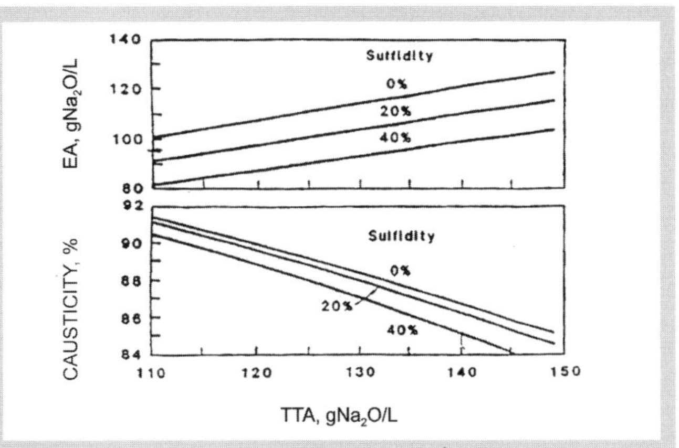

Figure 3. Effective alkali vs. white liquor TTA[4].

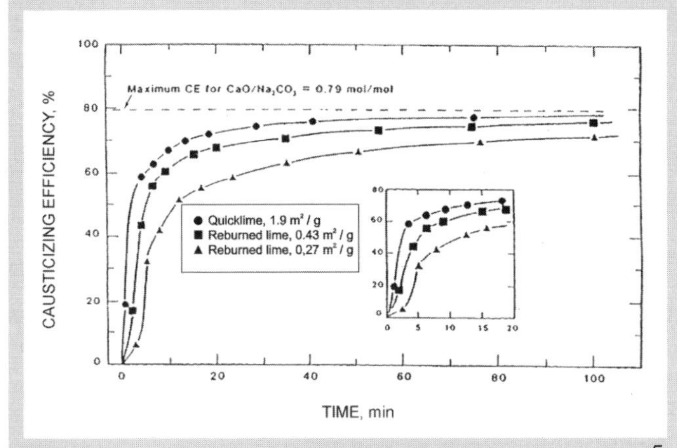

Figure 4. The dependence of causticizing efficiency on lime quality[5].

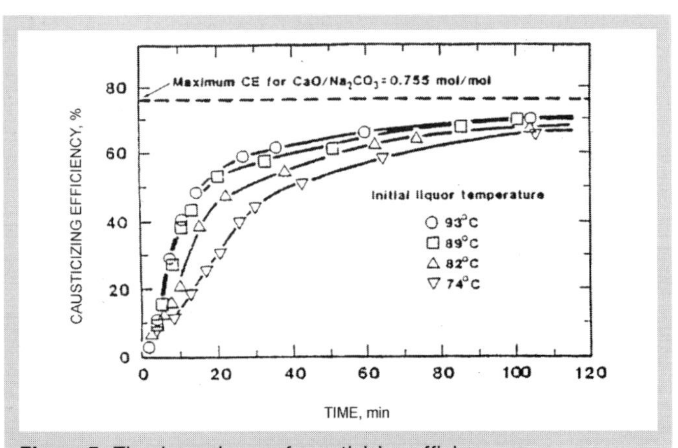

Figure 5. The dependence of causticizing efficiency on temperature[6].

amount of sulfide in proportion to the total alkali in liquor. The ratio of sodium sulfide and sodium hydroxide usually expressed as sulfidity can therefore vary countercurrently with causticizing efficiency.

1.3 Chemical reactions in lime reburning

By heating or reburning, lime mud converts to lime according to the following reaction:

$$CaCO_3 \rightarrow CaO + CO_2 \tag{3}$$

Dissociation of $CaCO_3$ to CaO and CO_2 begins when temperature goes above 820°C. Temperature increase greatly accelerates the reaction. To obtain sufficient reaction rate, reburning uses a temperature of approximately 1 100°C.

In addition to the main component, $CaCO_3$, lime mud entering the lime kiln also contains some unreacted lime as CaO, water, a small amount of alkali, and impurities. The amount of impurities in lime mud dry solids is typically about 7%–10% depending on the amount of impurities introduced in to the process with green liquor and makeup lime.

2 White liquor plant material balances

Table 3 presents an example of a mass flow in a modern recausticizing plant. The numbers refer to a similar modern plant shown in Fig. 1 having the following major equipment:

- Green liquor purification with a cross-flow filter
- Dregs washing with precoat filter
- White liquor filtration with a pressurized disc filter
- Lime mud filtration with a vacuum filter
- Lime reburning with a lime mud drier and a lime kiln
- Lime kiln equipped with an electrostatic precipitator and a scrubber
- Lime kiln fired with oil.

The case in Table 3 uses the following design values:

- Causticizing efficiency, 81.2%
- Sulfidity, 38.0%
- Reduction efficiency, 90.0%
- Total titratable alkali production, 865 t Na_2O/day
- Total titratable alkali, 123.5 g Na_2O/L
- Active alkali, 108.0 g Na_2O/L
- Effective alkali, 87.5 g Na_2O/L.

As Table 3 shows, the weak wash used in the dissolver increases the amount of TTA in green liquor by 2.5% (from 867 to 889 t Na_2O/day). When separated from green liquor, the dregs contain 4.9% of the TTA in the smelt. After washing, the final loss is only 0.1% of the total TTA. Dregs discharge is at 45% dry solids.

Before entering the slaker, green liquor undergoes slight dilution for density control. The lime fed into a slaker consists of two fractions: reburned and makeup lime. The share of the makeup lime in the total lime amount is 5.3% in this case.

After causticizing, the lime mud undergoes separation from the white liquor at 70% dry solids when using a disc filter as in this example. When using candle (ECO) filters, the separation consistency is lower at 38%–43%. Lime mud is then diluted for storage at 35% dry solids.

The lime mud is pumped from storage to the lime mud filter at 25% dry solids where it is finally washed before feeding into the lime kiln. The solid content of the lime mud is typically 75% dry solids. The filtrate and scrubber waters are pumped into the weak white liquor storage tank and finally used to dissolve smelt.

Cold and warm water and some steam are needed at the recausticizing plant. The cold water use is approximately 5 000 t/day, and the warm water use is approximately 7 000 t/day. The design capacity of the corresponding lime kiln installation with a lime mud dryer (LMD) is 530 t product /day. In a lime mud dryer, the mud dries from 75% dry solids to 100% dryness with kiln flue gases before feeding into the rotating kiln. Table 3 gives an example of a material balance for a modern recausticizing plant.

For new, modern kilns, the heat consumption is about 150 kg oil/t of product (Table 4). In many existing older kilns, the heat consumption figures can be considerably higher.

3 Sodium and sulfur balances

The design and the operation of the recausticizing plant also relate to sulfur and sodium balances of a mill. It is important to keep the process losses and fresh chemical consumption equal so that the sulfidity stays at a constant level and the liquor inventory at the mill does not change.

In the past, the chemical balance was controlled by sodium sulfate makeup. Today mills have severe restrictions on sulfur emissions. In some cases, the emissions are restricted to less than 1.0 kg S/t. This means that sodium sulfate or salt cake cannot balance the alkali losses and liquor inventory. The mole ratio of S/Na_2 in sodium sulfate is 1.0 compared with 0.3–0.4 in white liquor. Use of sodium sulfate in cases where sodium losses exceed sulfur losses will cause the sulfidity to increase rapidly.

This is especially true because mills have other natural sulfur makeup chemicals. These include small amounts of sulfur in wood and fuel, magnesium sulfate in oxygen delignification, chlorine dioxide waste acid or sesquisulfate, and sulfuric acid used in the tall oil plant. These streams cover more than enough of the sulfur makeup requirements.

Table 3. Material balance for a modern recausticizing plant with 7 000 m^3 white liquor/day.

Flow	Flow rate, t/day	TTA, t Na_2O/day
Smelt from the boiler		867
Green liquor from dissolver:		
Dregs	10.5	
Green liquor	8245	889
Dregs from green liquor filter:		
Dregs	10.5	
Green liquor	410	43.5
Dregs from dregs filter:		
Dregs including precoat	21.0	
Green liquor	25.7	1.1
Filtered green liquor	8 354	888
Reburned lime to slaker	492	
Makeup lime to slaker	25.9	
Lime mud to storage:		
Lime mud	834	
Weak wash	1548	23.1
Feed to lime mud filter:		
Lime mud	834	
Weak white liquor	2501	23.1
Feed into kiln		
Lime mud	834	
Weak white liquor	278	1.1
Filtrate from the lime mud filter	4 210	22.1
White liquor to storage	8137	865
Utilities:		
Cold water needed	5196	
Warm water needed	7081	
Steam needed	92	

Table 4. Material balance for a lime kiln with 530 t product/day.

Flow	Flow rate, t/day
Dry solids in lime mud feed	882.1
Moisture (water) in feed	293.8
Fuel oil	77.8
Combustion air	1078.8
Steam	7.8
Flue gas including leaks	2132.7
Product	529.6

At the present level of sulfur losses, the alkali makeup must consist of sulfur-free chemicals such as sodium hydroxide. One possibility is to use NaOH in the oxygen delignification stage. Normally, the alkali used in oxygen stages is oxidized white liquor. The use of an R8 type chlorine dioxide plant is common today, and the sodium coming with waste acid and salt from such plants fulfills most sodium makeup needs. Table 5 gives a typical example of sodium and sulfur balances.

Table 5. Sodium and sulfur balance of a modern softwood kraft mill (Post oxygen ECF-bleaching, ClO_2-charge 15 kg ClO_2 /t, ClO_2 plant of type R8).

	kg Na/t	kg S/t
Input:		
Wood and water	0.1	0.3
Fuels	–	0.5
$MgSO_4$	–	1.3
Crude tall oil plant	–	1.6
ClO_2 waste	6.0	4.5
Scrubber NaOH	4.0	–
Total input	10.1	8.2
Output:		
Emission to air	0.1	0.5
Dregs and grits	0.7	0.3
Washing losses	6.5	1.4
Spills	0.5	0.1
Total output	7.8	2.3
Difference between input and output	2.3	5.9

4 Environmental issues

The recausticizing plant as a part of a pulp mill process can today be built to perform without major environmental impact. When compared with the total balance of a mill, recaustizicing is a secondary source of emissions to air and water. Operated properly, it is not a significant source of odorous gases.

The liquor losses and spills in the recausticizing area are collected and returned to the process. If scrubbers are used, the scrubber water and collected chemicals can be returned to the process by using these waters as process water such as in the dissolver.

In the recausticizing area, gaseous emissions are generated in the smelt dissolver, in the lime kiln, and in the slaker. The slaker is of minor importance as an emission source, but vent gases undergo treatment in a simple scrubber to prevent local dusting. The same is also true with other liquor tanks. In some recent installations, vent gases from the tank farm are collected and introduced to the mill's lean gas collection system.

Major emissions originate from the dissolver and consist of droplets formed during smelt dissolving, gas leaks from the boiler smelt spouts, and evaporated impurities from the secondary condensates used in lime mud washing and then as a weak white liquor in smelt dissolving. Vent gases from the dissolver are washed in a separate scrubber and in the newest installations sent to the boiler.

The principal monitored emissions of the kiln flue gas are sulfur dioxide, TRS, dust, and in many countries today nitrogen oxides. To protect the environment, new kilns have an electrostatic precipitator and often a wet alkaline scrubber to ensure that sulfur dioxide and dust emissions are within acceptable limits. Whether the scrubber is required naturally depends on the sulfur content of the fuel.

Flue gases from the lime kiln occasionally contain minor amounts of hydrogen sulfide due to incomplete washing of lime mud. Otherwise, TRS emissions are not normally a problem if the lime mud washer shower waters are clean.

Many mills also burn collected odorous gases in lime kilns. When the burner has proper design and the combustion has correct control, this does not increase TRS emissions from the kiln. Nitrogen oxide emissions have been controlled by combustion air distribution adjustments without any special external gas cleaning devices.

Some solid wastes require discharge from the recausticizing plant. Green liquor dregs must be landfilled, and periodically some portion of the lime must be replaced because of its contamination. Landfilling is still the best way to handle these wastes in many countries. Some countries apply environmental taxes to landfilling. The amount of solid wastes should therefore be minimum. The amount of green liquor dregs is 5–20 kg dry solids/adt, and its pH is high. The dregs are therefore not suitable for landfilling. With waste lime mud, the amount of discharged material is 10–30 kg dry solids/t. There are already several new handling methods available for decreasing the amount of solid wastes. These include chamber filter press filtration.

5 Nonprocess elements in white liquor preparation

The need to close the chemical and water cycles of pulp mills has increased the interest in the accumulation of nonprocess elements (NPE) in white liquor production. In recausticizing, the impurities in lime mud have a negative effect on liquid-solid separation that also causes problems in lime reburning. These problems are often due to the low dry solids content of contaminated lime mud to the kiln and its high alkali content. The impurities in solid form accumulate especially in the lime cycle. Inert material in the lime mud also increases the solid material load of the kiln.

5.1 Nonprocess elements in recausticizing

Besides the main cooking chemicals, green liquor contains considerable amount of NPE as dissolved and suspended solids. They originate from wood raw material, bleaching, or makeup chemical or process water and enter the recausticizing through the recovery boiler. Weak white liquor often brings lime mud from recausticizing to the smelt dissolver. Fine carbon or char is a result of incomplete burning of black liquor in the recovery boiler bed.

The impurities fall into three categories:

- Very soluble in white and green liquor such as K and Cl
- Less soluble in white liquor than in green liquor such as Si and P
- Less or as soluble in green liquor than in white liquor such as Mg, Mn, and Fe.

Very soluble impurities neither accumulate in the lime cycle nor have a negative impact on the recausticizing process itself.

Impurities that are less soluble in white liquor than in green liquor may have a great importance in some cases. A good example is silica. This accumulates considerably in the lime cycle when pulping annual plants or mixed tropical hardwoods. Phosphorous accumulation may also cause problems when pulping softwood.

The impurities in the third group are present in all green liquors in quantities that will cause problems in recausticizing if not removed. Green liquor clarification concentrates on removing these impurities in solid form.

The removal efficiency of a green liquor clarification method for a specific NPE depends on the following:

- Solubility of the least soluble compound of the element in the green liquor
- Total solids removal efficiency
- Amount of impurity in the feed.

The removal efficiency is often 90%–99%. The amount of solids left in purified liquor is normally specific to the method applied. The removal efficiency for suspended solids therefore depends heavily on the feed concentration. The efficiency is not a good measure to judge functioning of a particular process. Better measures for separation efficiency are the total suspended solids, TSS, in purified green liquor or simply the concentration of a single NPE in the liquor.

The purification can improve in some cases by lowering the solubility of an element with the help of co-precipitating elements or compounds. For example, adding magnesium ions can precipitate aluminum and vice versa. Table 6 shows the removal of various elements from green liquor.

Table 6. Removal of various elements from green liquor[7].

	Raw green liquor	Filtered liquor
Dregs, mg/L	310	7
Potassium (K), mg/L	5.3	5.3
Magnesium (Mg), mg/L	53	1.1
Aluminum (Al), mg/L	60	40
Iron (Fe), mg/L	17	8.2
Manganese (Mn), mg/L	16	2.2
Copper (Cu), mg/L	1.3	0.88
Phosphorus (P), mg/L	43.8	41.2

The amount of NPE in white liquor depends on solubility and total input to the chemical cycle of the mill.

Only those nonprocess elements that are soluble in strong alkaline conditions occur in the white liquor because its preparation includes an efficient separation of suspended solids. Any solids remaining in clarified green liquor will separate with lime mud and accumulate in the lime cycle. If the separation of solids from white liquor is not operating satisfactorily, some calcium carbonate will remain in the white liquor and bring some other nonprocess elements in solid form to the digester.

5.2 Nonprocess elements in lime reburning

Limiting the input of impurities to the lime cycle is the best way to maintain the quality of lime. The quality of reburned lime depends heavily on the amount of impurities that enter the lime cycle via green liquor. Successful clarification of green liquor where dregs are efficiently removed is therefore essential. Some nonprocess elements dissolve in green liquor but precipitate with lime addition and then contaminate the lime mud. Replacing the reburned lime with better makeup lime is the only removal technique.

Most typical NPE in the lime cycle are magnesium (Mg), aluminum (Al), silicon (Si), phosphorus (P), iron (Fe), manganese (Mn), and sulfur (S). Silicon is mainly a problem with pulping of annual plants or tropical wood. Sulfur often comes from burning noncondensible gases in the lime kiln.

Table 7 provides an example of different lime compositions. The major differences in the mill limes is that SiO_2 in the lime is below 0.5% and over 5%, respectively.

Table 7. Examples of analysis of makeup lime and reburned lime.

Component (expressed as oxides)	Makeup lime	Scandinavian mill with softwood or hardwood as raw material	Mill with mixed tropical hardwood as raw material
CaO, wt.%	93	N/A	N/A
Na_2O, wt.%	0.06	N/A	N/A
SiO_2, wt.%	2.5	0.4	7.9
MgO, wt.%	1.3	1.3	1.2
Fe_2O_3, wt.%	0.5	0.06	0.4
MnO, wt.%	0.04	0.1	0.01

The physical properties of lime and lime mud also change when they contain more inert compounds. Lime mud with considerable impurities has an increased specific surface area as a result of finer particle size or more irregular particle shapes. Filtration cannot properly dewater this type of lime mud. The feed to the kiln will then contain more moisture, and the lime kiln energy consumption will increase because the heat required for evaporation is much higher than with uncontaminated lime mud.

In practice, the highest cost of burning impure lime mud can be due to poor kiln operation. This is particularly true if frequent shutdowns of the lime kiln begin to limit the entire mill production. Alkali that enters the lime kiln with moisture in the lime mud may cause frequent refractory failures. With sulfur input and variations in lime mud dry solids, this can also be a reason for ring formation. Some impurities can prevent sintering of lime to cause dust problems and limit capacity. Lime mud dry solids remain low, and available flue gas fans and ducts cannot handle the gas flow due to the increased water vapor content. Kiln capacity in terms of produced active lime is lower when lime mud contains more NPE.

Silica is the best known impurity in the lime cycle because of the problems it causes when pulping annual plants and tropical wood. The cell structure of annual plants contain solid silica particles that dissolve in the cooking liquor, proceed with the black liquor into the smelt, and end in the green liquor in dissolved form. Silica precipitates with lime mud during recausticizing because of the low solubility of calcium silicate in white liquor. Lime mud from annual plants often cannot be reburned at all. A large portion of tropical hardwood lime mud must be dumped to keep recausticizing in operation.

Lime mud containing silica has low dry solids content, and it contains alkali when it enters the lime kiln. Because of the high alkali content and silica, the lime mud does not sinter well but dusts. Some granulation is possible, but it requires higher temperature than normal. This introduces the risk of over burning the lime and overheating the kiln refractory. Lime quality is usually lower since silica will also reduce available lime.

Silica forms dicalcium silicate with CaO in the lime kiln. This compound resembles cement when it reacts with water. The reaction product of dicalcium silicate has a large surface area that binds moisture to lime mud. Figure 6 shows a clear correlation between high silica content in lime mud and dry solids content.

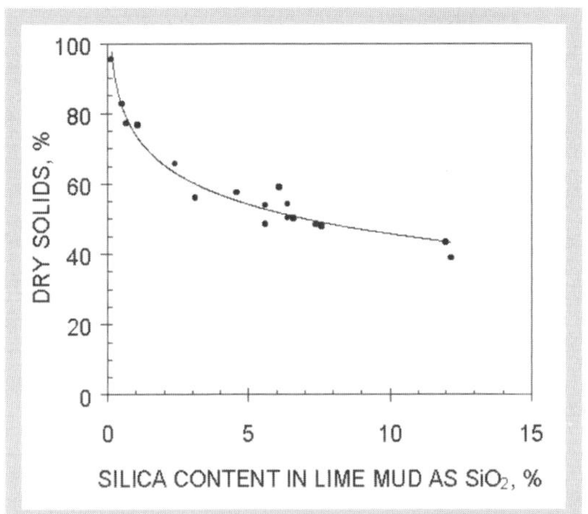

Figure 6. Lime mud dry substance and silica in lime mud.

Phosphate ($P_2O_5^{2-}$) also has high solubility in green liquor but poor solubility in white liquor. Preventing it from entering the lime cycle by green liquor filtration is therefore not possible. In lime, phosphates will bind available lime and increase inert load, since one mole of phosphorous combines with 1.5 moles of CaO[8]. Other effects have not been investigated as thoroughly as those of silica.

Magnesium in green liquor originates from wood and bleaching chemicals. It resembles lime very closely in its chemical reactions. Magnesium carbonate can be calcinated to MgO even at lower temperatures than limestone can be calcinated to lime. It can be slaked by water just like lime, but it does not causticize Na_2CO_3. For that reason, magnesium remains in lime mud as magnesium hydroxide. Magnesium in lime mud resembles free lime. Both occur in the form of hexagonal, plate-like particles that have a large surface area per unit mass. This leads to the same dewatering problems that dicalcium silicate causes.

Aluminum is relatively soluble in green and white liquor as sodium aluminate. It can, however, precipitate with Mg as hydrotalcite or with Si as aluminosilicates. This explains why aluminum sometimes occurs in green liquor dregs and in the lime cycle.

Iron and manganese have low solubility in green liquor like magnesium. In lime mud, they probably occur as hydroxides that lower the dry solids of lime mud. Figure 7 shows the influence of some impurities on lime mud dry solids

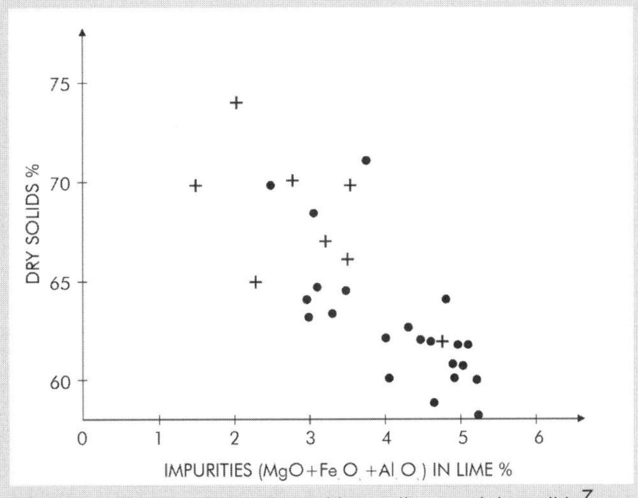

Figure 7. Influence of some impurities on lime mud dry solids[7].

Most sulfur that enters the lime kiln in fuel reacts with lime in the kiln to form calcium sulfate that dissolves in liquor during recausticizing and released calcium reacts to form calcium carbonate. Sulfur in oxidized form increases inert load in the lime and white liquor but does not otherwise harm the lime burning process. Sulfur takes part in ring formation in lime kilns.

Sulfur that enters the lime kiln in lime mud moisture may cause emission problems as odorous gases.

5.3 Reducing NPE in the lime cycle

Most impurities in the lime cycle usually come from green liquor. Many are removed as dregs. Green liquor filtration has given the best results in green liquor purification. Without any dregs separation, the need for lime makeup may be 30% of the lime circulation. With good purification of green liquor, the makeup required can decrease to only 3%– 5%.

In practice, the most difficult impurities are those that change their solubility during recausticizing such as silica and phosphate. They cannot be removed from green liquor with any economically suitable method. These impurities can be removed from the lime cycle only by dumping some contaminated lime mud and compensating the calcium loss with fresh makeup lime.

The quality of makeup lime varies depending on its origin. Makeup lime should contain as few inert substances that will accumulate in the lime cycle as possible. The impurities that enter the lime cycle in makeup lime are not usually a problem. Pure, fresh lime is available, and any impurities are often in the form of coarse sand removed at the slaker classifier.

If solid fuel such as bark or wood dust is fired in the kiln, the impurities in the fuel mix directly with lime. The impurities dissolve in white liquor until solubility limits are reached. The rest remains in the lime cycle.

Considerable savings can result with a more stable kiln operation, higher capacity, better emission control, lower energy costs, and better lime quality in recausticizing if the impurities in the lime cycle are constantly monitored and maintained below a reasonable limit.

6 Future challenges

At least three major drivers will probably change the level of future requirements[9] because of expected changes in the fiber line. The requirement of small process losses will make controlling of chemical balances and sulfidity more important. In the case when bleach plant filtrates return to the chemical recovery cycle, a separate recycling of alkali and sulfuric acid will be necessary[9]. In the first stage, a question of the recycling of alkaline filtrates may exist. In this case, the use of extended white liquor oxidation can be a partial solution.

The development of bleach plant closure will also introduce more impurities into the kraft recovery cycle. Those NPE's that are insoluble in alkaline solutions must be removed in the green liquor handling system. Those that coprecipitate with lime mud must be removed from the lime cycle[10].

The possible modifications and developments in cooking methods may require new types of alkalies with different compositions. The increased use of polysulfide requires selective sulfide oxidation at least for a certain fraction of recovered chemicals[11,12]. Alternative impregnation methods and split sulfidity cooking are other examples of possible trends that will require manufacturing of liquors with varying composition[13].

The importance of recaustizicing control will increase in the future. Further development efforts will accompany the general development of the kraft process.

7 Green liquor treatment

Green liquor (GL) consists of several inorganic chemicals in an aqueous solution. Chemicals such as Na_2CO_3, Na_2S, and Na_2SO_4 originate from kraft recovery boiler smelt. NaOH that usually occurs in green liquor analysis originates from weak white liquor (WL) returned from recausticizing to smelt dissolving. These compounds are the main components of green liquor as Table 8 shows.

Table 8. Typical green liquor composition.

Component	Concentration, g Na_2O/L
Na_2CO_3	80–100
Na_2S	40–50
Na_2SO_4	3–6
NaOH	5–20
Other dissolved	5–10
Suspended solids	0.6–1.5

Green liquor concentration depends on the operating strategy of the mill, e.g., sulfidity, white liquor active alkali, and efficiency in white liquor separation. Green liquor strength as total titratable alkali (TTA) is typically 110–140 g Na_2O/L. In green liquor treatment, the green liquor clarifying process used may be limited by the green liquor strength. This is because the increased liquor density or strength makes sedimentation of dregs difficult particularly when TTA is 135–140 g Na_2O/L or higher. Table 9 shows green liquor physical properties.

The volumetric flow of green liquor depends primarily on the amount of white liquor that must be produced. It is also influenced by the recausticizing process and equipment selection. Some white liquor separation processes recirculate more alkali in weak white liquor to the dissolving tank than others. The volumetric ratio between green liquor and white liquor from one process to another can be 1.05–1.25 m^3 GL/m^3 WL.

Table 9. Green liquor physical properties.

Temperature	80–100°C
Density	1.12–1.2 kg/L
Viscosity	0.9 cP (at 90°C)
Heat capacity	3.3– 3.8 kJ/kg • K

New cooking methods that use green liquor as impregnation liquor are available. This use may represent 10%–20% of the entire green liquor volume. Various methods for production of sulfide rich impregnation liquor have also been proposed. These mainly use the evaporation or cooling-crystallization of Na_2CO_3 with simultaneous preparation of large volumes of concentrated Na_2S liquor.

7.1 Separation of solid impurities

The basic purpose of green liquor treatment is to make the green liquor coming from the smelt dissolver into a proper feed for recausticizing. This treatment consists of removing solid impurities from the liquor, adjusting the temperature, and stabilizing system flow and concentration fluctuations. The green liquor treatment process in Fig. 8 includes the following main process stages:

- Separation of solid impurities and dregs from green liquor
- Cooling of green liquor for recausticizing
- Treating the dregs for proper disposal.

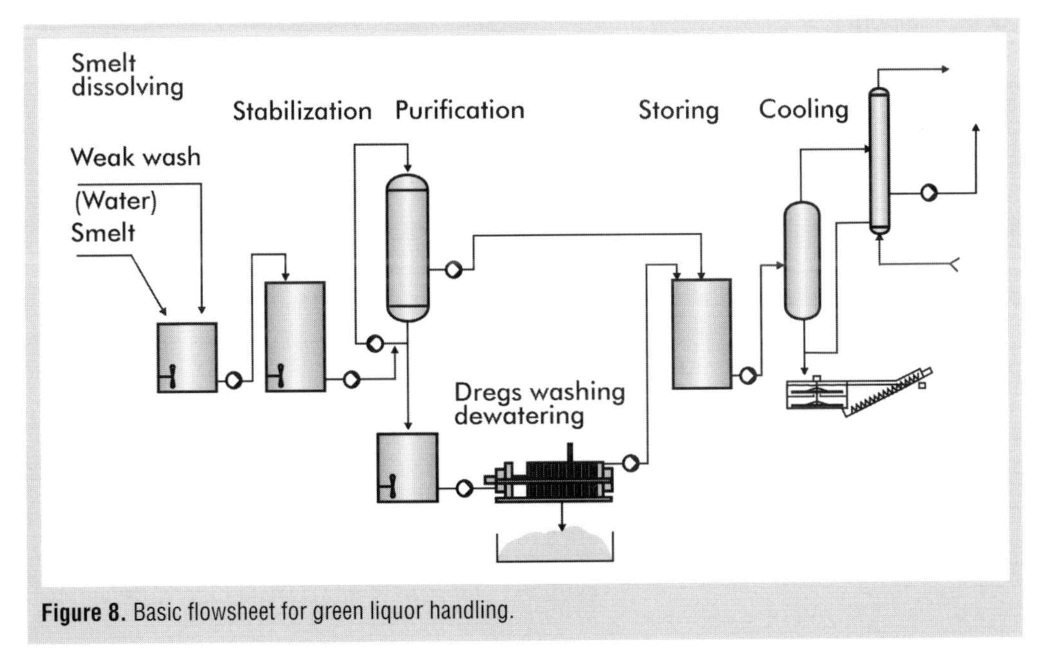

Figure 8. Basic flowsheet for green liquor handling.

To protect the lime cycle from contamination, the green liquor requires purification, i.e., removal of suspended solids. Although suspended solids can be separated effectively from green liquor, dissolved impurities will remain in it. These dissolved impurities may contaminate the lime cycle as discussed earlier if they form sparingly soluble compounds with reburned lime. Using precipitating agents[15] can reduce the solubilities of the impurities somewhat. The separation of the solids normally uses sedimentation or filtration methods.

Green liquor purification concepts include:

- Clarification
  - Sedimentation
  - Sludge blanket clarification
- Filtration
  - Cake filtration with or without precoat
  - Cross flow filtration[7].

The need to control the amount of nonprocess elements in recausticizing has increased the importance of green liquor purification. The extent to which green liquor needs to be purified depends on the technology used elsewhere at the mill. The green liquor used directly in cooking obviously needs the most effective treatment. If the green liquor is used in a recausticizing plant where the lime cycle is relatively open, green liquor of lesser purity is acceptable.

7.1.1 Clarification

A conventional method of green liquor purification is sedimentation of the dregs in a clarifier without any settling aids. When the clarifier load increases with mill production, a need for various flocculants to increase sedimentation rate may emerge. The use of so-called blanket clarifiers has achieved the highest loads. Clarifiers represent simple, proven technology, and their energy consumption is low.

Recovery boiler and dissolving tank outputs often fluctuate unpredictably. A clarifier dampens these fluctuations, but this may result in variation in overflow quality. Variations in liquor strength, temperature, and carbon content can have a negative effect on clarifier performance. Improper use of the stabilization tank and inadequate control of dregs removal can sometimes cause disturbances in the clarifier. Although clarifiers can adjust themselves to variations in flow and usually do not limit the production, the liquor purity cannot be guaranteed with high volume load.

Sedimentation

Clarifying by sedimentation uses the density difference between the solid material and liquid in green liquor. A clarifier is normally an open-top, cylindrical tank as Fig. 9 shows. The solid material that is heavier than the liquid settles to the bottom of the clarifier to form sludge. The clear liquor in the upper part of the clarifier is subsequently decanted.

Modern clarifiers usually have green liquor storage above the clarifier section. The overflow often contains 60–100 mg suspended solids/L. At the bottom of the clarifier, a rake moves the sludge sediment toward a well for sludge removal in the middle of the tank. Consistency of the withdrawn sludge is 2%–5%.

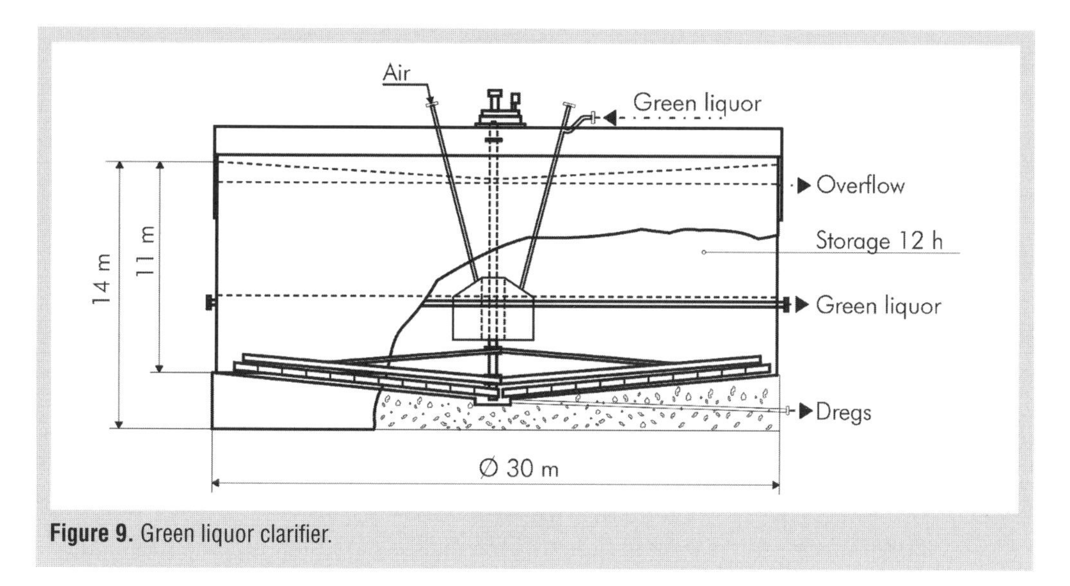

Figure 9. Green liquor clarifier.

Successful operation of a clarifier requires elimination of all causes for disturbing flow patterns inside the clarifier. Green liquor feed enters the clarifier via a distribution box located in the middle of the clarifier. The box design eliminates most kinetic energy of the feed and distributes it evenly in all directions. Large clarifiers especially cannot dampen the internal autonomous flows that density variations in the feed cause. A feed tank is usually in front of the clarifier to stabilize not only the density variation in feed but also temperature and flow variations that could hinder the solids sedimentation. Table 10 shows the design parameters for green liquor clarification.

Table 10. Design values for green liquor clarifying equipment.

Clarifier surface load	11–12 $(m^3/d)/m^2$ (diameter > 25–35 m)
Clarifier section height	3–4 m
Storage section retention time	10–12 h (diameter > 6–7 m)
Stabilization tank retention time	1–2 h

Of the various additives in modern bleaching processes, magnesium is used in amounts that influence the composition of dregs. Magnesium may cause problems in green liquor clarifying[19] because it is present as magnesium hydroxide in alkaline conditions. It forms very fine particles that are difficult to settle without using flocculation aids.

Sludge blanket clarification

The operation of a sludge blanket clarifier involves bringing the feed beneath a flocculated sludge layer so this layer acts as a filtering media through which liquor must pass. A heavy use of flocculating agents is normally necessary. Considerably high surface loads are possible with this method. The surface area of a sludge blanket clarifier is often only 10%–20% of the area of a conventional clarifier.

This method depends completely on flocculants. A mistake in flocculant dosing may cause the sludge blanket to mix into the system causing very high suspended solids in the overflow. If a green liquor storage tank is necessary, it is usually built separately instead of on top of the clarifier. Some Japanese mills use sludge blanket clarification for green liquor.

7.1.2 Filtration

Considerable efforts have recently been made to find a more effective green liquor purification method to replace clarification. During the 1990s, several green liquor filtration concepts entered the market. Maintaining a pressure difference through a filter media drives green liquor in the filtration. The media can be a filter cloth, a lime mud cake, a dregs cake, or a mixture of lime mud and dregs. In cake filtration, the dregs that separate from green liquor are collected in the cake. In cross-flow filtration, the dregs are collected in circulating liquor sludge outside the filter cloth.

Compared with clarification of green liquor, filtration ensures better liquor clarity since incoming liquor properties do not have any substantial effect on the filtrate quality. Liquor quality and variations can influence the filtration capacity. Plugging and wearing of filter cloths is impossible to avoid as they age. Occasionally, they will require washing by acid or complete replacement. These operations often require dividing green liquor storage volume into two separate tanks. One tank is for storage for 6 h before the filter. The other is the same size for the filtrate.

Cake filtration with precoat disc filter

Pressure disc precoat filtration uses lime mud as a filter precoat. Figure 10 shows that the filter resembles the disc filter generally used for white liquor separation. The difference is that the filter cake is not reslurried. It is discarded from the filter as a dry filter cake using a discard screw and pressure lock. The filter precoat forms on top of the filter cloth by filtering lime milk. After the precoat has reached a suitable thickness, lime milk feed is interrupted and replaced by green liquor. Green liquor dregs cake formed on top of the precoat is washed by wash showers and scraped from the surface of the cake. The screw removes the scraped cake to the pressure lock for removal from the filter. Scrapers move successively closer to the discs and remove the blinded surface of the precoat so filtration can proceed. A unit of this kind is in commercial use. This type of green liquor purification process does not require any separate treatment for dregs before disposal. A large amount of lime mud must be dumped with the dregs.

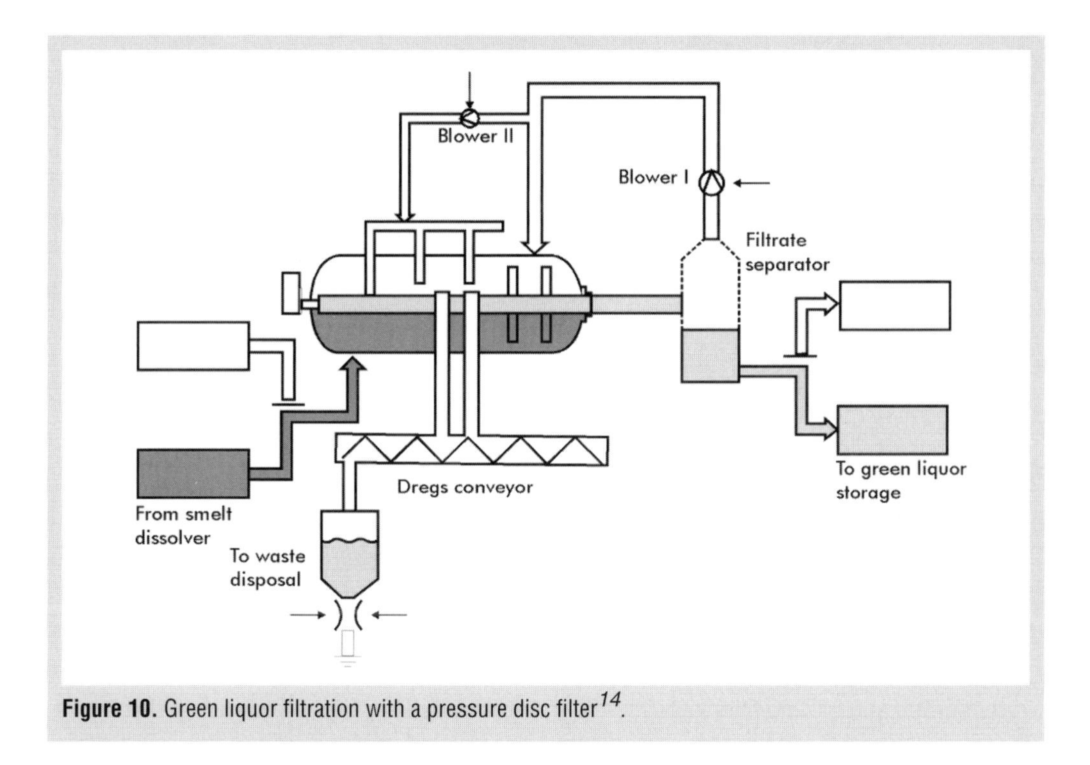

Figure 10. Green liquor filtration with a pressure disc filter[14].

Cake filtration with candle filter

Green liquor cake filtration uses over pressure. Some lime mud or lime is often added to green liquor as a filter aid. In green liquor filtration, 5–10 g/L of lime mud mixes with the green liquor feed. This slurry is pumped to the filter for a filtering sequence of approximately 1 h. When the pressure in the filter reaches its limit, the cake that has formed on top of the filter cloth is back flushed and allowed to settle inside the filter tank. A new filtration period begins after a short sedimentation time. Pumping removes dregs sediment from the bottom of the filter for dregs handling. The technique is simple, and it allows reuse of old white liquor candle filters. Unnecessarily large amounts of lime mud are wasted with this method. Using the filter as a polishing filter after a clarifier allows lowering lime mud consumption. Figure 11 shows a flowsheet for the process.

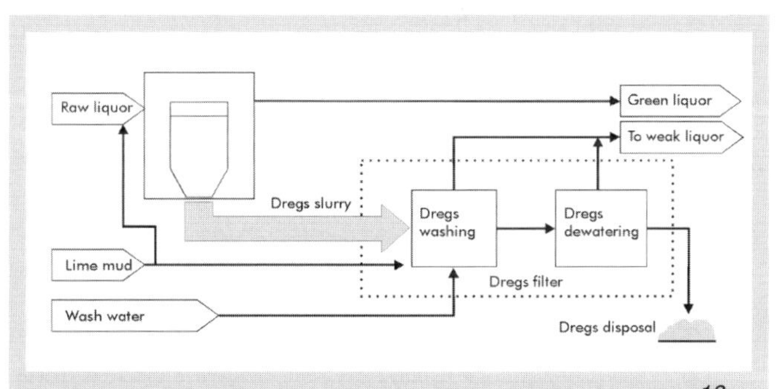

Figure 11. Flowsheet for green liquor filtration by a candle filter[18].

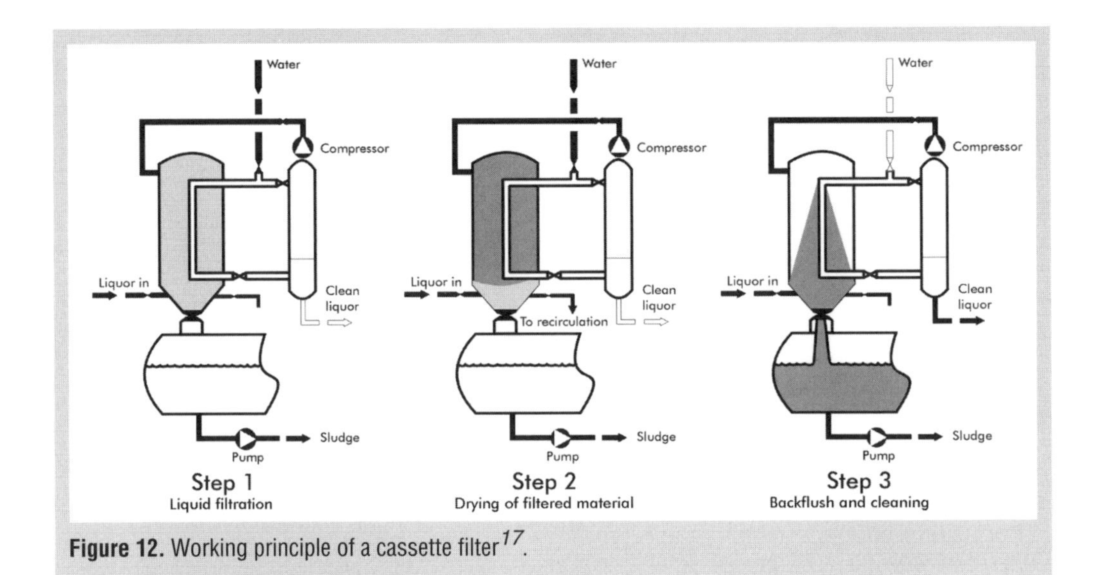

Figure 12. Working principle of a cassette filter[17].

Direct cake filtration

Direct cake filtration without lime mud filter aid uses the "cassette filter" (Fig. 12). This filter resembles the candle filters commonly used in white liquor filtration except the filtrate side is designed for over pressure to allow more effective back flushing. To speed filtration, a flocculant is added to green liquor before filtration. Green liquor is fed to the filter until full filtration pressure occurs. Then the filter is emptied using compressed air. Air pressure holds the filter cake on the filter cloth. When the filter is empty, the filter cloth is back washed with simultaneous reslurry and removal of dregs. The prewashed dregs are finally washed and dried using a dregs filter. No lime mud consumption occurs in the cassette filter. The high alkali removal efficiency is a positive factor. The removed dregs have a low consistency, and this may cause additional volumetric load on the dregs filter. An extra thickener is often necessary[17].

Cross-flow filtration

In cross-flow filtration, a strong tangential flow on the surface of the filter media prevents cake formation during pressure filtration. Green liquor cross-flow filter has an arrangement that makes a thin film of green liquor fall onto filter cloth. Some falling liquor is forced through the vertical filter cloth while most of the liquor with dregs continues in the film to the bottom for recirculation to the top of the filter to maintain the film. Filtration continues and the sludge thickens in the circulation until reaching the desired sludge density. The filtration resistance increases simultaneously, but increasing filter pressure maintains the filtration rate. Thickened sludge is discarded from the filter by the pressure inside the filter.

This method does not require lime mud as filter aid. The final density in the dregs is high, but a separate dregs washing and drying stage is still necessary[16]. Figure 13 outlines the process.

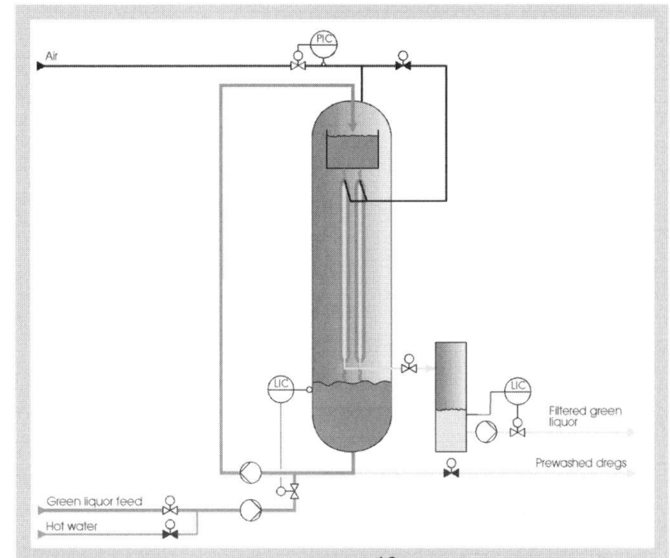

Figure 13. Cross-flow filter flowsheet[16].

7.2 Green liquor cooling

Green liquor cooling involves controlling the temperature of green liquor that enters the slaker in recausticizing to prevent boiling in the slaker. Boiling may occur due to the heat generated as a result of the exothermic slaking reaction. An uncontrollably over boiling slaker could cause a safety hazard due to the vapors and dusting occurring during intense boiling. Several mills control the lime charge with the help of temperature increase during slaking. This is not possible if the slaker has already reached the boiling temperature.

Cooling weak white liquor fed to the dissolving tank usually indirectly controls the green liquor temperature. Cooling of green liquor is almost impossible with indirect heat exchangers because of rapid scaling and plugging[20,21], and even weak white liquor coolers require regular cleaning. Weak white liquor temperature is typically about 50°C–60°C. This requires large amounts of cooling water.

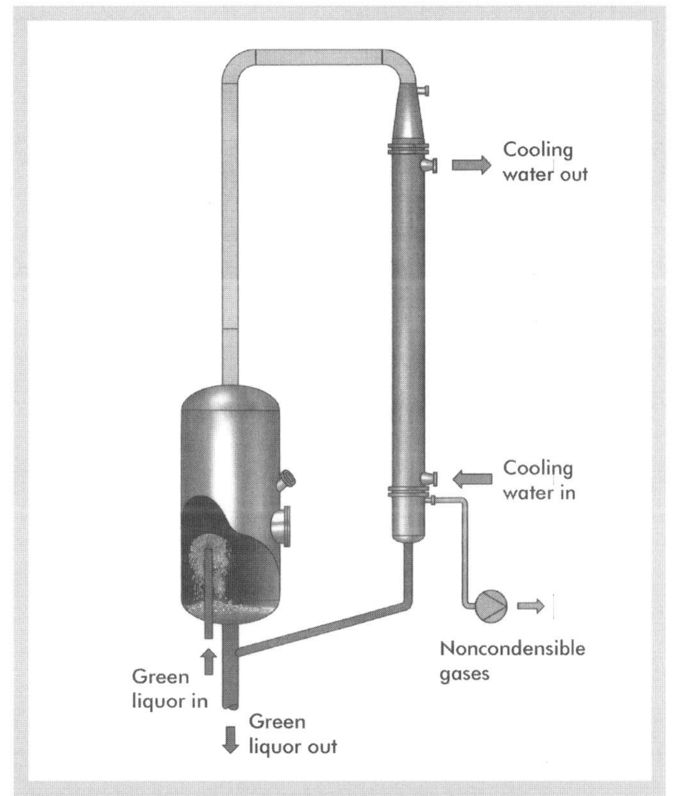

Figure 14. Green liquor cooling by an expansion cooler.

Figure 14 shows a new concept for green liquor cooling involving flashing hot liquid under vacuum. Liquor is normally stored at high temperatures and cooled immediately before feeding to the slaker. In flash cooling, green liquor is fed to a vacuum vessel where the pressure corresponds to the desired temperature. Water vapor formed during flashing condenses by indirect cooling so the cooling water is at 75°C– 80°C at the outlet. A vacuum pump removes noncondensable gases.

With this type of cooler, adjusting the green liquor temperature to 80°C–90°C by controlling the vacuum in the expansion vessel is easier. Since no direct contact exists between green liquor and heat exchanger surface, no plugging problems can occur.

7.3 Dregs handling

The amount of solids or dregs in green liquor varies considerably from one mill to another, but it is often 600–2 000 mg/L. Table 11 provides an example of dregs composition. The chemical composition of the dregs also varies depending on such items as closure of mill cycles, delignifying processes, and pulping raw materials. Other processes can also significantly influence the composition of dregs. Improper operation of the recovery boiler bed can result in fine carbon particles in smelt and green liquor. Difficulties in lime mud washing in recausticizing may also cause high mud carryover via weak wash to the dissolver.

Table 11. An example of dregs composition.

Component	Wt.%
CaO	15.2
Na_2O	15.5
MgO	13.2
MnO	2.9
Fe_2O_4	1.5
Al_2O_3	0.1
SiO_2	0.1
SO_3	12.8
P_2O_5	0.2
C	19.3

7.3.1 Dregs washer clarifier

Green liquor handling can have different goals. If recovering only the major alkali loss is important and dregs discharge concentration is not important, a washing clarifier will be suitable. One-stage or two-stage clarifier washing will work if it is followed by dregs removal at a consistency suitable for pumping. Dewatering of dregs is necessary before discarding them in a solid form to a landfill.

7.3.2 Vacuum precoat filtration

Lime mud is a common filtering aid either mixed with dregs or as precoat on a precoat filter, since filtering green liquor dregs as such is difficult. Lime mud can also be used as a filtering aid in belt filtration. Belt filtration can also be done without extra mud, but then the filtered dregs are removed by water showers and discarded as slurry.

Dregs slurry is most commonly filtered with a vacuum precoat drum filter (Fig. 15). With this type of filter, the cake becomes relatively dry. The process can be automatic, and the filter requires very little maintenance. Typical lime mud consumption with precoat filtration is about 1.2 times the separated dregs amount. The cake dryness is 30%–50% depending on dregs quality. Alkali loss is usually 4%–5% as Na_2O in the discharged cake.

Figure 15. Vacuum precoat dregs filter.

7.3.3 Pressure disc precoat filtration

As mentioned earlier, dregs washing and final dewatering can occur simultaneously with green liquor purification when using a pressure precoat disc filter for green liquor purification[14].

7.3.4 Chamber filter press filtration

In practice, the use of precoat filtration means that lime mud acting as a filter aid more than doubles the total amount of discharge for dumping. This mud can partially match the amount of mud produced by fresh makeup lime. The cost of landfilling and certain environmental quality issues have initiated a search for alternative methods. Dregs should be washed better than with the existing methods and dewatered effectively without any lime mud.

Chamber filtering presses have been widely used in the mining and mineral industries, but their applications in the pulp and paper industry are rare. Two different types of filtering presses are now commercially available for green liquor dregs filtration. The first one has the filtering chambers stacked horizontally, and the filtering media is a movable cloth. The other has a stationary cloth around vertically stacked chambers. The filtration procedure itself for both filters is very similar.

In both filters, pressurized sludge is fed to filter chambers. Filtrate passes through filter cloth that covers the chamber walls. Dregs cake forms in the chambers. After the cake forms, the feeding is discontinued, and the cake is washed by forcing water through the cake. Finally, the cake is dried by squeezing it with moving membrane chamber walls. The dried cake is discharged from the filter by opening the chambers. The process is cyclical, and a new cycle can begin by closing the chambers again.

The horizontal filter press of Fig. 16 with a moving filter cloth has filter chambers that lay horizontally in a stack closed by hydraulic cylinders. When the press opens, the filter cloth carries filter cakes from the chambers. All the chambers are discharged at the same time. The moving cloth is washed with a stationary washer.

Figure 17 shows a vertical type filter press with stationary cloths. This press has filter chambers stacked vertically next to each other. The stack is closed by a hydraulic cylinder. The cakes are discarded by opening the chambers singly and allowing the cakes to fall to a discard chute below the filter. A moveable washing device washes the filter cloth when necessary.

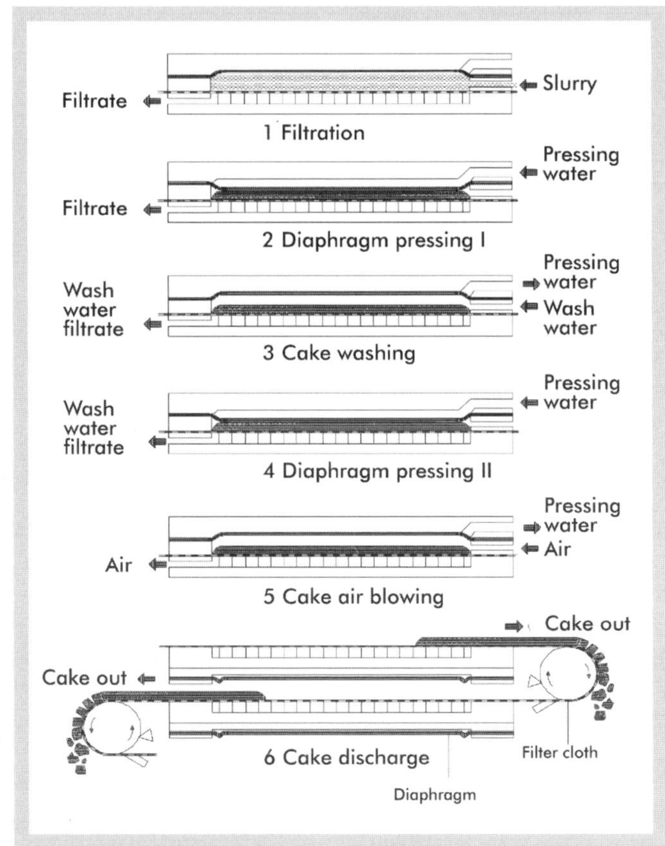

Figure 16. Working procedure in a horizontal type filter press.

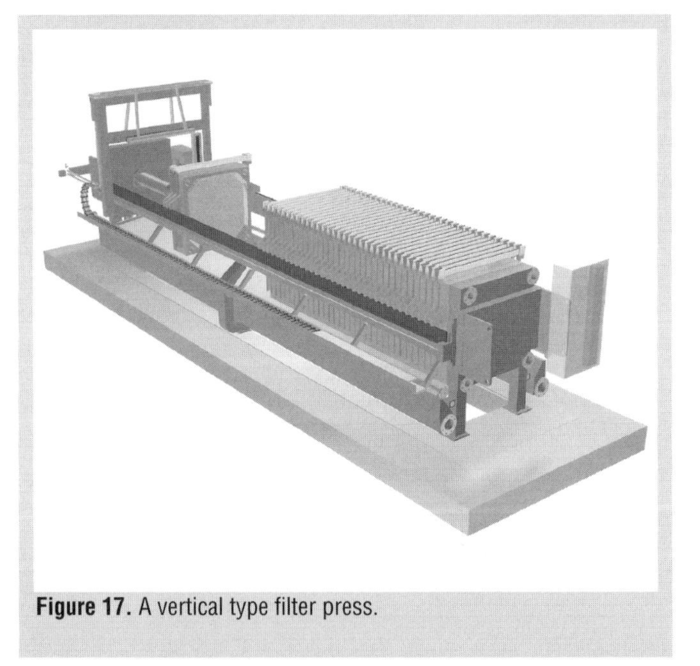

Figure 17. A vertical type filter press.

8 Recausticizing

8.1 General

Green liquor converts to white liquor during recausticizing when reburned lime reacts with sodium carbonate in green liquor. The reaction products separate from each other as white liquor in the solution and lime mud in the precipitate according to the following reactions:

$$CaO + H_2O \rightarrow Ca(OH)_2 \tag{4}$$

$$Ca(OH)_2 + Na_2CO_3 \rightarrow 2\ NaOH + CaCO_3\ (\downarrow) \tag{5}$$

The white liquor composition determines its properties. The major components of white liquor are sodium hydroxide and sodium sulfide that are active chemicals in the cooking process. White liquor also contains some inert compounds such as sodium sulfate and sodium carbonate that form during the recovery process. Some nonprocess elements may also enrich the white liquor. They enter the process with raw materials, with various chemicals used in the process, or as makeup.

The maximum concentration of green liquor occurs in the smelt dissolving tank. The liquor then becomes weaker by dilution during the recausticizing process.

8.2 Process

Recausticizing has three main steps:

- Reactions that include mixing the reactants and suitable reaction time: slaking and causticizing
- Solid-liquid separation where solid lime mud and liquid white liquor separate from each other: white liquor clarification by settling lime mud or filtering white liquor
- Lime mud washing and further dewatering where the remaining liquor in the mud is removed by displacement with water, settling, or filtration, followed by dewatering to increase lime mud dry solids fed into the lime kiln.

The aim of the recausticizing process is to produce as much strong white liquor as possible with as low sodium carbonate content as possible. These targets have both chemical and physical limitations. The chemical equilibrium sets limits to the causticity reached with a certain strength of white liquor. To avoid over liming and the subsequent changes in the filtration properties of lime mud due to $Ca(OH)_2$, a certain margin must cover unintentional changes in lime proportioning. In practice, the target for the causticizing efficiency is 3%–4% below the equilibrium value. With white liquor TTA at 125–130 g Na_2O/L, the causticizing efficiency will be about 80%–82%.

If higher TTA is required, the causticizing efficiency of the liquor will inevitably be lower, but the active alkali in the white liquor to the digester will increase. The clarification of green liquor by settling may be hindered by increasing strength and density of the liquor. A TTA of 140 g Na_2O/L is usually the practical limit for proper settling of dregs in green liquor. If suspended solids are removed from green liquor by filtration, no limitations for the strength of the liquor apply.

If the total alkali concentration of white liquor increases considerably, the amount of sodium carbonate will also increase. This may disturb the operation of the black liquor evaporation plant. This is especially true if combined with low reduction degree or when black liquor for some other reason contains sodium sulfate in considerable quantities.

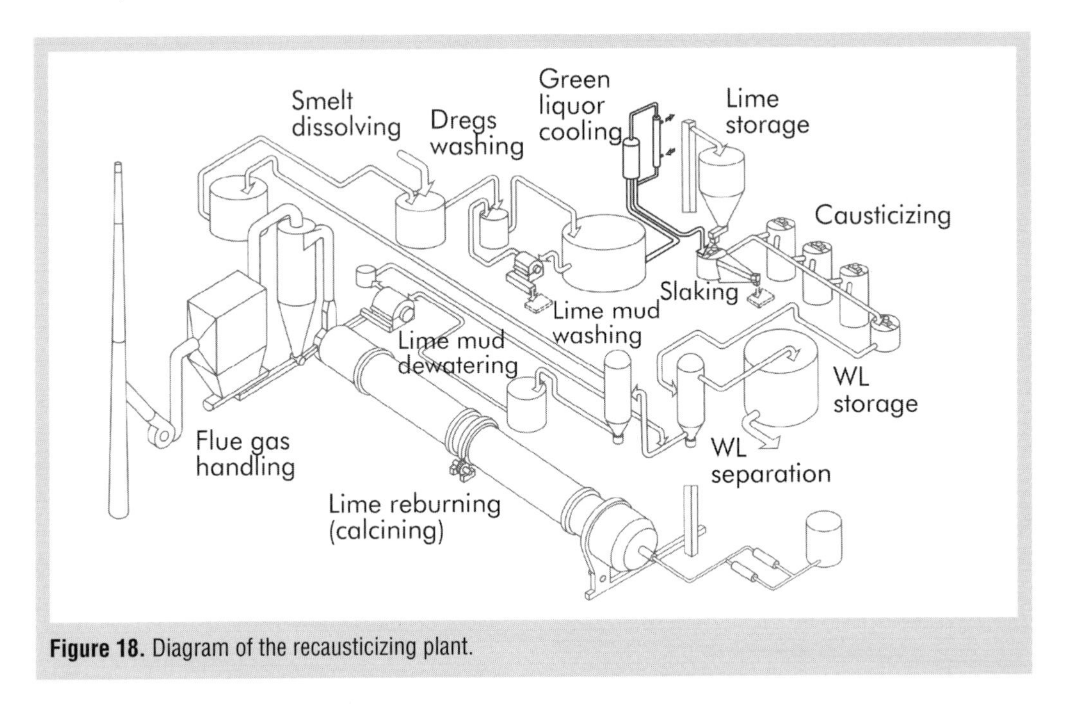

Figure 18. Diagram of the recausticizing plant.

Figure 18 shows a diagram of the recausticizing plant. If the plant uses a pressurized disc filter, lime mud can be washed in the same filter where white liquor is separated from lime mud. This avoids a separate piece of equipment for lime mud washing.

8.2.1 Slaking

Green liquor and lime mix in the slaker in a certain proportion by adding lime to green liquor. Lime that takes part in the causticizing reaction can be reburned lime from the kiln or makeup lime. When lime slakes with water, the heat release due to the slaking reaction must be distributed evenly to the liquor by agitation to avoid local overheating. The causticizing reaction will also begin and continue to 70% completion during the slaking.

The quality of reburned lime, especially the residual carbonate content, depends on the reburning conditions in the lime kiln. The residual carbonate content is normally 2%–4% as $CaCO_3$, but it may be higher if disturbances occur. The causticizing power of lime (percentage of lime that reacts with Na_2CO_3) normally remains at 85%–90%, but it can sometimes be less than 70% due to impurities in the lime. Low causticizing power results in problems in the solid-liquid separation of lime mud and white liquor.

Adequate mixing is necessary for the effective reaction of the lime feed to the slaker with green liquor. The mixture of liquor, lime mud, and the unreacted solids (sand, overburned lime) flows from the slaker tank to the classifier. The grits with typically larger particle size than the lime mud will settle in the classifier, and they will be removed with a conveyor. Slaked lime and liquor overflow to the causticizers where the causticizing reaction proceeds further. The mixture of slaked lime and liquor is sometimes called lime milk.

The heat release in slaking generates vapor that tends to carry out some lime dust from the slaker. The vapors are condensed with cold water showers that also scrub the gas from dust.

A plate feeder has traditionally proportioned the lime to the slaker, but today a proportioning screw is more common since its construction allows a more dust free operation. The amount of lime is adjusted by varying the rotating speed of the feeding screw.

The slaker is a mixing tank having an opening for lime mud and liquor to flow with unreacted lime to the classifier. Some slakers have two chambers equipped with rakes that have the same diameter as the chambers. Lime is fed to the outer periphery of the upper part of the slaker. The upper rake pushes lime to the middle of the tank. Lime then falls to the bottom level, and the bottom rake scrapes it to the outer periphery. Unslaked lime or grits flow from the outer periphery to the classifier where they are removed as waste from the process. Figure 19 is a diagram of a slaker.

This kind of slaker with two chambers is said to result in larger lime mud particles than the conventional slaker with traditional agitation. The rakes keep the entire bottom of each chamber clean. This makes the slaker insensitive to process disturbances that could otherwise cause scaling in the bottom and on the slaker walls.

The slaking reaction requires approximately 10–20 min for completion. This determines the residence time of reactants in the slaker. The reaction proceeds further to completion in a slaker with two chambers than in a slaker with only one chamber.

Grits settle to the bottom of the classifier for removal with a rake conveyor or classifier screw. Screw-like conveyors are common, since vapor-retaining rake conveyors are difficult to construct. Grits can be washed with water in the upper end of the classifier screw before discharging them. If grits with low alkali content are required, a separate additional washer such as a screw-type washer can be used. Grits are usually discharged with the dregs.

Vapors from the slaker are usually condensed and cleaned in a co-current scrubber that generates a slight vacuum in the slaker to avoid dust leakage at shaft penetrations or gaps in manholes. Both horizontal and vertical scrubbers are used. Scrubber water obtained from a horizontal scrubber is often reused in lime mud washing. Vapors can vent through the roof. In the vertical scrubber, vapors are introduced into a shower pointing downward. Water and vapors drain to a sump or to a lime mud washing stage.

To reduce dusting, reburned lime is introduced below the liquid surface through a pipe. This pipe tends to clog occasionally.

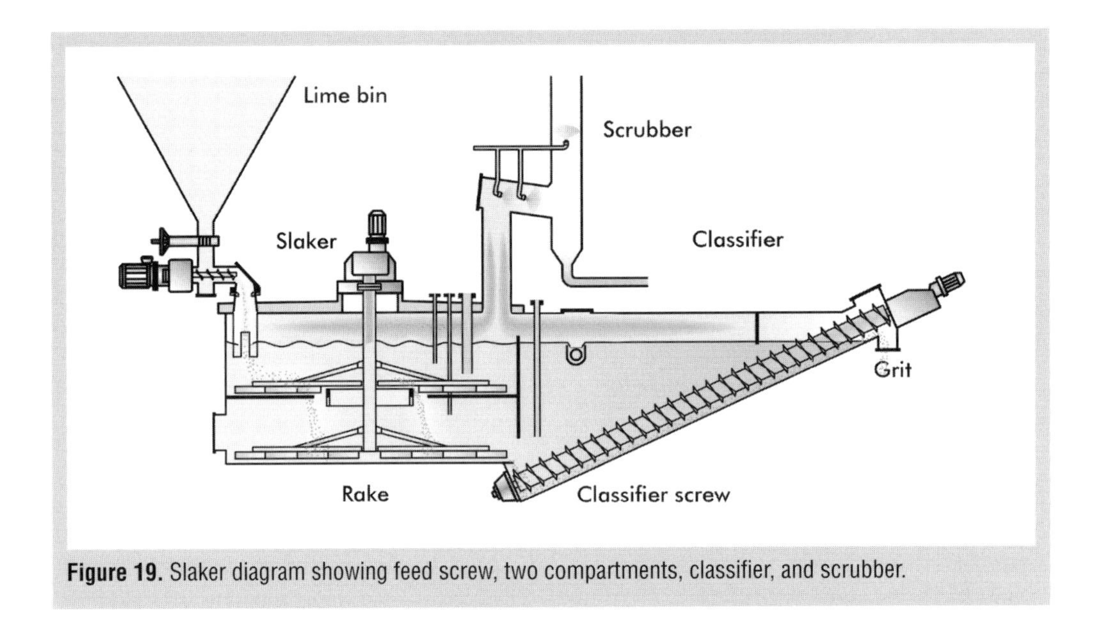

Figure 19. Slaker diagram showing feed screw, two compartments, classifier, and scrubber.

Some required lime can be fed to the last causticizer tank instead of the slaker to allow better control of the causticity and operation closer to the equilibrium without risk of over liming. The lime for this purpose must be clean and reactive because the retention time is only about 45 min before the lime mud separation.

8.2.2 Causticizer train

The main purpose of the causticizer is to complete the causticizing reaction that has already achieved up to two-thirds completion in the slaker. The remainder of the reaction requires more time because the product ($CaCO_3$) encapsulates unreacted lime particles. The diffusion of carbonate and hydroxide ions through the outer layer of the particles limits the reaction rate. Figure 20 shows lime mud particles during causticizing.

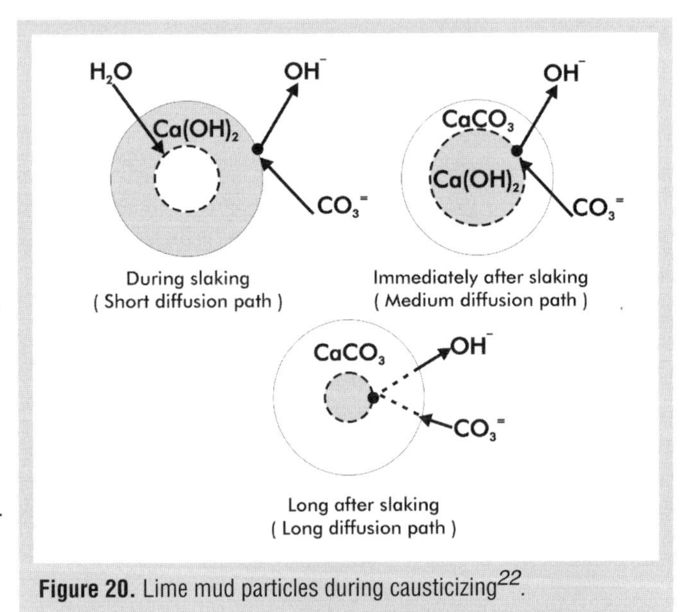

Figure 20. Lime mud particles during causticizing[22].

The causticizing reaction must proceed to completion to decrease the dead load (carbonate content) in white liquor and to avoid any unreacted calcium hydroxide in lime mud. Unreacted lime could make separation of white liquor difficult especially when using filtration. Filtration is more sensitive to unreacted lime than settling because calcium carbonate will begin to form within the filter media (cloth) and eventually blind the cloth.

The slaked lime slurry (lime milk) must remain in the causticizers sufficiently long, and the liquor must have proper contact with the lime mud particles at all times. The necessary residence time depends on the selected method for white liquor separation. If white liquor separates by settling, an approximate residence time of 1.5–2 h is adequate because reactions can continue in the clarifiers. If filtration is applied, the residence time in the causticizers is usually approximately 2.5 h.

The distribution of residence time in causticizers is of primary importance. The amount of lime that has passed quickly through the causticizers without reaction must be as small as possible. To minimize the effect of this shortcutting, the causticizer tanks must have independently operating chambers. Most causticizer trains have nine chambers. Earlier causticizers were constructed as separate tanks. Today, each causticizer tank usually has three chambers on top of each other, and a causticizer train typically has three such tanks in series. The agitators in the causticizers are propeller or turbine type. Figure 21 shows a causticizer.

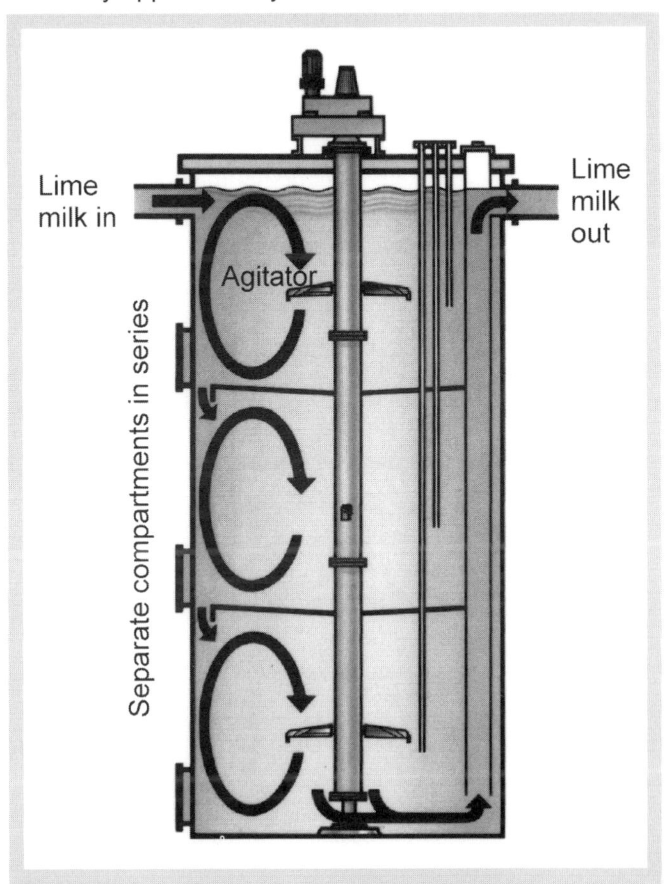

Figure 21. Causticizer.

8.2.3 White liquor separation

The reaction products of recausticizing, i.e., lime mud and white liquor, separate from each other in white liquor separation. The most important goal of separation is to produce clear white liquor without any residual lime mud. Efficient separation means less dead load of recirculated alkali in the recausticizing plant. The product should also be at high temperature and concentration. Cooling or dilution are undesirable.

Two principles apply to white liquor separation: clarification (settling) and filtration. Pressurized candle filters or pressurized disc filters are used today for filtration. Vacuum filtration with a belt type filter has been used previously.

White liquor clarification

White liquor clarification by settling is still commonly used. Lime mud settles to the bottom of a tank, and the clarified white liquor overflows to storage. Lime mud is collected from the bottom with a rake mechanism in the middle of the clarifier and pumped from there at 35%–40% suspended solids concentration to the lime mud washing stage. Mud washing is usually done by first diluting and then settling the slurry in a similar type of clarifier. The upper part of the clarifier can be used as storage to eliminate the need for a separate storage tank. Figure 22 shows a white liquor clarifier with storage tank above it.

The operation of the clarifier is stable and reliable, and it always reaches the required capacity with low energy consumption. If some disturbances occur, the suspended solids content in white liquor may increase. The separation efficiency of clarification is low. This results in a substantial need for washing and a high alkali circulation in the recausticizing, approximately 20% of the white liquor production. The clarifiers are large, requiring considerable space.

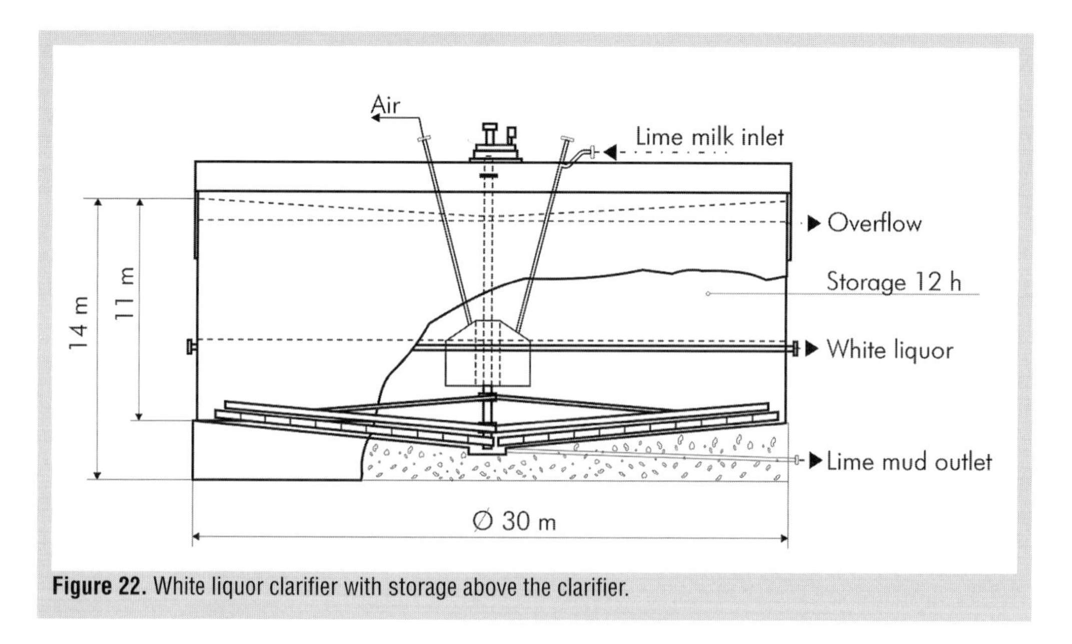

Figure 22. White liquor clarifier with storage above the clarifier.

Belt filter

Vacuum filtration with a belt filter ensures good separation efficiency and lime mud washing with the same piece of equipment. Due to the vacuum applied during the operation of the filter, the liquor cools approximately 15°C. This disadvantage is a primary reason for many mills to replace belt type filters with pressurized filters. Thanks to the efficient washing of the filter cloth, the belt type filter can tolerate minor upsets or changes in lime dosing well. Despite the rather complex structure that requires considerable maintenance work, belt filters are still used as lime mud washers following the pressurized filtration step.

Candle filter

During the energy crisis in the 1970s, the use of candle filters for white liquor separation became more common. This method avoided the cooling of liquor and ensured continuous good quality of white liquor, i.e., low suspended solids content of less than 20 mg/L.

In the candle filter of Fig. 23, lime milk is pumped to a pressure vessel in which liquor will pass through tube-like filter elements while lime mud remains on the surface of the filter cloth. After the cake thickness on the surface of the filter cloth is sufficient (approximately 3 min), the liquor is back-flushed though the filter elements for 1–2 s to release the lime mud cake from the cloth. After a short delay of about 30 s, lime mud starts to settle toward the bottom of the filter, and a new filtration sequence can begin.

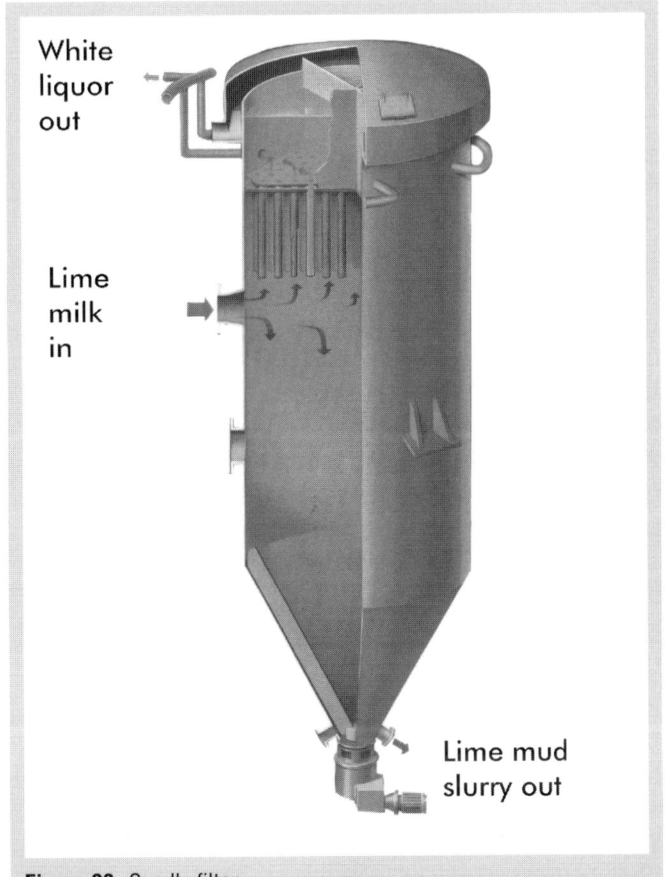

Figure 23. Candle filter.

Lime mud is continuously withdrawn from the bottom of the filter to be washed. The washing stage usually uses dilution washing. Lime mud is separated with the same kind of filter. Lime mud slurry is mixed with washing waters in a mixing tank before the lime mud wash filter. From the wash filter, lime mud slurry with 30%–35% suspended solids is pumped to the lime mud storage tank. Lime mud can also be separated by sedimentation or a belt type filter after dilution for washing.

The candle filter produces white liquor with no major heat losses and a low suspended solids content. Occasional operational disturbances in the recausticizing plant can influence capacity. Such disturbances can lead to blinding problems in the filter media. The separation efficiency of the candle filter is the same as that of the clarifier.

This means that the need for lime mud washing and the alkali circulation in the recausticizing are similar, and the lime mud washing stage is inevitable.

The filter cloths require maintenance so they will retain sufficient throughput. The cloths are washed with dilute acid in 1–3 months intervals and changed after one year of operation.

Pressurized disc filtration

Recent requirements for closure of the mill water balance have set new demands on lime mud washing. It should use less water and have greater efficiency. These requirements with the need for heat economy have led to the use of pressurized disc filtration for white liquor separation and lime mud washing. Figure 24 shows a pressurized disc filtration plant.

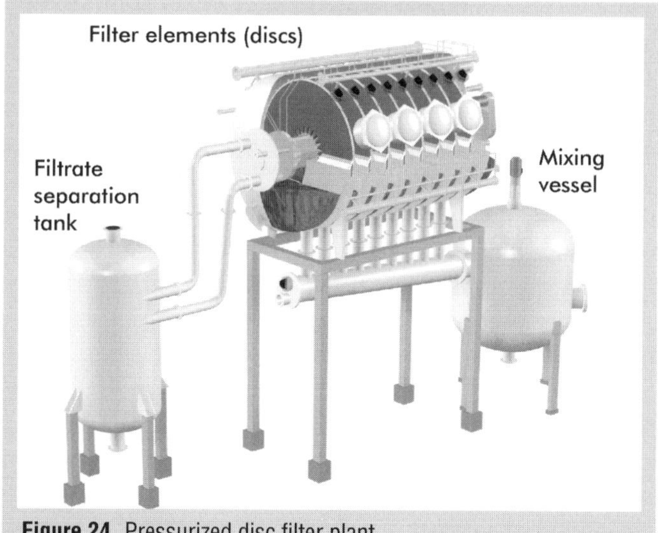

Figure 24. Pressurized disc filter plant.

The pressurized disc filter consists of disc shaped filter elements attached to a center shaft that rotates in a horizontally placed, pressurized vessel. A part of each disc passes through the lime milk in the bottom of the vessel. The pressure difference across the filter surface is maintained by withdrawing gas from the internal parts of the filter elements, compressing it, and returning it to the vessel outside the discs.

Lime mud remains on the filter cloth surface while white liquor passes through the cloth and inside the discs. The liquor goes from the discs via channels in the filter shaft to a pressurized vessel in which the filtrate and gas separate from each other. White liquor is pumped to storage, and gas is compressed back to the filter inside the pressure shell.

The lime mud cake formed on the cloth rises above the lime milk level when the discs rotate. The pressure difference across the cake and cloth drives the filtrate through. This results in about 60% dry solids content in the cake. The cake is further washed with water showers and then dewatered. The top layer of the lime mud cake is removed with a scraper before the discs again rotate partly under the lime milk surface. The remaining cake acts as a precoat during the next filtration cycle. Lime mud passes through chutes to a pressurized mixing vessel where it is suspended in water and pumped to the lime mud storage.

The lime mud precoat on the disc surface will slowly deteriorate. Shifting the scrapers closer to the discs for approximately one round intermittently removes material from the surface. Sometimes the whole precoat requires changing. This is usually combined with filter cloth washing. It occurs at 8–24 h intervals. Neither of these changes in the lime mud precoat causes any significant interruption to the production of white liquor. The filter cloths are washed with dilute acid at 1–3 months intervals. The cloths are replaced approximately once per year. All these procedures are automatic.

The pressurized disc filter is more complex and demanding than a clarifier technically and operationally. The need for maintenance is also greater as in most filtration methods. The power consumption is also higher than for clarifiers. Developments in process automation allow the programming of complex operating sequences with a distributed control system (DCS) to reduce operator work load.

Pressurized disc filtration offers many significant benefits including white liquor that has low suspended solids content (less than 20 mg/L), high separation efficiency for white liquor and lime mud separation, and the ability to perform lime mud washing with the same piece of equipment. The washing water dilutes the white liquor and decreases TTA approximately 2–5 g/L. Due to the high separation efficiency, the white liquor plants with disc filters have only minor alkali circulation of about 5%. With clarifiers, alkali circulation can be 20%.

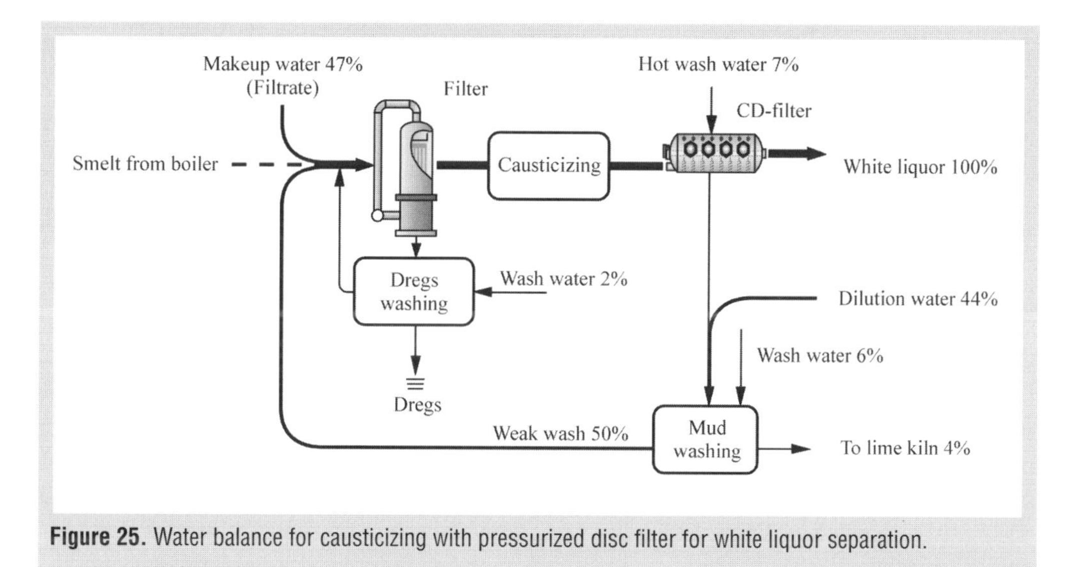

Figure 25. Water balance for causticizing with pressurized disc filter for white liquor separation.

The water balance of a pressurized disc filtration can be similar to that of belt filtration. Figure 25 shows the water balance for causticizing with a pressurized disc filter for white liquor separation. Lime mud washing occurs in the same piece of equipment as in belt filtration. The white liquor can be separated in the pressurized disc filter without any significant heat loss because of the applied pressure. Considering the total water balance of white liquor production, more than one-third of the water necessary for white liquor preparation can be added directly to the dissolving tank when using pressurized disc filtration. This will help water management and mill closure.

Other process options such as settling and candle filtration require that the filtrates from the lime mud washing stage recycle to the smelt dissolving. The production of weak wash matches the need for water for white liquor preparation. No bleach plant filtrates can be introduced in the dissolving tank. Figure 26 shows the water balance for causticizing with candle filters for white liquor separation.

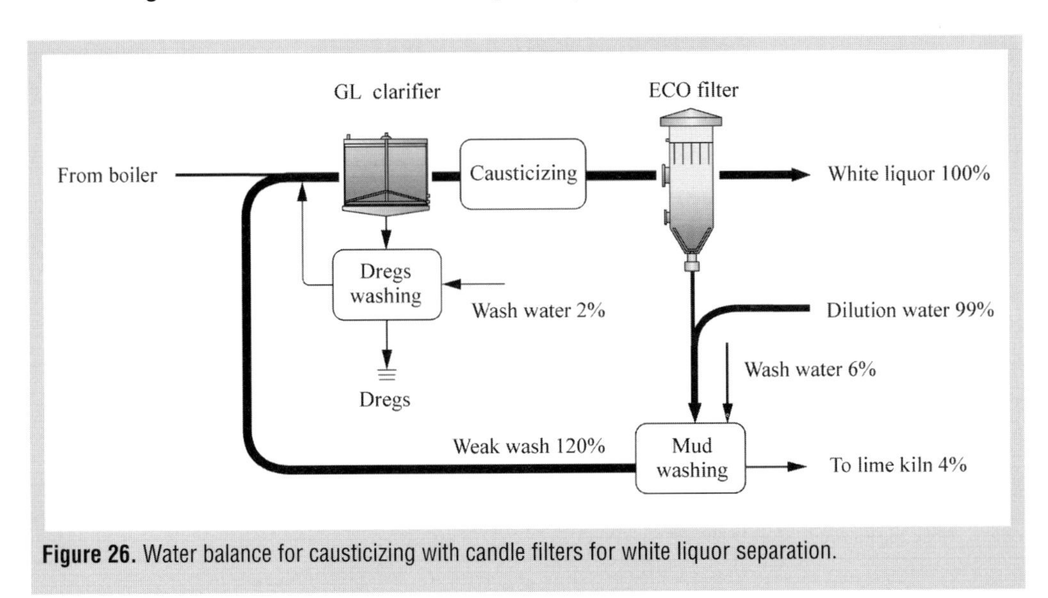

Figure 26. Water balance for causticizing with candle filters for white liquor separation.

The water balances in Figs. 25 and 26 give the amount of water in the process streams in proportion to the produced white liquor amount.

8.2.4 Storage tanks for white liquor and weak white liquor

The sizing of white liquor and weak white liquor tanks usually uses an average retention time of 12 h. The tanks often have a flat bottom and a conical top or a hanging roof if the diameter is over 19 m. The tanks have hatches, a manhole on the side, and piping overflow and drainage connections as necessary. The most common manufacturing material for the tanks is mild steel. White liquor tanks usually have a concrete lining. If the sulfidity is continuously low (below 25%) the concrete lining may be replaced with a suitable corrosion allowance.

The liquors in the recausticizing plant are usually stored in a tank farm centrally located inside a safety bund. All the channels, overflows, and drains inside the safety bund go into a sump. A tank usually overflows first to another storage tank instead of the sewer or drain. Clean water with low electrical conductivity flows from the sump to the sewer, and alkaline liquids are returned to the process, e.g., to the weak white liquor tank.

8.2.5 Lime mud dewatering

Lime mud is stored between recausticizing and the lime kiln to ensure stable and continuous operation of the kiln as a slurry at approximately 35%–40% suspended solids.

Keeping the lime mud suspended and ready for pumping requires continuous agitation in the lime mud tank with compressed air or a mechanical agitator.

The target in lime mud dewatering is to complete lime mud washing and increase the dry solids so that lime mud can be fed to the lime kiln. The moisture in lime mud has a considerable effect on the energy consumption of the kiln. The trend has therefore been to increase the dryness of lime mud after the filter. Dry solids contents of 80%–90% are often possible today. Water soluble alkali in the lime mud after the filter remains under 0.1% as Na_2O on dry mud. To obtain such a high dry solids content, lime mud requires proper washing before lime mud dewatering. Nonprocess elements should not contaminate the lime circulation. The specific load for lime mud filter is approximately 5–7 t/m^2 • day.

Vacuum drum filters are mostly used for lime mud dewatering, although other techniques including a belt filter have also been tried.

The same lime mud that is being dewatered is used as a precoat in lime mud filtration. In the beginning of the filtration, a precoat of lime mud with thickness approximately 10–15 mm forms on the filter drum. The scraper removes the outermost layer of lime mud cake to the belt conveyor. Lime mud is conveyed to the feeding screw and fed to the kiln. Figure 27 shows a lime mud filter plant.

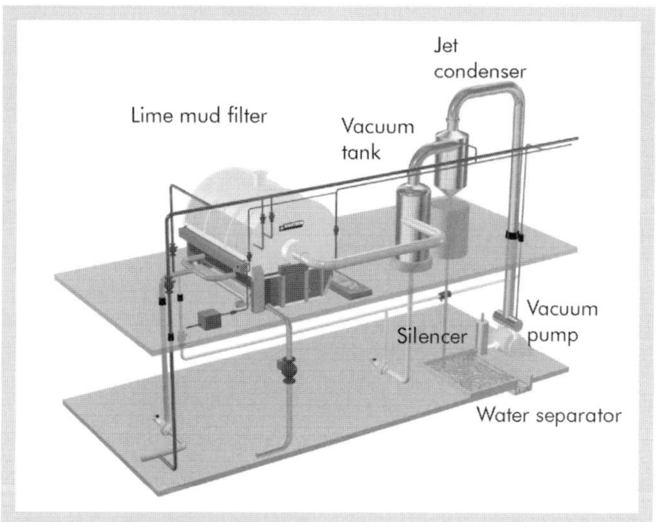

Figure 27. Lime mud filter plant.

In practice, the lime mud precoat gradually blinds. It was earlier completely changed at intervals of 8–24 h by dropping it into the lime mud filter vat and from there via the drain to lime mud storage. The precoat can also be blown onto the belt conveyor and then to the kiln. These methods significantly disturb the feed of lime mud to the kiln. The precoat can also be renewed by moving the scraper closer to the drum for one revolution and then returning it to the original position. This exchanges the outer layer of the precoat and the need to change the whole precoat decreases.

The most recent method for precoat renewal is a high pressure water nozzle that can accomplish partial or complete precoat removal depending on the water pressure. A small portion of the precoat is continuously changed, and the nozzle carriage passes the entire length of the filter drum approximately once every hour. The length of the drum surface that is completely renewed can be less than 0.5 m during each pass. This procedure ensures that the feed to the kiln will have almost no disturbance while the precoat is renewed.

Many techniques are possible to apply the high pressure precoat renewal technique. The water nozzle can be under the scraper or at the back of the filter drum below the wash sprays. In both cases, using a nozzle carriage that travels the whole length of the drum is possible. Other options are to manage part of the drum length using a nozzle tube that has several nozzles and oscillates. Figure 28 shows the high pressure precoat renewal apparatus.

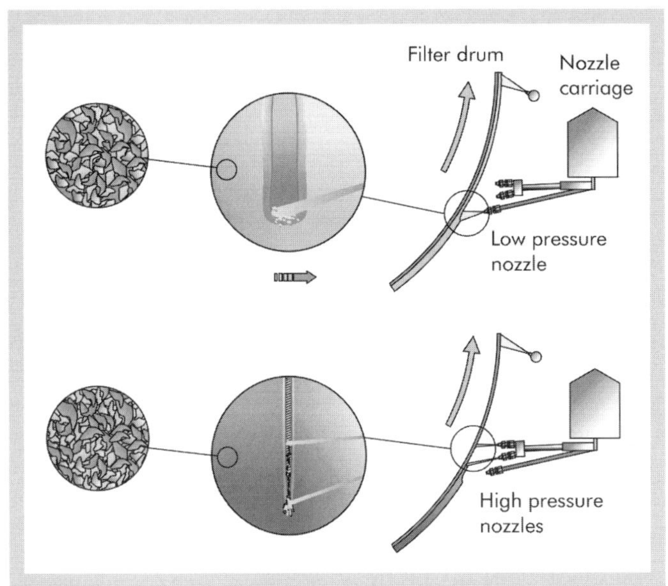

Figure 28. High pressure precoat renewal apparatus.

8.2.6 Modifications of white liquor

White liquor produced in the recausticizing plant is normally used for cooking in a kraft pulp mill. Modifications of the cooking process can require liquors that vary in strength or composition. White liquor can also be oxidized and used as an alkali source in the oxygen delignification stage of a bleaching plant.

Polysulfide liquors

The production of polysulfide liquors uses a catalyst that permits rapid oxidation of white liquor with air. One application is the MOXY process. It converts the sodium sulfide present in white and green liquors to sodium polysulfide (PS) and sodium hydroxide. The oxidation of white liquor occurs in a single step oxidation that results in an orange-colored pulping liquor. Air and white liquor are apportioned to the top of a vertical stainless steel vessel containing a packed bed of the catalyst and appropriate stationary distributors. The small amount of water evaporation occurring simultaneously with the exothermic reaction limits the increase in temperature through the reactor.

The reactions occurring in the reactor are the following:

$$4\ Na_2S + O_2 + 2\ H_2O \rightarrow 2\ Na_2S_2 + 4\ NaOH \qquad (6)$$

$$2\ Na_2S + 2\ O_2 + H_2O \rightarrow Na_2S_2O_3 + 2\ NaOH \qquad (7)$$

When treating white liquor by the MOXY process, about 60%–70% of the sulfide fed to the reactor oxidizes. Approximately two-thirds of the oxidized sulfide reacts to form polysulfide. One-third of the oxidized sulfide forms thiosulfate and caustic. The reactions are exothermic[23].

No change in active alkali occurs with the reactions to form polysulfide. The four moles of caustic formed are equivalent on an Na_2O basis to the two moles of sulfide oxidized. When the oxidation continues to produce thiosulfate, some active alkali is lost.

Pulping with polysulfide liquors results in regeneration of sodium sulfide early in the cook. Higher yields with no difference in bleachability have occurred when comparing polysulfide pulping to kraft cooking[24].

Oxidized white liquor

Although sodium sulfide is an active component in the cooking liquor like sodium hydroxide, sulfide will adversely influence the brightness and the viscosity of pulp when using white liquor as such for oxygen delignification. Certain advantages do exist for oxidizing white liquor and using it as an alternative alkali source to maintain optimum pH.

Oxidized white liquor preserves the chemical balance of the mill. If one used fresh caustic in the oxygen stage, sodium would accumulate in the recovery cycle leading to an imbalance of sodium and sulfur. The end result would be a continuous decrease in the sulfidity.

Oxidation of white liquor typically uses air or substantially pure oxygen as the source of oxygen. Air oxidation is usually at atmospheric pressure in a reactor by sparging air in white liquor. The chemical cost may be low, but the saturated vapors released to the atmosphere cause significant heat losses.

Oxidation with essentially pure oxygen is carried out at elevated temperatures and pressures. The mechanism of sulfide oxidation involves two primary processes. These are the absorption of oxygen from the gas phase into the solution and its reaction with the sulfide present in the solution. One can assume that increasing the partial pressure of oxygen will also increase the total oxidation rate. The partial pressure influences the solubility of oxygen. The increase in temperature enhances the oxidation reactions.

The reactions during white liquor oxidation include the formation of thiosulfate according to the following reaction[25]:

$$2\ NaHS + 2\ O_2 \rightarrow Na_2S_2O_3 + H_2O \quad (8)$$

and further oxidation of thiosulfate to sulfite:

$$Na_2S_2O_3 + O_2 + 2\ NaOH \rightarrow 2\ Na_2SO_3 + H_2O \quad (9)$$

or the production of sulfate:

$$2\ Na_2SO_3 + O_2 \rightarrow 2\ Na_2SO_4 \quad (10)$$

Sodium thiosulfate and sodium sulfate are the main reaction products in oxidized white liquor. Due to the consecutive reactions involved, the composition of the product depends on residence time in the reactor. The clear and sometimes even colorless appearance of oxidized white liquor excludes polysulfides as final reaction products. They may still take part in the reaction as intermediates.

8.3 Automation

8.3.1 Process requirements

The control of the causticizing reaction is important to ensure the quality of white liquor, i.e., maximum causticity degree, and to secure the operation of solid-liquid separation. If the amount of lime is inadequate in proportion to green liquor amount, all lime will react, but considerable unreacted sodium carbonate will be present in the white liquor. This will then increase the inert load in the liquor cycle and possibly influence the operation of the evaporation plant and digester. An excess of lime will result in "free lime" in lime mud that deteriorates the filtering of lime mud even in minor amounts. This also causes problems in solid-liquid separation by blinding the filtering media with precipitated calcium carbonate.

Mills usually operate below the equilibrium causticity. One reason is to avoid over liming due to normal quality variations seen in green liquor and lime. Another reason is to compensate for fluctuations in flow rates. The operation close to equilibrium requires very accurate control.

Recausticizing as a process is very slow. Only slight automation is possible. The situation has recently changed for several reasons. Today, mills operate with less personnel. This requires increasing the level of automation in all processes. Simultaneously, the main process equipment at the recausticizing plant has become more sophisticated. This also requires additional automation. The equipment has several fully automated sequences for operation such as automated acid washing and water flushing sequences. The use of automated sequences has increased the amount of instrumentation loops and especially the use of automated on-off valves.

In automation, operational sequence engineering has become increasingly important. An operational test of the DCS configuration before startup is a requirement today. This is a factory acceptance test (FAT) or staging. In this test, the DCS vendor, process equipment vendor, and customer operators witness the complete simulation of the operation of the sequences and interlockings programmed into the DCS.

8.4 Field instrumentation

Proper material and equipment model selections are essential for recausticizing field instrumentation to avoid erosion, blinding by sedimentation, and scaling. The standard measurement requirements and common practices are still valid in recausticizing instrumentation installations. The following text discusses some observations of applying the most common field instruments.

The main problem in using conductivity measurement with recausticizing process liquids is the frequent need of sensor cleaning. Some vendors have developed automatic signal compensation to allow continued measurement with dirty sensors. This compensation can help the situation only to a certain level, and sensor cleaning with acid is still necessary regularly.

All density measurement of process liquids uses nuclear density transmitters. In green liquor services, scaling is the main problem. Polytetrafluoroethylene (PTFE) lined spools and special cleaning can correct the problem. Temperature compensation connected directly to the density transmitter is also widely used. In vessels such as candle filters, pressure difference transmitters with capillaries, water purging, or both are used for density measurements.

Due to the nature of the process liquids, a magnetic flow meter is the main instrument used. PTFE is the most common material for the lining of magnetic flow tubes. A special protection flange in the incoming side of the tube is therefore necessary in lime mud and other erosive applications. Steam and air measurement uses standard orifice plates or vortex-type meters.

Sedimentation is the major problem with level measurement of recausticizing process liquids. Using water purging to clean the transmitter face can avoid this problem. Process liquid temperature is often near 100°C. This sometimes leads to use of capillary connections. Most main process equipment consists of pressurized vessels causing the need of minus-leg compensation in the level measurements. Dry and wet legs are used. In the areas where vessels are outside and freezing can be a problem, a capillary tube is a good choice for minus-leg compensation.

Standard pressure measurement devices with direct connection or capillaries are used.

Temperatures in the recausticizing area are near 100°C. Pt-100 resistance thermoelements can therefore handle all temperature measurements.

Erosion and scaling are the main problems for proper control and on-off valve selection. Using ceramic valves can solve the erosion problem. These valves are expensive and only have use in places where erosion is severe such as for lime mud flow control. Ball, segmented ball, and butterfly valves essentially cover all recausticizing control valve needs. To avoid leakage caused by scaling, ball valves with scraper type seals have wide use in green liquor services.

8.5 Process controls

8.5.1 Manual control

The control of the recausticizing department is still manual in many mills. Operators follow the process using laboratory analysis. The determination of total, active, and effective alkali per titration is the most common laboratory test method. Process delay between lime feeding to the slaker and the final product in the third causticizer vessel is approximately 2–3 h. Due to this long process delay, manual control is very difficult and can cause over liming. Over liming results in filter cloth plugging and requires more frequent filter acid and water washing. Automatic control and optimization packages for the recausticizing process area have recently become popular.

The following section summarizes the automatic control systems commercially available today.

8.5.2 Conductivity measurement and control

Special four electrode conductivity sensors with automatic compensation for scaling and contamination are available for this control system. The differential conductivity

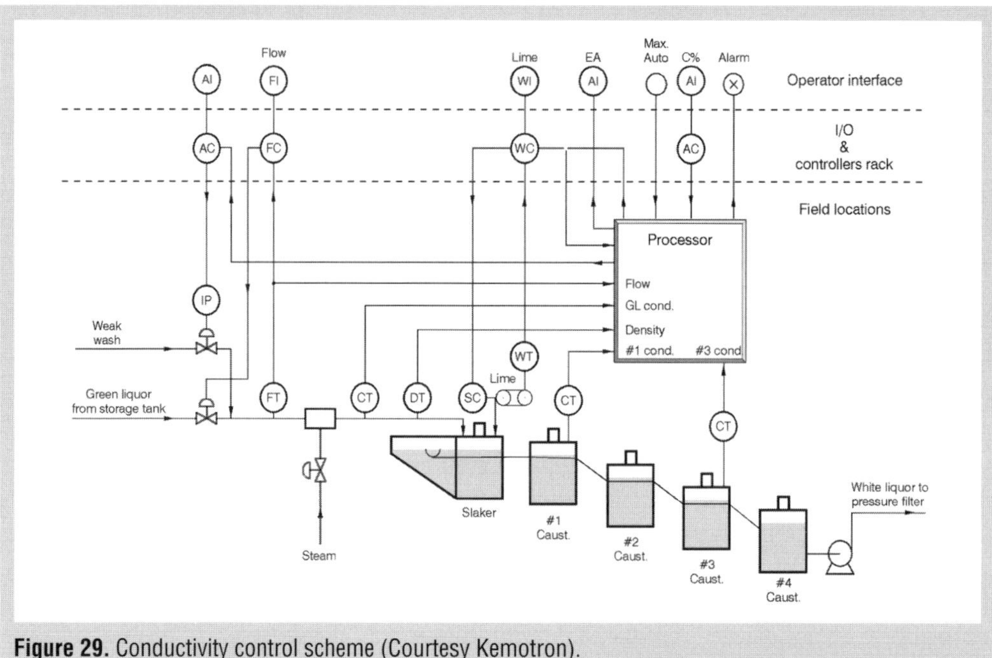

Figure 29. Conductivity control scheme (Courtesy Kemotron).

measured from the first and third causticizer corrected for TTA is the primary feedback for the lime feed control. Figure 29 shows such a system.

8.5.3 Automatic process titrator and control

In a typical installation, the automatic titrator takes samples from the incoming green liquor, the slaker, the first causticizing vessel, or combinations of these. Figure 30 shows the typical sampling points for causticizing process control. The system takes samples and analyzes them automatically. The analysis uses a modified titration procedure. For each sample, the analyzer reports the following:

- Effective alkali, EA
- Active alkali, AA
- Titratable alkali, TTA
- Sodium hydroxide, NaOH
- Sodium sulfide, Na_2S
- Sodium carbonate, Na_2CO_3
- Sulfidity, S%
- Causticizing efficiency, CE%.

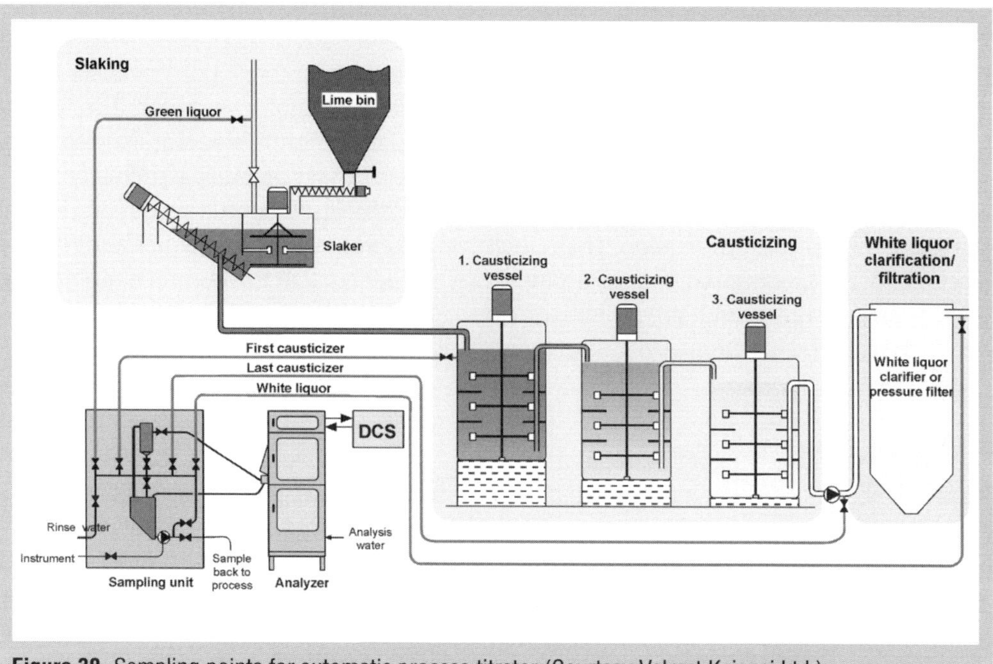

Figure 30. Sampling points for automatic process titrator (Courtesy Valmet Kajaani Ltd.).

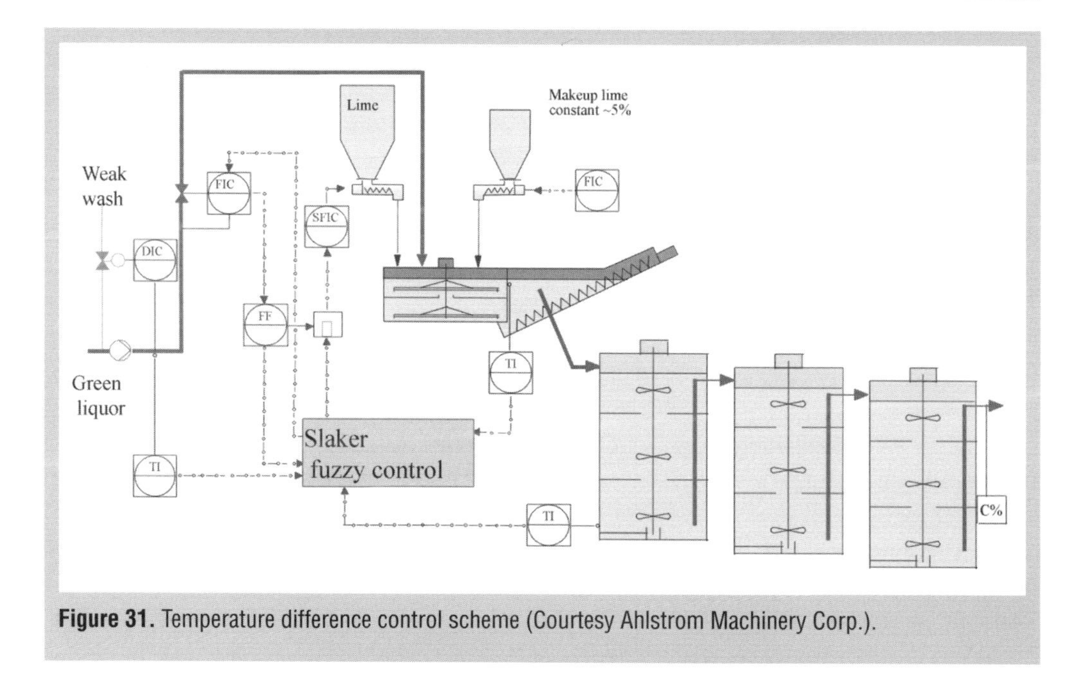

Figure 31. Temperature difference control scheme (Courtesy Ahlstrom Machinery Corp.).

8.5.4 Temperature difference control

The temperature difference control of Fig. 31 is a recent control technique. It measures the temperature increase between incoming green liquor and the liquor in the slaker. This control does not require any new field instrument or titrator investment. The necessary automation usually exists in the recausticizing department. Temperature measurement does not require the extra field instrument service that has been a problem with conductivity and titrator methods. The increase in temperature correlates directly with the extent of the slaking reaction. The causticizing degree measured by an automatic titrator or in the laboratory can have manual use as feedback to the control algorithm for its fine tuning. The control algorithm uses fuzzy control and is rule based.

The control methods presented above can use the DCS system or separate, personal computer hardware.

8.5.5 Fuzzy control

Today, almost all controls mentioned above use fuzzy control technology. Fuzzy control has some significant benefits compared to older proportional integral derivative (PID) controllers. It offers the possibility to control the process the way the best process operators would control it. Fuzzy control uses empirical statements such as the following: "If temperature of the green liquor in the first causticizer is high and increasing, then decrease the feed of lime into the slaker." Using these few clauses that are developed with process operators, fuzzy control can then be programmed to operate the process continuously using these rules.

The second main benefit of this new technology is that the control function programming can be nonlinear. For instance, the control gain can be programmed in different ways depending how close the control is to the equilibrium causticizing efficiency (Goodwin's curve). Causticizing efficiency can move closer to the curve slowly, but the control gain near the curve for the reverse control action can be much larger to avoid over liming in all process conditions.

9 Lime reburning

Lime reburning is part of the chemical circuit called lime cycle in a kraft pulp mill. After causticizing, lime is essentially in the form of calcium carbonate. The objective of lime reburning is to convert calcium carbonate back to calcium oxide. Reburned lime is a recirculating chemical used in converting green liquor to white liquor in the causticizing plant. Lime regeneration is called reburning because it involves treating lime mud at high temperatures in a lime kiln. Makeup lime or limestone compensate for lime losses in the lime cycle.

The lime reburning process uses a counter-currently operating, heat exchanging reactor where heat transfers from combustion gas to lime particles by direct contact. The heat source is heavy fuel oil or natural gas. The selection primarily depends on fuel price and environmental requirements.

The first lime kiln was built in the beginning of the 20th century. The number of kilns has now grown to more than 300. The total capacity of the kilns is about 100 000 t/day, and the highest capacity of a single kiln is about 1 000 t/day reburned lime. Early years saw use of open hearth and rotary kilns. After a short period, only rotary kilns were built.

A rotary lime kiln with a capacity of approximately 530 t reburned lime/day is typically 4–4.5 m in diameter and 100–140 m in length depending on the feed end construction. The production of a kiln of this size approximately matches the white liquor production of 7 000 m^3/day. The kiln is supported by three or four piers.

9.1 Lime reburning process and equipment selection

The lime reburning process of Fig. 32 consists of the following unit operations:

- Pumping lime mud from lime mud storage
- Mechanical dewatering of lime mud
- Thermal dewatering
- Heating and calcination
- Cooling of product
- Screening and crushing
- Conveying to storage.

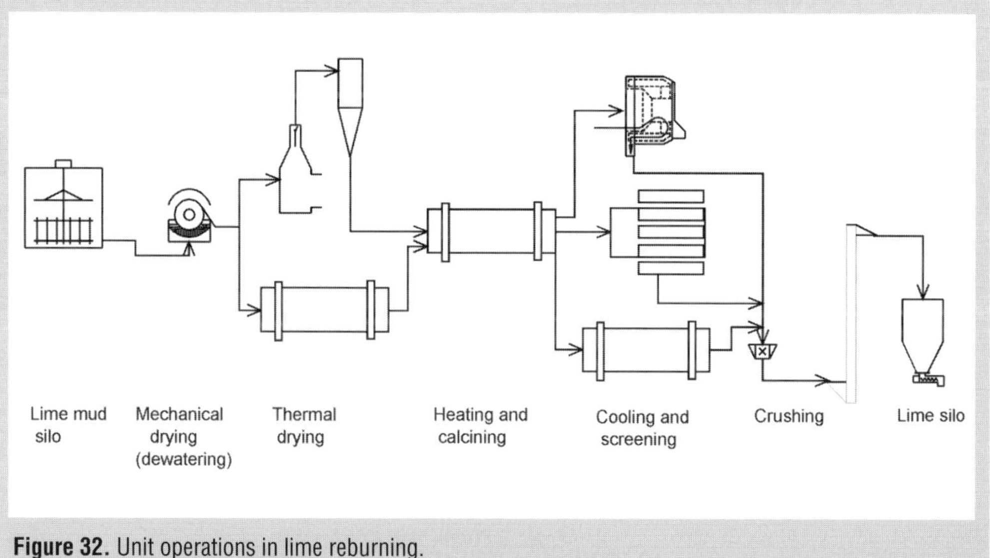

Figure 32. Unit operations in lime reburning.

9.1.1 Mechanical dewatering

The main requirements for the equipment used for dewatering are to give uniform flow and constant moisture. The most common device for this purpose is the vacuum drum filter described in detail in the recausticizing section. Lime mud flow to the filter determines the production of the kiln. Centrifuges have been used as an alternative, but they do not have wide application due to their lower dry solids content product.

Lime mud from the dewatering system collects on a belt conveyor that conveys the mud directly or via a screw feeder into the kiln. Figure 33 shows a screw feeder.

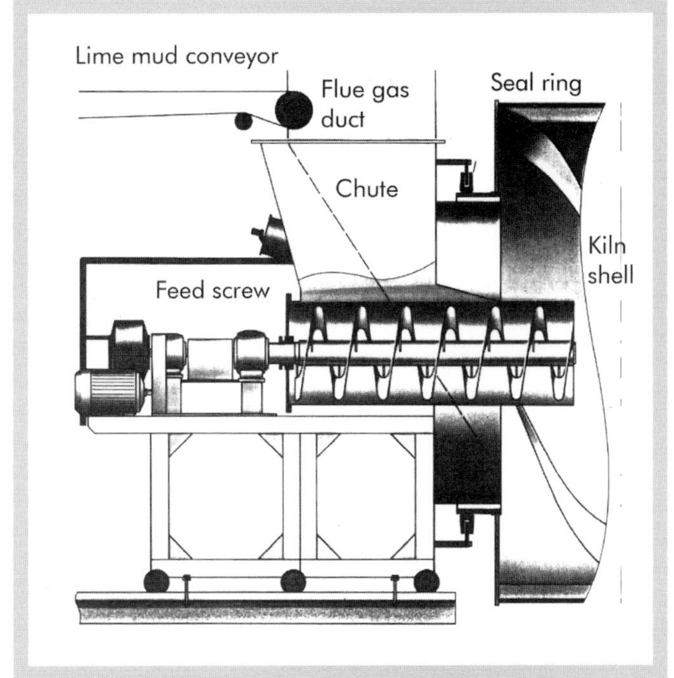

Figure 33. Screw feeder at the lime kiln feed end.

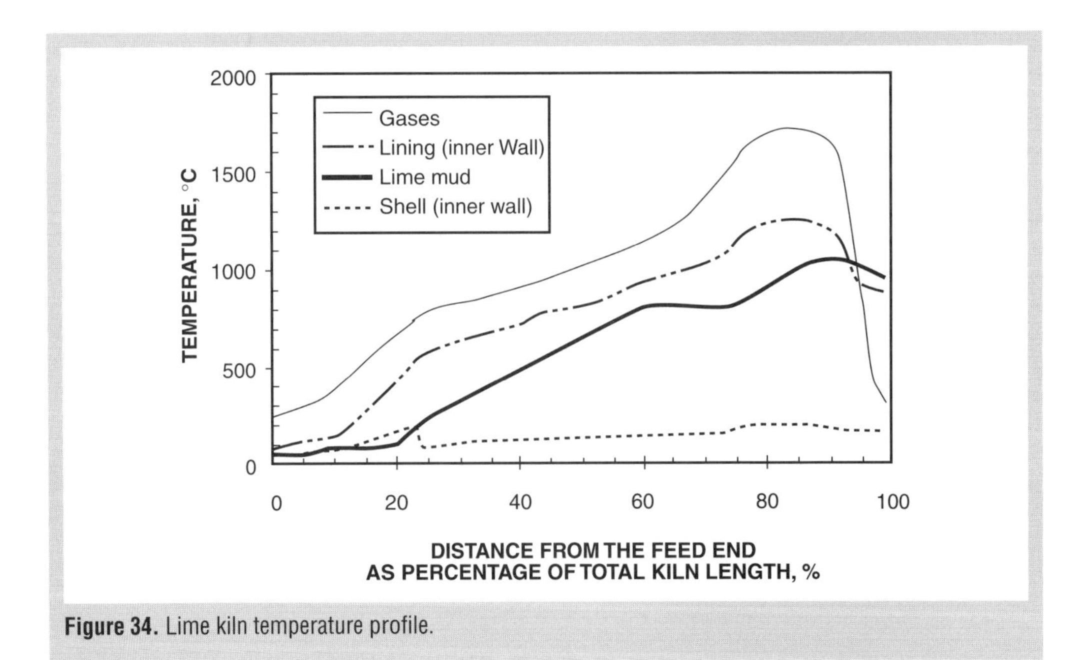

Figure 34. Lime kiln temperature profile.

9.1.2 Lime reburning process in the kiln

A rotary kiln slopes slightly toward the firing end. The rotation speed is typically 0.5–1.5 rpm. The lime retention time in the kiln depends primarily on kiln dimensions, rotation speed, and lime mud properties. Normally, it is approximately 2.5–4 h. When lime mud travels downhill toward the firing end in the kiln, moisture evaporates and the calcination reaction finally occurs in the actual burning zone where gas temperature increases to 1 100°C. The calcination occurs spontaneously at about 800°C in a gas atmosphere containing 20% CO_2. The gas temperature needs to be significantly higher because of the poor heat transfer in the kiln.

Figure 34 shows an example of a lime kiln heating profile. The process has four phases:

- Drying: moisture in the lime mud evaporates
- Heating: lime mud heats to the reaction temperature
- Calcination: calcium carbonate dissociates into calcium oxide and carbon dioxide
- Final treatment: lime cools before leaving the kiln.

The treatment of lime mud in the first three kiln zones requires external heat. The kiln therefore burns fuel oil or natural gas. Burning waste fuels to some extent is also possible in the kiln to use their heating value.

Heat transfer in the kiln depends primarily on radiation. Convection dominates only in the drying phase where the flue gas temperature has decreased considerably. Chains in the drying zone can improve heat transfer. The purpose of the chain section is to increase the heat transfer from the flue gas to the lime mud. The chain system absorbs heat from flue gases and transfers it to the lime mud.

Thermal drying

Two alternative methods are possible for the thermal drying of lime mud. The traditional method for drying is the chain section of the rotary kiln. Several patterns exist for hanging the chains. The most typical are the garland and curtain type arrangements. Figure 35 shows a chain system.

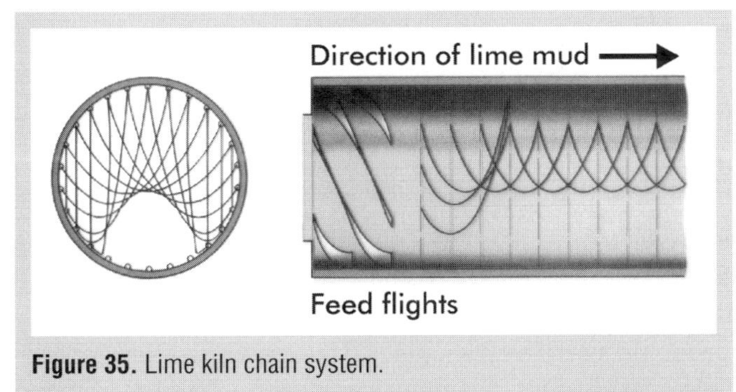

Figure 35. Lime kiln chain system.

One adjusts the chain length section so the mud is completely dry when leaving the section. The length is usually about 20% of the total kiln length.

Another common drying method is the pneumatic dryer of Fig. 36. Lime mud is fed to a flue gas stream where the heat of the gases dries the mud. A cyclone separates dry mud and feeds it to the kiln.

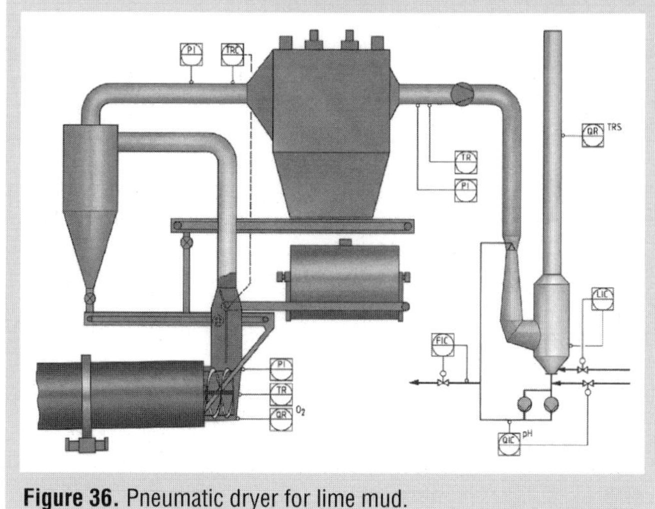

Figure 36. Pneumatic dryer for lime mud.

Heating and calcining

After drying, the mud is heated to the calcining temperature in the rotary part of the kiln. Various devices such as lifters, bars, cam lining, or baffles can improve the heat transfer in the heating zone. These devices are not very common because they damage easily during upset process conditions.

The objective of lime reburning is to obtain homogeneous and porous lime that will slake easily and produce lime mud that will separate easily from the liquor. Lime activity or causticizing power is a criterion for the slaking rate. It is proportional to lime porosity and surface area of the particles. Lime activity reaches its maximum value at a certain calcining temperature and then decreases. If the temperature in the kiln is too high, it causes changes in the lime crystal structure. This results in poorly slaking, hard, burnt lime.

The calcining occurs in a plain section, i.e., no internal devices are in the kiln. For heat transfer improvement, dams are installed at the hot end of the kiln near the discharge area. The dam forms by reducing the diameter of the shell or making the refractory lining thicker.

For calcination, a fluid bed reactor is a possibility. This has not become very common mainly because of its poor heat economy.

Product cooling, screening, and crushing

Most kilns have a product cooler for heat recovery. Separate cooling drums have also been used. The most common configuration is to attach the cooling device to the kiln itself. Figure 37 shows a sector and satellite cooler.

Burned lime exits the kiln via the cooler located at the kiln firing end. Cooling the lime with secondary combustion air recovers heat from the hot lime. This air then flows into the kiln. It accounts for more than 80% of the total air amount necessary for combustion. The secondary air proceeds through the cooler to the kiln with an induced draft fan.

The burned lime from the kiln has a wide particle size distribution. Large particles are crushed by a lump crusher or hammer mill. The cooling devices have screening arrangements so that the crusher will handle only oversized material.

Steady and trouble-free kiln operation is vital to ensure successful lime burning. To maintain stable production, the composition of lime requires continuous control. Lime mud is washed so the sodium content of the lime mud is less than 0.3% as Na_2O on a dry mud basis. This avoids operational problems in the lime kiln and limits flue gas emissions to an acceptable level.

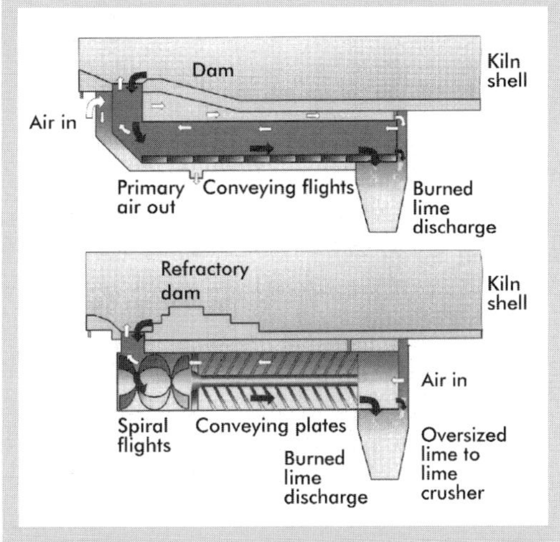

Figure 37. Sector and satellite cooler for burnt lime.

9.1.3 Brick lining

All lime kilns have a refractory lining that protects the kiln shell from overheating and limits the heat loss to an acceptable level. The temperature distribution of lime kilns is very wide from the colder feed-end housing to the hot firing hood. In conventional kilns in which the mud moisture is removed inside the kiln, the lowest temperature of the refractory lining is approximately 100°C. In the flame zone, the temperature can be 1 250°C as Fig. 34 shows. For very large kilns with shell diameter of 4.5 m and higher, the flame zone linings must tolerate even higher temperatures of 1 300°C–1 350°C. In modern kilns with separate mud dryer, the cold end temperature is higher at 400°C–600°C.

Selecting materials for the whole kiln according to the harshest conditions is impractical, because this would result in an expensive design. Normally, the kiln contains different zones. Figure 38 shows the lime kiln brick lining zones. Each zone has a lining of a certain material and thickness. The division into zones typically includes the following:

- Chain zone (not in all kilns)
- Heating zone
- Intermediate zone (sometimes combined with heating zone)
- Burning or flame zone
- Dam
- Discharge

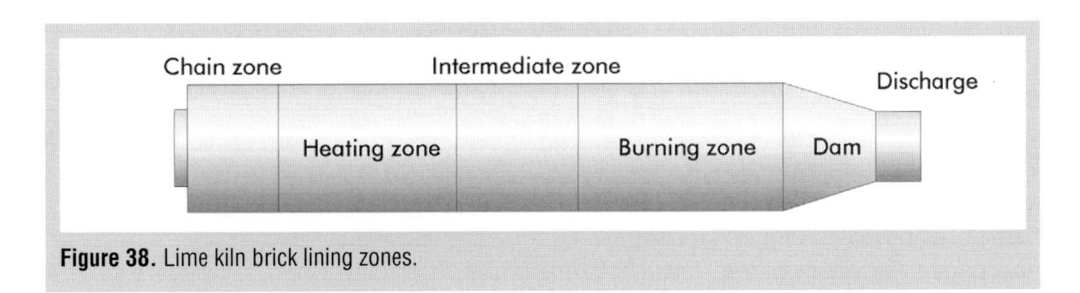

Figure 38. Lime kiln brick lining zones.

Lining materials

The chain or drying zone material must be hard to resist the abrasive contact of chains and must tolerate moisture[26]. Typical materials are low cement castable (LCC) and other abrasion resistant castable materials. Abrasion resistant fireclay bricks have also had successful use.

The heating zone, intermediate zone, and flame zone usually have back insulation. The hot face normally uses fireclay bricks. Insulation material bricks of diatomaceous earth have almost universal use due to their good combination of heat conductivity and strength.

The lining design of the intermediate zone is almost the same as in the heating zone. In larger kilns, the strength of the back insulation should be higher. Using light fireclay bricks instead of diatomaceous earth bricks can achieve this.

In the flame zone, the conditions on the hot face are harsh. The lining temperatures are typically 1 150°C–1 250°C in the presence of alkali compounds and burnt lime. Andalusite and sillimanite bricks have proven better in practice than other Al_2O_3 based materials. In large kilns with shell diameter 4.5 m and higher, the hottest area of the flame zone needs still better material such as magnesia spinel. This is due to alkali and lime attack at higher temperatures.

As back insulation, diatomaceous earth bricks are used. If the shell diameter is more than 3.2 m, a harder material such as lightweight fireclay bricks is necessary.

The dam area including the discharge zone is normally cast or gunned. Similar materials are used as in the chain zone. Because of the high working temperature of 1 000°C–1 150°C, the refractoriness under load must be at least 1 500°C.

Although casting or gunning has only been used in a few kiln sections, some totally cast designs have been built. The main drawback with this design is the time-consuming removal of old lining when replacing it with new lining. The development of casting and gunning materials was rapid in the 1980s and 1990s. Although some special characteristics of these materials may be suitable for this kiln area, experience of long time service and references are yet unavailable.

9.1.4 Firing equipment

A firing hood covers the discharge end of the kiln. The main burner is supported from the hood or is attached to a trolley. The main burner design can accommodate one fuel or many fuels. Besides fuel oil and natural gas ports, a combination burner can have ports for noncondensible gases and a

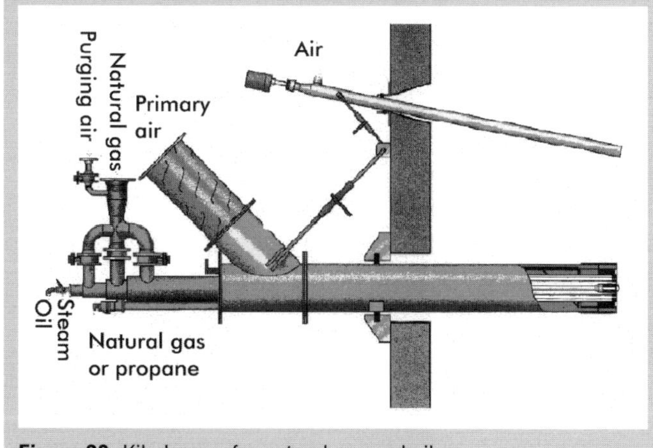

Figure 39. Kiln burner for natural gas and oil.

methanol feed arrangement. Primary air that is typically 10%–20% of the total amount of combustion air passes through the burner, cools it, and stabilizes the flame. The burner also has an igniter and hood flame scanning systems. Figure 39 shows a kiln burner.

9.1.5 Dust handling

The cyclone, electrostatic precipitator (ESP), or scrubber separates the lime dust escaping from the kiln with flue gases. The precipitator of Fig. 40 uses electricity to polarize the lime dust particles and remove them from the flue gases. The dust collects on plates in the precipitator that are rapped by impulse rappers periodically to remove the dust. It falls to the dust hopper at the bottom of the precipitator. Dry dust from the cyclone or ESP returns to the kiln on closed conveyors such as drag chain conveyors.

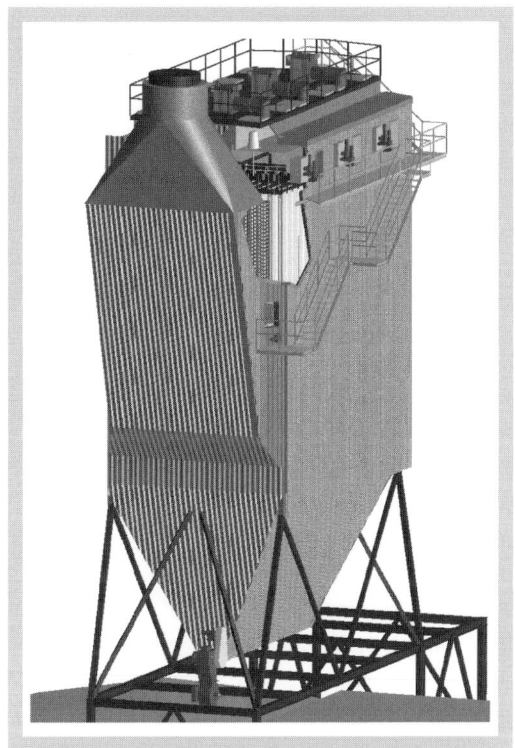

Figure 40. Electrostatic precipitator (Courtesy of ABB Power).

The electrostatic precipitators are efficient in removing dust particles. They offer very little resistance to the flow of gases. Replacing or cleaning elements is unnecessary. The advantage that electrostatic precipitators have over wet scrubbers is that they remove dust in the dry state without requiring pumps, valves, and associated equipment. The disadvantage is that the ESP does not reduce emissions of total reduced sulfur (TRS).

The hot lime conveying system has dust collection piping normally connected to the lime silo. A dust filter and a vacuum blower are on top of the silo.

9.1.6 Scrubber

Wet scrubbers have use for dust separation and gas absorption. A venturi type scrubber is the most common. The emission regulations are commonly so tight that wet scrubbers are today rarely used for dust separation. They are becoming more common for absorption of gaseous emission such as SO_2 in alkaline solutions. A high gas pressure drop in the venturi throat is used for liquid atomization or high liquid pressure atomization. Figure 41 shows an ejector type venturi scrubber.

9.2 Limestone burning

The need for makeup lime in a pulp mill usually comes from purchasing burnt lime or limestone that is then burned with lime mud in the lime kiln of the mill. Figure 42 shows a typical flowsheet of limestone handling.

Calcining properties of several limestones have been tested worldwide. Finnish recrystallized limestone and some limestone in Gotland (an island in the Baltic Sea) are extreme examples of such stones. The Finnish stone has compact, hard, and brittle construction. It requires a much higher burning temperature than lime mud. It is impossible to burn them together. The limestone of Gotland is soft and porous. When crushed to the correct particle size, it is suitable for burning with lime mud in a lime kiln.

Figure 41. Ejector type venturi scrubber.

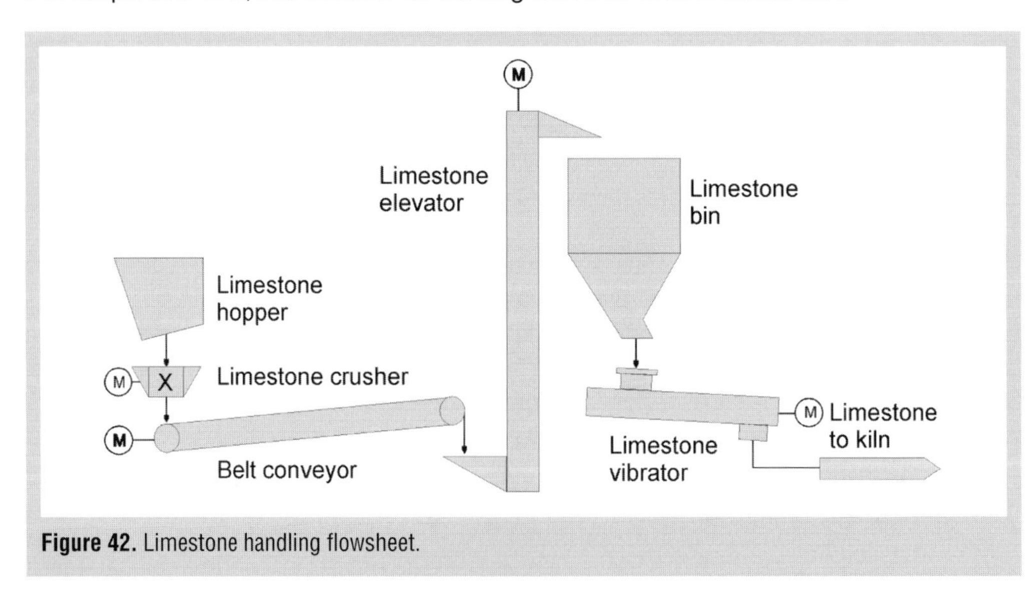

Figure 42. Limestone handling flowsheet.

Limestone should be calcined similarly to lime mud, i.e., at a temperature and retention time suitable for the lime mud. If this procedure is not successful, the residual carbonate remains high. If the burning temperature increases to the point at which the residual carbonate falls to an acceptable level, the material bed melts and the linings at the burning zone will incur damage.

A mill should use limestone crushed to the desired particle size. If this is unavailable, the mill must crush it. Besides the necessary conveyors, the handling system will require a crusher, a screen, and a system to return oversized lime particles to the beginning.

Storing of limestone either uncrushed or crushed to final particle size must consider keeping it dry under all conditions. Moisture in makeup limestone should be less than 1%. Otherwise, it is difficult to store, move, and feed. If the moisture exceeds 1%, the makeup limestone must undergo drying before crushing. If high moisture problems occur, screening the fines material from the limestone will improve the situation. Otherwise, various plugging problems in handling and storing can occur.

Limestone is finally fed to the lime kiln by mixing it with the lime mud after the lime mud filter or by feeding it directly to the kiln through a separate feed chute.

Hardness of the limestone usually relates to its reactivity. More reactive limestone usually causes less mechanical wear of linings. Burning of reactive limestone with the lime mud does not damage the lime kiln linings, because the stone is soft and does not require higher temperatures to calcine than lime mud.

9.2.1 Limestone quality

The proportion of limestone in the dry solids feed to the lime kiln can be as high as 50%. The particle size of the limestone should be 2–4 mm average with maximum particle size about 10 mm. If the stone used is very reactive, these particle sizes can be larger on a case-by-case basis. One should not exceed a maximum particle size of 20 mm.

Limestone should be fed to the lime kiln in a constant ratio to the dry solids flow of the lime mud. Feed equipment and its instrumentation should allow adjustment and measurement of the feed. If the feed ratio between the stone and lime mud dry solids varies continuously, varying moisture of the feed mixture will result, and the kiln will become more difficult to operate. This will cause problems especially if the need for makeup lime is great. This can happen at mills where much lime mud must be withdrawn from the lime circulation due to silica problems and replaced by makeup limestone.

9.3 Material and heat balance

Table 12 shows a typical material and heat balance for a lime kiln. The production is comparable to the production of 7 000 m^3 white liquor/day in recausticizing. The balance uses the following input values:

- The kiln is fueled with #6 fuel oil with lower heat value of 39.21 MJ/kg
- Dry solids content in lime kiln feed is 75%
- $CaCO_3$ content in lime mud is 92% resulting in approximately 85% lime availability

- Residual O_2 content in lime kiln feed end is 2.0%
- Residual $CaCO_3$ content in product is 2.0%
- The kiln has a two layer brick lining and a lime cooling system

Table 12. An example of a material and heat balance for a lime kiln.

	Flow, kg/s	Temperature, °C	Heat energy, kJ/kg product
IN			
Lime mud, dry solids	10.21	60	83.4
Water	3.40	60	139.4
Air	13.42	20	44.2
Oil	0.87	120	32.1
Fuel consumption			5548
Steam	0.09	220	6.0
Total in			5854
OUT			
Heat of reaction			2766
Evaporation	3.40		1388
Flue gas		181	756
Product	6.13	350	295
Heat losses			649
Total out			5854

9.3.1 Factors influencing heat balance

The heat consumption of a lime kiln depends on lime mud dry solids as Fig. 43 shows. Low lime mud dry solids content means high heat consumption due to excess evaporation of moisture. High lime mud dry solids content increases the flue gas outlet temperature when the excess heat energy in the flue gases cannot be completely used in the lime mud drying. Some heat energy is lost with the gas flow. The heat required for the calcination reaction remains the same.

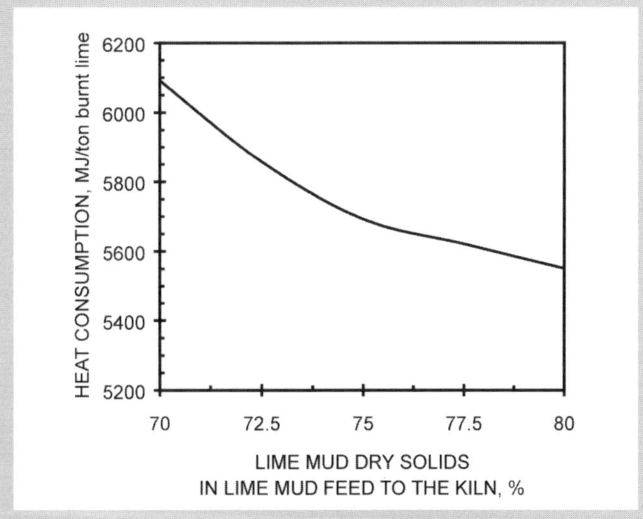

Figure 43. Lime kiln heat consumption in production of reburned lime vs. dry solids content in the lime mud feed to the kiln.

The residual O_2 content (excess O_2) in flue gas measured at the feed end of the kiln is an important controlling parameter of the lime kiln process. If O_2 content is too low, combustion is below stoichiometric and incomplete. This results in formation of CO and TRS causing high heat consumption. Excessively high O_2 content means that air is fed to the kiln in excess so the flue gas amount increases. The gas flow is heated in the kiln and results in an increase in heat consumption. Figure 44 shows the relation between residual O_2 content and heat consumption.

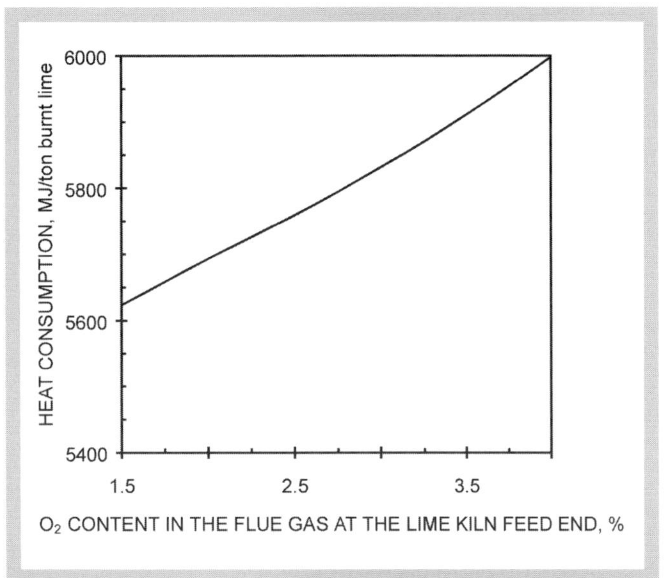

Figure 44. Lime kiln heat consumption in production of reburned lime vs. residual O_2 content in the flue gas from the kiln.

Flue gas outlet temperature is a result of many operational parameters. Higher flue gas temperature means higher heat loss with flue gas and also higher heat consumption. The ratio of kiln production rate to kiln diameter influences the temperature. Increasing burning load results in high flue gas temperature because heat transfer surface is constant at all production rates.

Residual O_2 content influences the flue gas temperature by changing the flue gas flow. High O_2 content means higher flue gas flow through the kiln and reduced heat transfer in the burning zone. If the fuel input does not simultaneously increase, the temperature profile of the kiln will change as heat goes from the burning zone toward the feed end. Excessively low O_2 content means incomplete combustion, lower burning zone temperature and reduced heat transfer from the flame to the lime. To maintain the burning zone temperature constant, fuel flow must increase[27].

The requirements of the causticizing reaction determine the target for residual $CaCO_3$. It is normally 2%– 4%. If the $CaCO_3$ content is lower, lime becomes harder and less reactive. Higher $CaCO_3$ content means higher inert load in recausticizing. Excessively low $CaCO_3$ means excess heat consumption in the kiln. If the residual $CaCO_3$ decreases, higher temperature is necessary in the burning zone. More fuel is also necessary because of the limited heat transfer surface. The flue gas outlet temperature will also increase simultaneously with the increase in the burning zone temperature[28].

The $CaCO_3$ content in lime mud and the residual $CaCO_3$ content in reburned lime determine the heat required for calcination reaction. Higher $CaCO_3$ content in lime mud requires more heat for reaction. The heat consumption per produced CaO amount is almost constant with different lime mud $CaCO_3$ contents, but heat consumption per production of burnt lime increases when lime mud $CaCO_3$ content increases. Figure 45 shows heat consumption at different lime mud $CaCO_3$ contents.

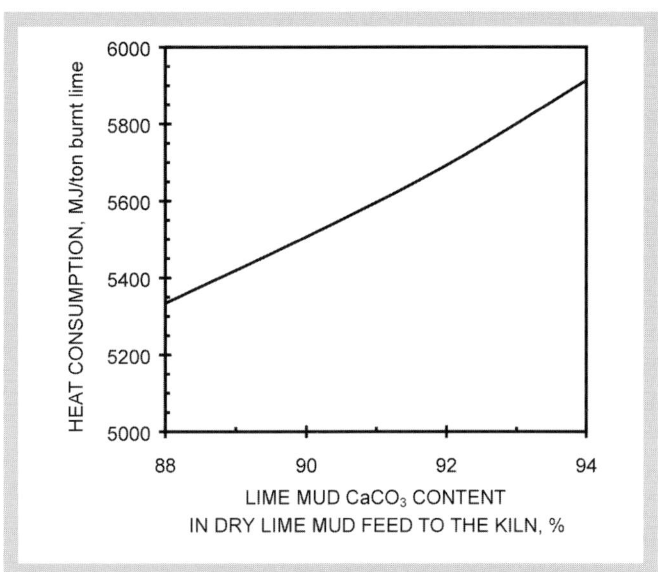

Figure 45. Lime kiln heat consumption per production of reburned lime vs. lime mud $CaCO_3$ content in the lime mud feed to the kiln.

The heat flow through the shell is approximately constant at all kiln loadings. This means higher relative shell heat loss at low capacities. The brick lining type (layers and materials) greatly influences the heat flow through the shell. Normally, new lime kilns use two-layer brick lining to reduce heat consumption. When using one-layer brick lining, shell surface temperatures are high, and more heat is necessary to maintain sufficient burning temperature.

Nonprocess elements (NPE) influence the heat consumption in two ways. Contaminated lime mud has a low $CaCO_3$ content meaning a high solids load to the lime mud filter. This results in low dry solids content and high solids load to the lime kiln. Both factors mean an increase in the heat consumption. Some NPE will also worsen lime mud dewatering properties causing low lime mud dry solids. In addition, some NPE lower lime smelting temperature. Then the kiln must operate at high residual $CaCO_3$ content of 4%–5%. This means that the CaO content in lime is low and the amount of lime into recausticizing must increase. This causes increased solids load to the lime mud filter and lime kiln.

9.4 Combustion

Heat transfers inside the kiln from the flame and the gas stream to the bed primarily by radiation. This requires a high fuel combustion temperature. Both stability and controllability of the combustion temperature are important because fluctuations in temperature easily change the heat transfer and cause subsequent loss of control with potential danger of refractory lining damage.

Limitations in formation of nitrogen oxides (NO_x) have recently become important. Achieving emission limits without reducing the flame temperature and the capacity of the kiln is often difficult.

9.4.1 Fuel oil and natural gas

Almost all lime kilns use heavy fuel oil or natural gas. The combustion temperature of both fuels is sufficiently high. Their critical properties such as heat value per unit volume and analysis per unit volume have the required stability. Nitrogen oxide emissions may be difficult to control especially if nitrogen content in the fuel is high (more than 1% nitrogen). This is normally a problem with heavy fuel oil.

Unlike cement kilns, lime kilns cannot use the pulverized coal. Cleaned, gasified coal is a potential fuel, but an acceptable heat value for the gas can only be reached by oxygen gasification.

The viscosity of heavy fuel oil is high at room temperature. The oil must therefore be preheated before entering the burner nozzle. The required viscosity for atomization depends on the burner design and the principle of atomization (pressure atomization or steam/air atomization). Typical atomization viscosity is 10–20 cSt. Table 13 provides typical properties of fuel oil and natural gas.

The adiabatic combustion temperature calculation uses stoichiometric combustion. Air temperature is 300°C corresponding to the true conditions in lime reburning kilns furnished with cooler for burnt lime.

Table 13. Typical properties of fuel oil and natural gas.

Elementary analysis of dry solids	Fuel oil	Natural gas
Carbon,% wt.	85.5	73.9
Hydrogen,% wt.	11.2	24.5
Sulfur,% wt.	2.4	0
Oxygen,% wt.	–	0.1
Nitrogen,% wt.	0.8	1.5
Ash,% wt.	0.04	0
Moisture,% wt.	0.3	–
Effective heat value in dry solids, MJ/kg	40.7	48.9
Density at 15°C, kg/m^3	970	
Viscosity at 50°C, cSt	380	
Adiabatic combustion temperature, °C	2 330	2 220

9.4.2 Noncondensible gases (NCG)

Handling systems for malodorous gases became necessary in the 1970s. Lime reburning kilns had use for destructive incineration especially for streams with high amounts of combustible compounds such as hydrogen sulfide, methyl mercaptan, and dimethyl sulfide. This first caused numerous failures of kiln refractories especially the linings of the flame zone. Moisture was the main cause for the failures. Because of the high risk of explosion, the gases were transported by steam ejectors. The moisture removal before feeding the gases into the kiln had poor design.

NCG typically contains considerable amount of transportation steam that decreases the heating value. Co-combustion with the main fuel reduces the flame temperature and makes ignition and flame stabilization more difficult. In practice, it has proven impossible to stabilize the heat value and volumetric flow. This has limited the amount of NCG to 15% maximum of the total heat input as a rule of thumb to avoid excessive fluctuations in the flame temperature and the temperature of bed and refractory linings in the flame area.

If NCG must cover more than 15% of the total heat input, the flow and heat value require control to correspond totally to the quality of combustion of natural gas or fuel oil.

Lean malodorous gases that are mainly air-rich vapors from various tanks and vessels of caustisizing plants can be used in lime kilns as combustion air by mixing them with secondary air before entering the lime cooler. One must be certain the amount of combustibles is well below the flammability limit[29].

9.4.3 Tall oil and methanol

Combustion of other liquids than fuel oil is possible in the lime kiln. They can replace the primary fuel partially or totally. The heat value and combustion properties of tall oil are very near those of fuel oil, and no significant difference exists in the heat transfer from flame to solid bed. The water content must be well below 5% when replacing a high proportion of primary fuel. The heat value of water-free methanol is much lower than the heat value of oil, but the combustion properties are sufficiently similar. The adiabatic combustion temperature in air is only 100°C lower.

The burner design must use the same principles as oil burners. Proper atomization is necessary, and the flame stabilization must be secured. Effective filtration of impurities is necessary because the openings of burner nozzles are often small.

Tall oil can be mixed with heavy fuel oil, but the mixing proportion must remain constant to avoid sudden deviations in density, viscosity and heat input. Methanol must be introduced into the kiln through a separate nozzle that can be part of the main burner or a completely separate burner. Table 14 shows the main differences of tall oil and methanol compared to fuel oil.

Table 14. Combustion properties of tall oil and methanol.

	Tall oil	Methanol	Fuel oil
Density at 15°C, kg/m^3	950–1 020	796	940–1000
Viscosity, cs	50–300 at 50°C	0.75 at 20°C	50–2000 at 50°C
Effective heat value (water free), MJ/kg	Approx. 35	21.1	40–41
Adiabatic combustion temperature,°C	2200–2250	Approx. 2200	Approx. 2300

9.4.4 Wood waste and other solid fuels

Solid fuels normally contain various amounts of ash. In the lime reburning process, fuel ash primarily occurs in the product. Aluminum, silica, phosphorus, and magnesium are the most harmful nonprocess elements of ash in a pulp mill. Coal ashes typically contain too much of these elements making combustion of coal in the kiln impossible.

Pulp mills produce considerable amounts of wood waste such as bark. The waste typically contains much smaller amounts of ash than coals and can be a potential fuel for lime reburning. The amount and composition of ash depend on the wood species. For example, mixed tropical hardwood contains considerable amount of silica.

Besides its ash composition, the moisture content of wood waste is also critical. The moisture content of bark is typically 50%–60% in Scandinavia. Properly controlling the direct combustion of fuel with high moisture content in the lime kiln is difficult. The main consequence of its use is the reduction of kiln capacity. The flame temperature is hundreds of degrees lower, and heat transfer to the lime bed is also lower. The moisture content is lower at 25%–50% in tropical conditions, but this is unacceptable as well.

Bark and other moist wood waste requires drying before combustion. The moisture content must be sufficiently low that the same flame temperature can result as in natural gas or oil combustion. In practice, the maximum allowable moisture content is approximately 15% on wet weight. At this moisture content, the flame temperature remains lower than in oil combustion, but the difference is acceptable.

A simple method to use wood waste as fuel for lime reburning is to dry it and then pulverize it. Depending on the dryer type, these process steps can be combined. Rotary drum dryers and pneumatic dryers have common use. The main problem with these drying methods is formation of emissions due to excessively high temperature of the drying gas. In addition, the nonprocess elements tend to enrich into the lime and liquor circulation. The closing of the mill cycles does not favor the direct combustion of wood waste in the future unless the removal methods for nonprocess elements improve.

Compared with direct combustion of dry pulverized wood waste, gasification offers a method to decrease the load of NPE in the fuel. This is because a significant part of the ash can separate in the gasifier or after gasification before the combustion. In the fluid bed gasifier, heavy ash compounds such as Fe, Al, and Si are removed primarily in bottom ash. The removal efficiency depends on the initial content in the dried fuel. Most Mg, Ca, and alkalis remain in the produced gas and become entrained in the kiln as fine dust.

The economy of wood waste combustion in lime reburning kilns whether direct or gasified depends mainly on the price of primary fuels, fuel oil, and natural gas. Since 1986, the prices have been inexpensive. Consequently, no plants for wood waste have been built.

9.5 Emissions

Emissions are today a main concern in the lime kiln process. Governing bodies usually set emission requirements for SO_2, TRS, CO, particulates, and NO_x. Requirements for emission of volatile organic compounds (VOC) also exist in some locations. Emissions monitoring occurs with continuous measurements or during scheduled measurement periods.

Controlling emissions continuously is easier when knowing their sources. In addition, the condition and maintenance of the monitoring equipment is essential for accuracy and reliability of the measurements.

9.5.1 Formation of SO_2, TRS, CO, particulate, and NO_x

SO_2 forms in combustion of fuel when the fuel contains sulfur. $CaCO_3$ absorbs some SO_2, but some SO_2 usually occurs in the stack[30,31]. The ability of lime itself to absorb SO_2 from the flue gas has often been discussed. Note that the absorption reaction

$$CaO + SO_2 + 1/2\ O_2 \rightarrow CaSO_4 \tag{11}$$

starts at 800°C. At a temperature above 900°C, calcium sulfate is not stable. $CaSO_4$ rapidly decomposes at temperatures over 1 000°C in the lime kiln environment.

TRS emission consists primarily of hydrogen sulfide. The emission may originate from two sources. If the fuel contains sulfur and insufficient air is present for complete combustion, the combustion is below stoichiometric. Carbon monoxide and hydrogen sulfide form and exit in the flue gas. The main origin of H_2S is sodium sulfide in the lime mud fed into the kiln. H_2S forms at the cold end of the kiln according to the following reaction[32]:

$$Na_2S + CO_2 + H_2O \rightarrow H_2S + Na_2CO_3 \tag{12}$$

Particulate matter consists of lime mud dust and alkali. The dust forms in the lime kiln through two different mechanisms: lime mud dust carryover from the feed end and alkali vaporization. In conventional kilns, the lime mud dust formation relates to the proper kiln feed end and chain section design and to the proper pelletizing process at the heating zone of the kiln. In kilns with a lime mud dryer outside the kiln, proper design of the cyclone controls the lime mud dust content.

Lime mud dust is almost totally collected with a scrubber or an electrostatic precipitator. Sodium components are usually so small that the scrubber cannot absorb them. If the dust emission limit is 150 mg/m^3n or lower, an electrostatic precipitator is necessary.

The mechanisms for alkali dust formation relate to high temperature at the hot end of the lime kiln. Sodium compounds vaporize in the burning zone and condense to form sub-micron sized particles when the flue gas cools.

In laboratory simulations, water soluble sodium is the principal origin of alkali fume. Alkali emission therefore relates to lime mud washing.

Various nitrogen oxides consist mainly of NO. The remainder is NO_2 and sometimes other nitrogen oxides. They form in high temperatures when nitrogen and oxygen are present. The formation of NO_x starts when the temperature is over 650°C and increases rapidly when temperature is over 1 400°C. Adjustment of lime kiln temperature distribution can avoid the temperature peaks. Burner design can also avoid NO_x formation.

Nitrogen oxides always form during combustion if nitrogen is present. Formation involves three different mechanisms[33]. The formation of thermal nitrogen monoxide starts at about 650°C, but becomes significant at temperatures above 1 300°C. Thermal NO_x forms when N_2 and O_2 from combustion air react according to the net reaction:

$$N_2 + O_2 \rightarrow 2\ NO \tag{13}$$

Prompt NO_x forms when oxygen and nitrogen from air react in the presence of hydrocarbon radicals in the flame pattern. The reaction time is in milliseconds. This reaction is minimal at temperatures below 1 500°C.

The fuel NO_x mechanism involves the decomposition of the organic nitrogen compounds from the fuel and their reaction with oxygen in the flame pattern. These reactions occur at much lower temperatures than thermal NO_x formation. Flue gas temperature in the lime kiln varies from 1 400°C at the burning zone to 150°C–300°C at the flue gas outlet. Fuel NO_x formation reactions start at 650°C. The forming section for thermal NO_x is very short. It is only in the hottest zone where gas remains for 1–2 s.

CO and VOC both relate to poor combustion. When sufficient air mixes properly with fuel, CO and VOC emissions remain low.

9.5.2 Emission control

SO_2 emission is usually less than 30 ppm for oil-fired lime kilns without a scrubber or NCG combustion. In other cases when the kiln is using natural gas or wood gas but not burning NCG, emission values are usually 10 ppm or less. When burning NCG in the lime kiln, SO_2 emission is considerably higher, and the actual SO_2 emission depends on the sulfur input to the kiln[34].

An alkali scrubber in one or two washing stages using fresh NaOH as makeup for pH control can remove SO_2 from line kiln flue gas. SO_2 emission can be under 10 ppm with the scrubber. The incoming level of SO_2 determines the amount of spraying nozzles in the scrubber and the alkali charge.

TRS emission is low with proper lime mud washing, i.e., the soluble alkali content is under 0.3% as Na_2O in lime mud on a dry solids basis. The lime mud dry solids content is also higher when lime mud contains less alkali.

TRS emission usually forms from Na_2S in the lime mud feed. In some cases, TRS emission was very high because the kiln was operated under less than stoichiometric conditions. A proper burner design and controlling the residual O_2 content in flue gas during operation can eliminate TRS formation in the flame. In practice, oxygen content of 1.5%–2.0% in wet gas at the feed end is sufficient.

Adjusting the kiln operating parameters will not eliminate the TRS emission originating from Na_2S in lime mud. A solution is to prevent Na_2S from entering the kiln with the proper operation of the lime mud precoat filter. With good washing and dewatering on a modern filter, a TRS level of 10 ppm should be easily achievable. Separating lime mud dust from flue gas is usually effective. The main components of the dust emission in the stack are the fine Na_2SO_4 or Na_2CO_3 with particle size below 10 µm. With the present limits for particulate emissions below 250 mg/m^{3}n, flue gas cleaning only with a wet scrubber is not acceptable. The plant must have an electrostatic precipitator. The actual particulate emission depends on the design of the ESP. Normally, particulate emission with ESP is less than 50 mg/m^{3}n.

CO and VOC emissions are usually no problem because kilns run with excess air. CO emission is normally less than 45 ppm. VOC emission is usually below 20 ppm as CH_4.

NO_x emission results during the combustion of fuel. This means that the burner provides the main control of NO_x emission. NO_x emission measurements have shown that the NO_x emission can often decrease by 50% with the adjustment of primary air distribution. The kiln temperature profile influences NO_x formation. Flame shaping controls the combustion temperature. NO_x emission normally is 50–200 ppm depending on fuel. It remains at the same level with oil and gas with correct flame shaping. When burning NCG, NO_x emission is somewhat higher.

9.6 Automation of rotating kiln

The lime kiln environment is very demanding for automation hardware and software. The kiln is a hot, continuously rotating piece of equipment with fine lime mud and dust. The process delay in the lime kiln from lime mud feed to final product output from the hot end is about 4 h. During this time, only few measurements can be taken from the product itself. The process changes can be reliably measured only indirectly. For example, changes of reburned lime or $CaCO_3$ content can be continuously measured only through reburned lime temperature. The air flow to combustion cannot be measured. The only way to control air flow is to measure residual O_2 content in the feed end. All kiln measurements are essential for kiln operation control, but regular visual monitoring is necessary by the operators who must know how these indirect measurements relate to the actual process changes.

After the final product quality has been checked and control adjustments made, four hours will pass before the results of the control can be seen. The process conditions will only occasionally remain constant during this time. This makes identification of the result of the control action is even more difficult. Taking representative lime samples and analyzing the residual carbonate accurately is also difficult. Conventional methods do not provide satisfactory replication.

The main process measurements and kiln control philosophy have remained the same for a long time. Following the common trend in the pulp and paper industry, higher level optimization control systems have recently been developed for this process area.

9.6.1 Lime kiln field instrumentation

The main process measurements include temperature, pressure, CO and O_2, lime mud filter feed density, and flow. The burner valve train has considerable instrumentation. Each fuel requires feed flow and pressure control. Damper or induced draft (ID) fan speed control usually controls gas flow. The main process measurements usually connect into the DCS system. The burner control either connects into the DCS or has its own programmable logic controller (PLC) depending on the automation strategy at the mill and local legislation[35]. Figure 46 shows typical lime kiln instrumentation at the cold end, and Fig. 47 shows the instrumentation at the hot end.

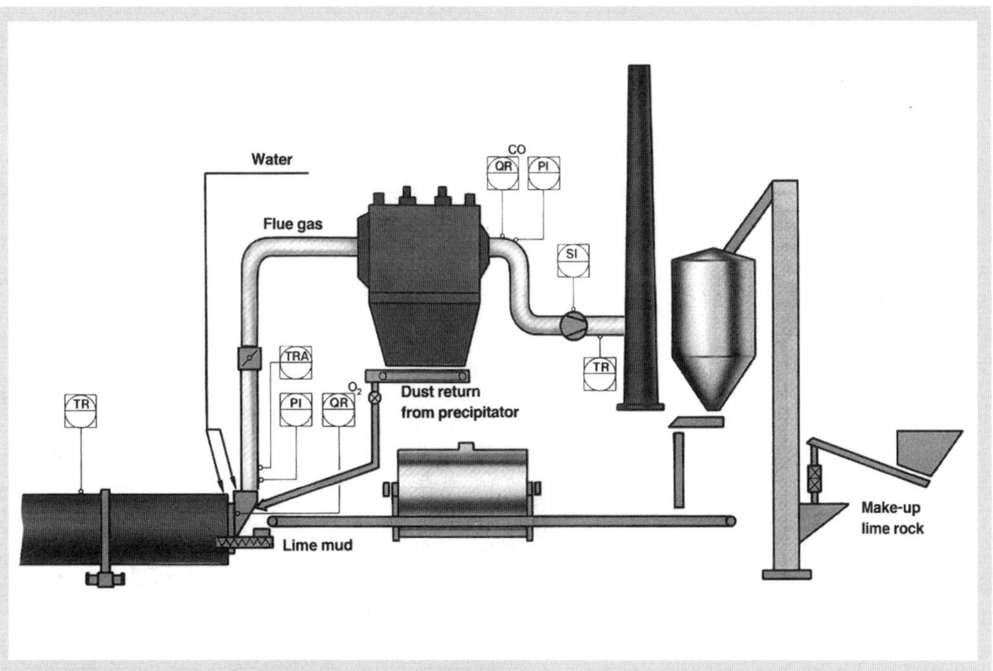

Figure 46. Typical lime kiln instrumentation at the cold end.

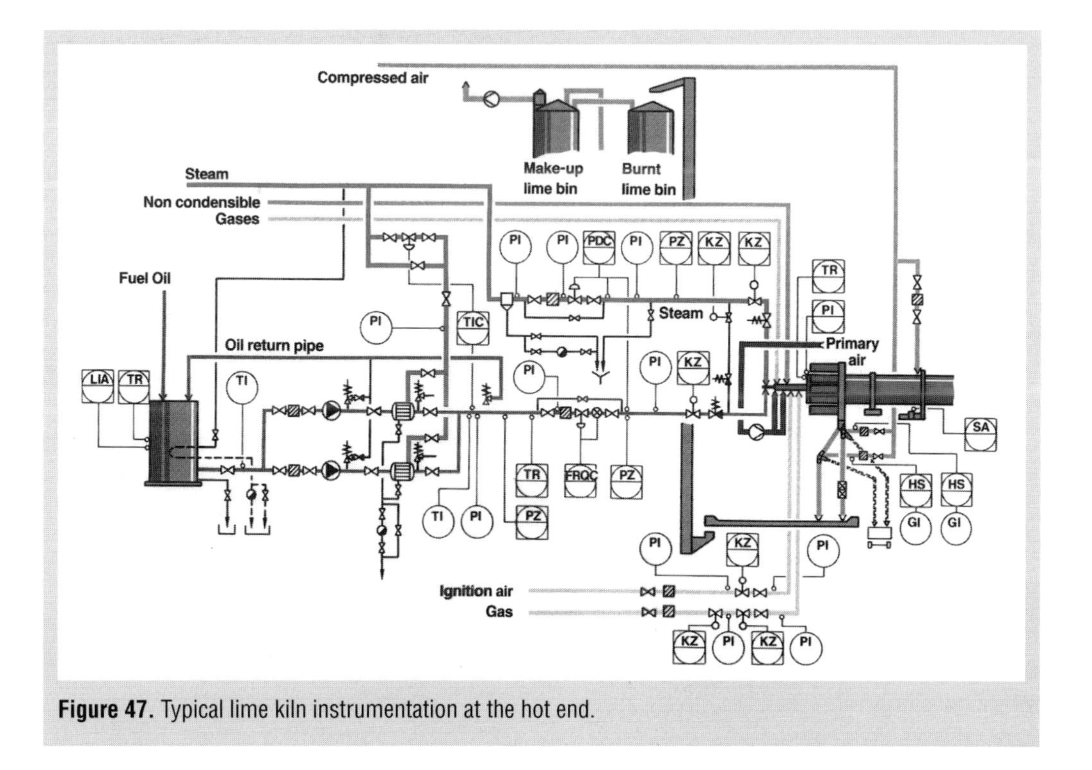

Figure 47. Typical lime kiln instrumentation at the hot end.

Kiln bed temperature is a very demanding measurement in the lime kiln process. The sensor is usually an infrared sensor located in the kiln firing end. The dust inside the kiln is the main problem in this measurement. With considerable dust in the kiln, the sensor cannot see the bed through the dust and cannot provide reliable measurement. Sensor manufacturers have developed compensation technologies to bypass this problem. Proper sensor installation is essential for the operation of the entire measurement system.

Measurement of the kiln shell temperature has recently become increasingly popular. In some areas, insurance companies even request it. Several technologies from the use of multiple Pt-100 sensors similar to kiln shell infrared sensors have been used. The infrared method is currently widely used. The system requires one or multiple sensors depending of the plant design and shell monitoring length. The unit usually comes with its own personal computer based hardware and has several graphical views to indicate hot spots in the kiln shell. Depending on the measuring distance, the resolution of this measurement in the screen can be about the size of one brick in the kiln shell. Weather influences this measurement if the sensors and kiln are outside.

Mid zone temperature measurement is not common today. Infrared sensors are used for monitoring the bottom of a tube in the kiln shell. These systems usually have so-called "peak picker" features that allow the instrument to remember the hottest spot during the entire kiln rotation period.

A color camera system often monitors the lime bed. This gives valuable on-line information of the kiln operation to the operators. Camera technology has improved lately, and the image quality has improved markedly.

O_2 measurement is another parameter of major importance for proper kiln operation. With long kilns, a sensor of the in-situ type has wide use. In the LMD kilns, the temperature at the kiln feed end exceeds the operational temperature of the sensor. Measurements that withdraw a gas sample and analyze it outside the kiln have become popular for the LMD kiln O_2 measurements. Correct sample taking technology is very critical for proper operation. Plugging is the major problem if O_2 is measured between kiln and electrostatic precipitator. Proper automated sample line back flushing may avoid this.

Local legislation often clearly specifies burner control instrumentation. Burner controls uses PLC and DCS equipment. Burner startup can often occur only from the local burner control panel, but capacity control and system shut down control can come from the DCS operator screen. A mass flow meter has wide use today for oil flow measurement. Flame scanners, K-type thermoelement, and sometimes special valve actuators are the special instruments in this area.

9.6.2 Lime kiln control

Manual control

Manual control is still widely used in kiln control. Operators monitor temperatures, residual carbonate, O_2, and CO measurements and control the kiln feed, kiln speed, and ID fan accordingly. The process delay as noted earlier makes the proper manual control very difficult. Overheating the kiln is the major risk to be avoided. If it occurs, it will damage the brick lining and cause a long, expensive kiln shutdown for repair. For this reason, there is considerable interest in developing an automated kiln higher level optimization control system.

Higher level optimization controls

Higher level optimization controls require considerable calculations[36]. A kiln filling degree calculation using the lime mud filter feed flow and density measurements is usually necessary. This calculated value can control the main fuel input.

Temperatures are measured from both kiln ends at least. The results control the temperature profile location inside the kiln by controlling the ID fan speed. O_2 measurement and residual carbonate test values can fine-tune the temperature profile and fuel control accordingly. A proper bed temperature reading is critical for the kiln control. Kiln optimization control usually requires a long fine-tuning period in the startup. This control uses standard proportional integral derivative (PID) and fuzzy controls. Figure 48 shows the lime kiln control principle.

9.7 Maintenance

Lime kilns require continuous maintenance to prevent unscheduled repairs. This includes daily monitoring of mechanical condition, inspection of the wearing components, lubrication, measurements and adjustments, systematic inspection during shutdown, and routinely performed service of wearing components. The daily routines include inspection for any unusual noise and visual observation. These are reported on daily monitoring records for scheduling the required inspections or repairs.

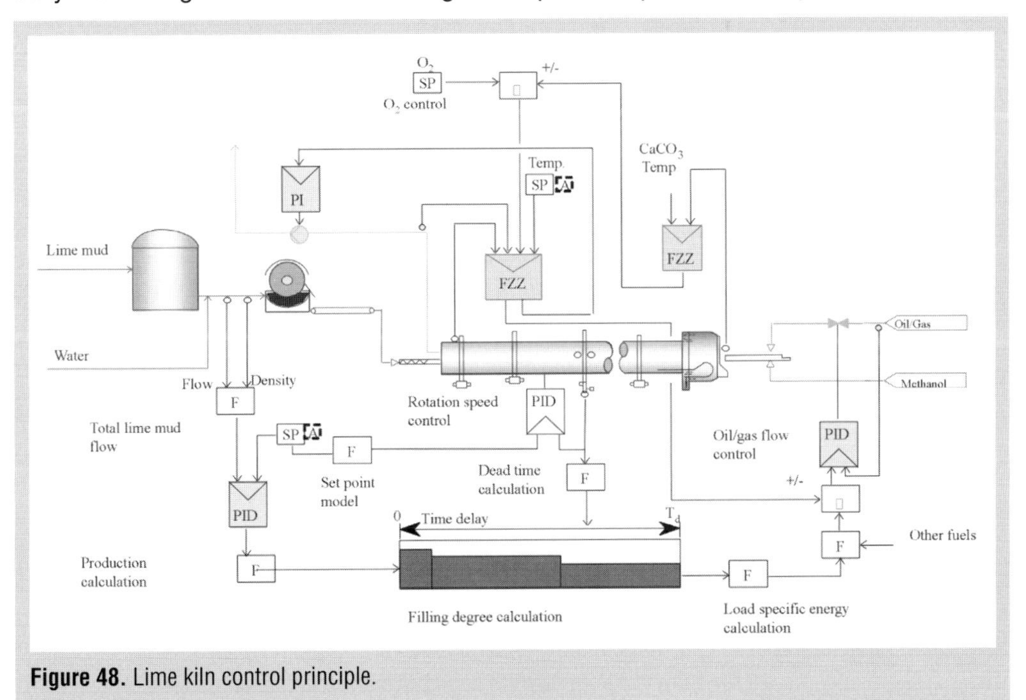

Figure 48. Lime kiln control principle.

Brick lining is vital to keep the lime kiln in continuous operation. To prevent shutdowns caused by sudden failure of brick lining, careful inspection during a scheduled shutdown is necessary. Shell surface temperatures are also measured regularly and recorded. The trend of the surface temperatures shows the wearing rate of the bricks. The condition of the brick lining is checked visually and by measuring the thickness in all zones during shutdowns. The thickness changes are recorded for scheduling of brick lining replacements.

Using the recorded inspection and repair requirements, maintenance is scheduled during a shutdown. The check list for inspections also reveals damage that is not visible during operation.

To prevent unscheduled shutdowns, the lime kiln condition is inspected by measuring straightness, degree of oval shape, and elevation of the supporters. Kiln adjustments are done using these measurements.

Burner inspection when it is not in operation is essential to maintaining stable and trouble-free operation of the kiln. Mechanical damages or plugging can cause severe problems in combustion and problems for brick lining, coolers, etc. Following the operational parameter trends of the instruments provides a monitor of the valve set and fuel piping condition. Flame shape monitoring is a component of the daily inspection routine by operators.

References

1. Virkola, N-E., Pikka, O., Keitaanniemi, O., et al., in Puumassan Valmistus II (N-E. Virkola, Ed.), Suomen Paperi-insinöörien Yhdistys, Turku, 1993, p. 296.
2. Lindberg, H. and Ulmgren, P.,J. Pulp Paper Sci. 9(1):TR7(1983).
3. Daily, C. M. and Genco, J. M., TAPPI 1988 Pulping Conference Proceedings, TAPPI PRESS, Atlanta, p. 181.
4. Dorris, G. M., TAPPI 1990 Pulping Conference Proceedings, TAPPI PRESS, Atlanta, p. 245.
5. Dorris, G. M. and Allen, L.H., TAPPI 1986 Kraft Recovery Operations Seminar, TAPPI PRESS, Atlanta, p. 7.
6. Dorris, G. M. and Allen, L.H., Operating Variables Affecting the Causticizing of Green Liquors with Reburned Limes, PPR 548, Pulp and Paper Research Institute, Quebec, 1985, p. 4.
7. Keskinen, H., Engdahl, H., and Beer, C., 1995 International Chemical Recovery Conference Proceedings, CPPA, Montreal, p. A305.
8. Ulmgren, P. and Rådeström, R., J. Pulp Paper Sci. 22(2):J52(1997).
9. Kiiskilä, E.,Paperi Puu 76(8):574(1994).
10. Kiiskilä, E. and Lindberg, H., 1995 International Chemical Recovery Conference Proceedings, CPPA, Montreal, p. B159.
11. Jiang, J. E., Tappi J.77(2):120(1994).
12. Varhimo, A., Pekkala, O., Ranua, M., et al., TAPPI 1997 Pulping Conference Proceedings, TAPPI PRESS, Atlanta, p. 909.
13. Varhimo, A., Pekkala, O., Ranua, M., et al., TAPPI 1996 Pulping Conference Proceedings, TAPPI PRESS, Atlanta, p.957.
14. Sandvold, B. and Pettersson, B., Pulp Paper Can. 92(5):T130(1991).
15. Ulmgren, P., Nordic Pulp Paper Res. J. 2(1):4(1987).
16. Engdahl, H. and Tormikoski, P., Paperi Puu 76(5):327(1994).
17. Wimby, M., Gustaffson, T., and Larsson P., 1995 CPPA International Chemical Recovery Conference, CPPA, Montreal, p. A299.
18. Lumikko, J., Fin. Pat. FI 65918 (May 29, 1980).

19. Lidén, J., 1995 CPPA International Chemical Recovery Conference Proceedings, CPPA, Montreal, p. A291.

20. Frederick, W. J, Jr., Rajeev, K., and Ayers, R. J., Tappi J. 73(8):135(1990).

21. Salmenoja, K. and Kosonen, J., TAPPI 1996 Engineering Conference Proceedings, TAPPI PRESS, Atlanta, p. 793.

22. Blackwell, B., Pulp Paper Can. 88(6):T181(1987).

23. Anon., Chemical and Physical Reactions during the Oxidation of Filtered Kraft White liquor and during Pulping, Mead Chemical Systems, Dayton, 1975, p. 3.

24. Anon., Comparison of Polysulfide and Kraft Pulping of some Canadian Wood Species, Mead Chemical Systems, Dayton, 1975, p. 1.

25. Thring, R.W., Uloth, V.C., Dorris, G.M et al., "White liquor oxidation in a pilot-plant pipeline reactor," paper presented at the AIChE Annual Meeting, St. Louis, Nov. 7–12, 1993.

26. Bartha, P., TIZ-Fachberichte 110(12):835 (1996).

27. Mullinger, P.J., World Cement 24(6):20 (1993).

28. De Leo, V.C., "Control of underburning in lime kilns using a reburned lime analyzer," 1981 TAPPI Pulping Conference Proceedings, TAPPI PRESS, Atlanta, p. 249.

29. Banks, D. and Horne, G., Pulp Paper Can. 97(10):57 (1996).

30. Holman, K. K., Golike, G. P., and Carlson, K. R., AIChE Symposium 92(311):99 (1996).

31. Springer, A. M. and Arceneaux, D., in Air Pollution in the Industry (A. M. Springer, Ed.), TAPPI PRESS, Atlanta, 1993, pp. 565–581.

32. Steen, B. and Stijnen, T., Svensk Papperstid. 87(3):R14 (1984).

33. Ahola, A., "Typpioksidien vahentaminen maakaasun vaiheistetulla lisapoltolla," M.Sc. thesis, Lappeenranta University of Technology, Lappeenranta, 1991.

34. Klafka, S. J., "Sulfur dioxide control in a rotary lime kiln", 1993 86th Annual Meeting & Exhibition Proceedings, Air & Waste Management Association, Denver, p. 93-TA-173.01P.

35. Blevins, T. and Rice, R., Tappi J. 66(3):103 (1983).

36. Crowther, C., Blevins, T., and Burns, D., Appita 40(1):29(1987).

CHAPTER 15

Combustion of bark

Markku Huhtinen, Arto Hotta

Combustion of bark

1 Introduction

This section provides a thorough discussion of the combustion technology of wood based waste fuels in the pulp and paper industry. It presents physical and chemical properties and combustion calculations of wood based fuels. A section on combustion technologies of bark first shortly describes grate firing systems that were the only combustion technologies for bark and wood wastes in the pulp and paper industry until the 1970s. Fluidized bed technologies that have since then gradually replaced grate firing, are then presented more thoroughly. Fuel handling systems, air and flue gas systems, and the water and steam cycle of natural circulation boilers are discussed in the section on boiler processes. This is followed by a section on the structure and thermodynamic design of boiler heat exchangers including economizer, evaporator, superheater, and air preheater. The final discussion covers the determination of boiler efficiencies and losses in energy production.

1.1 Operating principle of steam boilers

A steam boiler produces steam from water that is fed into the boiler. When considering the water and steam cycle, a steam boiler is simply a long heated tube. Water feeds into one end of the tube and exits from the other end as superheated steam.

Figure 1 presents the steam and water cycles in a natural circulation boiler. The feed water is first heated at or close to the evaporation temperature corresponding to the operating pressure in the economizer. After the economizer, the feed water enters the steam drum and the evaporator section. Water is heated further in the evaporator where a portion of it evaporates. After the evaporator, a

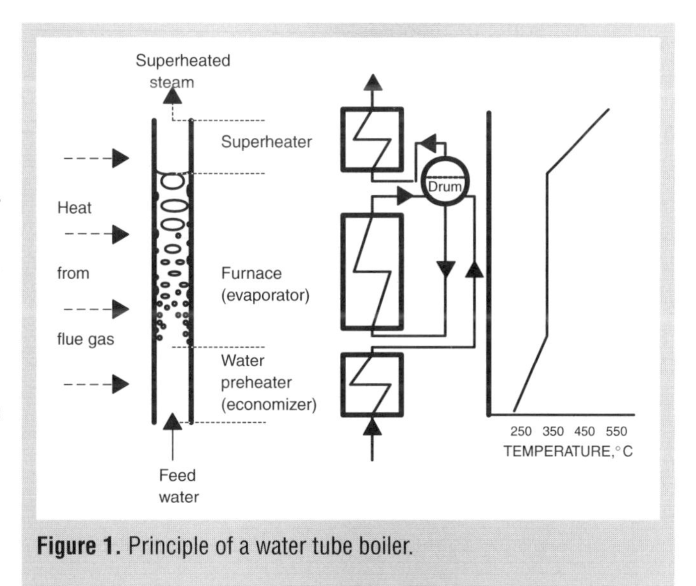

Figure 1. Principle of a water tube boiler.

mixture of saturated steam and water returns back to the steam drum. The saturated steam separates from the mixture in the steam drum and enters the superheater section where it is heated to the desired temperature. The steam pressure of a bark boiler in a pulp and paper mill is typically 7–10 MPa, and the steam temperature is 450°C–540°C.

The energy for heating and evaporating feed water and superheating steam in a boiler comes from firing fossil fuels such as coal, oil, gas, peat, etc. In a pulp and paper mill, the main fuels are bark, wood chips, and sometimes sludge from effluent treatment. These provide a large resource of cheap energy for the wood processing industries. If bark and wood based fuels are not sufficient to produce the desired steam capacity oil, gas, or coal is also fired in the bark boiler.

The fuel and the required amount of combustion air to complete the burning of the fuel are fed into the combustion chamber of the boiler. The fuel reacts with the oxygen in the combustion air. The chemical energy in the fuel transfers into thermal energy bound in the flue gases. The flue gases undergo cooling in various heat exchangers in the boiler (evaporator, superheater, economizer, and air preheater) before they go to a flue gas cleaning system and to the stack.

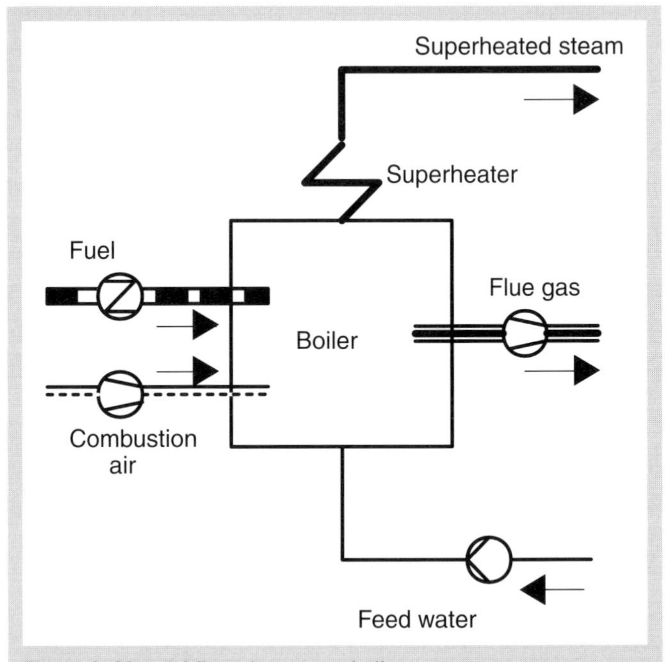

Figure 2. Material flows in a steam boiler.

Figure 2 shows the material flows in a steam boiler. A boiler plant has the following components or systems:

- Fuel handling, conveying, and feeding equipment for feeding the fuel into the boiler
- Air fans, ducts, and air preheater
- Flue gas cleaning
- Combustion chamber or furnace where the combustion of fuels occurs
- Tubing for steam and water that forms heat exchangers to cool the flue gas
- Control system.

Until the late 1970s, grate or stoker-fired boilers were the most common boiler types for combustion of bark. The bark boiler today is typically a bubbling fluidized bed boiler (BFB) or a circulating fluidized bed boiler (CFB). The fluidized bed boilers have replaced other technologies due to their better fuel flexibility and improved emissions performance.

1.2 Cogeneration steam cycles

The wood wastes in the pulp and paper industry today commonly find use in cogeneration power plants. The chemical energy in the wood waste is released by burning in steam boiler. With the released chemical energy the steam boiler produces steam with pressure and thermal energy for a steam turbine. When the steam goes through the turbine its pressure and temperature decrease and the released energy is converted into mechanical rotation energy of the turbine. The shaft of the turbine is connected to the shaft of a generator where the mechanical rotation energy is converted into electricity. Steam is taken for heating purposes from turbine extraction or turbine exhaust to various consumer points in the process. Cogeneration allows production of cheap electricity as a by-product, since the main purpose is to produce steam for process heating. In the consumer points steam releases its latent energy into the process by condensating. Normally most of the steam is brought back to the power plant as condensate. The condensate is pumped as feed water into the boiler again. Figure 3 shows the flow diagram.

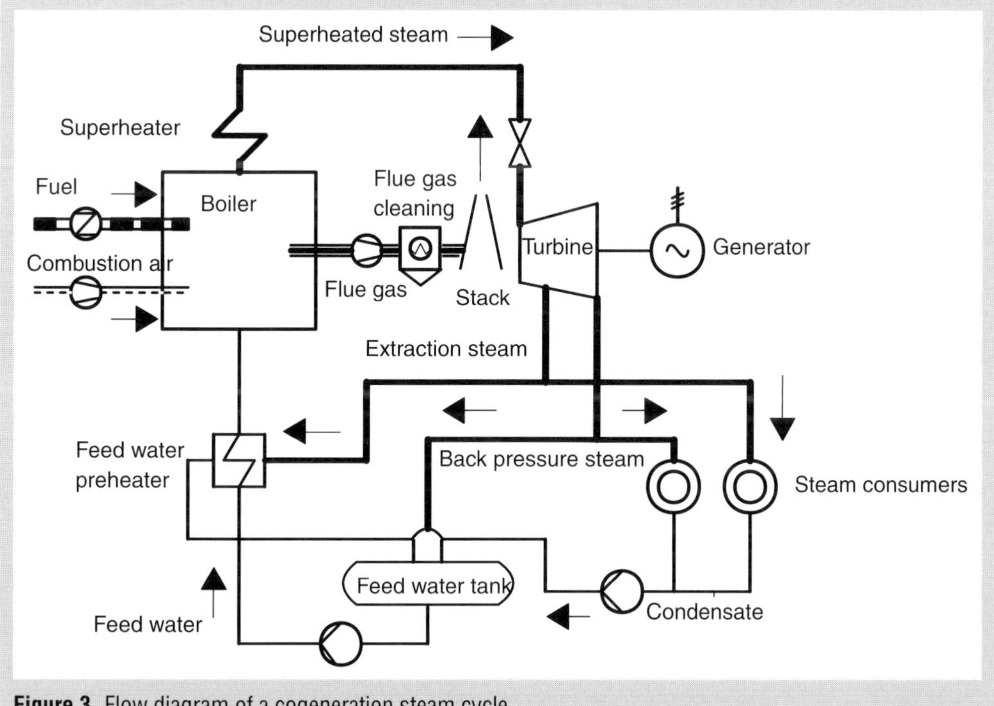

Figure 3. Flow diagram of a cogeneration steam cycle.

2 Physical and chemical properties of fuels

2.1 Fuels and their composition

Bark, wood, and hog fuel are the main fuels burned in the power boiler in a pulp and paper mill. Oil or natural gas is also needed for boiler startups, and for supporting firing during disturbances in solid fuel feeding or when the moisture content of the solid fuel is too high to support stable burning. Coal or peat is often cofired in power boilers if bark and wood waste produced in the mill is not sufficient to meet the energy needs of the mill.

Physical and chemical properties of fuel influence the fuel handling system design, combustion behavior, heat transfer surface design, flue gas cleaning, and ash handling. This section concentrates on the properties of the fuels that are the by-products in pulp and paper mills, i.e., bark, wood, hog fuel, and sludge from the water treatment plant.

2.2 General terminology and properties of fuels

Ultimate analysis

The ultimate analysis or elementary analysis gives the composition of carbon (C), hydrogen (H), oxygen (O), nitrogen (N), and sulfur (S) expressed in percent by weight for the fuel. The analysis is done by completely burning the fuel sample. Ultimate analyses are necessary to compute air requirements and the amount of combustion gas produced.

Proximate analysis

The proximate analysis gives information on the behavior of fuel when heated. It tells how much fuel vaporizes as gas and tar (volatile matter) and how much remains as fixed carbon. The volatile matter and fixed carbon are expressed in percent by weight. The proximate analysis is useful when determining the combustion behavior for coal. For wood based fuels the amount of volatile matter is much bigger than for coal so the proximate analysis is normally not necessary.

Table 1. The composition of dry substances (% by wt.) and typical moisture content for fuels[1,2,3].

Fuel	C	H	S	O	N	Ash	Moisture
Bituminous coal	73.2	4.7	1.0	9.1	1.0	11.0	9
Heavy fuel oil	88.3	10.1	1.0	0.2	0.4	0.04	0.3
Light fuel oil	86.2	13.7	0.1	-	0.02	0.01	0.01
Wood	50.4	6.2	-	42.5	0.5	0.4	55
Bark (pine)	54.5	5.9	-	37.7	0.3	1.7	60
(spruce)	50.6	5.9	-	40.2	0.5	2.8	60
(birch)	56.6	6.8	-	34.2	0.8	1.6	55
Black liquor							
(pine)	39.0	4.0	4.5	33.4	0.1	19	40
(birch)	37.0	4.0	4.0	33.9	0.1	21	40
Sod peat	55.0	5.5	0.2	32.6	1.7	5.0	50
Natural gas	CH_4	C_2H_8	O_2	CO_2	N_2		
(Russian)	98.92	0.11	0.05	0.02	0.9		

Figure 4 shows the differences in volatile matter and fixed carbon for fuels. The figure indicates the progressive stages of transformation of vegetal matter into coal.

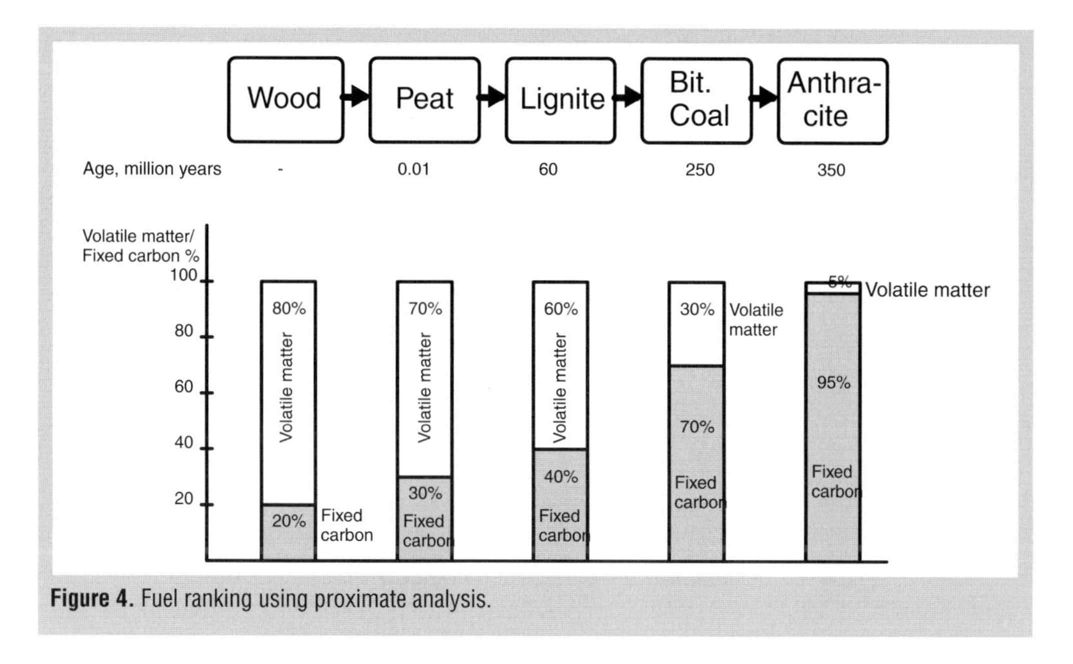

Figure 4. Fuel ranking using proximate analysis.

The volatile matter influences the ignition temperature of the fuel. Table 2 shows that fuels with high volatile matter content ignite at low temperatures compared with fuels having low volatile matter content. Due to the high reactivity of the volatile matter, burning of high volatile fuels is faster and more complete than burning of low volatile fuels.

Table 2. The volatile matter content in dry, ash free fuel and ignition temperature for fuels.

Fuel	Volatile matter,%	Ignition temperature, °C
Wood	70–85	200–400
Peat	70–75	200–250
Lignite	40–60	300–400
Bituminous coal	28–37	300–500
Anthracite	3–9	500–600
Charcoal	1–3	600–700
Heavy fuel oil	82–100	255–265

Moisture content

Moisture content is the amount of water in the fuel expressed as weight percent. Bark, wood wastes, and sludge have high moisture content. It is typically 55%–65%. Moisture is therefore a major characteristic influencing the boiler design for those fuels. High moisture content will result in high flue gas volume and large boiler size. Since the evaporation of water in fuel takes energy from the burning substance, the heating value decreases rapidly with increasing moisture content. At moisture of approximately 62%–65%, stable combustion becomes difficult to maintain, and supporting the firing with some fossil fuel is necessary. Table 3 shows the effect of moisture on the heating value of bark.

Table 3. Relationship of bark moisture to heating value[4].

Moisture content,%	Heating value ,HHV, MJkg
0	20.4
20	16.3
40	12.2
50	10.2
60	8.1
70	6.1
80	4.1
90	2.0

Dry substance

Dry substance is normally the basis for reporting fuel and ash analyses. Dry substance (kg) is the total mass of the water free portion of the fuel. It contains combustible and inherent or noncombustible materials such as ash. The combustible material consists primarily of carbon, hydrogen, and sulfur.

Ash content

Ash is the solid, inorganic residue from combustion. Ash content of fuel is expressed as percent weight of the dry substance.

Ash fusion temperature

Fusion temperature varies considerably with different fuels. Heated ash becomes soft and sticky. As temperature increases, ash becomes fluid. If the ash fusion temperature is low, the ash may be in a molten state when it reaches the first heat transfer surfaces in the boiler. The molten ash will then adhere on the heat transfer tubes creating slagging. This slagging will decrease the heat transfer and can even plug the gas path completely. Boiler design must consider ash fusion tendency. Ash fusion temperature is typically low for wood and sludge from water treatment due to the high alkali (Na and K) content.

The ash fusion tendency is measured by observing the temperatures at which triangular cones prepared from fuel ash attain and pass through certain defined stages of fusing . Four stages of fusion temperature are commonly reported:

- Initial or first rounding of the cone
- Softening when the cone height has diminished to equal the width at the base
- Hemispherical when the height of the lump equals half the width of the base
- Fluid when the height is no higher than 1/16 in.

Heating value

The heating value of fuel is expressed in MJ per kg of fuel on as-received, dry, moisture-free, or ash-free basis. The calorific or high heating value is the amount of heat recovered when the products of complete combustion of a unit quantity of fuel cool to a base temperature of 25°C. This includes the latent heat of the water vapor in the products of combustion. The water vapor produced in the combustion and the fuel moisture are in a liquid state at the reference temperature. An adiabatic bomb calorimeter is used to determine the calorific heating value.

In the actual conditions of a boiler, water vapor does not cool below its dew point, and the latent heat is not available for making steam. Heating value is therefore often expressed as low heating value (LHV) or net heating value. This is the latent heat of the water vapor subtracted from the calorific heating value, i.e. the water vapor produced in the combustion and the fuel moisture remain as vapor at the base temperature. Low heating values are standard in Europe, and high heating values (HHV) are standard in America.

Because the fuel moisture especially with bark and wood can change considerably, the calorific heating value is determined on dry fuel. When the moisture content is known, the low heating value for the as-fired fuel can be calculated.

Calculation of the low heating value for the dry fuel uses the following formula:

$$LHV_{dry} = HHV_{dry} - 8.935 \cdot l_{25} \cdot m_h \qquad (1)$$

where LHV_{dry} is low heating value for dry fuel
HHV_{dry} high heating value for dry fuel
l_{25} the evaporation heat for water at 25°C
m_h hydrogen in fuel (kg H/kg fuel)

Calculation of the low heating value for as-fired fuel can be calculated using the following formula:

$$LHV_{af} = LHV_{dry} \cdot (1 - m_w) - l_{25} \cdot m_w \qquad (2)$$

where LHV_{af} is low heating value for as-fired fuel
m_w mass of water in fuel (kg H_2O/kg fuel).

Heating values for solid and liquid fuels can be calculated with error below 2% using the reaction heats of the combustible components and oxygen with the following formula:

$$LHV_{af} = 34.8m_c + 93.8m_h + 10.5m_s + 6.3m_n - 10.8m_o - 125m_w \qquad (3)$$

where m_c is mass of carbon in fuel (kg C/kg fuel)
m_s mass of sulfur in fuel (kg S/kg fuel)
m_n mass of nitrogen in fuel (kg N/kg fuel)
m_o mass of oxygen in fuel (kg O/kg fuel).

Table 4 lists the heating values for most common fuels. The low heating value for moist fuels is calculated with the moisture content from Table 1.

Table 4. Heating values for common fuels[1,2,3].

Fuel	High heating value, MJ/kg	Low heating value for dry fuel, MJ/kg	Low heating value for moist fuel, MJ/kg
Bituminous coal	29.8	28.8	26.0
Heavy fuel oil	43.1	40.9	40.8
Light fuel oil	46.0	43.0	43.0
Wood	20.5	19.1	7.25
Bark (pine)	21.3	20.0	6.5
(spruce)	19.9	18.6	6.0
(birch)	24.2	22.7	8.9
Black liquor			
(pine)	15.5	14.6	7.8
(birch)	14.7	13.8	7.3
Sod peat	22.0	20.8	9.2
Natural gas	54.6	49.2	49.2
(MJ/m^3n)	39.3	35.3	35.3

2.3 Bark, wood, and hog fuel

Most wood based fuel in a pulp mill is bark from tree trunks. Wood chips and hog fuel are also available for power production.

Wood

Table 5 shows the chemical composition of several different wood species.

Table 5. Typical analysis of dry wood (% by wt.)[4].

Fuel	C	H	S	O	N	Ash	Heating value (HHV), MJ/kg
Softwoods							
Cedar, white	48.8	6.4		44.5		0.37	19.6
Cypress	55.0	6.5		38.1		0.40	23.0
Fir, Douglas	52.3	6.3		40.5	0.1	0.8	21.1
Hemlock, western	50.4	5.8	0.1	41.4	0.1	2.2	20.1
Pine, pitch	59.0	7.2		32.7		1.13	26.4
white	52.6	6.1		41.3		0.12	20.7
yellow	52.6	7.0		40.1		0.31	22.4
Redwood	53.5	5.9		40.3	0.1	0.2	21.0
Hardwoods							
Ash, white	49.7	6.9		43.0		0.30	20.8
Beech	51.6	6.3		41.5		0.65	20.4

Table 5. Typical analysis of dry wood (% by wt.)[4].

Fuel	C	H	S	O	N	Ash	Heating value (HHV), MJ/kg
Birch, white	49.8	6.5		43.5		0.29	20.1
Elm	50.4	6.6		42.3		0.74	20.5
Hickory	49.7	6.5		43.1		0.73	20.2
Maple	50.6	6.0		41.7	0.25	1.35	20.0
Oak, black	48.8	6.1		45.0		0.15	19.0
red	49.5	6.6		43.7		0.15	20.2
white	50.4	6.6		42.7		0.24	20.5
Poplar	51.6	6.3		41.5		0.65	20.8

The moisture content of freshly cut wood is 30%–50%. This value as received in the furnace will depend on the extraneous water from the source, storage and handling, species, and the time of the year the wood is cut.

Wood fuel can be chips, shavings, and saw dust. Each form has its specific combustion problems that require consideration in the design of fuel feeding equipment and the furnace.

Wood typically has a higher content of potassium than bark. The high potassium content with low ash content can result in slagging of heat transfer surfaces when the amount of wood fuel is high in a boiler.

Bark

Table 6 shows ultimate analyses of bark from some wood species.

Table 6. Ultimate analyses of bark.

Fuel	C	H	S	O	N	Ash	Heating value, MJ/kg
Bark (pine)	54.5	5.9	-	37.7	0.3	1.7	6.5
(spruce)	50.6	5.9	-	40.2	0.5	2.8	6.0
(birch)	56.6	6.8	-	34.2	0.8	1.6	8.9

With a wet debarking process, the bark as received from the barking drums contains 80% or more moisture and has no value as a fuel. To make the bark suitable for burning in a boiler, a mechanical press may be used to reduce the moisture to 55%–60%.

The bark comes in long, rope-like strips. The shape, size, and high moisture content make handling the fuel difficult. The handling properties of bark also vary with different wood species.

Bark naturally contains a low amount of potassium, sodium, and chlorine. It does not have a high potential for slagging or corrosion. If logs have floated in sea water, the salt in the water can cause severe slagging and superheater corrosion when burning such bark in a steam boiler.

Hog fuel

The term hog fuel covers the mixture of all log waste material from its harvesting to processing. Hog fuel can therefore contain slabs, edge pieces, trimmings, bark, saw dust, and shavings. Since the availability of these waste materials is very mill dependent, hog fuel has no typical analysis. The mixture of different waste materials in hog fuel constantly changes at a specific mill. This makes the fuel handling and control of the combustion process a very challenging task.

3 Calculation of combustion

In the combustion process, a substance combines chemically with oxygen. In fuel, the inflammable materials or materials that react with oxygen are carbon (C), hydrogen (H) and sulfur (S). The combustion reactions of these subjects are exothermic, i.e., the specific reaction releases energy as heat.

Knowing the composition and combustion reactions of a fuel allows calculation of the demand for combustion air and the amount of fuel gases produced. The fuel gases from a boiler plant undergo constant analysis to determine whether combustion has been complete and if the fuel to air ratio is correct.

3.1 Combustion reactions

To calculate combustion air demand and the formation of fuel gases, the reactions between inflammable components carbon (C), hydrogen (H_2) and sulfur (S) and oxygen (O_2) must be known. The combustion reactions that follow are net reactions where the reactants and the end products can be obtained. In reality, the reactions are more complex and contain reaction chains with multiple intermediates. The net reaction formulas give sufficient information for calculating the combustion air demand and the amount of flue gases. Carbon is the most important burning component in all common fuels. The net reaction formula for its combustion is the following:

$$C + O_2 \rightarrow CO_2 \qquad (+411 kJ/kmol) \tag{4}$$

Note from the formula that the reaction of C with O_2 produces carbon dioxide (CO_2) and 411 kJ/kmol of heat. One mole of C requires one mole of O_2 for complete combustion to one mole of CO_2. If the combustion of C is incomplete, the end product is carbon monoxide, CO. This will be discussed later in this section with other emissions.

The net reaction formula for combustion of hydrogen is the following:

$$2H_2 + O_2 \rightarrow 2H_2O \qquad (+242 kJ/kmol) \tag{5}$$

According to the formula, one mole of O_2 reacts with two moles of H_2 to produce two moles of H_2O. When H_2 combusts, the result is pure water vapor.

The net reaction formula for combustion of sulfur is as follows:

$$S + O_2 \rightarrow SO_2 \qquad (+9.2 kJ/kmol) \tag{6}$$

Combustion of sulfur results in sulfur dioxide (SO_2) formation. In the boiler, a small quantity of SO_2 usually oxidizes further to sulfur trioxide, SO_3. This reacts with water in flue gases to form sulfuric acid. Sulfur contained in the fuel is harmful to the boiler and the environment even if it provides energy during combustion.

Fuels also contain nonflammable components that still have an effect on the process. These components are oxygen, nitrogen, moisture, and ash. Oxygen in the fuel reduces the demand for combustion air. Before fuel burns, its moisture evaporates and reduces the heating value of the fuel. Some nitrogen in the fuel forms nitrogen oxides that are harmful to the environment. The incombustible ash increases the dust content in the fuel gases and increases the need for flue gas cleaning. In the following calculations, the nitrogen, moisture, and ash contained by the fuel are assumed to exist in the flue gases in their original form.

3.2 Combustion air demand

The following is the calculation of combustion air required for one kilogram of fuel. The results, given as kg air/kg fuel (kga/kgf) can be used to calculate the combustion air demand if the amount of fuel to be burned is known.

The calculation of combustion air demand begins with fuel composition and the reactions between the fuel components and oxygen. The composition of fuel is usually known as percentages by weight. See Tables 1 and 5. For calculation the composition is presented either in grams or kilograms per kilogram of fuel. As Table 7 shows.

Table 7. Fuel composition for calculating fuel combustion.

Component	Quantity, g/kgf	Molecular weight, kg/kmol	Quantity, mol/kgf
C (carbon)	w_c	12.01	$n_c=w_c/12.01$
H (hydrogen)	w_{h2}	2.016	$n_{h2}=w_{h2}/2.106$
S (sulfur)	w_s	32.06	$n_s=w_s/32.06$
O (oxygen)	w_{o2}	32.00	$n_{o2}=w_{o2}/32.00$

The quantities of different components must be in moles, since the quantities of the reactants in the combustion reactions are in moles. This is done by dividing the quantity given in grams by the molecular weight of the component. See Table 7.

Reaction equations can provide the amount of oxygen needed for combustion. The total oxygen demand in theoretical or stoichiometric combustion with the excess air ratio, $\lambda = 1$, can be calculated with the following equation:

$$\frac{N_{O2(theor)}}{m_f} = n_c + 0.5 \cdot n_{h2} + n_s - n_{o2} \tag{7}$$

i.e. by subtracting the quantity of oxygen contained by the fuel from the total oxygen demands of the burning components. The oxygen comes from air that also contains other gases (Table 8).

Table 8. Composition of dry air.

Substance	Concentration, vol.%	Molecular weight, kg/kmol
N_2, Nitrogen	78.03	28.02
O_2, Oxygen	20.99	32.00
Ar, Argon	0.94	39.94
CO_2, Carbon dioxide	0.03	44.01
H_2, Hydrogen	0.01	2.02
Total	**100.00**	**28.96**

Air contains mostly nitrogen and oxygen. Since argon is an inert gas it is often included in the percentage of the other inert gas nitrogen, this gives nitrogen the value 79% and leaves 21% for oxygen. With this and the total oxygen demand calculated as explained above, the equation for calculating the demand for dry combustion air is as follows:

$$\frac{N_{a(theor)}}{m_f} = \frac{N_{O2(theor)}}{m_f} \cdot \frac{1}{0.21} = \frac{N_{O2(theor)}}{m_f} \cdot 4.76 \quad (8)$$

In the equation the amount of air is in mol/kgf. Multiplying this by the molecular weight for air, 28.96 g/mol, transforms it to grams (g/kgf). To obtain the amount in L/kgf, multiply $N_{a(theor)}/m_f$ by the molecular volume for air, 22.40 L/mol.

Table 8 shows the composition of dry air. In practice, air is not dry. It contains a small amount of water vapor. In exact calculations the water vapor must be considered, although in practice it is usually ignored.

In practice complete combustion does not result with the theoretical (or stoichiometric) minimum amount of oxygen. The furnace must have slightly more air available. The ratio between the actual and the theoretical amount of air is excess air ratio, λ. Solid fuels require a larger excess air ratio than the more readily combustible liquid and gaseous fuels. Table 9 gives some typical values for excess air ratio in power plant boilers.

Table 9. Typical excess air ratios for different fuels used in power plant boilers.

Fuel	Excess air ratio, λ
Gas	1.02–1.10
Heavy oil	1.03–1.10
Coal	1.15–1.35
Peat (moisture content 50%)	1.15–1.35
Bark, wood (moisture content 60%)	1.15–1.50
Black liquor	1.10–1.25

One can calculate the actual amount of necessary combustion air by multiplying the theoretical amount of air with the excess air ratio according to the following equation:

$$\frac{N_{a(real)}}{m_f} = \frac{N_{a(theor)}}{m_f} \cdot \lambda \tag{9}$$

3.3 Flue gas composition

The amount and composition of flue gases can be determined from the fuel composition and the combustion equations in the same manner as the required amount of combustion air. According to the reaction equations presented earlier in this chapter, the amount of products in moles equals the amount of components in the fuel.

Flue gases also contain nitrogen. The ratio between nitrogen and oxygen in air is 3.76. Other incombustible components such as moisture transfer to flue gases in their original form. The composition of flue gases or the amount of each component in moles is obtained as described above. Adding these gives the total amount of flue gases.

If the amount of flue gases is necessary in kilograms instead of moles, multiply the molecular amounts of components by their molecular weights. If the volume of the flue gases is necessary, multiply the amounts in moles by the molecular volumes. For most gases, the approximate molecular volume of 22.4 L/mol (the molecular volume of an ideal gas) is satisfactory.

Flue gases may be expressed as dry or wet. The difference between these is that wet flue gases contain water vapor, but dry flue gases do not.

The following table gives the combustion air demand and the amount and composition of flue gases with excess air ratio, $\lambda = 1$, for some common fuels in pulp and paper industry. When the air demand and the amount of flue gases in theoretical combustion $(\lambda = 1)$ are known, the amount of flue gases with other excess air ratios results from the following equation:

$$\frac{N_{fg(real)}}{m_f} = \frac{N_{fg(theor)}}{m_f} + (\lambda - 1) \cdot \frac{N_{a(theor)}}{m_f} \tag{10}$$

The equation is based on the fact that the theoretical flue gas flow stays the same when the excess air ratio increases, but the excess air transfers to flue gases.

The excess air ratio can be calculated according to the carbon dioxide ratio or the oxygen ratio in the flue gases. In theoretical combustion when $\lambda = 1$, the oxygen concentration in the flue gases is zero. When the excess air ratio and the amount of combustion air increase, the amount of oxygen in the flue gases starts to increase. This excess O_2 from dry flue gas is usually measured in the boiler, and based on this measured oxygen concentration $(X_{O2(meas)})$, the excess air ratio can be determined using the following equation:

$$\lambda = 1 + \left(\frac{V_{fg,\,dry(theor)}}{V_{a(theor)}}\right) \cdot \left(\frac{X_{O2(meas)}}{20.9 - X_{O2(meas)}}\right) \tag{11}$$

Because for all fuels the ratio between theoretical dry flue gas flow $(V_{fg,dry(theor)})$ and theoretical air flow $(V_{a(theor)})$ is near one, the following approximate formula can be used:

$$\lambda = \frac{20.9}{20.9 - X_{O2(meas)}} \qquad (12)$$

Determining the excess air ratio using the oxygen concentration is easy because the oxygen concentration in the fuel gases does not depend on the fuel. When determining excess air ratio using the carbon dioxide concentration measurement, the maximum amount of carbon dioxide that forms in theoretical combustion of the fuel must be known. In theoretical combustion, the CO_2 concentration is highest $(CO_{2(max)})$. As the amount of air in the flue gases increases, the CO_2 concentration begins to decrease. By measuring this real CO_2 consentration $(CO_{2(meas)})$ the excess air ratio can be determined by using the following formula:

$$\lambda = 1 + \left(\frac{V_{fg,\,dry(theor)}}{V_{a(theor)}}\right) \cdot \left(\frac{X_{CO2(max)}}{X_{CO2(meas)}} - 1\right) \qquad (13)$$

Because the theoretical dry flue gas flow is almost equal to the theoretical air flow, also this formula can be simplified and the approximate formula is:

$$\lambda = \frac{X_{CO2(max)}}{X_{CO2(meas)}} \qquad (14)$$

For most common fuels in pulp and paper industry, Table 10 provides the numerical values for $X_{CO_2(max)}$.

Table 10. Gas flows in combustion of wood (bark) and black liquor $(\lambda = 1)$.

		Wood, humidity 60%	**Wood, humidity 30%**	**Black liquor, humidity 40%**	**Black liquor, humidity 20%**
Oxygen demand	kmol/kgf	0.0193	0.0338	0.0202	0.0269
	kg/kgf	0.618	1.082	0.645	0.860
Air demand	kmol/kgf	0.092	0.161	0.096	0.128
	Nm^3/kgf	2.06	3.61	2.15	2.87
	kg/kgf	2.67	4.67	2.78	3.72
Amount of wet flue gas	kmol/kgf	0.136	0.197	0.128	0.152
	Nm^3/kgf	3.05	4.41	2.87	3.41
	kg/kgf	3.64	5.65	3.55	4.36
Amount of dry flue gas	kmol/kgf	0.091	0.159	0.0939	0.125
	Nm^3/kgf	2.04	3.56	2.21	2.80
	kg/kgf	2.84	4.96	2.92	3.89

Table 10. Gas flows in combustion of wood (bark) and black liquor $(\lambda = 1)$.

		Wood, humidity 60%	Wood, humidity 30%	Black liquor, humidity 40%	Black liquor, humidity 20%
Flue gas molecular weight, kg/kmol	Wet gases Dry gases	26.8 31.2	28.7 31.2	27.7 31.1	28.7 31.1
Flue gas mole fractions	N_2 (wet)	0.535	0.649	0.593	0.665
	N_2 (dry)	0.801	0.802	0.809	0.811
	H_2O (wet)	0.332	0.191	0.267	0.178
	H_2O (dry)	0.0	0.0	0.0	0.0
	CO_2 (wet)	0.132	0.161	0.140	0.157
	CO_2 (dry)	0.199	0.199	0.190	0.192
	SO_2 (wet)	0.0	0.0	0.0002	0.0002
	SO_2 (dry)	0.0	0.0	0.0002	0.0002
Flue gas density,	Wet gases	1.20	1.28	1.24	1.28
kg/Nm^3	Dry gases	1.40	1.40	1.39	1.39
X_{CO2max}, mole fraction	Wet gases	0.132	0.161	0.140	0.157
	Dry gases	0.199	0.199	0.190	0.192

3.4 Adiabatic temperature of combustion

The highest temperature theoretically achievable when combusting fuel with no heat transfer to the surroundings is the adiabatic temperature of combustion. In this case, all heat that releases heats the flue gases. Adiabatic temperatures for different fuels are unequal. Figure 5 shows examination of the adiabatic temperature by preparing an energy balance for the combustion chamber.

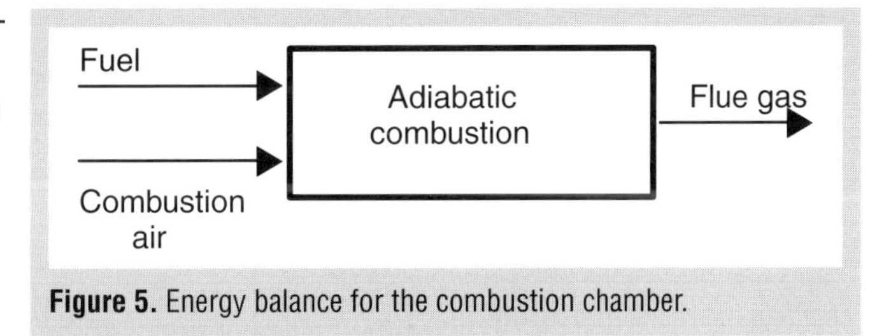

Figure 5. Energy balance for the combustion chamber.

According to the energy balance, a fuel amount, m_f, with the low heating value of LHV enters the combustion chamber. Combustion air amount, m_a, at temperature, t_a, also enters the combustion chamber. Flue gas amount, m_{fg}, leaves the combustion chamber at the adiabatic temperature of combustion. This gives the following equation for determining the adiabatic temperature of combustion:

$$m_f \cdot LHV + m_a \cdot c_{pa} \cdot t_a = m_{fg} \cdot c_{pfg} \cdot t_{fg} \tag{15}$$

where m_f is amount of fuel (kgf),
LHV low heating value (MJ/kgf)
m_a amount of combustion air (kga)

c_{pa}	average specific heat (kJ/kgK) of air between temperatures 0°C and t_a,
t_a	temperature of air (°C),
m_{fg}	amount of flue gas (kgfg),
c_{pfg}	average specific heat (kJ/kgK) of flue gases between temperatures 0°C and t_{fg},
t_{fg}	flue gas temperature or adiabatic temperature of combustion

When the equation is divided by fuel mass m_f it follows

$$LHV + c_{pa} \cdot t_a \cdot m_a / m_f = c_{pfg} \cdot t_{fg} \cdot m_{fg} / m_f \tag{16}$$

where $m_a/m_f = \mu_{a(real)}$ is actual air amount per kilogram fuel
$m_{fg}/m_f = \mu_{fg(real)}$ actual flue gas amount per kilogram fuel.

These give the following formula for calculating the adiabatic temperature of combustion:

$$t_{ad} = \frac{LHV + \mu_{a(real)} \cdot t_a \cdot c_{pa}}{\mu_{fg(real)} \cdot c_{pfg}} \tag{17}$$

The specific heat of flue gases is required to calculate the adiabatic temperature of combustion (see section 3.6.). Figure 6 shows the adiabatic temperature of combustion for a moist wood using the above method. The adiabatic temperature of combustion depends on the excess air ratio and the heating value of the fuel. Adiabatic temperatures of combustion are normally 1 500–2 000°C depending on the fuel and excess air ratio.

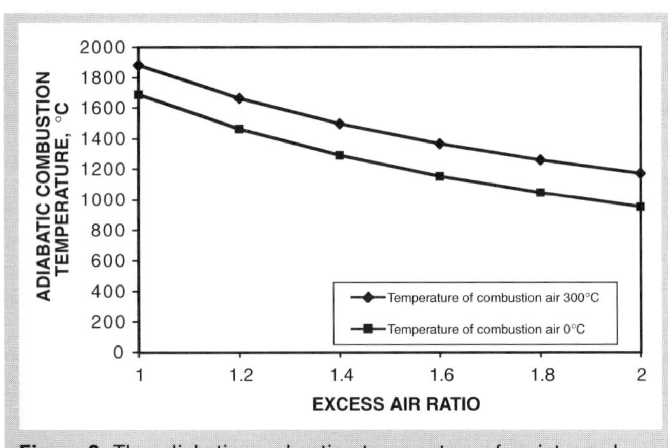

Figure 6. The adiabatic combustion temperature of moist wood.

3.5 Emissions from combustion

Because of the combustion reactions of the main components of fuel, flue gases contain water from the combustion of hydrogen, carbon dioxide from the combustion of carbon, and sulfur dioxide from sulfur combustion. Flue gases also contain nitrogen that originates from the fuel or combustion air, argon from the combustion air, oxygen from the excess air, and particulate (fly ash).

Many components in the flue gas are harmful for the environment and their content in the flue gas has to be reduced. These components are e.g. sulfur dioxide (SO_2), nitrogen oxides (NO_x), carbon monoxide (CO), carbon dioxide (CO_2), hydrocarbons (C_xH_y) and particulate. The only harmless components in the flue gas are water, oxygen, nitrogen and argon. Today most countries have stringent regulations for the emissions of power plants. The emissions can be reduced in the combustion process (fluidized bed combustion) and with additional flue gas cleaning equipment after boiler.

Carbon monoxide

In incomplete combustion with a shortage of oxygen, carbon reacts with oxygen as follows:

$$C + 1/2O_2 = CO \qquad (+110kJ/kmol) \tag{18}$$

When CO forms, only a quarter of the amount of heat releases compared to the reaction where CO_2 forms. So when CO is formed a part of heating value of fuel is not releasing. This is one loss that may decrease the efficiency of boiler. As Fig. 7 shows, the CO concentration will decrease with a sufficiently large excess air ratio. Excess air does increase the air flow through the boiler and so flue gas loss of the boiler is increasing. Figure 7 shows that minimum total loss occurs with a certain excess air ratio. Section 7 provides more information of boiler efficiency and losses.

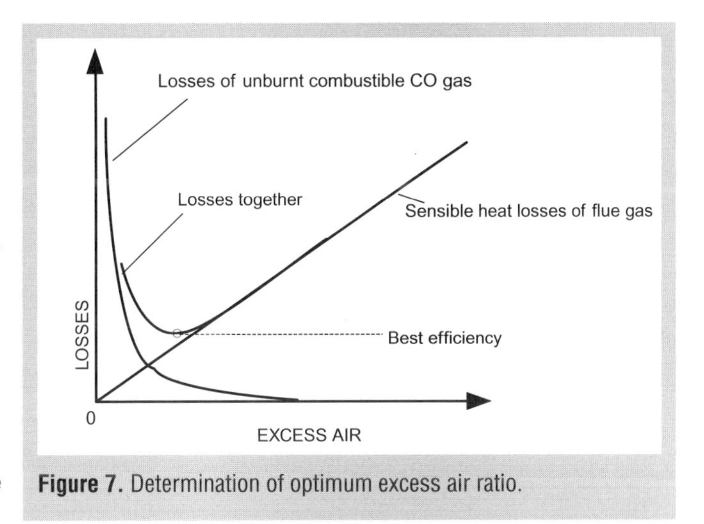

Figure 7. Determination of optimum excess air ratio.

Due to the imperfect mixing of fuel and combustion air, flue gases usually contain some carbon monoxide even when using excess air. The amount of carbon monoxide in flue gases of power plant boilers that function properly is usually 30–100 ppm. In the atmosphere, carbon monoxide oxidizes to carbon dioxide.

Hydrocarbons

Hydrocarbons (HC) in flue gases are fuel that has not combusted. HC emissions usually occur with CO emissions. Like CO emissions, HC emissions are the result of imperfect mixing of fuel and combustion air. Low temperature decreases burning velocity in the furnace and contributes to the formation of hydrocarbons. HC emissions from power plant boilers and from all oil and gas fired boilers are usually very small. They are often below the measurable range. The greatest emissions come from small domestic boilers that use wood as fuel.

Particulate emissions

The ash with the unburned solid fuel (carbon) leaving the furnace with flue gas forms particulate emissions. By analysing the amount of carbon in the fly ash it can be determined the effect that unburned solids have on the boiler efficiency (see section 7). High carbon concentration indicates the need for adjustment or maintenance of burners or fuel grinders. The particulate emissions can be cleaned efficiently with electrostatic precipitator or fabric filters .

Sulfur dioxide

During combustion sulfur (S) contained in fuel oxidizes to sulfur dioxide (SO_2). A level of 1/40–1/80 of sulfur dioxide further oxidizes to sulfur trioxide (SO_3) that reacts with water (H_2O) to form sulfuric acid (H_2SO_4). The best way to prevent corrosion of the heat exchanger of the boiler by sulfuric acid is to keep the flue gas temperature in the cold parts of the boiler higher than the acid dew point of the flue gas.

Even if protecting the heat exchanger is possible, sulfuric acid emissions are harmful to the environment. Sulfur fallout causes acidification of soil and water.

The largest sulfur dioxide emissions are due to industrial boilers and power plants. Authorities therefore control these sulfur dioxide emissions. Coals and fuel oils contain large amounts of sulfur. Usually new coal-fired power plants must have combustion technic (fluidized bed combustion) or flue gas cleaners with which sulfur dioxide emissions can be reduced effectively.

Nitrogen oxides

When nitrogen reacts with oxygen in air, nitrogen monoxide (NO) and nitrogen dioxide (NO_2) form:

$$N_2 + O_2 = 2NO \qquad NO + 1/2O_2 = NO_2 \tag{19}$$

In the furnace, primarily NO forms. In the atmosphere, NO later oxidizes to NO_2. NO and NO_2 emissions together are called NO_x emissions. Like sulfur dioxide, NO_x gases also acidify the environment. Restrictions covering their emission therefore exist. Power plants and industrial boiler plants are not the biggest producers of NO_x emissions. The emissions per consumed fuel energy (mg/MJ) from combustion engines are far greater. That is why traffic (including cars, ships etc.) is the greatest producer of NO_x emissions.

Nitrogen that reacts in the furnace with oxygen in air comes from the fuel or combustion air. During combustion, nitrogen oxides primarily form in three ways:

- At high temperatures, nitrogen in combustion air reacts with oxygen (thermal formation of nitrogen oxide)
- Nitrogen in the combustion air first reacts with hydrocarbon radicals in the part of the flame rich with hydrocarbons and then formed substancies (HCN, NH, N) may react to form NO (prompt formation of nitrogen oxide)
- Nitrogen in the fuel reacts with oxygen (formation of fuel based nitrogen oxide).

Figure 8 illustrates how amounts of NO_x emissions from different sources change with temperature. NO_x emissions typically increase as the flame temperature in the furnace increases because the thermal NO_x formation is highly temperature sensitive.

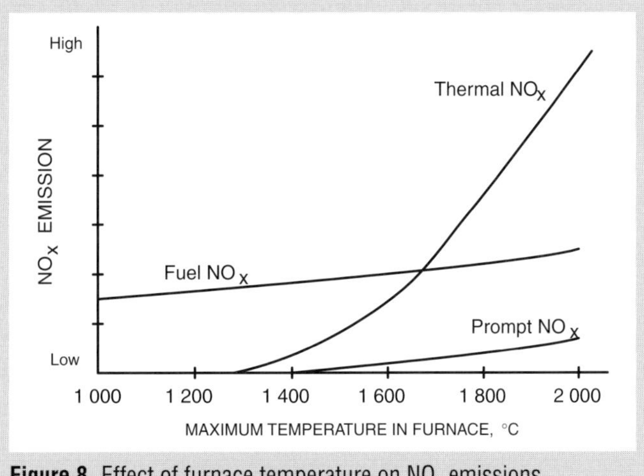

Figure 8. Effect of furnace temperature on NO_x emissions.

Eliminating NO_x gases from flue gases with various cleaning methods or reducing the formation of NO_x in the furnace can decrease emissions of nitrogen oxide. Lowering the temperature of the flame can reduce the formation of thermal NO_x. Reducing the oxygen concentration in the flame can reduce the formation of NO_x from fuel.

Carbon dioxide

Carbon dioxide is an important greenhouse gas. The others are nitrous oxides (N_2O), methane, and water vapor. Greenhouse gases allow the heat energy transferred by the radiation of the sun to reach the earth but reflect the heat radiation from earth back to earth. The temperature of the globe surface therefore depends considerably on the amount of greenhouse gases in the atmosphere. Because of increasing fossil fuel use, carbon dioxide concentration in the atmosphere has risen 20% in the previous 100 years. It is presently approximately 350 ppm. This has resulted in restriction of carbon dioxide emissions.

All commercially used fuels contain carbon that forms carbon dioxide upon combustion. Carbon concentrations vary for different fuels. The most favorable fuel is natural gas because it gives 35% less carbon dioxide than coal that has the same fuel energy. One of the best ways to reduce carbon dioxide emissions is however energy efficient energy production and energy efficient use of produced energy.

3.6 Heat transfer properties of flue gases

The most important heat transfer properties of flue gases are the following:

- Specific heat value
- Thermal conductivity
- Viscosity
- Density.

Values of these parameters are necessary for various calculations concerning heat and flow. Flue gases contain several mixed gases that have different heat transfer properties. Values for these properties are available in heat transfer literature. Using these allows determination of the properties of the flue gas in question. In Section 6.8 the basic heat transfer calculation methods are described.

4 Combustion technologies of bark

4.1 Grate firing

In grate or stoker fired boilers, the combustion of solid fuel occurs in a bed at the bottom of the furnace. Grate firing is the oldest method for combustion of solid fuels. Numerous different applications of grate or stoker firing systems exist for burning of different solid fuels. In all cases, the fuel burns in some form of grate through which some or all the air for combustion passes. The grate surface can be stationary or moving.

Grate firing was the only combustion technique for burning bark and wood waste in the pulp and paper industry until the 1970s. Since then, fluidized bed technology has replaced grate firing as the combustion technique for power boilers. The fluidized bed boilers provide better fuel efficiency, lower emission, less maintenance, and better fuel flexibility. The major disadvantage with the grate boilers in the pulp and paper industry is that they cannot burn the waste sludge from the mill in large quantities.

Grate constructions

The grate construction can be stationary or moving. Stationary grates are more common in small boilers. Larger boilers such as the power boilers in the pulp and paper industry have mechanical or moving grates with automatic fuel feed and ash removal. The other choice for classifying grates is the method of cooling the grate. Combustion air can cool the grate. Integration into the water circulation of the boiler can also cool the grate. A third choice for classifying grate boilers is the method of feeding fuel to the grate. The three general classes in use are under feed stokers, over feed stokers, and spreader stokers. Numerous combinations of all the above constructions can exist depending on the fuel used, boiler size, boiler manufacturer, etc. This text next describes only the most common designs used in burning bark and wood wastes in the pulp and paper industry:

- Stationary inclined or step grate
- Traveling grate
- Mechanical inclined grate (Kablitz grate).

Stationary inclined or step grate

When the grate has a 30°–50° inclination from horizontal, the fuel will move on the grate by gravity. The inclination of the grate depends on the fuel and its ability to flow. The inclination can change at different locations of the grate. It is typically higher at the upper end of the grate. To complete the burning of fuel, many inclined grates have a small horizontal grate after the inclined section. The grate is constructed of water cooled membrane tubes that are integrated in the evaporator circuit of the boiler.

In a step grate, the step construction is made of cast iron grate bars. Air is horizontally introduced between the grate plates.

Traveling grate

A traveling grate or chain grate has small (10 cm x 40 cm) grate elements joined from the shorter sides in chains that move between two sprocket wheels, of which one is motorized. Larger boilers have several parallel chains with parallel grate elements between them. The design of the traveling grate is such that when the grate elements move on the upper side from the idle wheel to the motorized wheel, they are in a horizontal position forming a slowly moving level for feeding and burning fuel.

Figure 11 shows feeding fuel on the grate from the other end of the grate. A vertically moving gate situated in the feeding side of the grate can alter the thickness of the fuel layer. Changing the thickness of the fuel layer and the speed of the grate regulates the burning effect of the boiler. For coal, a suitable thickness is 10–20 cm, and for wood it is 30–90 cm.

The traveling grate can have a spreader stoker (Fig. 9) that throws fuel over the grate against its moving direction. Heavy particles fly on the grate, but light particles burn while flying through the furnace. Spreader stokers make possible adding the fuel effect of a traveling grate, but the amount of unburned fuel in the fly ash increases. Feeding fly ash back from the fly ash separator to the furnace is then useful to minimize the losses of unburned fuel.

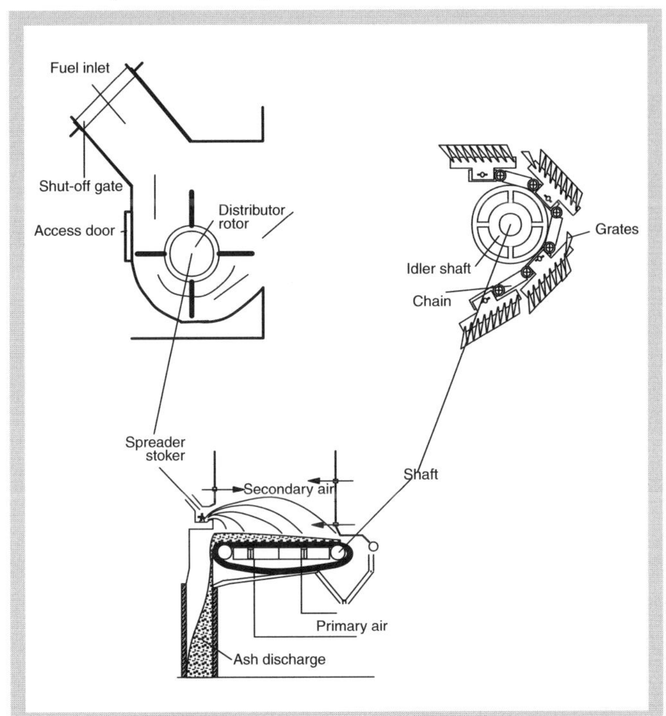

Figure 9. A traveling grate with spreader stoker and front ash discharge.

Mechanical inclined grate

On the mechanical inclined grate (Fig. 10), fuel moves not only by gravity (as with the stationary inclined grate) but also by horizontal back and forward moving grate elements. A mechanical inclined grate therefore does not have as deep an inclining angle as the stationary grate. A suitable angle is 15°. Regulating the moving speed of fuel on the grate is possible by changing the speed of the grate. The speed can be different at different sections of the grate.

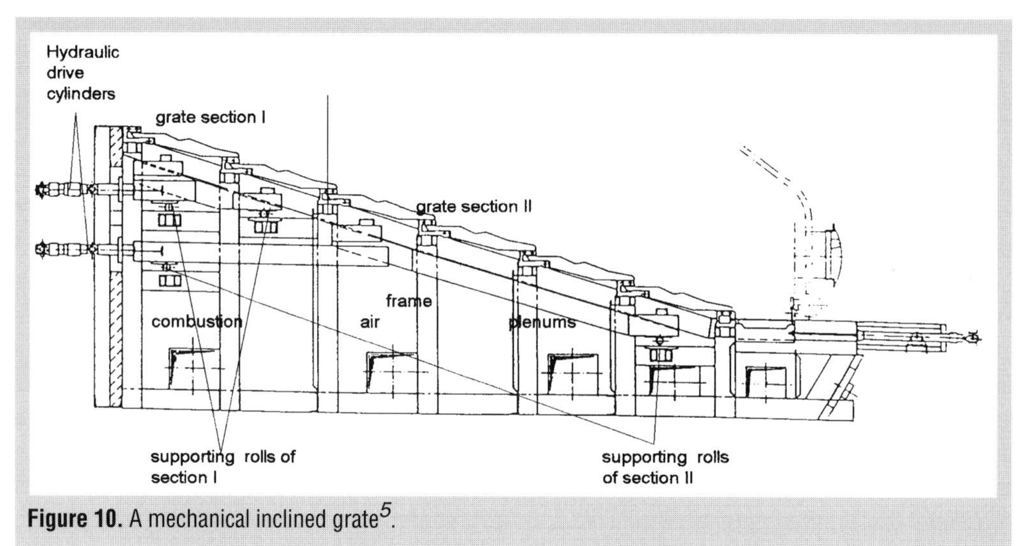

Figure 10. A mechanical inclined grate[5].

Combustion of bark and wood on a grate

The burning of bark on a grate follows the same principles as any combustion technique. Burning has the following phases:

- Drying of the fuel
- Pyrolysis and combustion of volatile matter
- Combustion of char.

All these phases occur in parallel for a single fuel particle, although fuel particles are simultaneously at different burning phases on the grid. Large fuel particles can have fresh fuel in the core while the char on the surface is burning. Figure 11 shows the phases of burning on a traveling grate.

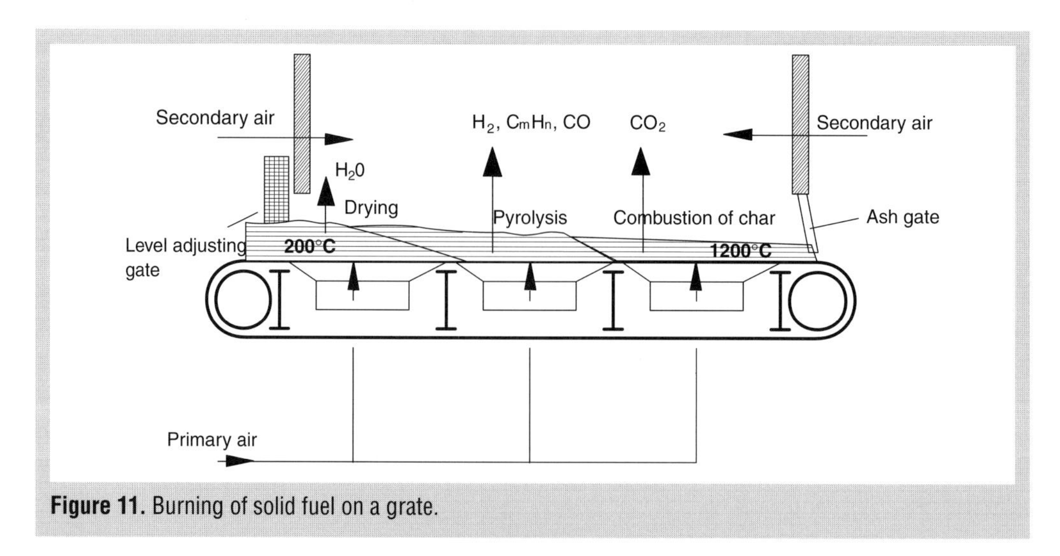

Figure 11. Burning of solid fuel on a grate.

The achieved fuel effect per grate area is typically the following:

- Coal (traveling grate) 1.6 MW/m^2
- Bark and wood (moisture content 60%) 0.4 MW/ m^2
- Bark and wood (moisture content 30%) 0.8 MW/m^2.

The maximum fuel effect also depends on emission limits. Particulate emissions increase as the burning effect increases. This is because air velocity through the grate is higher and fuel particles entrain more unburned in flue gases from the furnace.

4.2 Introduction to fluidized bed combustion

Background

For decades, fluidized bed reactors have been used in noncombustion reactions in which the thorough mixing and intimate contact of the reactants gave high product yields with improved economy. The applications of fluidized bed technology for combustion of solid fuels were commercially available in the 1970s. Since then, fluidized bed combustion has become a widely accepted combustion technology. Fluidized beds are very suitable for combustion of low grade fuels with high moisture or ash content. These are normally difficult to burn with other combustion methods. The benefits of fluidized bed combustion are the possibility to use different fuels simultaneously, simple and cheap sulfur removal by injecting limestone into the furnace, high combustion efficiency, and low emissions of NO_x.

Two types of fluidized bed systems are available today for combustion of bark and wood wastes: the bubbling fluidized bed boiler (BFB) and the circulating fluidized bed boiler (CFB). The question of which system is best for a given application is often difficult to answer. In certain cases such as the need for low SO_2 and NO_x emission limits or the requirement to burn coal in the boiler, CFB may be the only choice. More commercial experience is available with CFB units on a wide range of boiler capacities, steam cycles, and fuels. For most applications, BFB and CFB are both technically feasible. Defining the lowest cost option requires estimating capital and operating costs. The comparison will vary depending on unit capacity, steam cycle, fuel, sorbent space requirement, and emission limits.

General description of fluidized bed combustion

Fluidization refers to a condition in which fine solids transform into a fluid-like state through contact with gas or liquid. In a fluidized bed boiler, fluidizing results by blowing air through the solid bed material lying on an air distributor grid.

Figure 12 shows the operating principle of a fluidized bed. It illustrates the behavior of a bed of sand particles as a gas flows through the bed at different velocities, and the gas pressure drop through the bed as a function of gas velocity. For a fixed bed, the gas pressure drop is proportional to the square of velocity. As velocity increases, the bed becomes fluidized. The velocity at which this transition occurs is the minimum fluidizing velocity, U_{mf}. The minimum fluidizing velocity depends on many factors including particle diameter, gas and particle density, particle shape, gas viscosity, and bed void fraction. The following formula calculates the minimum fluidizing velocity[6/]:

$$U_{mf} = \frac{\mu_g}{d_p \cdot \rho_g} \cdot \left[\sqrt{33.7^2 + 0.0408 \cdot \frac{d_p^3 \cdot \rho_g \cdot (\rho_p - \rho_g) \cdot g}{\mu_g^2}} - 33.7 \right] \quad (20)$$

where μ_g is dynamic viscosity
d_p particle diameter
ρ_g gas density
ρ_p particle density
g acceleration of gravity.

At velocities above U_{mf}, the pressure drop through the bed remains constant and equals the weight of solids per unit area as the drag forces on the particles barely overcome gravitational forces. The following equation shows the pressure drop:

$$\Delta p = (\rho_p - \rho_g) \cdot (1 - \varepsilon) \cdot g \cdot H \quad (21)$$

where ρ_g is gas density
ρ_p particle density
ε ratio of empty volume in bed
g acceleration of gravity.

Figure 12 shows the gas pressure drop through a fluidized bed and the state of fluidization vs. gas velocity.

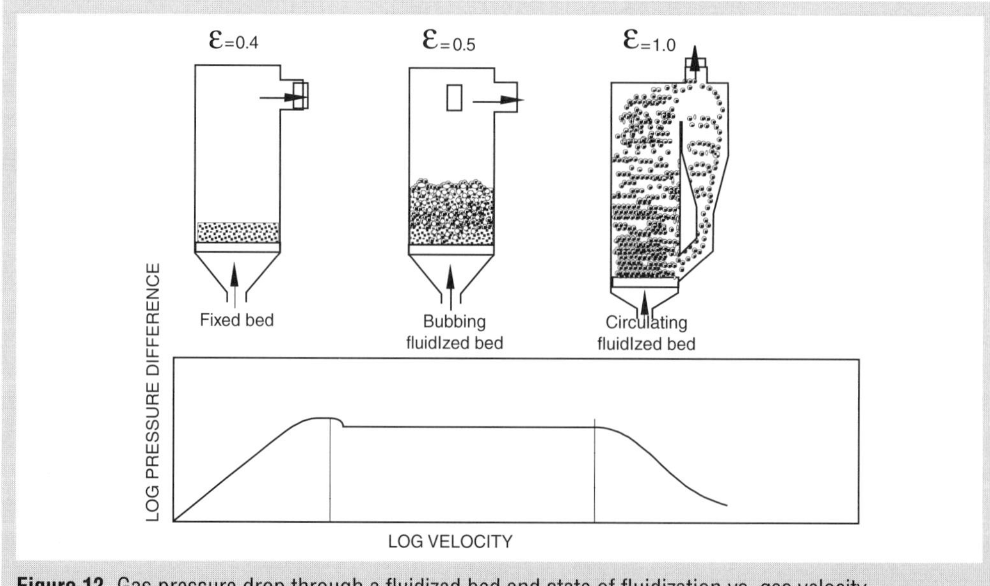

Figure 12. Gas pressure drop through a fluidized bed and state of fluidization vs. gas velocity.

When the fluidizing velocity becomes higher than the minimum fluidizing velocity, the gas in excess of that required to fluidize the bed passes through the bed as bubbles. This system is a bubbling bed, and boilers that use this system are bubbling fluidized bed boilers (BFB). The BFB has a modest solids mixing rate and low solids entrainment rate into the flue gas. The bubbling bed has a clear, visible surface level where the bed ends and freeboard begins.

As the fluidizing velocity increase the fluidized bed surface becomes more diffuse and solids entrainment increases. A defined bed surface no longer exists, and recycle of entrained material to the bed is necessary to maintain bed inventory. A fluidized bed with these characteristics is called a circulating fluidized bed, because of the high rate of bed material circulating from the furnace to the particle separator and back to the furnace. Boilers that use this system are circulating fluidized bed boilers (CFB).

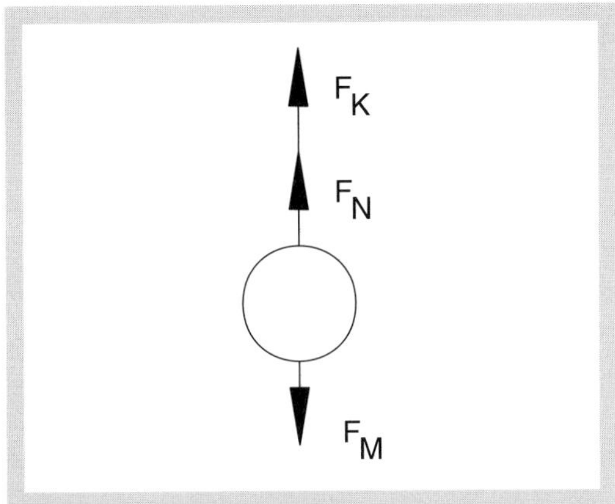

Figure 13. The force balance of a particle in a fluidized bed.

Gravity, buoyant force, and a drag force by the fluidizing medium influence a solid particle in a fluidized bed (Fig.13). If air is the fluidizing medium, the effect of the buoyant force is negligible. Forces that influence the particles can be calculated using the following formulas:

Drag force:

$$F_k = C_d \cdot \frac{\pi(d_p)^2}{4} \cdot \frac{1}{2}\rho_g \cdot U^2 \tag{22}$$

where C_d is drag coefficient
d_p particle diameter
U fluidizing velocity

Buoyant force:

$$F_N = \frac{\pi(d_p)^3}{6} \cdot \rho_g \cdot g \tag{23}$$

Gravity force:

$$F_M = \frac{\pi(d_p)^3}{6} \cdot \rho_p \cdot g \tag{24}$$

In a fluidized system, the force balance is follows:

$$F_K = F_M - F_N \tag{25}$$

A force ratio, n, is defined as:

$$n = \frac{F_K}{F_M - F_N} \tag{26}$$

$n = 1$ in a fludized bed system, $n > 1$ in a pneumatic transport system, and $n < 1$ in a fixed bed system. In the situation where $n = 1$, the terminal velocity, U_t, is:

$$U_t = \sqrt{\frac{4}{3} \cdot \frac{d_p \cdot (\rho_p - \rho_g) \cdot g}{\rho_g \cdot C_d}} \tag{27}$$

As the fluidizing velocity becomes higher than terminal velocity, U_t, the solids start to separate from the bed.

Heat transfer in fluidized bed boilers

An advantage of the fluidized bed technique is the good heat transfer between the gas and heat transfer surface. Figure 14 illustrates the phenomenon. The heat transfer between the bed and heat transfer surface rises rapidly when the velocity of gas exceeds the minimum fluidizing velocity. The explanation for this is that the hot bed material comes continuously in contact with the heat exchanger due to the mixing when the bed starts to fluidize. Hot bed material releases heat to the surface and immediately after that new hot particles come into the contact with the surface due to the mixing of the bed.

Several correlations are available for calculation of the maximum achievable heat transfer coefficient. An early one is the following Zabrodskies correlation[7]:

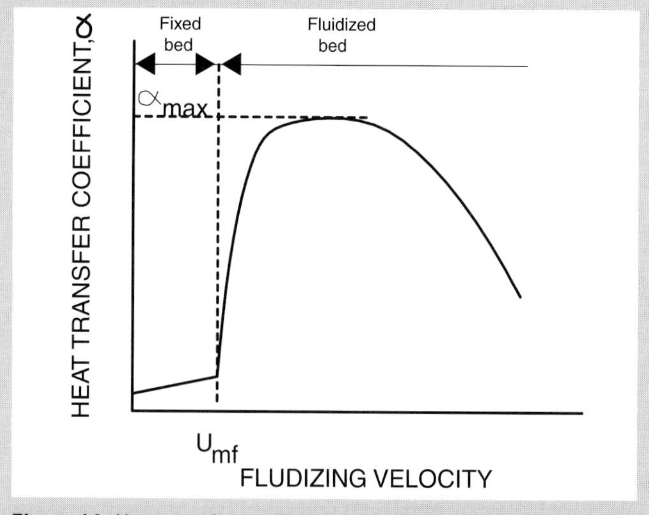

Figure 14. Heat transfer coefficient between the fluidized bed and the heat transfer surface.

$$\alpha_{max} = 35.7 \cdot \lambda_g^{0.6} \cdot d_p^{-0.36} \cdot \rho_p^{0.2} \tag{28}$$

where λ_g is thermal conductivity of gas.

4.3 Bubbling fluidized bed (BFB) combustion

Figure 15 shows a bubbling fluidized bed boiler, and Table 11 presents typical operating parameters for a BFB boiler. The height of the fluidized bed is 0.4–0.8 m resulting in a 6–12 kPa pressure drop over the bed. The bed material is sand and fuel ash. If sulfur capture is necessary, limestone is also fed into the furnace, and the reaction products (limestone) will form part of the bed material. The optimum bed material particle size is 1–3 mm when the fluidizing velocity is 0.7–2 m/s.

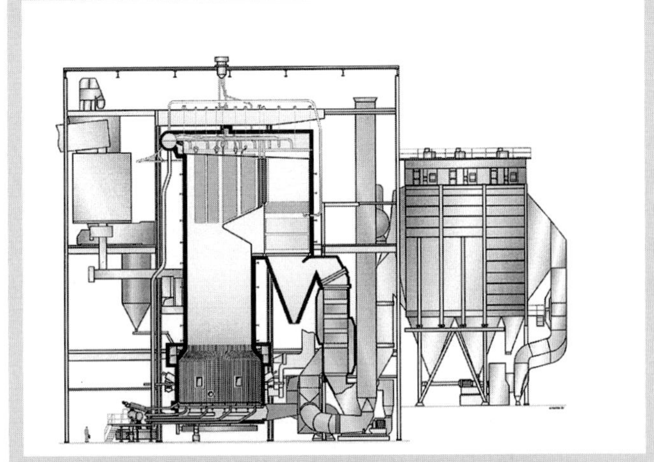

Figure 15. A bubbling fluidized bed (BFB) boiler[8]. (133 MWth, 40 kg/s, 112 bar, 525°C)

Crushed fuel is fed to the top of the bed. Fluidizing air is supplied to the bottom of the bed through a plenum and an air distributor. The combustion of heavy fuel particles occurs in the bed. Light fuel particles and volatile matter burn above the bed in the freeboard. The bed temperature is 700°C–1 000°C depending on the fuel quality and load.

Table 11. Typical operating parameters for a BFB.

Volume heat load	0.1–0.5 MW/m^3
Cross section heat load	0.7–3 MW/m^2
Pressure drop over the bed	6.0–12 kPa
Fluidizing velocity	1–3 m/s
Height of the bed	0.4–0.8 m
Temperature of primary air	20–400°C
Temperature of secondary air	20–400°C
Bed temperature	700–1 000°C
Freeboard temperature	700–1 000°C
Excess air ratio	1.1–1.4
Density of bed	1 000–1 500 kg/m^3

The high heat capacity of the bed material makes the BFB suitable for burning wet fuels such as bark and wood wastes. No external drying other than mechanical pressing of bark is necessary. The fuel dries and ignites rapidly in the bed due to the efficient mixing of the hot bed material. The high heat capacity of the bed material also helps even the effects of the changes in fuel properties.

Figure 15 shows a typical fuel feeding system for a BFB having a day bin or silo, a discharge device in the day bin, and fuel conveyor and feed chutes to the furnace. Air or screw feeders assist the fuel transport in the feed chutes depending on the type of fuel. A rotary feeder is usually between the feed chute and the fuel conveyor to seal the fuel transport system from the hot flue gases in the furnace. Several feed chutes are necessary to distribute the fuel evenly across the full bed area. The fuel system typically causes most operational disturbances in bark boilers. A redundancy in the fuel feed system is often necessary to ensure good boiler availability.

During the boiler startup, the fluidized bed first heats to 400°C–600°C with oil- or gas-fired startup burners to ensure the safe ignition of solid fuel. The startup burner can be above the bed, inside the bed, or in the primary air plenum or primary air duct. The fastest startup time and lowest oil or gas consumption result from heating the fluidizing or primary air in the air plenum or duct. Above-bed burners are often necessary to provide load carrying capability for situations when the solid fuel is unavailable for some reason.

The furnace enclosure is often constructed of welded water-cooled tubing (membrane wall) and does evaporative duty. The tubes in the lower furnace are covered with refractory lining up to approximately 2 m from the grid level. The refractory is necessary to protect the tubes from erosion and assist in maintaining sufficient bed temperature when burning fuels with high moisture content.

The fluidizing air is distributed evenly at the gross section of the bed with air distributor or grid that forms the furnace floor. The grid can be a plate construction or made of the tubing and integrated with the evaporator circuit similar to the furnace walls. The entire floor area has many closely pitched air nozzles. The pressure drop over the fluidizing nozzles must be at least 30%–50% of the bed pressure drop to ensure adequate fluidizing air distribution and avoid back shifting of bed material into the air plenum during low load operation.

The coarse fraction of the fuel ash, stones, and iron will accumulate in the bed material and require removal from the bed. Alkalis such as potassium and sodium in the fuel will also accumulate in the bed material, decrease the melting point of the bed material, and finally fuse or sinter the entire bed unless the bed material changes continuously. The spent bed material or bottom ash is typically removed through drain pipes that are located evenly in the grid. The bottom ash falls to water-cooled screws that cool the ash to an acceptable temperature. The drain pipes may have classification to return the finer fraction of the bottom ash to the furnace. The bottom ash is sometimes classified after the cooling, and the usable bed material is returned to the furnace.

The fine fraction of the ash and bed material entrains from the bed as fly ash. Since the ash content of bark and wood is low, some makeup material such as sand must be added to the bed to maintain sufficient bed inventory.

As mentioned earlier, the alkalis in the ash accumulate in the bed material. In addition to continuous change of bed material to flush the alkalis from the bed material, a good control of bed temperature is essential to avoid sintering of the bed. In bark and wood firing, the bed temperature should be below 900°C to avoid sintering. If the bed sinters, the boiler must be shut down to remove the sintered bed material manually. In most cases, controlling the ratio of fluidizing air vs. total air flow will give sufficient bed temperature control. If large variations in fuel moisture might occur or fuels with different moisture content or heat value are burned, a flue gas reinjection system will control the bed temperature.

Some oxygen for combustion comes through the grid with the fluidizing or primary air. The remaining combustion air enters through the secondary air ports located above the bed level. The amount of secondary air can be 30%–70% of total air flow depending on the fuel burned and boiler load. The secondary air ports can be installed at different levels to decrease NOx emissions by staged combustion. Centrifugal fans are used to supply primary and secondary air due to the high air pressures.

The turndown of a BFB is typically 30%–40% of the maximum continuous rating (MCR) load with solid fuels. Lower loads require use of support firing by oil or gas. The decrease in bed temperature primarily limits the minimum load. This temperature should be above 700°C. Fuel moisture therefore strongly influences the minimum load achievable without support fuel. Above MCR conditions, the maximum bed temperature, entrainment of bed material with flue gas and increased emissions, and unburned matter limit the load. The BFB is most suitable for combustion of fuels with high volatile content such as bark and wood wastes.

4.4 Circulating fluidized bed (CFB) combustion

The main feature of the CFB combustion process is that the fluidization velocity is increased past the bubbling regime and a large fraction of the bed mass becomes entrained with the gas stream. The entrained material requires collection in a solids separator for return to the lower furnace to maintain bed inventory. Due to the entrainment of particles, a clear distinction does not exist between the bed and the freeboard as in a BFB process. The pressure drop from the bottom to the top of the furnace follows a declining gradient as Fig. 16 shows.

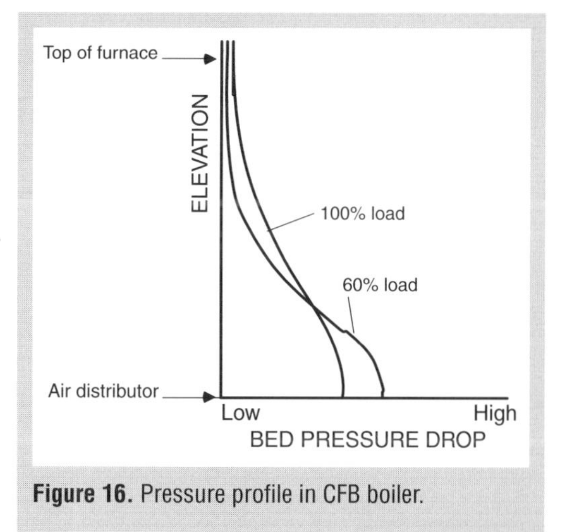

Figure 16. Pressure profile in CFB boiler.

When the gas velocity becomes higher than the entrainment velocity, the solids density in the upper part of the furnace increases. Some entrained solids form locally higher concentrations or clusters that decompose, reform, and move up and down within the furnace. The formation of clusters causes a large internal recycling of solids in the furnace. With the external recycle from the solids separator, the CFB process leads to excellent mixing and gas solids contact. Because of the high solids loading in the entire furnace volume, the gas and solids temperatures are essentially uniform throughout the furnace and solids separator. The good mixing and uniform temperature in the CFB furnace result in high combustion efficiency, effective use of limestone if sulfur capture is required, and low emissions of NO_x. The high solids loading and heat capacity ensures stable combustion of moist fuels like bark, wood, and sludge. In addition, the CFB process design can easily allow multi fuel burning capability. Table 12 shows typical operating parameters for CFB boilers.

Table 12. Typical operating parameters for CFB boilers.

Volume heat load	0.1–0.3 MW/m^3
Cross section heat load	0.7–5 MW/m^2
Total pressure drop	10–15 kPa
Bed material particle size	0.1–0.5 mm
Fly ash particle size	< 100 μm
Bottom ash particle size	0.5-10 mm (individual particles up to 100 mm)
Fluidizing velocity	3–10 m/s
Temperature of primary air	20–400°C
Temperature of secondary air	20–400°C
Bed temperature	850–950°C
Temperature after the cyclone	850–950°C
Excess air ratio	1.1–1.3
Density of bed	10–100 kg/m^3
Recircle ratio	10–100

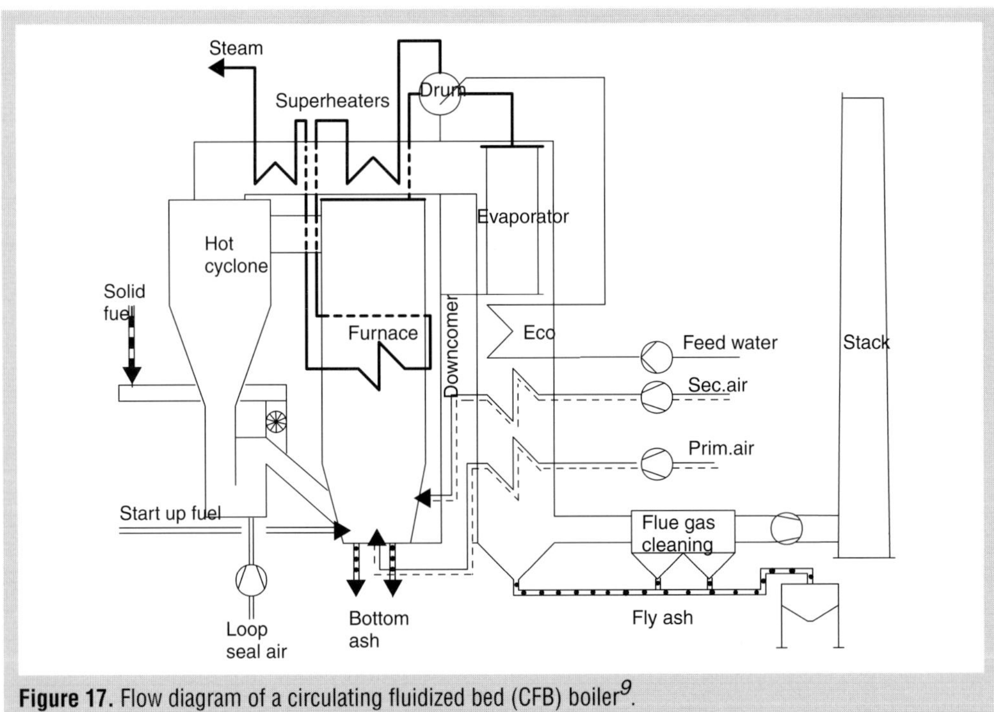

Figure 17. Flow diagram of a circulating fluidized bed (CFB) boiler[9].

Figure 17 shows the flow diagram of a circulating fluidized bed boiler. The main components of a CFB-boiler are the furnace and the solids separator that is typically a cyclone unit. The solids separator captures entrained bed material, sorbent, and unburned fuel particles and returns them to the lower part of the furnace through a non-mechanical valve or loop seal. After the cyclone, the flue gases flow through convective heat exchangers or superheaters, feed water, and air preheaters. Each boiler has an individual design for the fuel being burned. The order and placement of heat exchanger surfaces may therefore vary in different boilers. For example, a part of the evaporator is in the convective section after the solids separator in a boiler designed for low steam pressure and moist fuel such as bark. For high steam pressure boilers, some superheater surfaces are usually located in the furnace.

The CFB furnace walls (membrane walls/pannels) are made of together welded tubes (see Fig. 48). Furnace wall tubes are water-cooled which means that inside the tubes water is evaporating. The tubes in the lower furnace have a refractory lining cover up to at least 2–3 m height from the grid. The refractory is necessary to protect the tubes from erosion and to assist in maintaining sufficient bed temperature when burning fuels with high moisture content. The dimensions of a CFB furnace for a given fuel and steam output depend on the following:

- Gas velocity
- Minimum burning time of fuel
- Heat transfer in furnace.

The fuel feeding in a CFB furnace has a similar arrangement to the BFB furnace explained earlier (Sec. 4.3). Feeding the fuel into the solids return in the loop seal can give good mixing and uniform distribution of fuel in a CFB boiler. Limestone for sulfur capture is fed pneumatically into the lower furnace through several feed points.

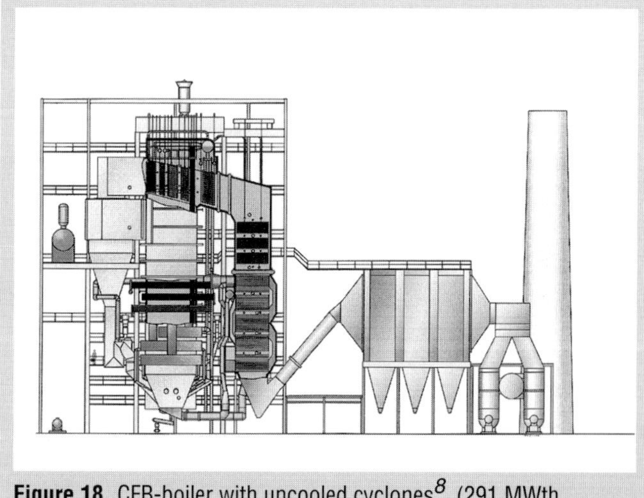

Figure 18. CFB-boiler with uncooled cyclones[8]. (291 MWth, 103/89 kg/s, 156/20 bar, 540/540°C)

A CFB boiler typically has one or more high temperature solids separators. The maximum practical physical size of the separator, even distribution of returning solids into the furnace, and furnace layout determine the number of solids separators. A cyclone is the most common design for a solids separator in CFB boilers. Figure 18 shows a CFB boiler equipped with hot cyclones. The cyclones are typically constructed of steel plate with a multiple layer of refractory lining to provide erosion protection and thermal insulations. Alternate cyclone constructions made of membrane tube walls with water or steam cooling are also available. The cooled cyclone has higher capital cost, but it will provide lighter construction and less refractory maintenance due to the thinner refractory layer.

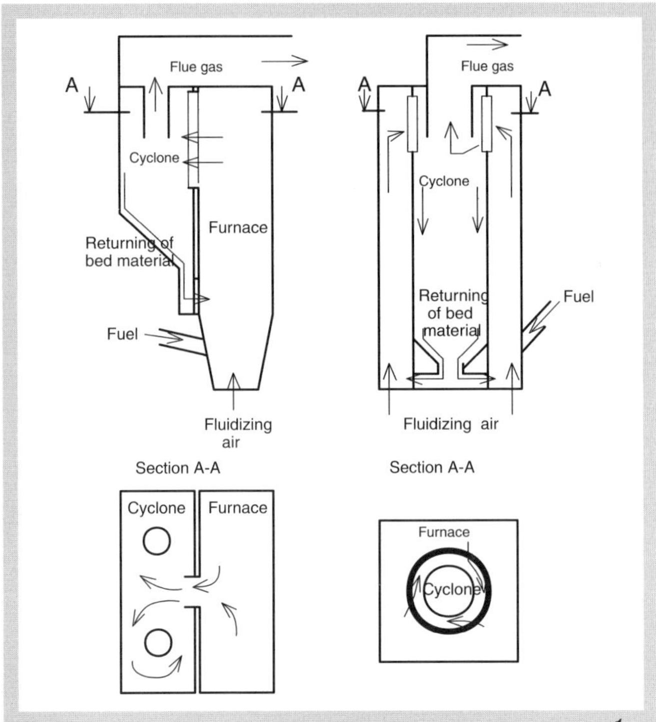

Figure 19. New types of CFB boilers with water cooled cyclones[1].

Alternate solids separator designs are also commercially available. The goal in developing these designs has been to replace the cyclone with a solids separator design that is more compact and lighter, has less refractory, and is easier to manufacture and erect. Figure 19 shows some new CFB-boiler designs with water-cooled solids separator. The drawing on the left shows a CFB-boiler with a compact separator. The compact separator has a square or rectangular cross section, and the separator walls are made of membrane wall similar to the walls in the furnace. The drawing on the right illustrates a CFB-boiler with a multi inlet cyclone located inside the furnace. Both boilers are faster to regulate and have shorter startup time compared with CFB boilers with uncooled cyclones because of thinner refractory lining.

The solids collected in the solids separator must be moved back into the furnace against furnace back pressure. Besides the back flow of gases through the solids return must be prevented. A nonmechanical valve does this. It is commonly a loop seal in which the sealing and solids movement is accomplished by the fluidization of the solids in the loop seal. Several alternate designs for a loop seal exist. All typically use fluidization of the returning solids to provide the sealing against the furnace back pressure. The loop seal provides an ideal feeding point for fuel and sorbent by assisting their mixing and distribution into the furnace.

Proper operation of the CFB process requires that the particle size of the bed material and the circulated solids are within proper size distribution. The large ash and other inert particles therefore require continuous removal from the CFB furnace. Continuous bottom ash removal is also important if the fuel contains alkalis such as potassium and sodium that can accumulate in the bed material and lower the melting point of bed material. Figure 20 shows a bottom ash classifier in which the bottom ash removed from the furnace is classified with air or flue gas. The fine particles return to the furnace, and the coarse material is removed through the drain pipe to a conveyor in which additional cooling can occur. The bottom ash can also be removed through drain pipes, that penetrate the grid floor. The drained bottom ash is then cooled in a water cooled screw conveyor.

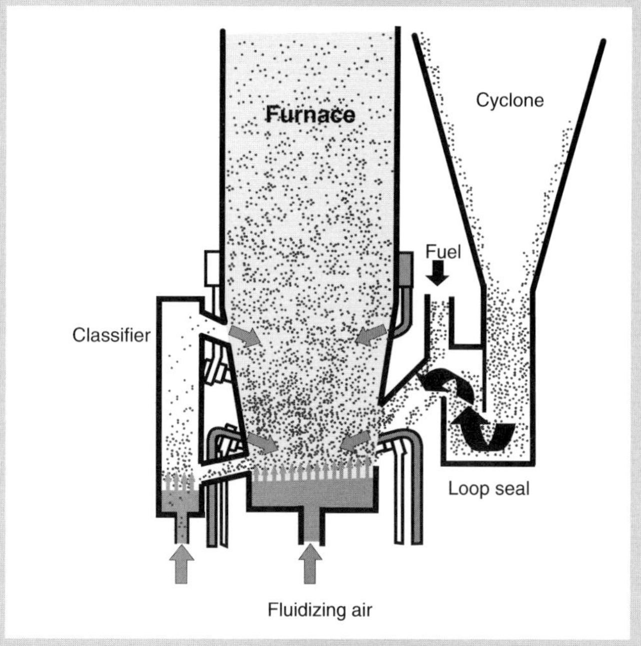

Figure 20. Return of bed material from the cyclone via a loop seal, fuel feeding, and bottom ash classification in a CFB boiler[8].

The combustion air system in CFB boilers is similar to that of BFB boilers using primary and secondary air. Primary air is blown through the grid nozzles located in the bottom of the furnace. The construction and design requirements of the grid are similar to those for a BFB boiler. The pressure of primary air is 15–20 kPa, and the amount is 30%–60% of maximum total combustion air (see Fig. 29). Secondary air is supplied through several secondary air ports in one or two levels located 2–5 m from the bottom. The amount of secondary air can be regulated from 10% to 50% of maximum total combustion air (see Fig. 29). A certain minimum pressure drop over the grid nozzles is necessary to avoid uneven fluidization, improper temperature distribution in the bed, and back shifting of bed material into the wind box. The minimum acceptable grid pressure drop determines the minimum primary air flow. This is typically about 50% of the total primary air flow rate. At low loads, the excess air starts to increase as the minimum primary air is reached and the furnace temperature drops. The solid fuel burning temperature allowable sets the minimum load without support oil or gas firing. Depending on the fuels used and the grid design, the turndown of a CFB boiler is 25%–40% of full load.

4.5 Environmental aspects of fluidized bed combustion

Sulfur capture

In the fluidized bed process and especially in the CFB process, the sulfur that releases in the combustion of fuels can be easily captured with low capital costs by simply adding limestone in the furnace. The capture of sulfur in fluidized bed combustion uses the chemical reaction between calcium oxide (CaO) and sulfur dioxide (SO_2). Calcium is introduced into the fluidized bed as limestone $(CaCO_3)$. Once introduced into the bed, limestone will change into calcium oxide and carbon dioxide in the high temperature of the furnace according to the following reaction:

$$CaCO_3 \rightarrow CaO + CO_2 \quad (29)$$

CaO reacts with SO_2 and oxygen and form calcium sulfate:

$$CaO + SO_2 + 1/2\ O_2 \rightarrow CaSO_4 \quad (30)$$

Due to incomplete mixing and entrainment of the sorbent, not all CaO react with SO_2. Excess limestone must be fed into the bed to achieve the desired reduction rate of SO_2 emission. The excess amount is expressed as Ca/S molar ratio, which represents the actual calcium input compared to theoretical. The Ca/S ratio depends on the limestone and fuel properties, combustion temperature, and the sulfur removal rate. The optimum temperature for sulfur capture in a fluidized bed is 750°C–900°C. Figure 21 shows the effectiveness of sulfur capture in BFB and CFB processes.

The proper size of lime particles is also important to achieve proper SO_2 reduction rate. Small particles have a larger surface area for reaction, but they entrain from the bed before they react. A smaller Ca/S-ratio occurs in CFB boilers due to recycle of fine particles and the efficient use of the whole furnace volume for the sulfur capture reactions.

NO_x formation and capture

Due to the low combustion temperature in fluidized bed combustion, "thermal" NO_x formation by oxidation of molecular nitrogen in the air is negligible. Figure 8 shows the effect of combustion temperature on the formation of thermal NO_x. The NO_x emission in fluidized bed combustion comes primarily from nitrogen in the fuel. Staged combustion reduces the formation of NO_x from nitrogen in the fuel. In the lower part of the bed, combustion occurs under reducing conditions that lead to the formation of molecular nitro-

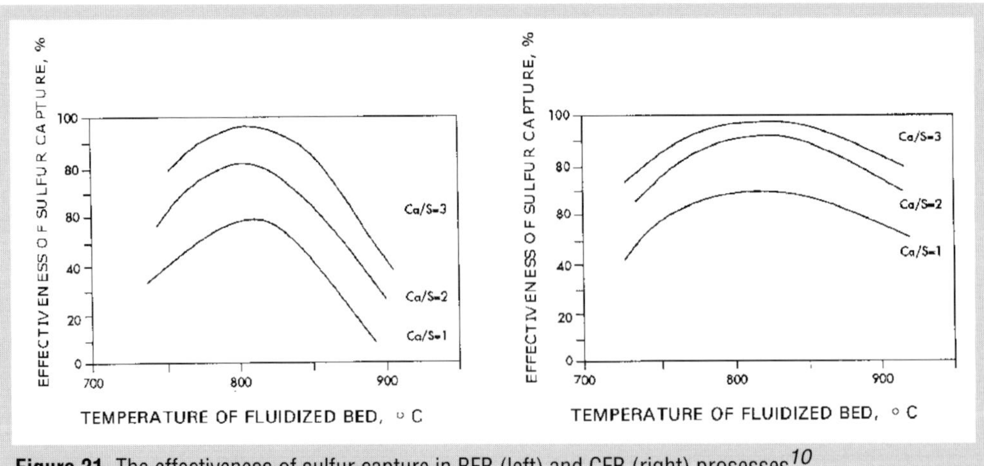

Figure 21. The effectiveness of sulfur capture in BFB (left) and CFB (right) prosesses[10].

gen, N_2, instead of NO as in the case with oxidizing conditions. Additional secondary air to complete the combustion is introduced at higher levels. NO_x emissions in fluidized bed combustion depend primarily on the fuel. They are typically 50–200 ppm in CFB combustion and 100–250 ppm in BFB combustion.

Although NO_x -emissions are low in fluidized bed combustion, additional NO_x control measures might sometimes be necessary to meet the lowest emissions standards of some areas. In fluidized bed boilers, one can inject ammonia (NH_3) into a solids separator in a CFB or into a furnace in a BFB. The efficiency of the ammonia injection is especially good in CFB due to efficient gas mixing in the solids separator. Ammonia reacts with nitrogen oxides and forms nitrogen and water vapor according to the following equations:

$$4NO + 4NH_3 + O_2 \rightarrow 4N_2 + 6H_2O \tag{31}$$

$$2NO_2 + 4NH_3 + O_2 \rightarrow 3N_2 + 6H_2O \tag{32}$$

In a CFB, a 50% NO_x reduction is possible with NH_3/NO_x molar ratio of 1.5–2.5 depending on the base NO_x emission.

4.6 Gasification

Gasification of solid fuels such as bark and wood waste has commercial use in the pulp and paper industry to produce combustible gas for lime kiln as a substitute for oil. Figure 22 shows the process diagram for a circulating fluidized bed gasifier connected to a lime kiln.

In a lime kiln application, bark or wood requires drying before gasification to produce gas that has sufficiently high heating value to achieve the desired flame temperature in the lime kiln. The drying can be done by firing a side stream of the product gas at the drying plant. The other option is to do the drying with flue gases taken after the lime kiln or a bark or recovery boiler. After drying, the fuel is fed to the CFB gasifier, and the product gas is burned in a gas burner connected to the lime kiln.

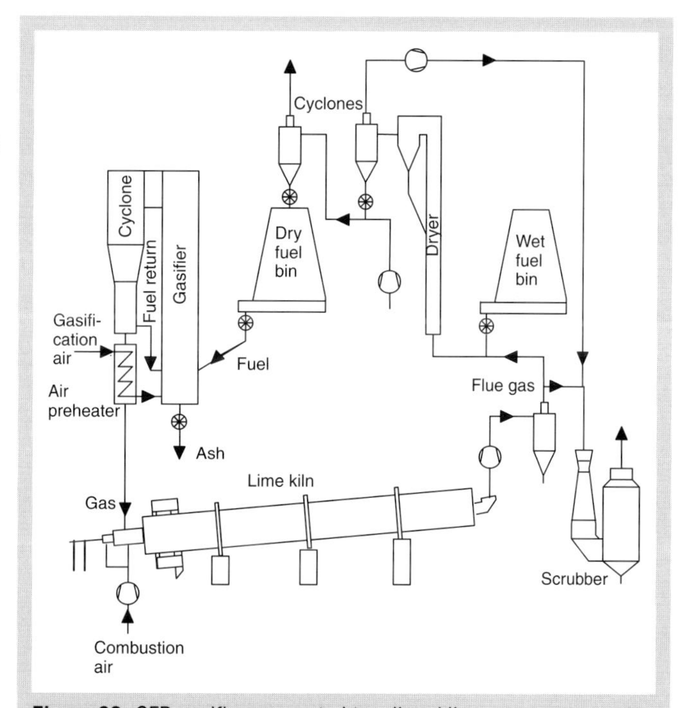

Figure 22. CFB gasifier connected to a lime kiln.

Figure 23 shows a CFB gasifier. It has a gasifier reactor that is a vertical refractory lined steel cylinder. The fuel is fed into the reactor at a level where the upward flowing gas stream does not contain free oxygen. Similar to a CFB boiler, the dense solids suspension in the reactor provides high heat capacity and creates even temperature throughout the reactor. In the hot gas at 850°C–950°C with a gas and solids suspension, the fuel dries and pyrolyzes. During the pyrolyzing, the volatiles liberate from the fuel and form the combustible gas. The fixed carbon remains in solid form as char. Table 13 gives typical gas composition on a dry basis.

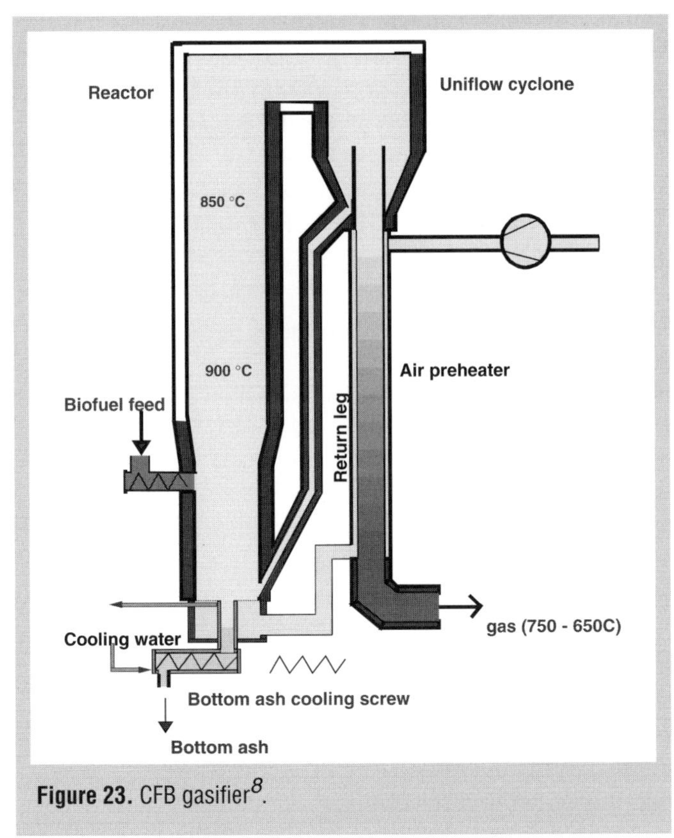

Figure 23. CFB gasifier[8].

Table 13. Typical gas composition on dry basis for a CFB wood gasifier.

Component	Amount, vol.% (dry basis)
CO	21–22
CO_2	10–11
CH_4	5–6
H_2	15–16
N_2	46–47

After the gasifier reactor, the product gas and the suspended solids enter the hot cyclone where solids separate from the gas and return to the bottom of the gasifier. Preheated air is introduced through the grid floor into the gasifier reactor to fluidize the bed material and provide oxygen to burn the carbon in the returned solid material. The combustion of the carbon residue generates sufficient heat to maintain the gasifier at the desired temperature.

5 Boiler processes

5.1 Handling of bark

Bark stripping and treatment of bark at a debarking plant

Fuel wood used in the industry is primarily bark disposal from wood used as raw material for pulping.

Wood is usually debarked in a debarking drum. The drums are slowly rotating hollow cylinders into which wood is fed. The diameter of these drums is 3–5 m, and the length is tens of meters. Barking irons are welded in the longitudinal direction in the drums. As the drum rotates, the barking irons lift the pieces of wood along the walls of the drum where wood falls on top of the other pieces of wood. Debarking of wood is the result of the friction and blows that the pieces of wood give each other. Debarking is wet or dry depending on the amount of water used in the debarking drum. The bark recovered from the drum undergoes shredding and pressing to make it suitable for burning.

The dry substance of a growing tree is 40%–60%. If wood floats to the factory, the dry substance falls to 17%–28%.The dry matter content of bark that comes from the wet debarking process and washing is 20%–35% depending on wood type. Pressing can increase it to 35%–50%. If the wood is transported by land and debarked by the dry debarking process the amount of dry substance in the bark cannot be increased by pressing.

Handling of bark in a power plant

If debarking occurs near a power plant, the treatment of bark only involves transport (usually with a belt conveyor) and intermediate storage. The size of the intermediate storage depends on the differences between production and consumption that some disturbances in the process may cause. Preparing for these interruptions usually means having a storage bin designed to store sufficient fuel for use during a few days.

Figure 24 shows a typical fuel feeding system for a CFB having a storage bin, a discharge device in it, and fuel conveyor and feed chutes to the furnace. A rotary feeder is between the feed chute and the fuel conveyor to seal the fuel transport system from the hot flue gases in the furnace. Several feed chutes are necessary to distribute the fuel evenly across the full bed area.

If the debarking occurs far from the power plant, bark is usually transported by rail or car. If bark produced outside the mill is also used at the power plant, building a bark handling plant that consists of a receiving station equipped typically with bar dischargers, an intermediate storage silo, a magnetic separator for iron, a screen, and a crusher is also necessary.

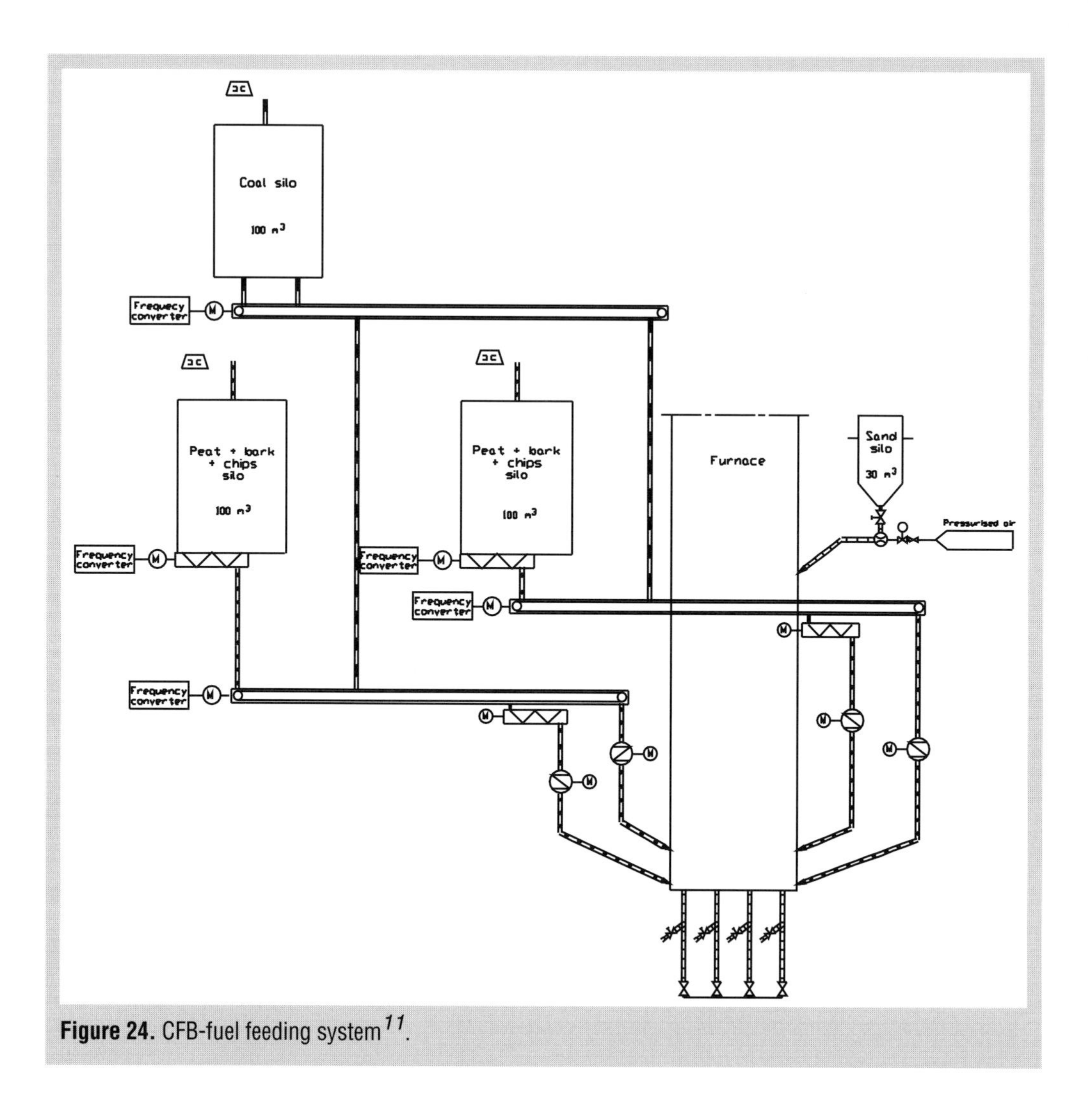

Figure 24. CFB-fuel feeding system[11].

5.2 Handling of auxiliary fuels

Handling of oil

Figure 25 presents a typical pipeline for oil fuel from storage tank to burners in an industrial power plant. The essential components of the system are the following:

- Heated storage tank for heavy oil
- Suction heater for oil in storage tank
- Pipeline with accompanying electricity or steam heating
- Filter before pumps
- Screw type pumps

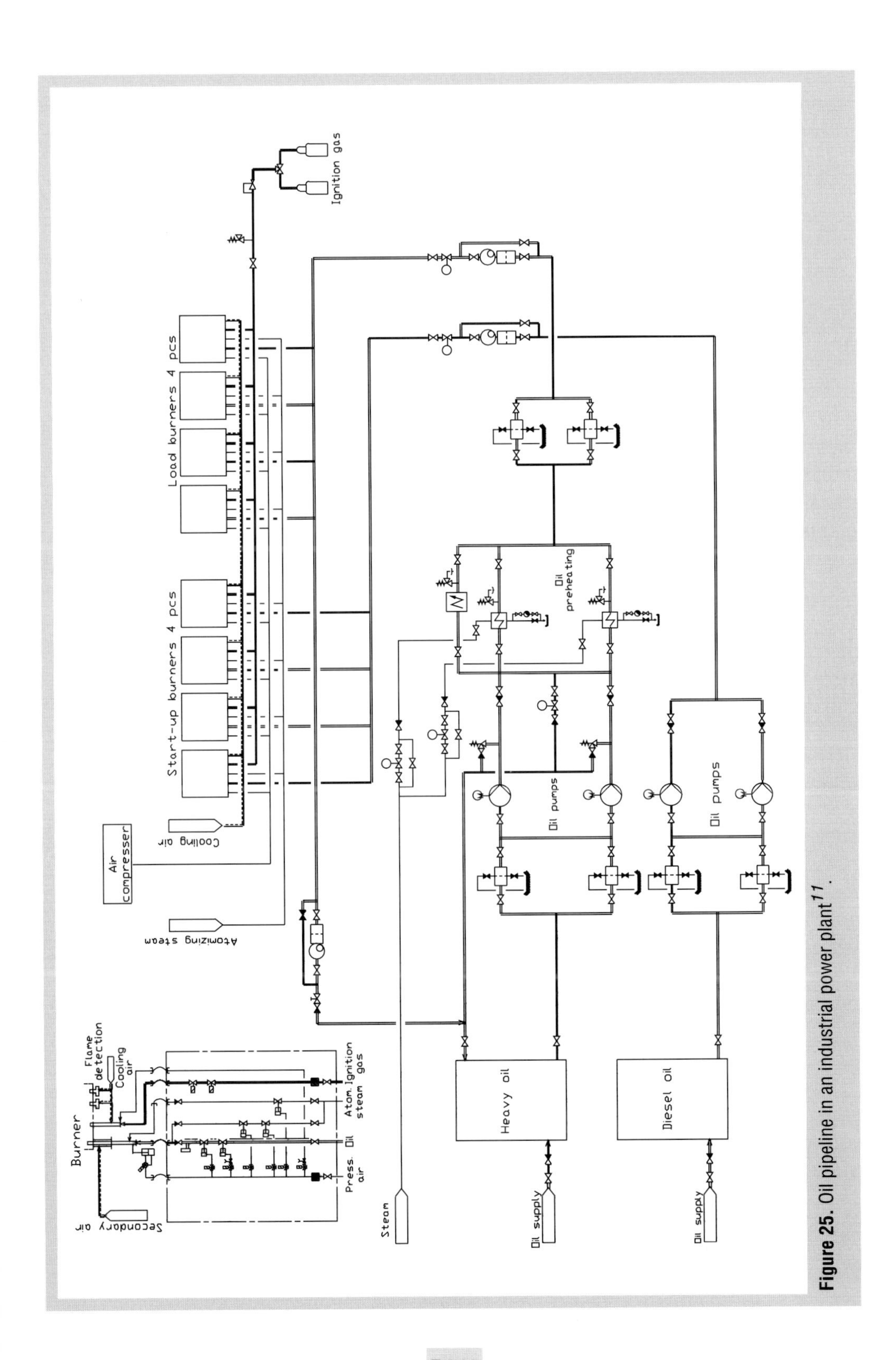

Figure 25. Oil pipeline in an industrial power plant[11].

- Steam heated preheater for oil before burners
- Burners equipped with
  - Liquid gas fired startup burners
  - Steam pipeline connections for the atomizing steam of oil
  - Quick action cut-off valves
  - Flame detection devices
- Reverse pipeline to lead excess oil back to the tank
- Valves to control the oil flow
- Light (diesel) oil facilities.

The temperatures necessary in the storage tank and for combustion depend on the fuel. The viscosity of oil being pumped should be under 500 mm^2/s. This means temperature values of 30°C–50°C for the commonly used heavy fuel oils. In burners used in power plant boilers, steam normally atomizes the oil. The viscosity for proper atomizing of the oil should be 10–20 mm^2/s. This requires a temperature of oil before the burner at 100°C–150°C.

Equipping the system with possibilities to fill the pipelines with light fuel oil in case of a longer standstill is also useful. Light fuel oil may also be used in startup burners.

Handling of gas

Natural gas comes to the power station from a pressure reduction station at a pressure of 0.4–0.8 MPa. Figure 26 shows a typical gas pipeline for an industrial power plant. The pipeline has the following features

- Filters to remove mechanical impurities
- Gas vents to empty the pipeline from natural gas and fill it with inert gas in case of repair of the pipeline.
- Pressure reduction valves (PRV) to reduce pressure to a level suitable for burners, i.e., 0.1–0.2 MPa
- Safety valve situated before PRV that automatically closes when the pressure after the PRV is too high or too low (too low means that there may be a leakage in the pipe line)
- Pressure and temperature indicators

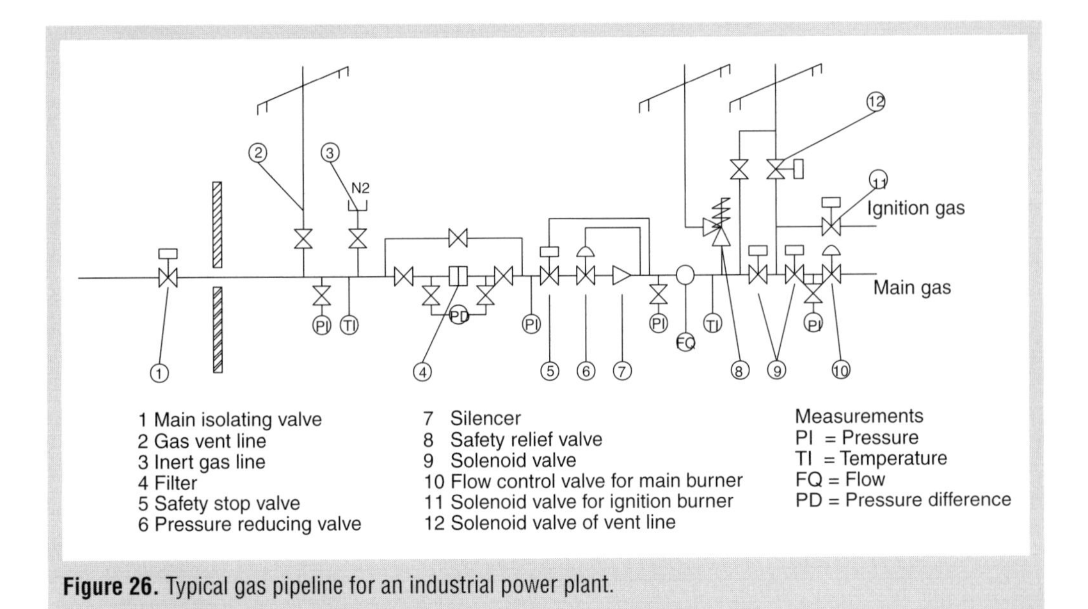

Figure 26. Typical gas pipeline for an industrial power plant.

- Flow meter
- Safety valve that opens and blows the gas into the atmosphere if the pressure of gas is too high

Normally integrated to the burner:

- Quick action cut-off valves
- Valve for regulation of gas flow
- Startup burner

5.3 Air and flue gas circuit, fans

In boiler plants, fans supply primary and secondary air to the furnace. The air is primarily used for combustion of fuels, but can also be utilized in pneumatic transport of fuels and other solid materials to the furnace. The induced-draft (ID) fans exhaust combustion gases from the boiler. In operation, they normally cause a small under pressure into the furnace. Fluidized bed boilers can be equipped with flue gas recirculation fan as shown in Fig. 27. The recirculated flue gas is used for bed temperature control.

Air and flue gas channels must be gas tight and be able to endure over and under pressure. This tightness depends on the pressures in the channel (at least ±5 kPa sometimes +20 kPa). Flue gas channels must be well isolated so the sulfur in the flue gases will not cause damage to the structures.

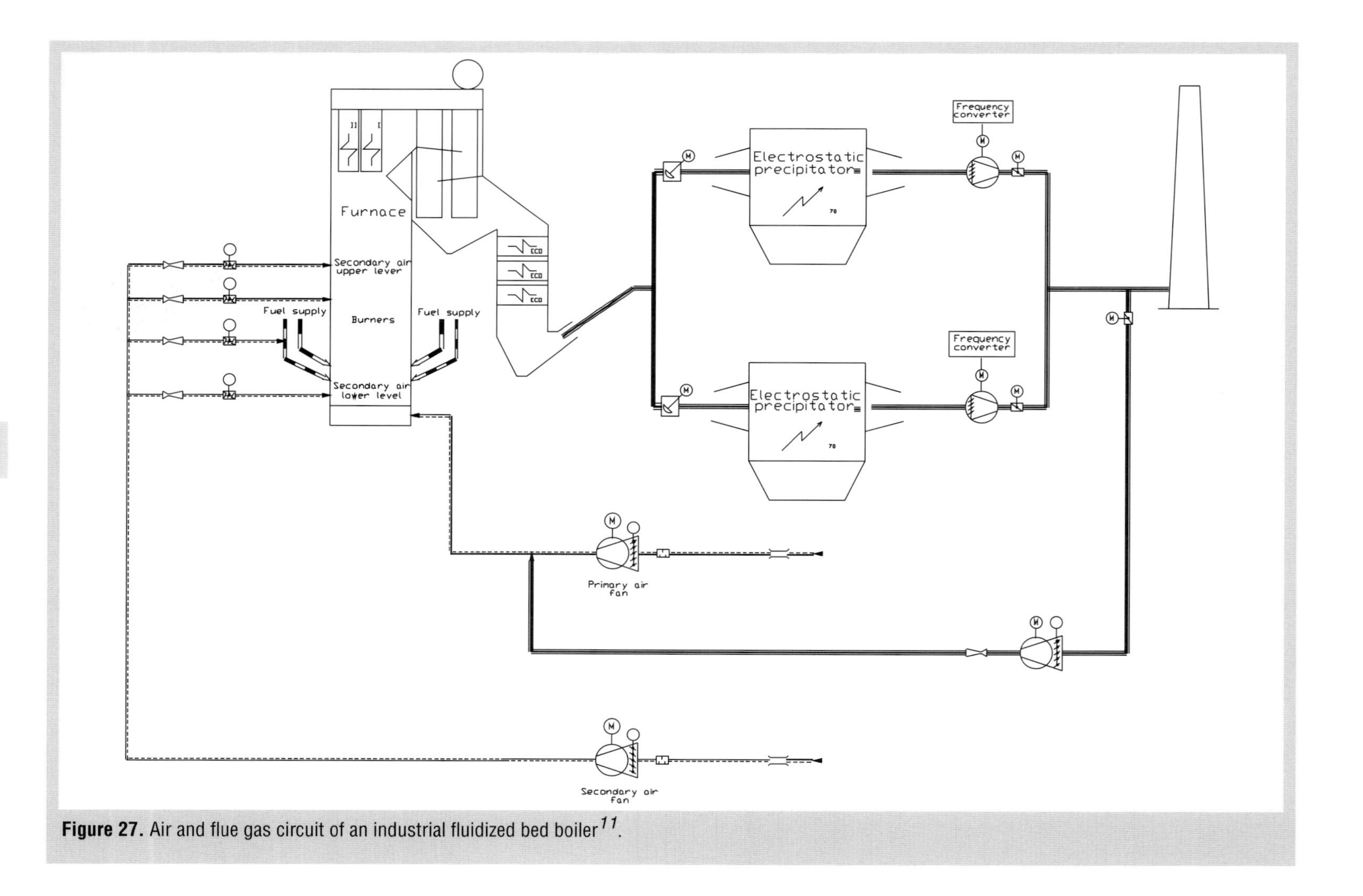

Figure 27. Air and flue gas circuit of an industrial fluidized bed boiler[11].

Flue gas velocity must be at least 8–10 m/s even with minimum load to prevent accumulation of fly ash in the ducts. To reduce pressure losses and fan power consumption, the flue gas velocities at full load must not exceed 30–35 m/s.

The selection of fan is made with the performance curves provided by the fan manufacturer. The curves are based on experiments that the manufacturers have made in laboratories for different types of fans. The curve illustrates the change in the total pressure created by the fan as a function of volume flow and speed of rotation .

When choosing a fan, the required volume flow and pressure difference must be known. Other factors influencing the choice are the following:

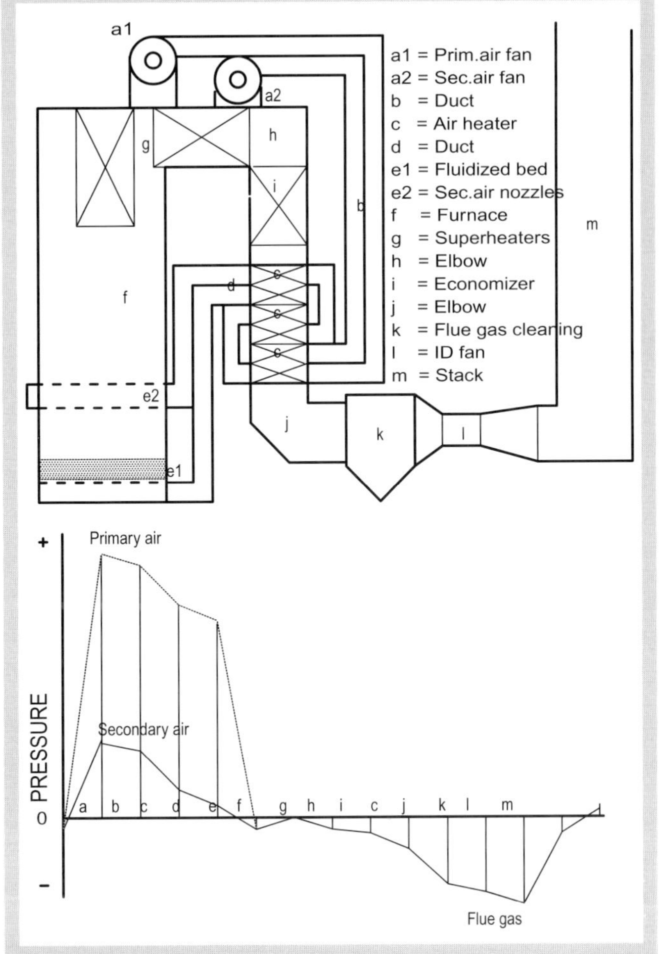

Figure 28. Pressure profile in the boiler.

- Efficiency
- Required space
- Shape of the characteristic curve for the fan.

The following formula provides calculation for the pressure losses caused by flow resistance when air or flue gases flow in the boiler channels:

$$\Delta p = \left(\frac{\lambda \cdot L}{d} + \Sigma\zeta\right)\frac{1}{2}\rho w^2 \qquad (33)$$

where L is channel length
d channel diameter
ρ density of the flowing matter
w velocity of the flowing matter
ζ single resistance coefficient
λ friction coefficient

To minimize the pressure losses in ducts sudden changes in direction, narrowings, and enlargements must be avoided. This allows minimization of the single resistance coefficient in the above formula.

Figure 28 shows a typical pressure profile in the boiler. Primary air flow through the fluidized bed causes the greatest pressure loss and second biggest are due to flue gas and air flow through the dense tube bundles in heat exchangers. ID fan adjustment gives a slight under pressure in the upper part of the furnace. The air fan regulates the oxygen content in the combustion. The stack causes an uplift that reduces the static head of the ID fan.

Fan control

Fig 29 shows the percentile amount of primary and secondary air of a CFB boiler when the capacity of the boiler is changing. Especially the amount of secondary air is changing a lot, at full load it is 45% of total air flow and at minimum load only 20% of maximum secondary air flow. In order to maintain proper fluidization in the bed also at minimum load it is not possible to decrease primary air flow so much. Typically minimum primary air flow is 50% of its maximum value.

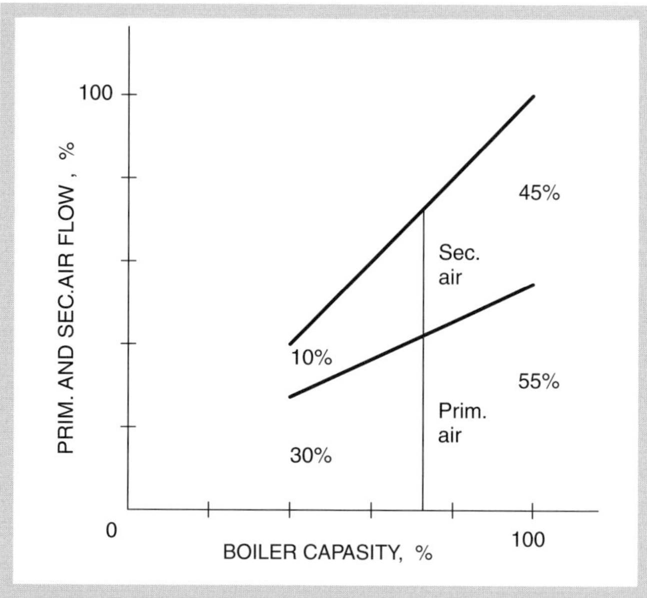

Figure 29. The percentile amount of primary and secondary air of a CFB boiler.

The most common methods for regulating the fans are throttling, inlet vane control, blade pitch control, and speed control. When choosing the method or system for requlating the fan, the following factors require consideration:

- Investment costs
- Durability
- Energy consumption.

Throttling

Dampers installed in the channels can reduce the air flow of the fan. This method is inexpensive and applicable to all types of fans. The power that is comparable with the pressure loss of the damper is wasted. The power consumption is therefore larger than with other control methods. This is the reason that throttling is not common.

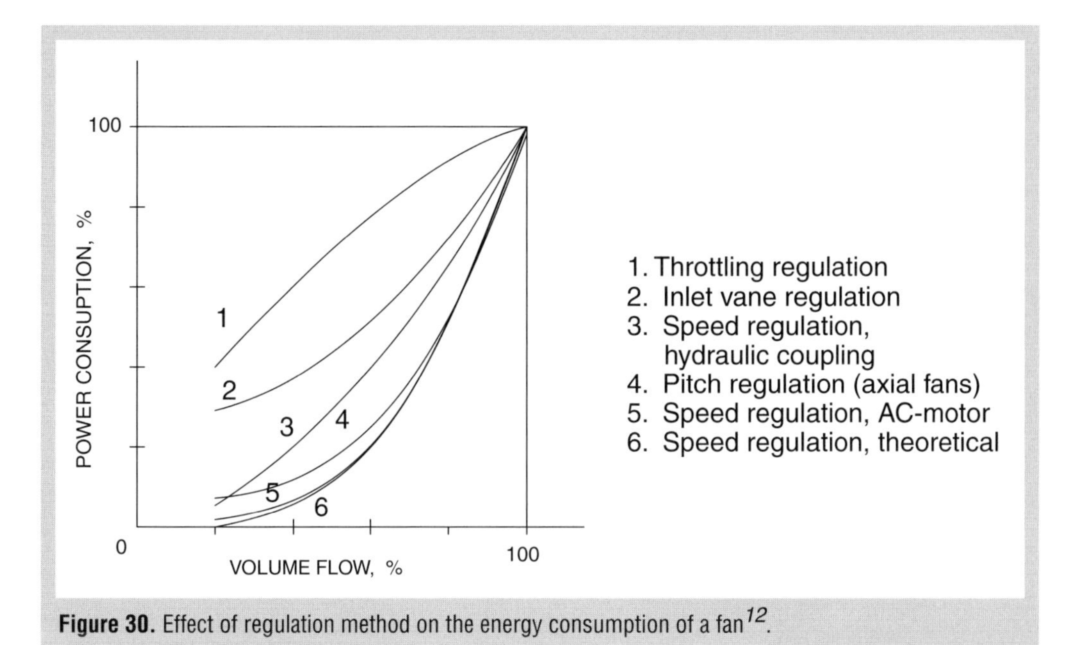

Figure 30. Effect of regulation method on the energy consumption of a fan[12].

Inlet vane regulation

With inlet vane regulation, the gas receives a rotating motion in the inlet cone. The volume flow decreases depending on the speed of rotating motion. The power demand of the fan decreases simultaneously. The investment costs for this method are low, and the energy losses are smaller than in throttling regulation. Inlet vane regulation is commonly used in centrifugal fans.

Blade pitch control

Blade pitch adjustment is only suitable for axial fans (see Fig. 32). Adjusting the angle of the blade controls pressure difference and volume control. This method is energy efficient, and the losses are small.

Speed control

Altering the rotation speed of fans can control their flow and pressure. This method is ideal for conserving energy. Speed control is suitable for all types of fans. Section 5.5 describes this method in greater detail. Figure 30 shows how different methods of regulation influence the energy consumption of fans.

Centrifugal fans

Centrifugal fans are used when high pressures up to 50 kPa are necessary. Figure 31 shows a typical power plant centrifugal fan. The suction inlet of a centrifugal fan is situated around the axis. The air comes from the outer edge of the blade wheel at the spiral casing surrounding the impeller. When desired, an inlet vane regulation can be installed

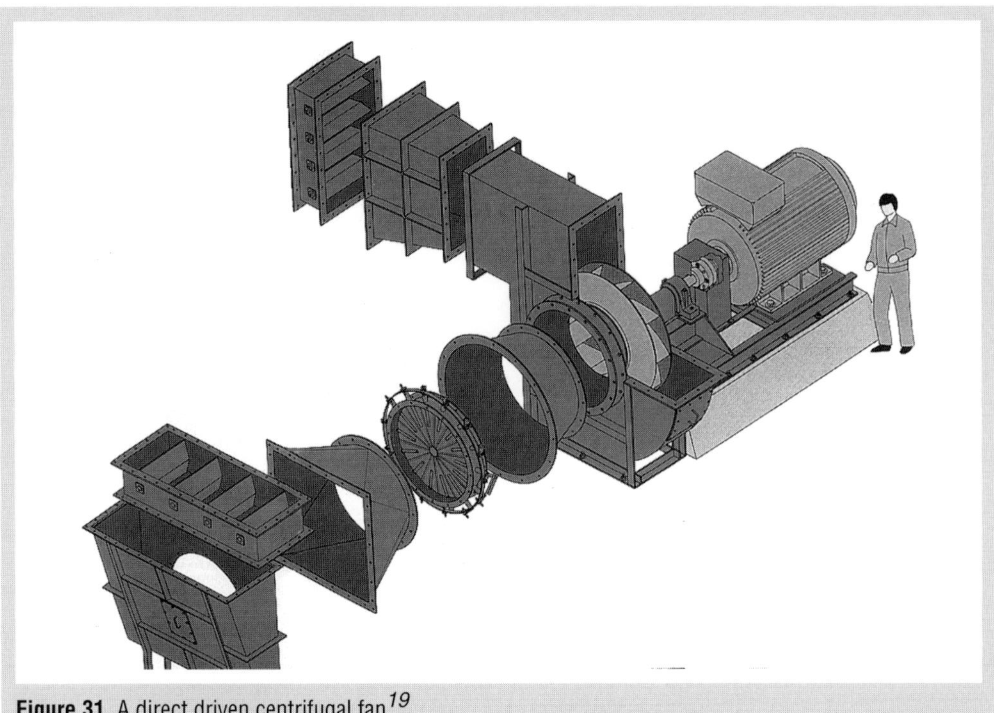

Figure 31. A direct driven centrifugal fan[19].

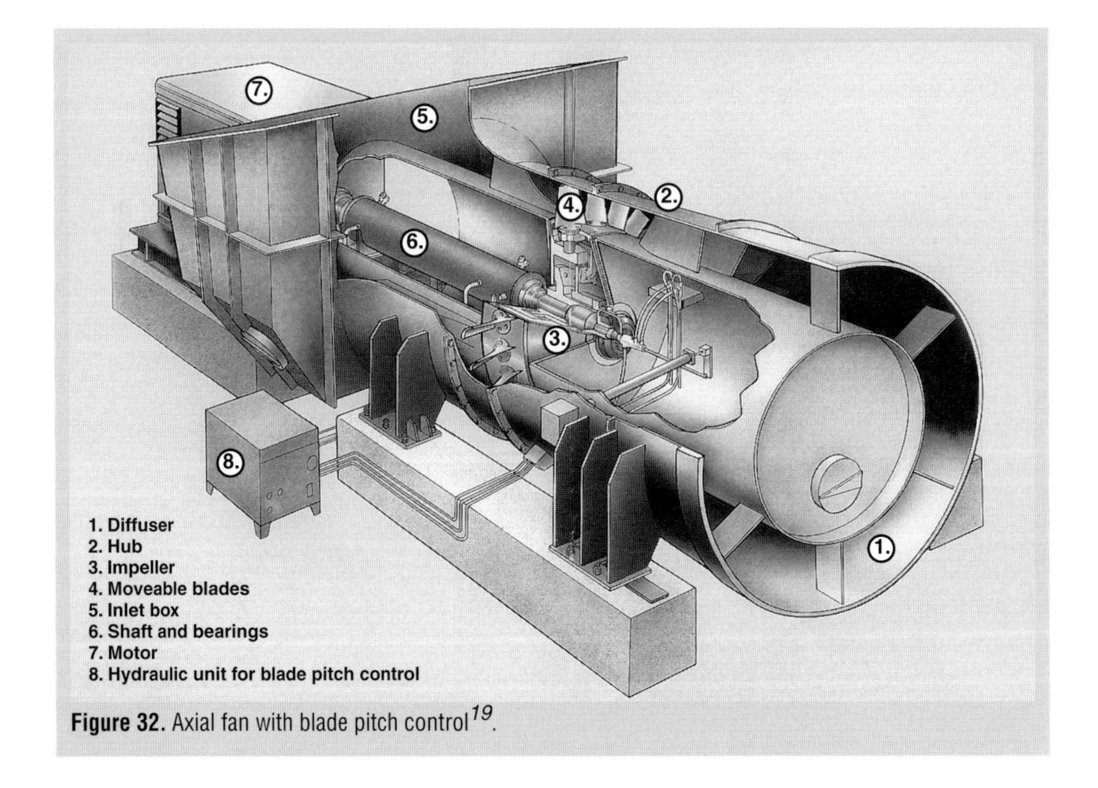

Figure 32. Axial fan with blade pitch control[19].

on the suction inlet of the fan. Flue gas fans are made of corrosion proof material. The areas exposed to erosion caused by dust are protected using wearing plates.

Axial fan

In an axial fan, the gas flows through the fan parallel to the impeller shaft. Axial fans are used in situations where low air pressures are adequate. Normally, axial fans do not have leading blades. This makes them simple, small, and easily adjustable to the system. Axial fans with leading blades are suitable for higher pressures than the fan shown in the figure (single-phase fans up to 15 kPa and quarter-phase fans up to 30 kPa). Blade pitch can be adjusted during operation. This type of fan is often used as ID fan. Because of the complexity of the adjustable pitch, axial fans need more frequent maintenance than centrifugal fans. Figure 32 shows an axial fan with blade pitch control.

Stack

A stack creates the draft reducing the head of ID fan. Also the purpose of the stack is to dilute and disperse the flue gas emissions into the atmosphere. The structure of the stack depends on the size of the boiler plant. The stack of small and medium size boiler plants uses steel construction. In large boiler plants, the load-bearing structure is a slip-form, cast reinforced concrete pipe or a steel framework structure with a duct supported by a load-bearing structure. With large stacks, a load-bearing outer casing is always necessary. In smaller, single duct systems, the outer casing is not always used.

Important factors for designing a stack are the following:

- Flue gas velocity on full load must be 20–30 m/s
- Creation of sufficient draft
- Dispersing the harmful emissions that form in combustion into the atmosphere.

The height of the stack can be estimated from emissions and the allowable concentration in the environment using the diagrams of Figs. 33 and 34[13]. The emissions from the boiler ,Q (g/s),and the hourly average maximum permissible concentration, C_m (mg/m^3),for the emission component in question must be known. From the set of curves in Fig. 33, the reference height, h_{ref}, for the stack is obtained after the amount of emissions has been calculated

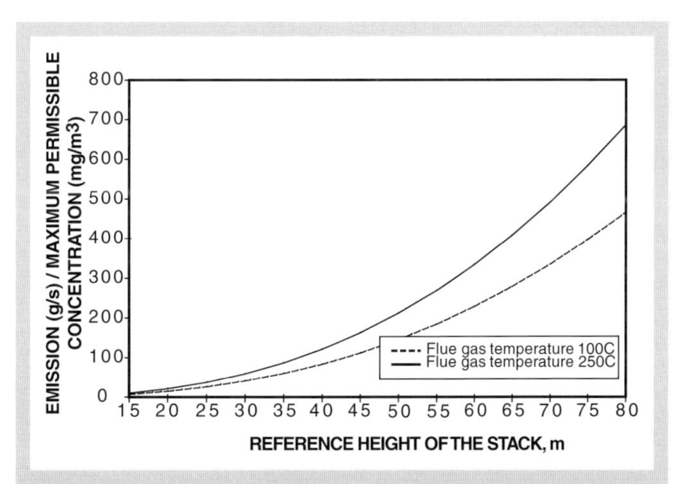

Figure 33. Reference height of the stack using flue gas temperature, emissions, Q (g/s), and maximum permissible concentration, C_m (mg/m^3)[13].

and divided by the maximum permissible concentration. After doing this, the actual height of the stack can be determined from the set of curves in Fig. 34 when the height of the tallest building in the surroundings is known.

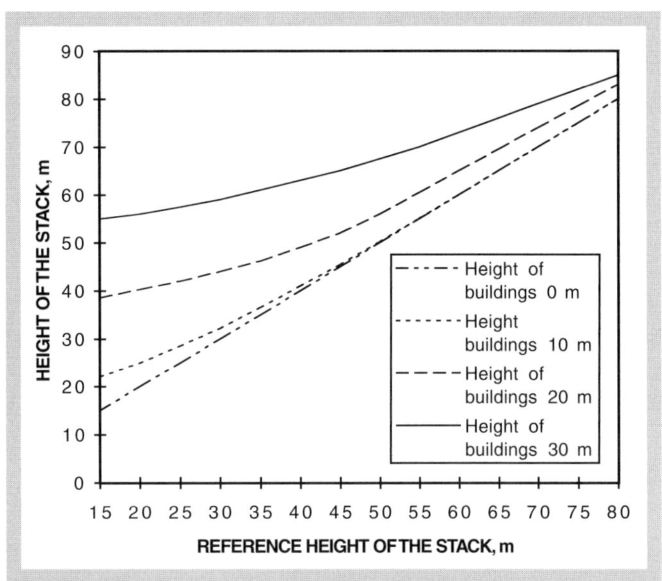

Figure 34. Determination of the height. h, of the stack with h_{ref} obtained from Fig. 33 and the heights of the buildings in the surrounding 500 m[13].

5.4 Dust separation methods for flue gas cleaning

Separating and cleaning solid impurities from flue gases is possible using different types of separators such as electrostatic precipitators, dynamic separators, bag filters and flue gas scrubbers. Local authorities regulate the maximum allowable particle emissions in every country. The acceptable particle emissions are typically 30–50 mg/m^3 .

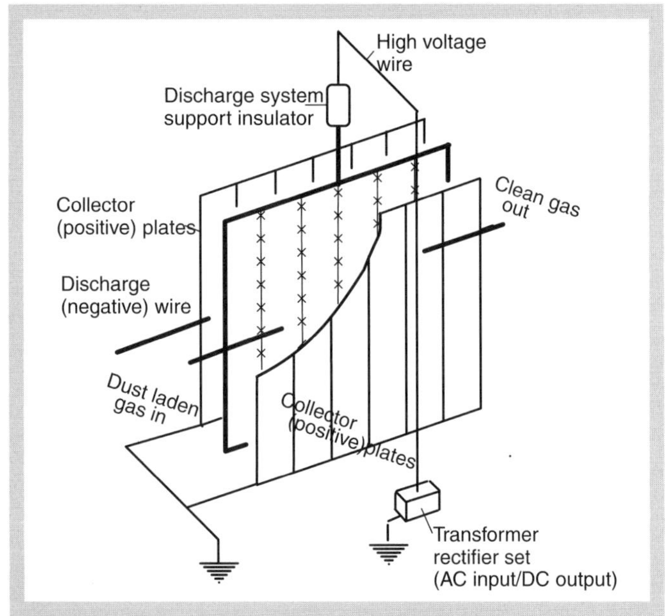

Figure 35. Structure of electrostatic precipitator.

Electrostatic precipitator

In the electrostatic precipitator of Fig. 35, particles receive a negative charge as they pass through an ionizing zone. A strong electric field then separates the charged particles from the gas. An electrostatic precipitator has two sets of electrodes: a collection electrode and a discharge electrode. The high voltage of 30–70 kV between the electrodes generates a strong electric field between them. The high voltage electric field produces a phenomenon called corona around the discharge electrodes. This creates a lot of gas molecules with negative and positive charges. When the discharge electrode has a negative charge and the collection electrode has a positive charge, the ions with a positive charge migrate to the discharge electrode. They surrender their charge and turn into neutral gas molecules.

As the negative ions migrate toward the collection electrode, they collide with dust particles and charge them with negative potential. The strong electric field between the electrodes causes the negatively charged dust particles to migrate toward the collection electrodes. The particles collect on the electrode. Rappers or water flushing removes the dust adhering to the electrode.

Advantages of using an electrostatic precipitator as the dust separator are the following:

- Large amounts of gas are suitable
- Long operating life
- Suitable for almost all processes according to the dust properties up to a temperature of 420°C
- Low operating costs due to the small pressure loss and power consumption
- High collection efficiencies are possible even for particles smaller than 1 μm.

Dynamic separators

The functioning of dynamic separators uses mass action. Many different structures for dynamic separators exist. The most common type is the cyclone. Other types of dynamic separators such as descending chambers do not have use in power plant boilers.

Gas containing dust goes to the cyclone either tangentially or axially. With the help of leading blades, the gas receives a spiral path toward the bottom of the cyclone. The mass action causes the dust particles to be thrown to the cyclone wall and then slip down to the outlet on the bottom of the cyclone. The purified gas flows from the middle of the cyclone.

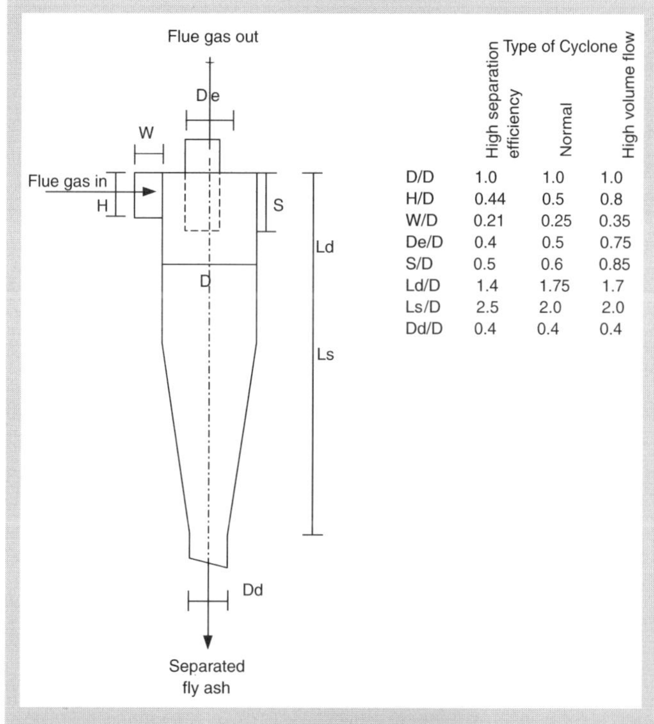

	High separation efficiency	Normal	High volume flow
D/D	1.0	1.0	1.0
H/D	0.44	0.5	0.8
W/D	0.21	0.25	0.35
De/D	0.4	0.5	0.75
S/D	0.5	0.6	0.85
Ld/D	1.4	1.75	1.7
Ls/D	2.5	2.0	2.0
Dd/D	0.4	0.4	0.4

(Type of Cyclone)

Figure 36. Structure and dimensions of a cyclone.

The collection efficiency of the cyclone depends on the particle size, gas velocity, viscosity, and cyclone geometry. The flue gas inflow velocity depends on the size of the inlet opening. This is usually 15–20 m/s. It is slightly slower if the dust is abrasive. Dividing a large flue gas flow into several small cyclones (multi cyclone) diminishes the cyclone diameter and increases a centrifugal force, F. The value for the force is $F=mv^2/r$. The division improves the collection efficiency. Figure 36 shows the dimensions of a cyclone.

Fabric filters

In a fabric filter, the flue gases undergo cleaning by passing them through cloth. Filter cloth is natural fiber, mineral fiber, or synthetic fiber. Recent development of cloth materials allows their use to filter gases at temperatures as high as 200°C–250°C. This reduces the need for cooling the gases. The most common type of fabric filter is the bag filter of Fig. 37. The cloth bags fit on cylindrical frames that are open on one end. Gas containing dust proceeds through the cloth at 0.5 m/s. The dust separates from the flue gases by stratifying on the cloth. The stratified dust improves the collection efficiency of the filter. The collection efficiency of filters is good, and the pressure drop is 500–1500 Pa depending on the cloth and the load. The dust collected on the filter cloth is removed by rinsing the cloth with air that flows in the opposite direction of the gas undergoing cleaning.

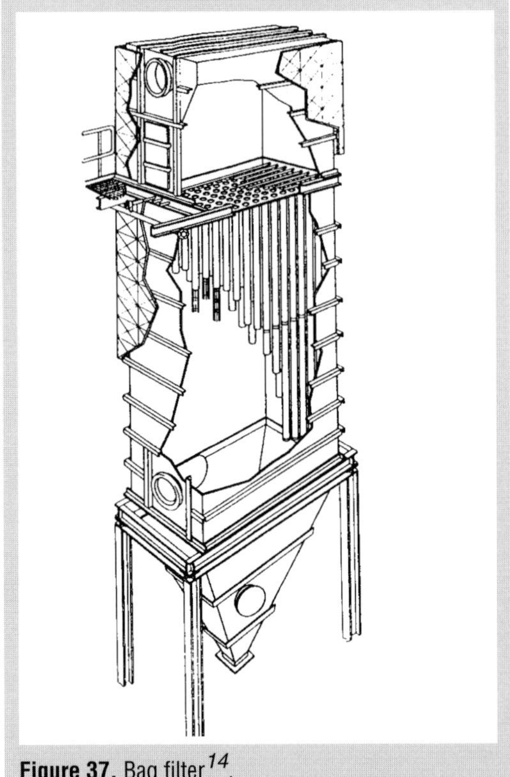

Figure 37. Bag filter[14].

Since air pollution control regulations have become more strict, using bag filters with half-dry desulfurization installations are common to filter the flue gases leaving the desulfurization plant. Bag filters are very suitable for this application. After desulfurization, the flue gases contain small amounts of unreacted lime. The lime accumulates on the cloth to form an active layer where an additional absorption of the sulfur dioxide occurs.

Scrubbers

In a scrubber, flue gases contact water. Water disperses into small drops in a scrubber and agglomerates with dust particles from the flue gases. Water with dust particles that form this way separates from the flue gases in a drop separator or flows by gravity to the bottom of the scrubber. Scrubbers are not common for separating dust from flue gases of bark boilers since the waste waters created in the process often require expensive water purifiers.

Comparsion of diffferent methods

Table 14 compares the collection efficiencies of the different dust separators described above. The collection efficiency depends primarily on the size of the solid particles. All methods can easily remove the largest particles. As the values in the table show, the cyclone has the lowest collection efficiency. Constructing a multi cyclone instead of a single cyclone can improve the efficiency. Scrubbers can reach a collection efficiency that is slightly better than that of electrostatic precipitators. Bag filters have the best collection efficiency.

Table 14. Collection efficiencies for different dust separators,%.

Dust separator	Particle size 0.5 μm	Particle size over 0.5 μm
Cyclone	Under 40	50–97
Multi cyclone	Under 60	75–100
Electrostatic precipitator	70	97–100
Scrubber	90	98–100
Bag filter	99.5	100

5.5 Feed water pumping

The main duty of the feed water pump is to transport water from the feed water tank into the boiler. Figure 38 shows a typical arrangement for feed and makeup water system. When selecting a pump, the amount of water to be pumped and the pressure difference or delivery head required must be known.

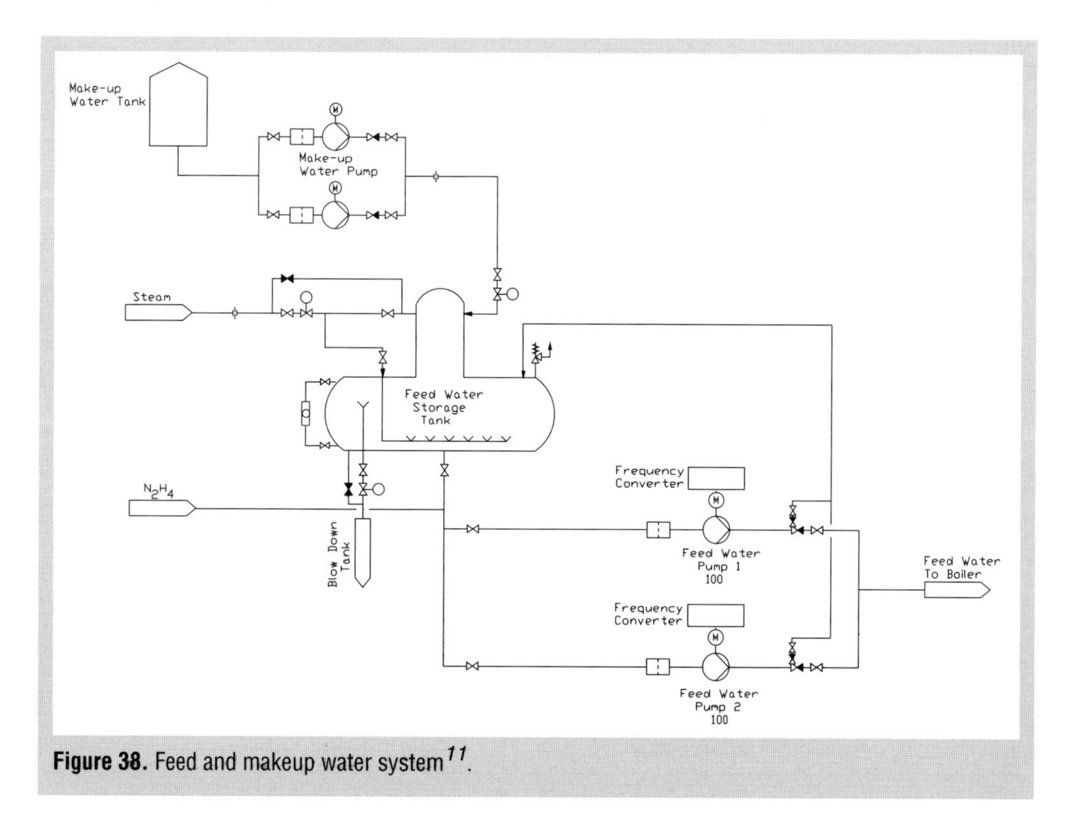

Figure 38. Feed and makeup water system[11].

Delivery head, H

The pressure difference created with the pump given as the head of water column that causes an equivalent pressure is the delivery head of a pump. The following formula gives the relation between the pressure difference and the delivery head:

$$\Delta p = \rho \cdot g \cdot H \quad (34)$$

where Δp is pressure difference created with the pump (Pa)
ρ water density (kg/m^3)
g gravitational acceleration (9.81 m/s^2)
H delivery head as a water column (m)

As Fig. 39. shows, the total head of a pump depends on the geodetic difference of altitude between the water levels of the feed water tank and the drum. The pressure difference between the feed water tank and the drum, the resistance in the piping, and the change in dynamic pressure i.e. velocities, also have an effect on the total head. Using these factors, the following formula gives the delivery head:

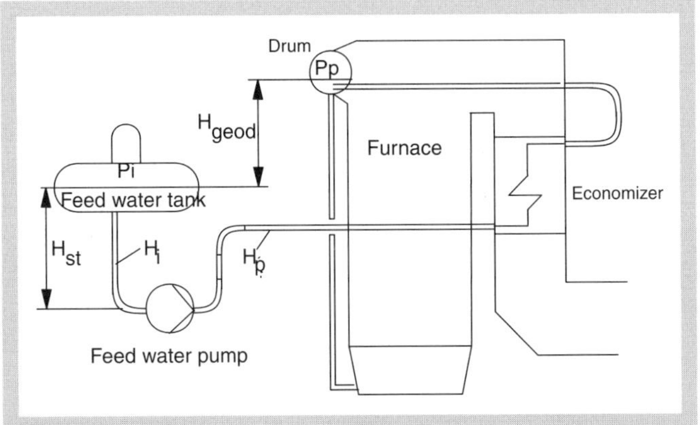

Figure 39. Factors necessary to determine total head of feed water pump.

$$H = H_{geod} + \frac{p_p - p_i}{\rho g} \cdot 10^5 + \frac{w_p^2 - w_i^2}{2g} + H_p + H_i \quad (35)$$

where H_{geod} is difference of altitude between the water levels in the incoming and outgoing sides (m)
p_p pressure in the drum (bar)
p_i pressure in the feed water tank (bar)
w_p velocity of the water into the drum (m/s)
w_i velocity of the water out of the feed water tank (m/s)
H_p pressure loss after the pump including pressure losses in economizer (m)
H_i pressure loss in the inlet pipe (m).

The following formula gives the pressure losses in the pressure and inlet tubes:

$$\Delta p = \left(\frac{\lambda \cdot L}{d} + \Sigma\zeta\right)\frac{1}{2}\rho w^2 \tag{36}$$

where L is tube length
d tube diameter
ρ density of the flowing water
w velocity of the flowing water
ζ single resistance loss coefficient
λ friction factor.

The tube side pressure drop, p_{ts}, of a heat exhanger such as an economizer is the sum of individual pressure drops:

$$\Delta p_{ts} = \Delta p_b + \Delta p_i + \Delta p_o + \Delta p_f \tag{37}$$

where Δp_b is bend pressure losses
Δp_i inlet pressure losses
Δp_o outlet pressure losses
Δp_f friction losses

since all pressure losses depend on dynamic pressure, p_d,

$$p_d = \frac{\rho w^2}{2} = \frac{\rho \cdot 8 \cdot Q^2}{\Pi^2 d_i^4} \tag{38}$$

$$\Delta p_i + \Delta p_o = \zeta_{io} p_d$$

$$\Delta p_b = n_b \zeta_b p_d$$

$$\Delta p_f = \lambda \frac{L}{d_i} p_d$$

Then

$$\Delta p_{ts} = \left(\zeta_{10} + n_b\zeta_b + \lambda\frac{L}{d_i}\right)\Delta p_d \tag{39}$$

where Q is the volume flow (m^3/s)
λ the friction factor
L the length of the tubes (m)
d_i the inside diameter of tubes (m)
n_b the number of bends
ζ the loss coefficient.

For steam generator conditions, inlet and outlet loss coefficient, ζ_{io}, is approximately 1.5.

For laminar flow (Re < 1000), the friction factor, λ, is calculated from the Hagen-Poiseuille law:

$$\lambda = \frac{64}{Re} \tag{40}$$

For turbulent flow λ must be iterated from the following equation:

$$\frac{1}{\sqrt{\lambda}} = -2\log_{10}\left[\frac{2.51}{Re\sqrt{\lambda}} + \frac{K/d_i}{3.7}\right] \tag{41}$$

Tube roughness, K, is 0.05 for typical boiler applications.

When the delivery head , H, and volume flow, Q, for the pump are known, the pump can be chosen using the performance curves of different pumps that the manufacturers have created as Fig. 40 shows. The performance curve gives information on how the delivery head depends on the volume flow of the pump.

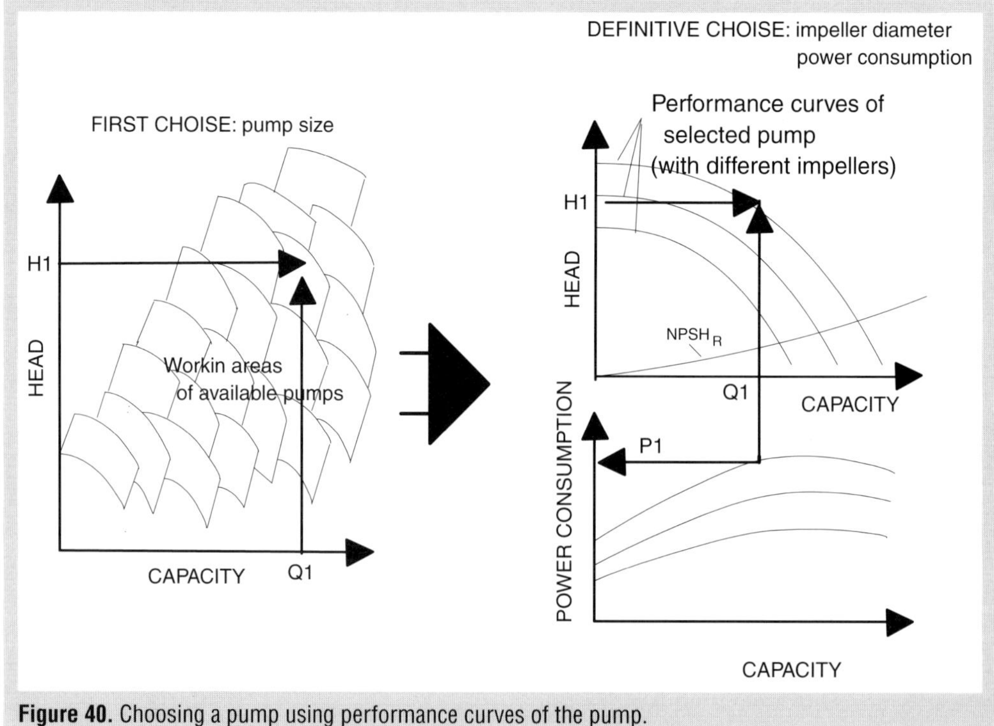

Figure 40. Choosing a pump using performance curves of the pump.

Power required

The required power, P, can be calculated from the following equation:

$$P = \frac{\rho g Q H}{\eta} \tag{42}$$

where ρ is density of the water to be pumped (kg/m^3)
g gravitational acceleration (9.81 m/s^2)
Q volume flow (m^3/s)
H delivery head (m)
η pump efficiency (%/100).

Cavitation and net positive suction head (NPSH)

When the pressure of a water decreases to the vaporizing pressure, steam bubbles form. The steam bubbles suddenly disappear as they move with the flow and the pressure rises again. The phenomenon is called cavitation. It causes strong blows on the blades of the pump. The typical logation where cavitation occurs is the inlet edge of the blades. The required *NPSH* value of the pump $(NPSH_R)$ gives the minimum suction head necessary to prevent cavitation.

Manufacturers give the net positive suction head necessary in the performance curves (see Fig.40). The placement of the pump and the piping influence the available NPSH value in the suction inlet $(NPSH_A)$, which can be calculated using:

$$NPSH_A = \frac{p_i}{\rho g} - \frac{p_v}{\rho g} - H_{st} - H_i \tag{43}$$

where $NPSH_A$ isthe available *NPSH* formed with the system
p_i absolute pressure above the surface in the suction tank
p_v vapor pressure of the water to be pumped in its temperature
ρ density of the liquid to be pumped in its temperature
g gravitational acceleration
H_{st} static suction head or altitude difference between the pump center line and the liquid level in the suction tank. (H_{st} is negative, if the liquid level in the tank is above the pump center line.)
H_i friction losses in the inlet pipings.

To prevent cavitation, one must place the pump so that $NPSH_A > NPSH_R$. When pumping saturated water, the pressure in the suction tank (p_i) equals the vaporizing pressure (p_v). To prevent cavitation, the pumps must be several meters beneath the suction tank.

Methods for controlling the pump

The pressure and volume flow created by the pump can be requlated by throttling a valve in the pipeline or by adjusting the speed of rotation of the pump.

Throttling

Throttling is the simplest way of regulating the pump. The valve situated in the pipeline after the pump does the throttling. By closing or throttling the valve, the tube resistance increases with concomitant increase in the delivery height, and with decrease in volume flow. The operating point moves left following the performance curve of the pump. In excample in Fig. 41 the new operation point for smaller capasity Q_2 is point 2B with head H_{2B}. Throttling is not economical since the pressure losses ($\Delta p = \rho g(H_{2B}-H_{2A})$) in the throttling valve is wasted. The valve also wears quickly because of the high velocity. Note that the valve in the suction tube should never be throttled because it might decrease the entrance pressure so the pump would cavitate.

Speed control

Controlling the speed of rotation alters the volume flow, delivery head, and required power necessary for pumping according to the so-called affinity rules as follows:

$$\frac{Q_1}{Q_2} = \frac{n_1}{n_2}, \qquad \frac{H_1}{H_2} = \left(\frac{n_1}{n_2}\right)^2, \qquad \frac{P_1}{P_2} = \left(\frac{n_1}{n_2}\right)^3 \tag{44}$$

where Q is the volume flow (m^3/s)
H delivery head (m)
n speed of rotation (rpm)
P power needed for pumping (kW).

Using the affinity rules or the performance curves provided by the manufacturers, one can determine the speed at a new operating point. In rotation speed control operation point is moving along the system head curve. In excample in Fig. 41 the new operation point for smaller capasity Q_2 is point 2A with head H_{2A}. The required power in throttling is:

$$P = \frac{Q_2 \cdot P_{2B}}{\eta} \tag{45}$$

and in the rotation speed regulation it is:

$$P = \frac{Q_2 \cdot P_{2A}}{\eta} \tag{46}$$

Therefore the required power for pumping when using the rotation speed regulation is smaller due to the lower pressure created by the pump. In addition, the operating point of the pump in throttling is usually in the area of lower efficiency than in speed control.

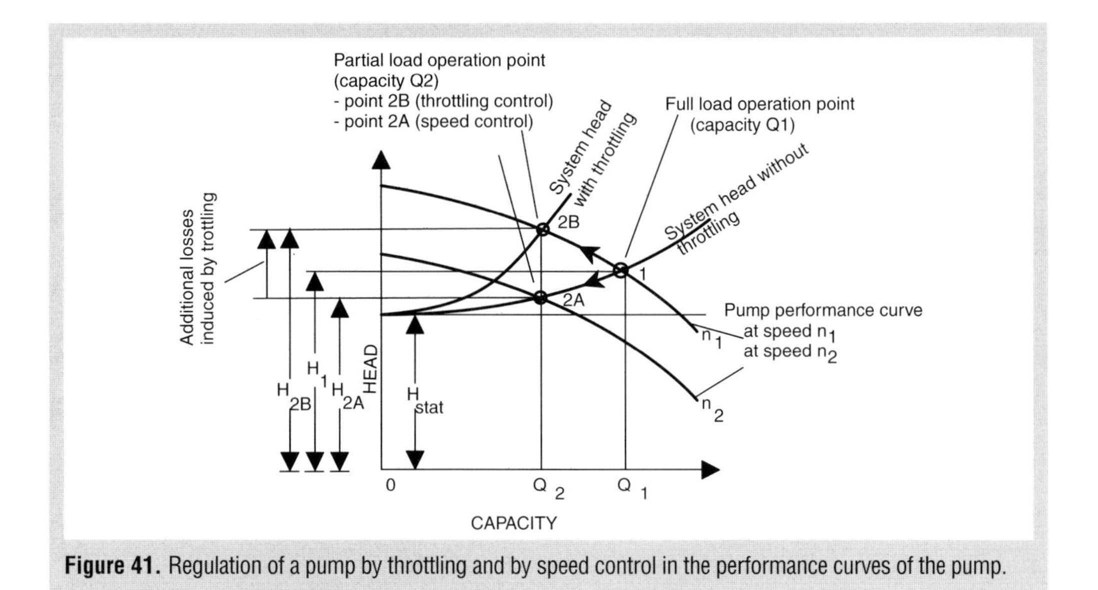

Figure 41. Regulation of a pump by throttling and by speed control in the performance curves of the pump.

The following observations apply to each technique:

- Throttling
  - Simple method with low investment cost
  - High operating costs
- Speed control
  - High investment costs
  - Low operating costs

Speed control can be attained with fluid drives, two-speed motor drives, variable-speed drives, thyristor controlled direct current motors and inverter controlled alternating current motors.

Feed water pumps

The feed water pumps must be able to create pressures up to 12 MPa in industrial boilers and operate at 100°C–200°C. Regulations concerning the number of boiler plant pumps are given in standards. At least two feed water pumps are often required in a boiler plant as Fig. 38 shows. Sometimes three pumps designed to operate at 50% capacity are used, with one pump in reserve.

Feed water pumps are multi stage centrifugal pumps as Fig. 42 shows. The pump sucks the water from the feed water tank through the suction connection on the right to the first impeller. This gives the liquid a certain pressure increase. In a multi stage pump, the liquid enters through the diffusors to the next impeller, etc., until it exists from the discharge connection.

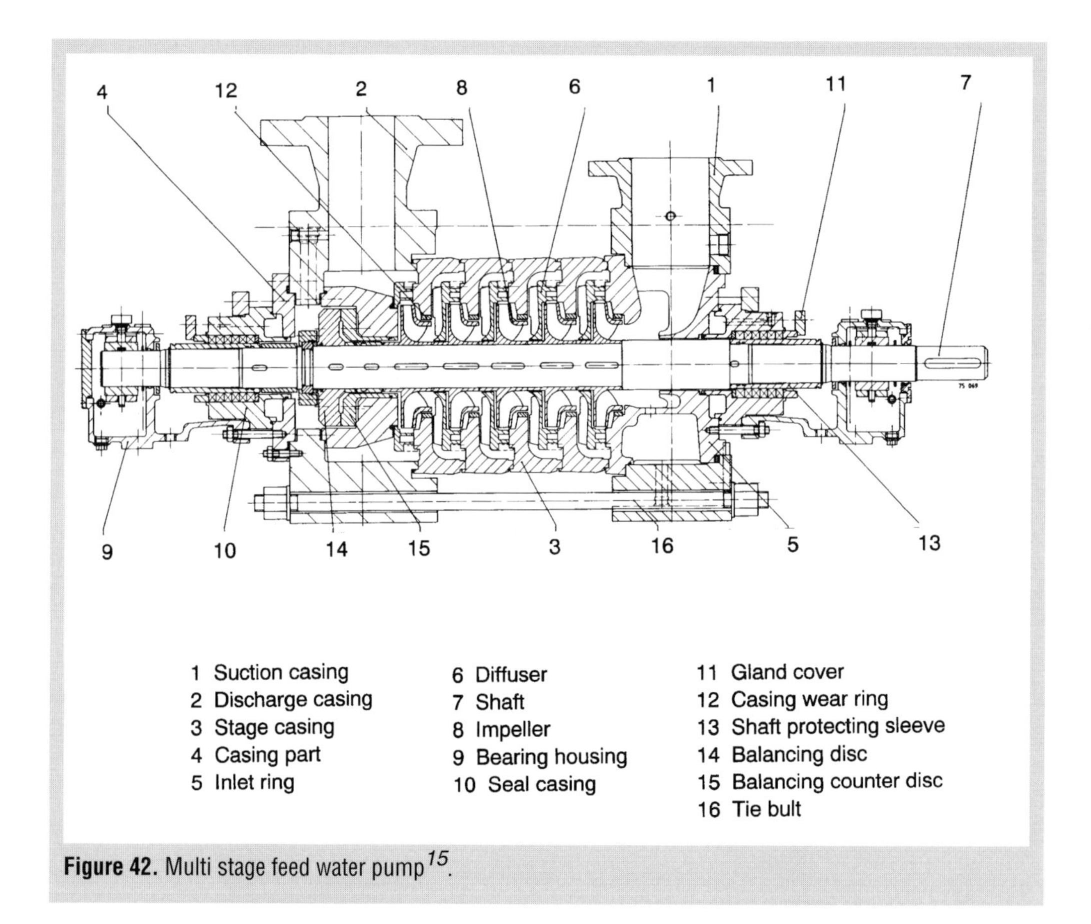

Figure 42. Multi stage feed water pump[15].

The multi stage pump has a hydraulic balancing disc. Due to the pressure difference between the suction and the pressure connections, large axial forces towards suction connection are formed inside the pump. These forces are balanced with the balancing disc with which pressurized water forms an axial force in opposite direction. Leakage water that goes through the balancing disc is returned to the feed water tank or led to the suction pipeline. Leakage water must be running during operation or the pump will definitely become disabled. A minimum flow valve is situated after the pump. A so-called minimum flow circulates through the valve. This minimum flow prevents the pump from overheating, since the power that the pump consumes transforms into heat if the valve in the pressure pipeline is closed.

The suction pipeline must have a sieve. This prevents the solid particles that have come loose from the piping or the feed water tank from entering the pump when the plant undergoes startup or after the feed water system or tank have been repaired.

5.6 Water-steam circuit

The first boilers built in the 18th and 19th centuries resembled cooking pots. They were closed units half filled with water and heated from below with a flame. The first modern boiler type was the so-called fire tube boiler. Developed in the 19th century, this unit allowed better cooling of the flue gases to improve the efficiency of steam production. This type of boiler is still in use today in low pressure steam production. The invention of the steam turbine and the need to raise the pressure in boilers to improve the efficiency of electricity production led to creation of water-tube boilers in the beginning of the 20th century. It has been typical for the development of boiler water-steam circuits during the 20th century to maximize the efficiency of electricity production in power plants. Equipment that allows higher pressure has been under constant development. Service pressure in boilers today varies from 0.1–24 MPa according to the purpose of the unit. Different pressures have different water-steam circuits.

Boilers are fire tube boilers or water-tube boilers according to the structure of their water-steam circuit. In fire tube boilers, the flue gases first go through a flame tube and then through flue tubes. The water vaporizes around the tubes. In water-tube boilers, the water vaporizes inside the tubes. The water-tube structure is more suitable for higher pressures. Power plant boilers are therefore water-tube boilers. The fire tube boilers have primary use in low pressure process steam production when electricity is not produced. A classification of water-tube boilers uses water circulation. The boilers can be natural circulation boilers, controlled circulation boilers or forced circulation boilers, and once-through boilers. In a natural circulation boiler, water and steam move from a drum into the pipes surrounding the furnace and back to the drum as a result of the density difference of water and steam. In forced circulation and once-through boilers, water and steam flow through the furnace wall tubes because of the pressure created by a pump.

Principle of natural circulation boiler

A natural circulation boiler is a water-tube boiler where water to be vaporized flows in tubes. Figure 43 shows the principal circuit diagram for a natural circulation boiler. The essential components of a water-steam circuit are the economizer, the steam drum, the evaporator, and the superheater.

The feed water enters the boiler from the feed water tank using the feed water pump. The water first goes to the economizer that is heated by the flue gases where it is heated to almost the saturation temperature. The economizer lowers flue gas temperature and therefore improves boiler efficiency.

From the economizer, the heated feed water goes to the steam drum, through the downcomers to the headers, and to the risers, i.e., the piping surrounding the furnace. Some water vaporizes in the risers when it receives heat from the combustion gases. The water and the vaporized steam return through risers to the steam drum where steam and water separate. Steam rises to the top of the drum and flows to the superheater. The unvaporized water mixes with new water fed into the steam drum and goes through the downcomers back to the risers.

In a natural circulation boiler, the circulation of water and steam between the steam drum and the evaporator uses the density difference of water and steam. The downcomer and the risers around the furnace form a continuous tube system. In the risers, some water vaporizes due to the heat from the furnace. The density of the water-

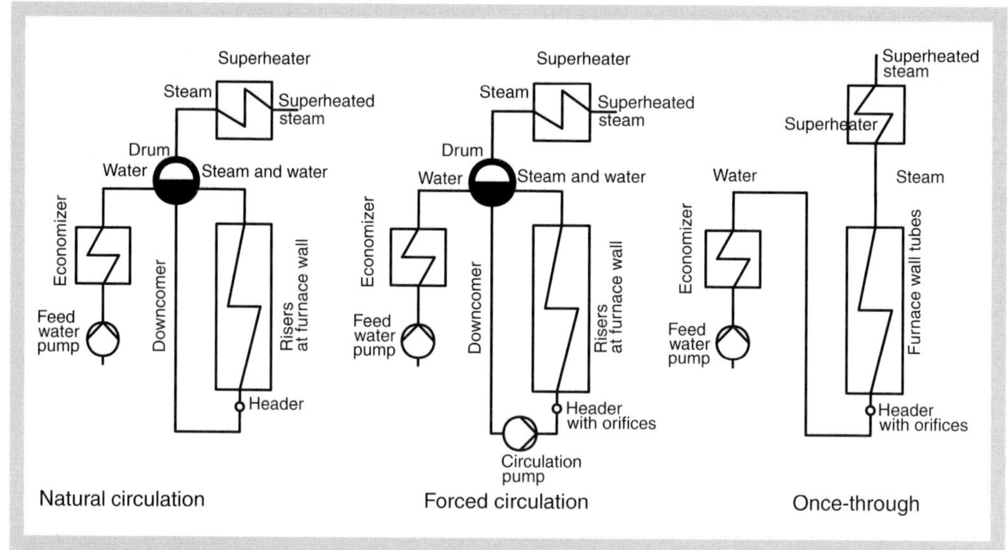

Figure 43. The principle of the water-steam circuit for natural circulation, forced circulation, and once-through boilers.

steam mixture in the risers is lower than the density of the water in the downcomer. Because of this density difference the mixture of water and steam starts to rise and flows back into the steam drum. From the steam drum, saturated water with a higher density flows down to replace the water-steam mixture.

Because a circulating pump is not necessary in a natural circulation boiler, the electric power consumption is smaller than with a forced circulation or a once-through boiler.

The density difference between water and steam decreases as the pressure rises. At the critical pressure, p = 22.1MPa, the densities of water and steam are equal at 315 kg/m^3. Natural circulation boilers are therefore not suitable for very high pressures. For the natural circulation to work, the pressure of steam that leaves the superheater must be under 17.0 MPa. In this case, the density of water is still approximately five times the density of steam. In industrial power plants, the pressure of the steam produced is normally 8–11 MPa. The boilers are therefore usually natural circulation boilers.

As the steam separates from the water in the steam drum, the impurities that came with the feed water that could cause deposits inside the boiler tubes primarily remain in the boiler water. Blowing down some water from the steam drum and replacing it with makeup water can keep the amount of impurities to a suitable level. In this way, boilers with steam drums do not require as extensive water treatment as once-through boilers where it is not possible to remove the impurities of feed water from the boiler by blow down.

Water circulation in evaporator

The pressure loss due to the circulation of water and steam is equal to the pressure difference caused by the density difference of water in the downcomer and the mixture of water and steam in the risers. The pressure loss is a result of the friction resistance of the piping and the increase of dynamic pressure caused by the change in the flow rate. The balance equation for this is the following:

$$\Delta p_{st} = \Delta p_R + \Delta p_B \tag{47}$$

where Δp_{st} is driving pressure
Δp_R friction resistance
Δp_B acceleration resistance

The following formula calculates the driving pressure, Δp_{st}, that causes the circulation of water and steam, is:

$$\Delta p_{st} = g \cdot \Delta H \cdot (\rho_w - \rho_G) \tag{48}$$

where ρ_G is average density of water-steam mixture in the riser
ρ_w density of saturated water in the downcomer
ΔH difference in levels of the steam drum and the point in the risers where the vaporizing begins.

Steam drum structure

The most important function of a steam drum is to separate the saturated steam formed in the risers of the boiler from the saturated water. The separation efficiency must be as high as possible because harmful salts dissolved in the water will cause unwanted deposits on the inner walls of the superheaters and turbine.

Water and steam separate in the steam drum due to their density differences. The simplest way to separate steam and water in the drum would be to use gravity. Gravitational separation, however, is not commonly used because it is not sufficiently efficient.

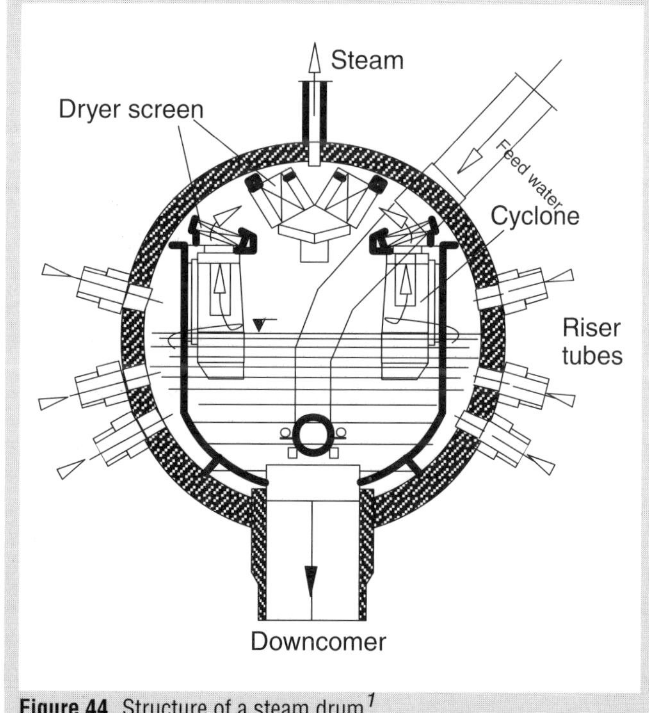

Figure 44. Structure of a steam drum[1].

Figure 44 shows a steam drum of a large, modern, natural circulation boiler where the separation of water and steam uses cyclones and drop separators. The water-steam mixture enters from the risers to numerous cyclones connected in parallel where steam and water separate due to the centrifugal force. Cyclone separators are effective, although they have a large pressure loss. After the cyclones, the steam flows further through the drop separators. A drop separator has a screen structure containing corrugated plates. As the steam flows through the separator, heavy water drops hit the plates in curves and separate from the steam flow.

Because the steam drum and the tubes are large in natural circulation boilers, the water volume is also large. Shutdowns, startups, and large load changes of the boiler require longer time. On the other hand the large water volume works as a barrier for sudden changes in load. For excample when the need for steam suddenly increases, then the steam pressure and the vaporizing temperature drop, and an amount of energy proportional to the fall of temperature vaporizes water and increases steam production.

Forced circulation boilers

In a forced circulation boiler, the water circulation in the evaporator is assisted with a pump. Figure 43 shows the principle of a forced circulation boiler. Because of the forced circulation, the boiler is suitable for pressures higher than for the natural circulation boilers. Because the separation of water and steam in the steam drum uses the differences in density, boilers that work on this principle are not suitable for supercritical pressures ($p > 22.1$ MPa). In practice, the live steam pressure in forced circulation boilers must not exceed approximately 19.0 MPa. Forced circulation boilers are not common in industrial power plants.

5.7 Main controls of boilers

The purpose of boiler regulations is to ensure that the boiler produces steam at desired pressure and temperature safely and economically. The primary controls of natural and forced circulation boilers are similar. Figure 45 illustrates the main controls of a natural circulation boiler.

As the figure shows, the main controls include:

- Control of live steam pressure by altering the fuel flow
- Control of live steam temperature by spraying water in the steam if the steam temperature rises too high.

These controls ensure that the boiler produces steam that has the desired temperature and pressure. The main controls also include control of the water flow into the boiler.

Control of the feed water ensures that water is always in the boiler and that it enters the boiler according to the consumption of steam. The simplest way to control the feed water flow is to control it according to the water level in the drum. Normally, the so-called three point regulation that also uses steam flow and the amount of water fed into the boiler is common.

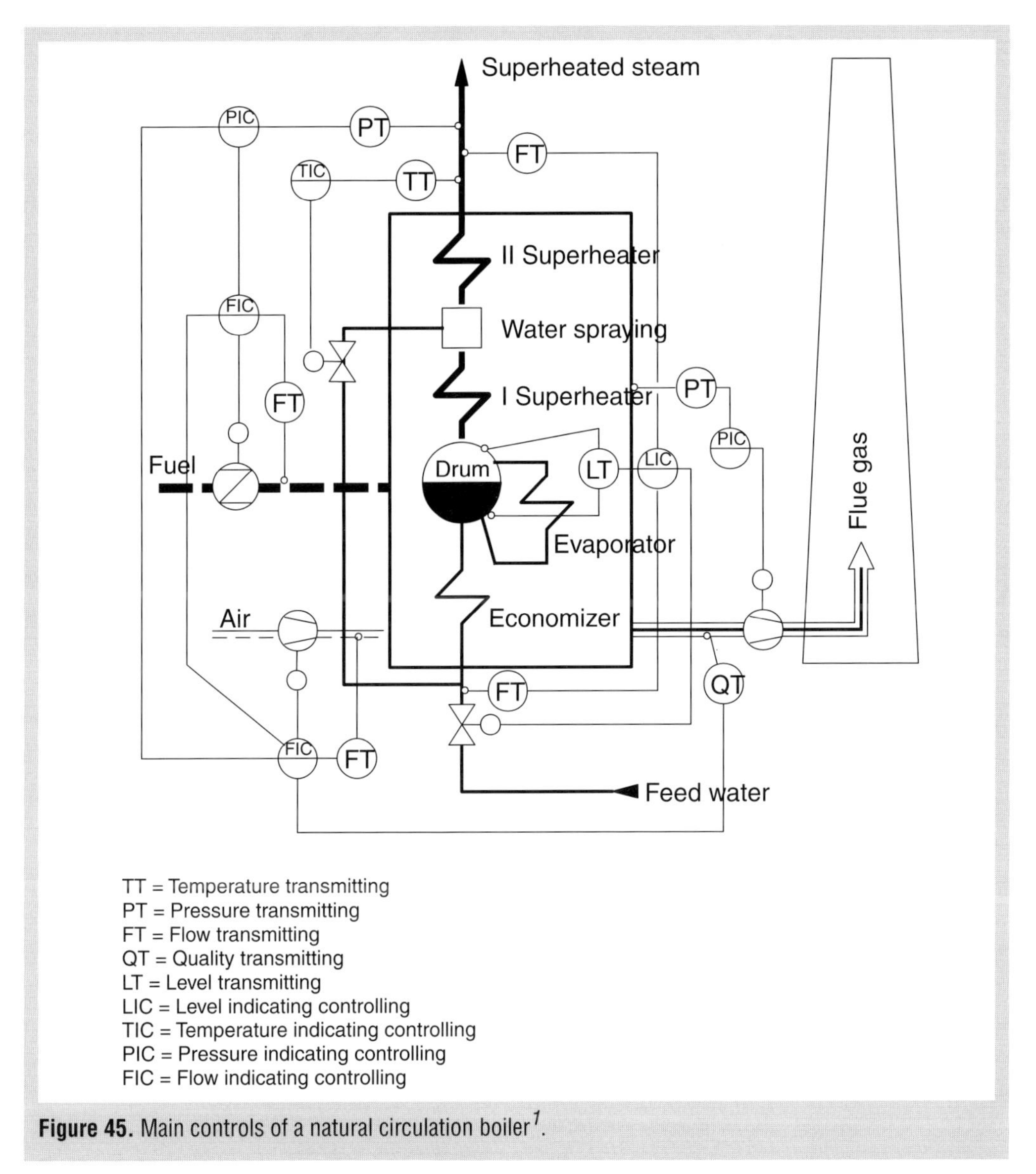

Figure 45. Main controls of a natural circulation boiler[1].

The main control circuit also includes:

- Control of the combustion air supplied to the boiler
- Control of the pressure of the flue gas in the furnace.

Regulating the combustion air ensures sufficient air for complete combustion. Combustion air enters the boiler in proportion to the amount of fuel fed into the boiler. Measurement of oxygen content in the flue gases can verify the proper ratio of air and fuel and also provide correction of the fuel to air ratio. The purpose for controlling the pressure in the furnace is to exhaust the flue gases from the furnace by controlling the flue gas fan.

6 Structure and design of boiler heat exchangers

6.1 Introduction

The purpose of boiler heat transfer surfaces is to transfer as efficiently as possible the heat energy of the flue gases to the water-steam circuit of the boiler. All the heat transfer surfaces heated with the flue gases and cooled with heat recovering mass flows (water, air, and steam) are boiler heat transfer surfaces.

Boiler heat exchangers have the following classifications according to their purpose:

- Evaporator
- Superheater
- Reheater
- Economizer
- Air preheater.

Figures 15, 17,18, and 28 show the typical locations of heat delivery surfaces in boilers. The flue gases cool from the furnace temperature of 800°C–1 300°C to 150°C–200°C after the air preheater. The correct placing of the heat delivery surfaces influences the following:

- Durability of heat exchanger materials
- Slagging and fouling of heat exchanger materials
- Temperature of live steam
- Exit temperature of flue gases.

Figure 46 gives an idea of the temperature levels and temperature differences in a boiler and of the amount of heat transferred by different heat exchangers. In the furnace of bark boilers about 40% of heat released in burning is transferred to the evaporating water in wall tubes so furnace temperature does not rise nearly as high as the adiabatic temperature. The furnace temperature is usually 800°C–900°C in fluidized bed boilers and 900°C–1 400°C in other boilers. Due to the high temperature in the furnace, effective cooling of boiler tubes is necessary. The vaporizing water is an excellent cooling medium and the temperature of the heat exchanger walls is near the saturation temperature of water. This depends on the pressure that varies from 179°C at 1.0 MPa to 374°C at 22.1 MPa.

The temperature of steam leaving the superheaters is typically 450°C–550°C. The superheated steam does not cool the heat transfer surfaces as effectively as water or vaporizing water. The superheaters should therefore not be in the lower furnace where the tube materials might overheat. The superheaters are normally in the convective pass after the furnace or on top of the furnace. The flue gases are still sufficiently hot for the high superheating temperatures, but the risk of overheating the tube materials is considerably smaller. About 20% of heat released in burning is transferred in superheaters.

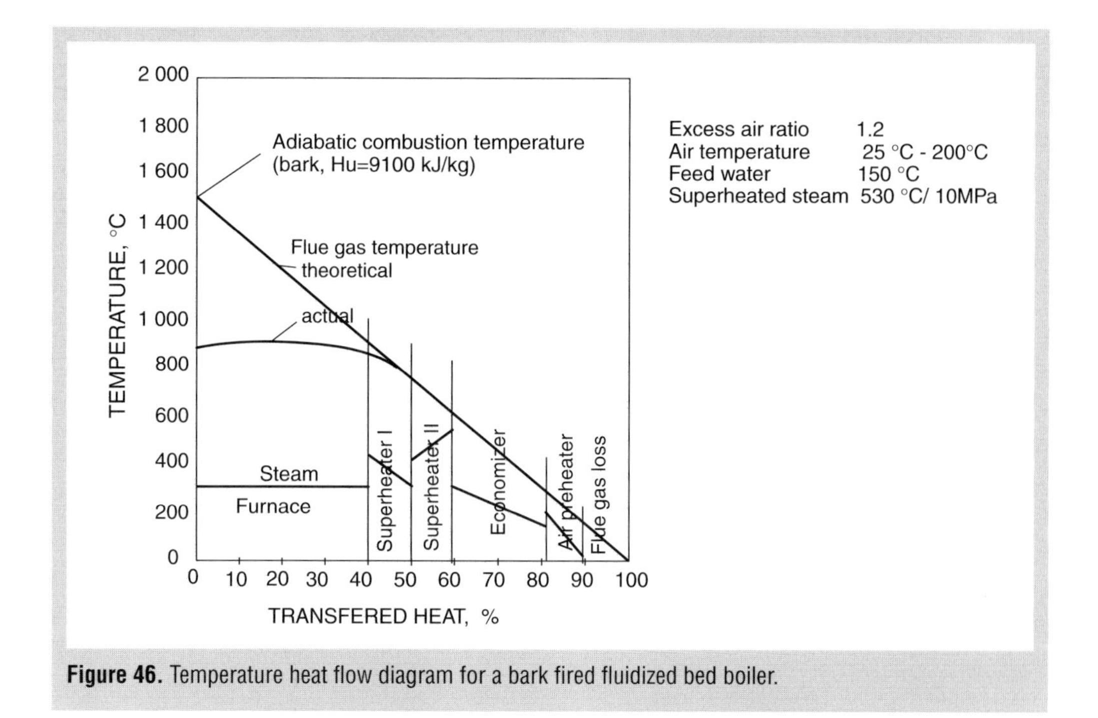

Figure 46. Temperature heat flow diagram for a bark fired fluidized bed boiler.

The flue gas temperature is still high, 600°C–800°C, after the superheaters. So high flue gas temperature would cause big flue gas losses and poor boiler efficiency (see sect. 7). So flue gases are further cooled with economizer and combustion air preheaters to gain good efficiency. In a power plant, the temperature of the feed water entering the boiler is approximately 150°C–250°C, and combustion air enters the air preheater at 25°C–45°C. In order to receive maximum cooling it is advantageous first cool flue gases with economizer and then finally with air preheater. The flue gases must not cool below their sulfuric acid dew point. This limits the gas exit temperature to approximately 130°C–150°C depending on the fuel moisture and sulfur content. With this exit temperature the flue gas loss is normally about 10%.

6.2 Corrosion of heat delivery surfaces

Corrosion of heat transfer surfaces can cause additional maintenance and operating costs in steam boilers. Selection of proper designs and materials for the heat transfer surfaces are therefore important as are correct operating procedures and preventive methods. Corrosion can occur on the gas side (outside) and water or steam side (inside) of the tubes. Corrosive elements in the flue gas induce fireside corrosion. Water and steam side corrosion can occur if the boiler water contains impurities, or if the boiler tubes are overheated.

Proper cleaning of the boiler make-up water and condensates that return to the boiler from the mill can prevent water and steam side corrosion. Continuous analysis of the quality of boiler water, steam, and condensates is important to prevent any failures of the boiler tubes. Long operation with dirty boiler water can result in complete damage of evaporator tubes and cause excessive maintenance cost and operational losses.

The material temperature of boiler tubes may exceed the maximum allowable temperature due to improper design or operational problems. Overheating of tubes is short-term or long-term depending on the duration of the temperature excursion. Any overheating is always detrimental to the tube and can result in sudden rupture of the tube or shortening of tube lifetime. The failure mechanics that can occur due to tube overheating are thermal oxidation and creep rupture.

Two fireside corrosion phenomena that can normally occur in bark boilers are chlorine induced high temperature corrosion and sulfur induced low temperature corrosion. Besides these, other types of corrosion such as CO induced corrosion, may occur locally under reducing conditions. Oil-fired boilers may suffer from high temperature corrosion induced by vanadium.

High temperature chlorine corrosion

High temperature chlorine induced corrosion can occur in superheater tubes when the fuel contains chlorine with alkalis (potassium or sodium), heavy metals, or both. Examples of metals are lead and zinc. Chlorine and alkalis or heavy metals form low melting compounds that condense on the tube surfaces. The corrosion of tube metal can be very aggressive especially when the chlorine containing deposits are in a molten state on the tube surfaces.

At severe conditions such as those in refuse fired boilers, high temperature corrosion can start at 350°C–400°C steam temperature. In bark boilers, the fuel always contains alkalis, and chlorine can be present when the logs are floated in sea water. A risk for high temperature corrosion is evident when chlorine bleaching is used in the pulp mill and the sludge from the pulp mill is burned in the boiler. In bark boilers, high temperature chlorine corrosion does not typically cause metal wastage in the superheater tubes when the steam temperature is lower than 480°C–500°C. If the steam temperature is higher, high temperature corrosion requires prevention. Preventive methods are necessary in boiler design and operation if the corrosion risk exists.

Selecting high alloy tube materials for the hot parts of the superheaters and placing the hottest parts of superheaters in lower gas temperature can reduce corrosion rate. By using proper preventive designs, the corrosion rate can decrease considerably but not be completely eliminated. In the case where conditions for high temperature chlorine corrosion are present, the lifetime of superheater tubes will decrease.

Low temperature corrosion

When fuels containing sulfur are burned, the sulfur forms corrosive reaction products, primarily sulfur dioxide (SO_2), with a small amount of sulfur trioxide (SO_3). When the flue gas temperature falls below the dew point, moisture condenses and forms sulfuric acid with the SO_3 in the flue gases according to the following equations:

$$S + O_2 \rightarrow SO_2 \tag{49}$$

$$SO_2 + 1/2O_2 \rightarrow SO_3 \tag{50}$$

$$SO_3 + H_2O \rightarrow H_2SO_4 \tag{51}$$

The amount of sulfur trioxide depends primarily on the following:

- Sulfur content of the fuel
- Residence time of flue gases in the furnace
- Excess oxygen
- Level of catalysts such as V_2O_5.

Figure 47 shows the effect of flue gas temperature on acid corrosion of heat delivery surfaces. Sulfuric acid corrosion occurs in air preheaters and in the flue gas ducts. Corrosion begins when the temperature is below the sulfur acid dew point at approximately 150°C and the acid drops have started to condense. Corrosion rate is also high when the temperature falls below the water vapor dew point of approximately 50°C. At this temperature, sulfurous acid forms due to the SO_2 and water droplets in the flue gas. At these low temperatures sulfurous acid corrodes steel even more than sulfuric acid.

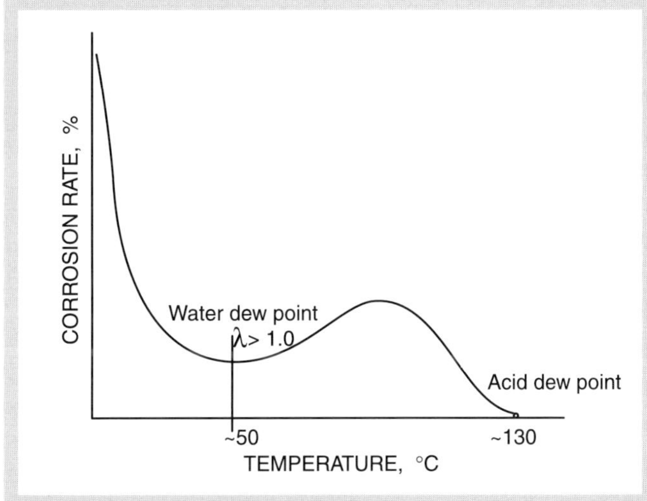

Figure 47. The effect of flue gas temperature on the acid corrosion of heat delivery surfaces.

6.3 Furnace tubes

Water that comes from the economizer vaporizes in the evaporator. In modern boilers, the furnace wall tubes form the evaporator. Furnace wall tubes form a gas tight membrane wall enclosure around the furnace as Fig. 48 shows. In low pressure boilers, an additional evaporator and evaporative tube bank situated in the convective pass is necessary.

By placing the evaporator in the furnace, the hottest area of the boiler receives good cooling. The risk of overheating the furnace wall tubes decreases. The temperature of the tube material of the evaporators increases slightly over the saturation temperature but always remains below 450°C. Normally, the construction material of furnaces can be carbon steel.

Boilers may be bottom supported or top supported by their structure. In the bottom supported boilers, the supporting level is below the boiler so heat expansion goes upward. Modern power plant boilers today are top supported boilers. Their supporting level is above the boiler, and the heat expansion goes downward. Vertical steel members at the top and rods for hanging the boiler form the actual supporting structure. Because of the pressure loads of the furnace and the vibrations of the large wall areas, the furnace walls of a top supported boiler have vertical buckstays as Fig. 48 shows.

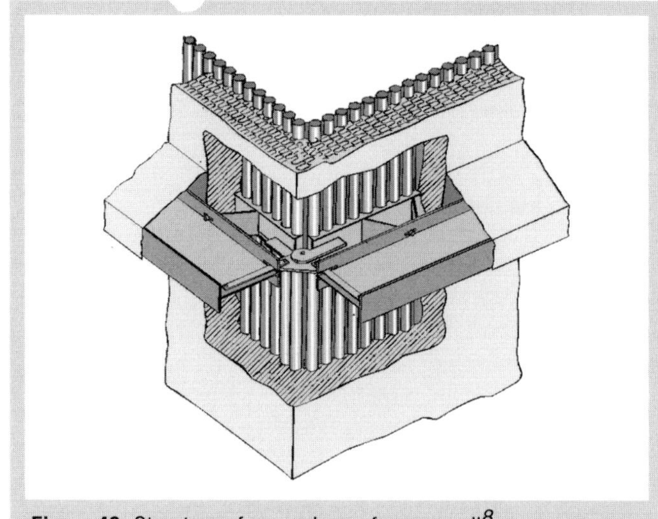

Figure 48. Structure of a membrane furnace wall[8].

Flue gas temperatures in the lower furnace are 800–1 300°C. The heat in modern boilers primarily transfers through radiation. They are radiating boilers to distinguish them from the previously used boilers with refractory linings. In those units, the heat is primarily transferred in the convection heat exchangers after the furnace.

Heat transfer through radiation is considerably more effective than through convection. In the furnace of a radiating boiler, the heat flux is approximately 200 kW/m^2.

6.4 Superheaters

For a given steam flow rate, steam with higher superheated temperature produces more electricity in turbine. All power plant boilers therefore have a superheater. The superheating temperatures do not normally exceed 550°C because higher temperatures would require too expensive materials.

A reheater can further improve electricity production of a steam turbine. The steam that has flowed through the high pressure turbine is led back into the boiler for reheating at lower pressure. Exit temperatures of steam after the reheaters are approximately the same as with superheaters. The investment costs of turbine and boiler increase with the use of reheaters. They are only profitable in the largest power plants. Reheaters do not have normal use in industrial power plants.

The superheaters are in the upper part of the furnace as Fig 49 shows. In this area, the flue gases are still sufficiently hot to reach the desired high superheating temperatures. The heat transfer through radiation is not as strong as in the furnace, and the risk of overheating the superheater tubes is smaller.

A stable temperature is necessary for the superheated steam for the following reasons:

- If temperature decreases, the electricity production decreases.
- If temperature increases, the superheater tubes may overheat.

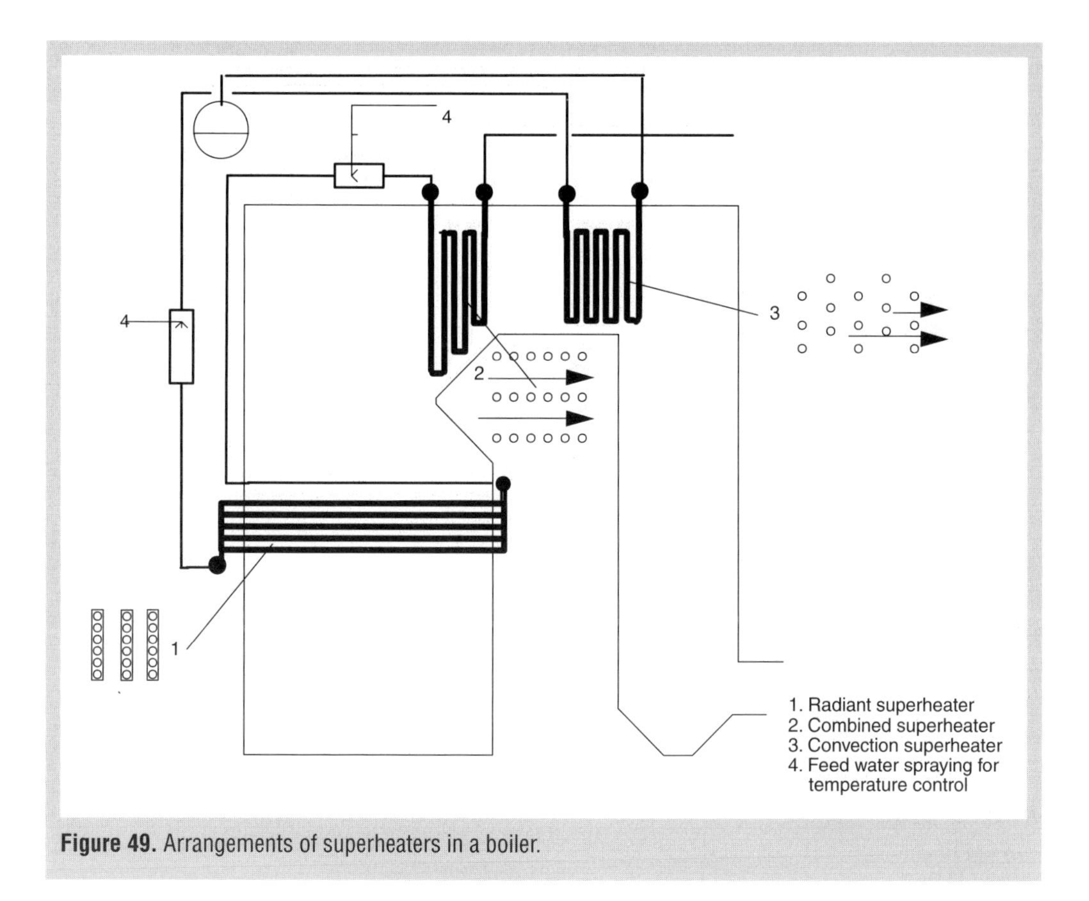

Figure 49. Arrangements of superheaters in a boiler.

A typical demand for the maximum allowable variation of the steam temperature is ±5°C–10°C after the final superheating phase. A superheater has several sections as Figure 49 shows. The superheating temperature can be controlled as the figure shows by spraying water to the superheated steam between different sections if the temperature of the steam increases too high. Other ways of adjusting the superheated steam temperature are also possible, but they are not as common as spraying.

Superheater types

Superheaters are the following types depending on the way they are placed in the boiler as Fig. 49 shows:

- Radiant superheaters
- Convective superheaters
- Combined superheaters.

A radiant superheater is in the upper part of the furnace, and the heat energy transfers into it primarily through radiation from the flames. A radiant superheater is very often a wall superheater. Because the heat transfers through radiation, the heat flux is high requiring high steam flow velocities in the superheaters to cool the superheater tubes. The temperature of the materials of radiant superheaters is approximately 60°C higher than the steam temperature.

A convective superheater is the most common superheater type in boilers. Its location is after the furnace so it has protection from the radiant heat of the flames. Heat from flue gases primarily transfers through convection. The temperature of the material in convection superheaters is 10°C–40°C higher than the temperature of the steam.

A combination where one section of the superheater functions as a radiant superheater and another section as a convection superheater is another possibility for superheaters.

Superheater characteristic

The relation between the temperature of steam leaving the superheater and the boiler load is the superheater characteristic. Figure 50 shows this. The superheater characteristic is different for different types of superheaters.

The heat flux or furnace radiation that comes to the radiant superheater increases insificantly as the boiler capacity increases.

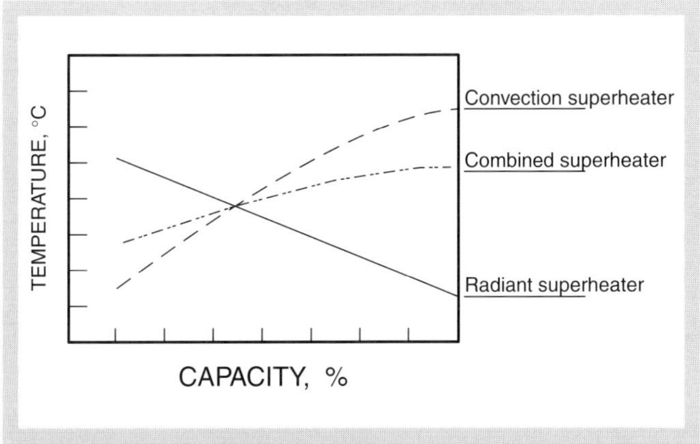

Figure 50. Superheater characteristic.

The temperature of the steam leaving the radiant superheater is therefore at its highest point at low loads. As the load and the steam flow increase, the radiant superheater cannot heat the steam flow to as high a temperature as at small loads. The characteristic curve is descending as Fig. 50 shows.

In a convection superheater, the temperature of superheated steam increases as a result of an icrease in load because the coefficients of heat transfer for convective heat transfer grow as flue gas velocity increases. The characteristic curve for convection superheater is therefore ascending. A slightly ascending curve similar to combined superheaters results from placing both radiant and convection superheaters in a boiler.

6.5 Economizers (feed water preheaters)

The economizer warms the feed water before the evaporator. Economizers can be nonevaporative or evaporating depending on their mode of operation. In the nonevaporative economizers, the temperature of the outflowing water must be approximately 20°C below the boiling point to prevent the risk of boiling due to load changes at any point. The evaporating economizer design allows the water to reach its boiling point and part of the feed water to evaporate at the end of the economizer.

Flue gas temperatures after the superheaters are 600°C–800°C. The efficiency of such a boiler would be low if the flue gases would not be cooled more. The flue gases can be cooled almost to the incoming temperature of the feed water by placing the economizer after the convective superheaters. In power plants, the efficiency of electricity production improves by heating the feed water with extraction steam from the turbine. With larger power plants, heating the feed water to a higher temperature is more economical, and there is normally at least one (sometimes even two or three) with extraction steam heated high pressure feed water preheater after feed water tank. The temperature of water entering a flue gas heated economizer therefore varies in different power plants at 100°C–250°C, and the temperature of flue gases after the economizer is 250°C–450°C. The economizer consist of tube bundles as convective superheaters. Normal carbon steels are primarily used for economizer tubes.

6.6 Air preheaters

By preheating the combustion air more heat is fed into the furnace and the furnace temperature is higher than without air preheating. In Fig. 6 is presented the effect of air preheating on the adiabatic combustion temperature. Air preheating accelerate ignition and combustion of the fuel. The temperature of combustion air is 100°C–400°C depending on the fuel and the method of combustion. The importance of preheating the combustion air grows as the moisture content of the fuel increases and it becomes less homogeneous. Heat transfer rate is also bigger when air preheating is used because of bigger temperature difference between flue gases and heat exhangers. Air preheater is the last heat exchanger in a boiler and it improve boiler efficiency by cooling flue gases more than it would be possible with other means. Flue gas temperature after the air preheater depends on the acid dew point of the flue gas. For heat exchanger durability, the temperature should be above the acid dew point. Typical flue gas temperatures after the air preheaters are therefore as follows:

- 120°C–150°C in coal combustion
- 130°C–150°C in combustion of industrial waste
- 100°C–120°C in natural gas combustion
- 140°C–160°C in oil combustion.

If concerns for the temperature falling below the acid dew point exists, air coming to the flue gas heated preheater can be preheated with a steam heated air preheater.

Materials used in air preheaters are carbon steel and cast iron. Carbon steel is excellent as an air preheater material when no risk of the temperature falling below the acid dew point exists.

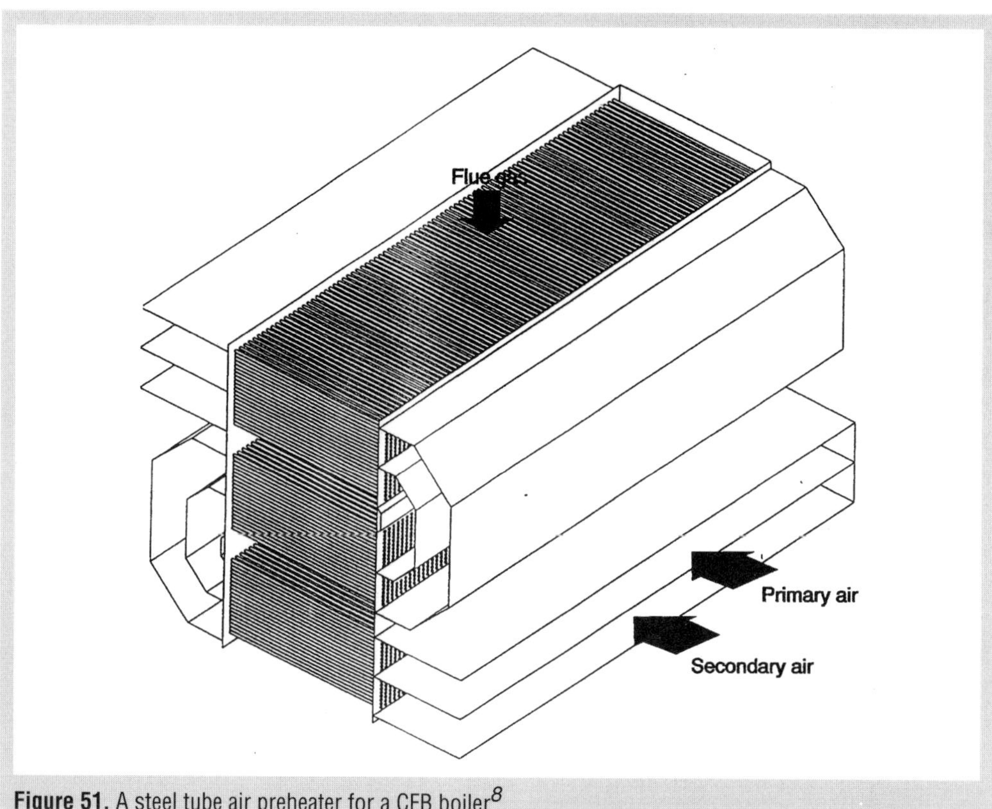

Figure 51. A steel tube air preheater for a CFB boiler[8].

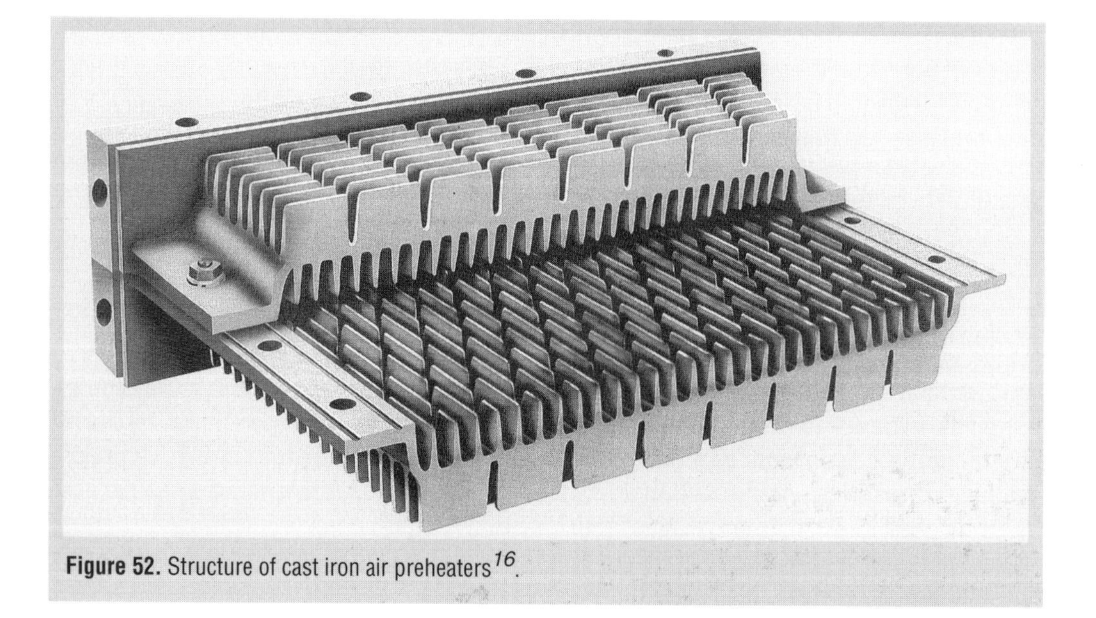

Figure 52. Structure of cast iron air preheaters[16].

Figure 51 illustrates the steel tube air preheater of a circulating fluidized bed boiler where primary and secondary air are heated separately. Air flows in tubes placed perpendicular to the flue gas flow.

Cast iron endures corrosion caused by sulfur acid that can form in the flue gases better than steel. A heat exchanger construction can be cast iron tubes or plates connected using bolts as Fig. 52 shows. A broken element is easily replacable with a new one. Heat transfer surfaces usually have fins on both sides to enhance the heat transfer. If a higher temperature on the heat exchanger surface in the flue gas side is necessary, no fins are on the combustion air side of the heat exchanger.

Like the steel tube heat exchanger in the figure, the cast iron heat exchanger is the cross-flow type. The flue gases flow downward in a canal formed by elements, and air flows horizontally straight through the heat exchanger. Several cast iron heat exchangers can connect one on top of each other.

Steam heated air preheaters are used when a danger of the flue gas final temperature falling below the acid dew point exists. To increase the flue gas final temperature the combustion air is first heated by steam heated air preheater and then by flue gas heated air preheater.

Usually, low pressure steam at 0.4–1.0 MPa from the turbine extraction is used. Special cases can use hot water. A steam air preheater construction has tubes with fins on the outside. Steam cools in these tubes. Fins are profitable because the heat transfer properties of condensing steam are much better than for flowing air. Figure 53 shows the diagram for a steam air preheater.

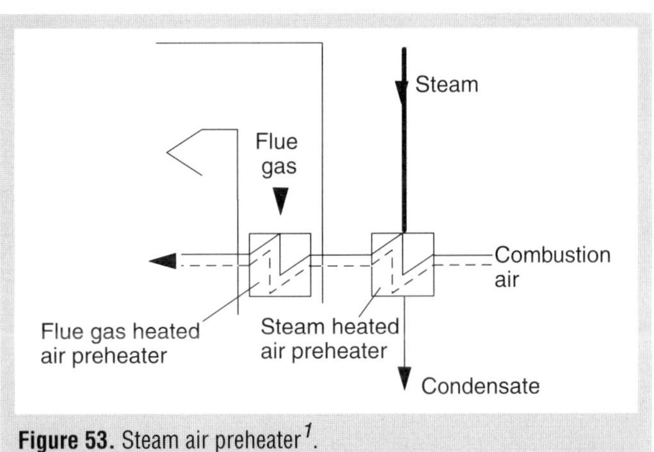

Figure 53. Steam air preheater[1].

Steam air preheaters have use especially on small boiler loads and in startup because then the temperature of the flue gas could drop below the acid dew point.

6.7 Cleaning of heat delivery surfaces

An important boiler operation is the on-line fireside cleaning of heat transfer surfaces. The cleaning is important for proper heat transfer. It prevents the tube bundles from becoming severely plugged. Plugged sections can restrict gas flow and cause load limitations. Plugging can also cause locally high velocities and result in tube erosion.

All boilers that burn solid fuels must have fireside cleaning equipment. In bark boilers, cleaning of the furnace wall is usually not necessary. The convective heat transfer surfaces of the boiler must always have a fireside cleaning system. The type of deposits in the radiant and convective sections of the boiler can vary from very hard, tenacious slag to a dry powdery coating. The most important fuel properties influencing the severity and rate of ash accumulation are ash softening and fluid temperatures, ash content in the fuel, and content of alkalis (sodium and potassium) in the fuel. Fuels with high ash content, low ash fusion temperature, and high alkali content are the most difficult to keep clean.

In bark boilers, the accumulation in the hot section of a superheater typically forms from the molten ash components. This makes cleaning more difficult. In the cooler section of a superheater, in an economizer, and in an air preheater, the gas in the solid state is not sticky and does not cause slagging. The accumulation in the colder section of the boiler is therefore powder-like and becomes easier to clean as the gas temperature decreases.

Several possible systems or equipment are available for cleaning the boiler heat transfer surfaces. The following are types of cleaning systems used in boilers:

- Soot blowers
- Acoustic cleaning
- Mechanical rapping
- Shot cleaning
- Water washing.

In modern boilers, the most common cleaning equipment is the soot blower that operates with steam. Another system that is gaining more popularity is the acoustic cleaning system.

Soot blowers

Soot blowers clean the heating surfaces with steam or compressed air. Both are equally effective in deposit removal. The high velocity steam or air jets emitted from special nozzles clean the tubes. The most common blowing medium is steam since it can come directly from the boiler. The steam is taken after the high pressure superheater through a reduction valve or directly from the cold reheater. The pressure of the steam is 20–30 bar, and the temperature is approximately 100°C higher than that of saturated steam to avoid water droplets that could cause erosion of the heating surfaces. In the case of air, large compressors must be installed. This results in high investment and operating costs.

In modern boilers, an automatic control system takes care of the soot blowing operation. A proper automatic sequential operation of the system can be programmed after ash accumulation patterns are established under actual operating conditions. With a properly programmed blowing sequence, ash deposits on the boiler tubes and the consumption of blowing media will be minimum.

Wall blowers clean ash deposits on furnace walls. The blower has a short-stroke, 50–200 mm retractable lance with one or two nozzles on the tip. Convective sections of the boiler are cleaned with long lances that are fully or partially retractable. These penetrate the cavities between heat transfer sections. Cooler sections are usually cleaned with nonretractable blowers with multiple nozzles that rotate and allow each nozzle to clean a tube row.

The retractable blower shown in Fig. 54 is used at high temperatures of 1 500°C maximum to clean the superheaters. The lance is pushed in the boiler only for the time it is in use and when it is used the through flowing steam is cooling the blower tube. It usually has two nozzles. The retractable blowers can be 16 m long. With a horizontal blower, the bending of the lance due to gravity has to be taken into the consideration when designing the blower.

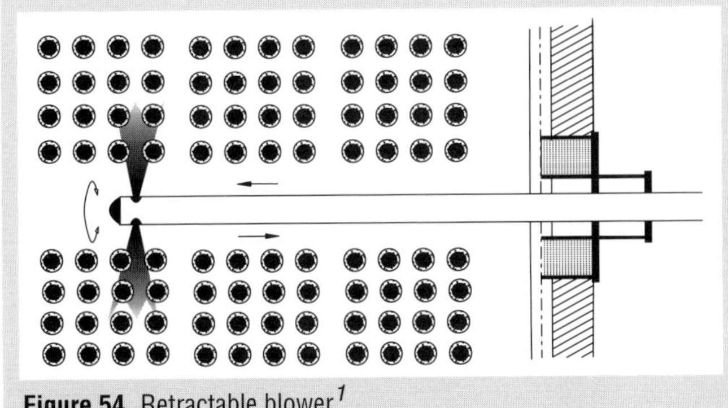

Figure 54. Retractable blower[1].

Figure 55 shows a rotary multi nozzle blower used for cleaning convection surfaces with flue gases at temperatures below 800°C. The lance is supported inside the boiler from the convection surfaces. No supports are on the wall. The distances between nozzles are such to allow the blowing to impinge between the tube surfaces to be cleaned. The maximum length of a lance is 5 m. Convective heat exchangers normally have blowers on the inlet and outlet sides.

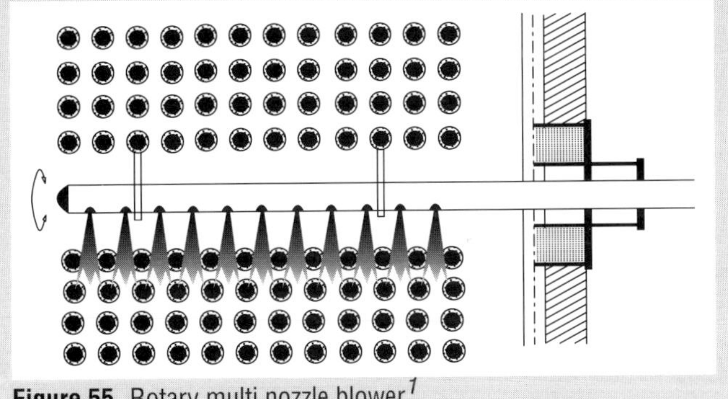

Figure 55. Rotary multi nozzle blower[1].

Acoustic cleaning

Acoustic soot cleaners create low frequency infrasound or audible, very loud (150 dB) sound that causes the heat delivery surfaces to vibrate. This removes the dirt. Figure 56 shows the principle of the acoustic soot removal system. The acoustic cleaning is effective in cleaning powder-like, friable deposits but it is not so effective in removing of hard slag deposits. The acoustic cleaning system requires low investment and maintenance and has low operating costs. Acoustic cleaning is often a good solution to clean the economizer and air preheater sections in a bark boiler.

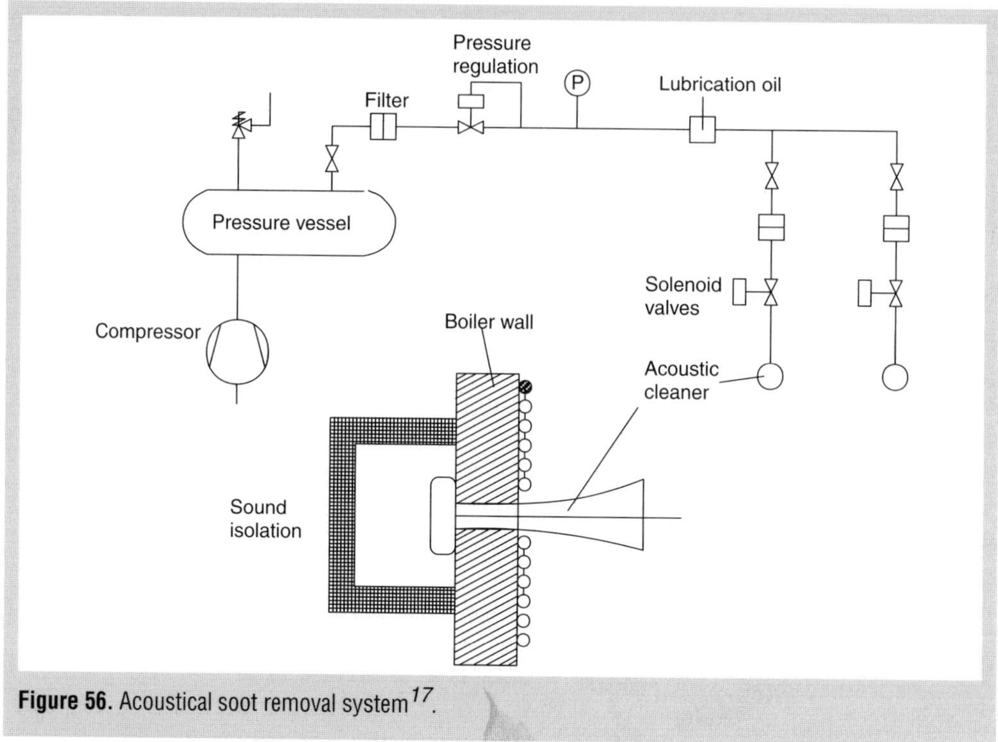

Figure 56. Acoustical soot removal system[17].

Water washing

Water can clean a boiler that is not operating. Cleaning must begin immediately after the shutdown while the heat delivery surfaces are still hot. After washing, the boiler requires drying to prevent corrosion. Cleaning with water is possible during operation but only for particularly troublesome deposits. Thermal shocks from water may reduce

tube life due to cyclical fatigue and increase the risk for corrosion. This is because the sudden temperature drop may break the layer of oxide protecting the metal surfaces. Water may also be blown in through soot blowers.

Shot cleaning

In a shot cleaning system, steel, cast iron, or aluminum shots are dropped on the horizontal convection surfaces. As the shots fall in the heat transfer section, they mechanically clean the tubes. The shot cleaning is not common today because it causes additional wear of boiler tubes and demands high system maintenance.

Mechanical rapping

Mechanical rapping is good for very difficult slag deposits such as in waste heat boilers in the metallurgical industry and boilers that burn municipal waste. Bark boilers do not use rapping systems. In a rapping system, cleaning occurs by vibration of the tube banks. Hitting the tube or its support structure with a special hammer initiates the vibration.

6.8 Thermodynamic design of heat exchangers

Calculation of required heat delivery surface

Figure 57 shows the mass flows of heat exchangers. The heat transferred to the heated flow or from the heating flow can be calculated with the formulas:

$$\Phi_1 = m_1 c_p (t_2 - t_1) = m_1 (h_2 - h_1) \tag{52}$$

$$\Phi_{II} = m_3 c_p (t_4 - t_3) = m_3 (h_4 - h_3) \tag{53}$$

where Φ is is transferred heat (W)
m mass flow (kg/s)
c_p specific heat capacities (J/kgK)
t temperature (°C)
h enthalpy (J/kg)

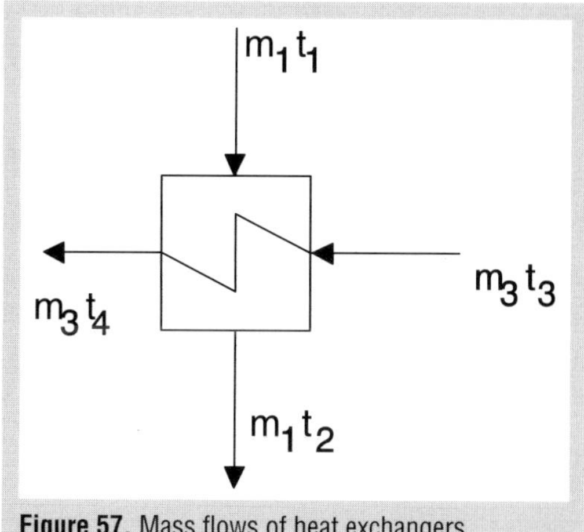

Figure 57. Mass flows of heat exchangers.

The heat flow, Φ, transferred in a heat exchanger can be calculated with the following formula:

$$\Phi = k \cdot A \cdot \Delta T \tag{54}$$

where A is heat exchanger area (m^2)
k total heat transfer coefficient (W/m^{2}K)
ΔT mean temperature difference between the cooling and the heated mass flow (K)

This leads to the following for calculating the required heat delivery surface:

$$A = \frac{\phi}{k \cdot \Delta T} \tag{55}$$

Calculation of mean temperature difference

Calculating the mean temperature difference between the cooling and the heating mass flows usually uses arithmetic, ΔT_{arit}, or logarithmic mean temperature difference, ΔT_{log},:

$$\Delta T_{\log} = \frac{\Delta T_1 - \Delta T_2}{ln \frac{\Delta T_1}{\Delta T_2}} \tag{56}$$

$$\Delta T_{arit} = \frac{\Delta T_1 - \Delta T_2}{2} \tag{57}$$

Arithmetic is easier to calculate, but logarithmic is more accurate. Figures 58 and 59 show the meaning of ΔT_1 and ΔT_2 in the above formulas.

A heat exchanger can be one of the following types depending on its operating principle:

- Counter flow
- Parallel flow
- Cross flow.

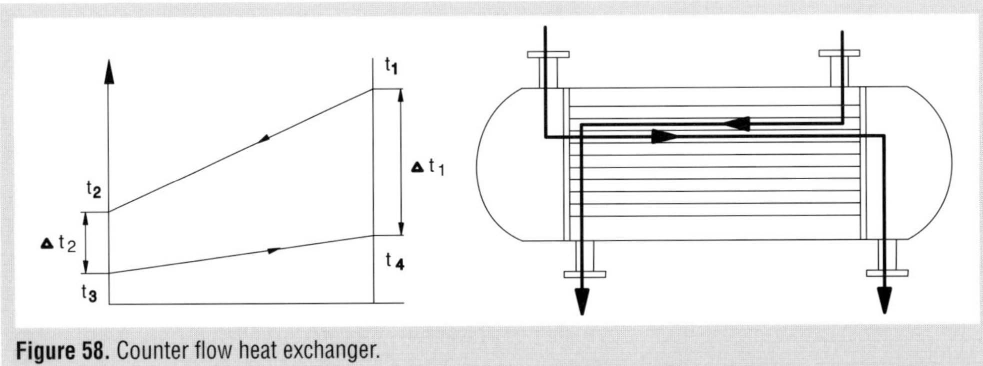

Figure 58. Counter flow heat exchanger.

In the counter flow heat exchanger of Fig. 58, the flows are in opposite directions. In the parallel flow heat exchanger of Fig. 59, the materials flow in the same direction. Figure 60 shows a cross flow heat exchanger where the flows are at right angles.

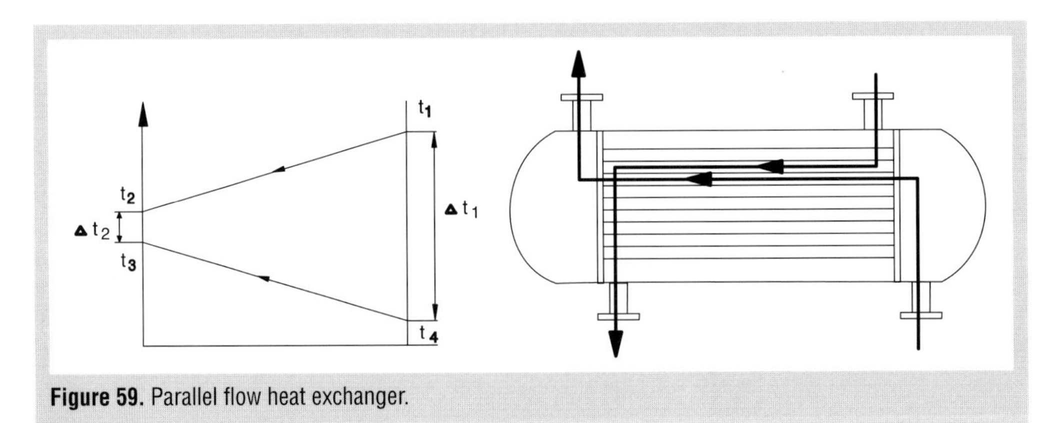

Figure 59. Parallel flow heat exchanger.

In a parallel flow heat exchanger, the largest temperature difference occurs when the flows enter the unit. The heated fluid cannot warm to a temperature higher than the exit temperature of the heating fluid. In a counter flow heat exchanger, this is possible. In a counter flow heat exchanger, the heat can transfer using the smallest surface area. The heat exchangers in a boiler are usually cross flow heat exchangers. If several cross-flow heat exchangers connect one after another, the heat exchangers can be thermally regarded as a counter flow or parallel flow heat unit. A disadvantage with the cross flow heat exchangers are the cold and hot corners where the material may overstrain.

For calculations with the counter flow and the parallel flow heat exchangers, one can use the logarithmic temperature difference. With cross flow heat exchangers, an adjustment factor is necessary:

$$F = \frac{\Delta T_{crossflow}}{\Delta T_{\log arit}} \tag{58}$$

Adjustment factors for cross flow and other types of heat exchangers are available in heat transfer literature.

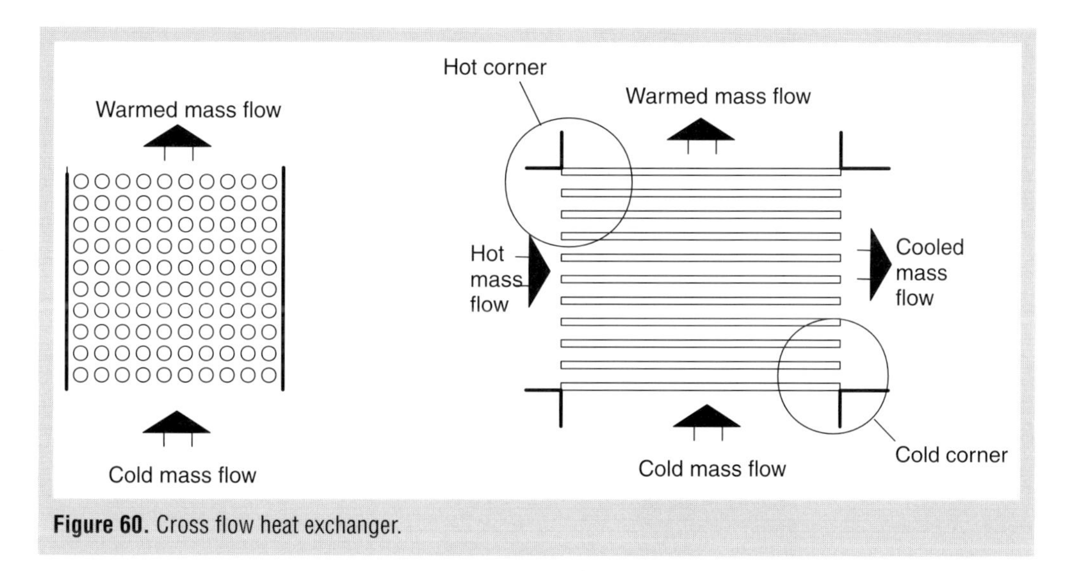

Figure 60. Cross flow heat exchanger.

Determination of total heat transfer coefficient

The required heat transfer surface area depends on the total heat transfer coefficient. This represents the ability of the heat exchanger to transfer heat between the mass flows. Following are three different ways to transfer heat:

- Conduction
- Convection
- Radiation.

Heat flow caused by the temperature difference through a solid material is conduction. Heat flows in the direction of the lower temperature.

The following formula gives the heat flow through conduction:

$$q = \frac{\lambda}{s} \cdot A \cdot (t_1 - t_2) \tag{59}$$

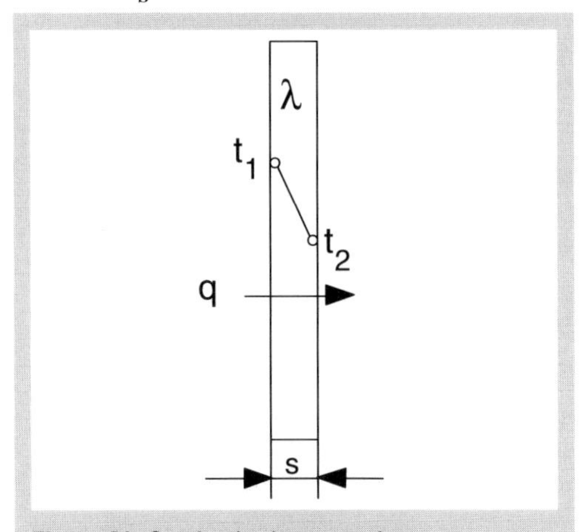

Figure 61. Conductive heat transfer.

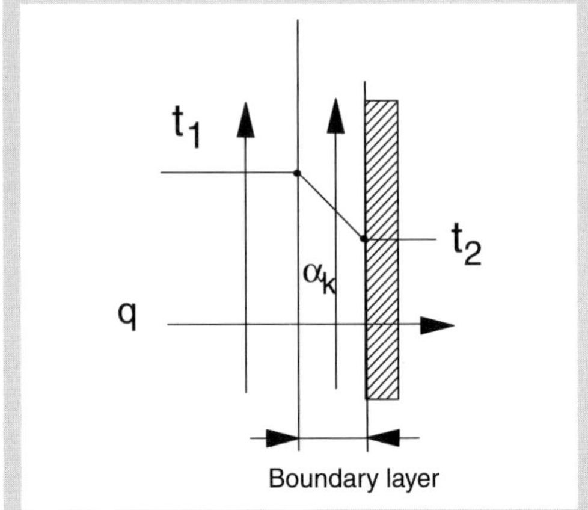

Figure 62. Convective heat transfer from a fluid to a plate.

where λ is thermal conductivity (W/m°C).
A surface area of the plate (m^2)
s thickness of the plate (m)
t_1 and t_2 are surface temperature (°C).

Different materials have different thermal conductivities. Metals are excellent thermal conductors, while gases have very low thermal conductivity. The insulating capacity of insulators uses the poor thermal capacity of air in a closed space. Table 15 gives the thermal conductivities for some materials.

Table 15. Thermal conductivity (W/m^2K) for selected materials.

Material	Thermal conductivity, W/m^2K
Copper	370
Aluminum	210
Steel	45
Stainless steel	20
Deposit on boiler tubes (water or steam side)	0.08–2.3
Insulators	0.03–0.1

Heat flow from a moving liquid or gaseous matter to a solid surface or vice versa is convection.

The following formula calculates heat flow transferred from the flowing gas to the plate:

$$q = \alpha_k \cdot A \cdot (t_1 - t_2) \tag{60}$$

where α_k is convection coefficient of heat transfer (W/m^2°C).
A surface area (m^2)
t_1 the temperature of the fluid (°C)
t_2 the temperature of the surface (°C)

The coefficient of heat transfer, α_k, depends on the quality, flow rate, surface roughness, and similar factors. Correlation equations in heat transfer literature provide its calculation. Table 16 gives typical convection heat transfer coefficients.

The temperature of the flowing fluid changes to the temperature of the surface, and its velocity falls to zero in a boundary layer. Because heat transfers through the boundary layer through conduction, convection heat transfer coefficient is thermal conductivity divided by the thickness of the boundary layer. Note that liquids have large heat transfer coefficients because their heat conductivity is good. If pure steam condenses on the heat surface the only thermal resistance that forms is from the condensed layer of liquid. The heat transfer coefficient is high. If the liquid contains air or another non-condensed gas, the heat transfer coefficient decreases considerably.

Table 16. Convective heat transfer coefficients (W/m^{2}K).

Intermediate agent	Heat transfer coefficient, W/m^{2}K
Flowing water	500–10 000
Boiling water	1 000–60 000
Condensating water	6 000–17 000
Flowing air	10–100

All materials with a temperature above absolute zero (– 273.15°C) radiate energy as electromagnetic radiation according to their temperature. The following formula gives heat flow radiating from an item:

$$q = C \cdot A \cdot \left(\frac{T_1}{100}\right)^4 \tag{61}$$

where C is coefficient of radiation (W/m^{2}K^4).
T_1 the temperature of the item measured in Kelvin (K) (0°C = 273.15K)
A the surface area (m^2),

Value of coefficient C depends on the quality of the surface, chemical properties of the item, etc. The value cannot exceed 5.67 W/m^{2}K^4, which is the value for totally black surface. For boiler surfaces, the value of coefficient of radiation is 3.5–5 W/m^{2}K^4. Because the surroundings radiate heat to an item, its compensation requires consideration when calculating the heat transfer flow:

$$q_{12} = C_s \cdot A_1 \cdot \frac{\left[\left(\frac{T_1}{100}\right)^4 - \left(\frac{T_2}{100}\right)^4\right]}{\frac{1}{\varepsilon_1} + \frac{A_1}{A_2}\left(\frac{1}{\varepsilon_2} - 1\right)} = C_{12} \cdot A_1 \cdot \left[\left(\frac{T_1}{100}\right)^4 - \left(\frac{T_2}{100}\right)^4\right] \qquad (62)$$

where T_2 is temperature of the surroundings (K)
ε_1 emissivity of material 1
ε_2 emissivity of material 2.

The importance of heat transfer through radiation increases as the temperature rises. The main radiation areas in a boiler are in the furnace and the upper part of the boiler.

One can also write the formula for calculating the radiation heat transfer in a form parallel with the formulas for calculating conduction and convection:

$$q = \alpha_r \cdot A \cdot (t_1 - t_2) \qquad (63)$$

where α_r is radiation heat transfer coefficient.

To determine the total heat transfer coefficient, examine the situation in Fig. 63.

In the figure, heat transfers from the left to the right. From fluid 1, heat transfers with convection, radiation, or both. In the wall, the heat transfers with conduction according to the thermal conductivity of the material. From the wall, heat transfers with convection to fluid 2. The heat flows must be equal because no heat losses occur. The following are therefore true:

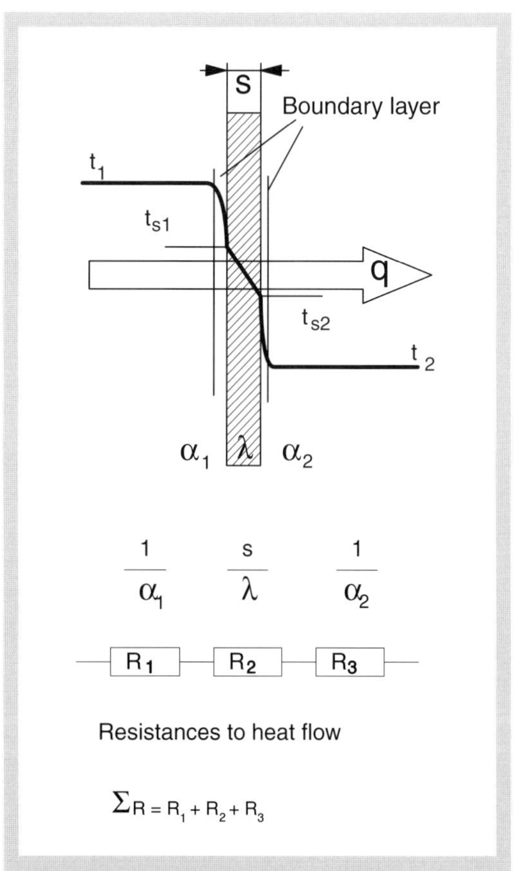

Figure 63. Resistance of heat transfer.

$$q = \alpha_1 \cdot A \cdot (t_1 - t_{s1}) \quad (64)$$

$$q = \frac{\lambda}{S} \cdot A \cdot (t_{s1} - t_{s2}) \quad (65)$$

$$q = \alpha_2 \cdot A \cdot (t_{s2} - t_2) \quad (66)$$

When including heat transfer with radiation, $\alpha_1 = \alpha_c + \alpha_r$. The heat transfer coefficient, α_1, is the sum of convection and radiation heat transfer coefficients. Solution of the set of three equations above involves eliminating the unknown temperatures, t_{s1} and t_{s2}, to give the following:

$$q = \frac{A \cdot (t_1 - t_2)}{\frac{1}{\alpha_1} + \frac{S}{\lambda} + \frac{1}{\alpha_2}} = A \cdot k \cdot (t_1 - t_2) \quad (67)$$

$$k = \frac{1}{\frac{1}{\alpha_1} + \frac{S}{\lambda} + \frac{1}{\alpha_2}} \quad (68)$$

where k is total heat transfer coefficient (W/m^{2}K).

The formula also has use when calculating the heat transfer in thin wall tubes. Total heat transfer coefficient for tubes with thick walls results from the following formula:

$$q = \frac{t_1 - t_2}{\Sigma R} \quad (69)$$

$$\Sigma R = \frac{1}{\alpha_1 D \Pi L} + \frac{ln(D/d)}{2\Pi\lambda L} + \frac{1}{\alpha_2 d \Pi L} \quad (70)$$

where ΣR is sum of thermal resistancies (K/W)
d tube inner diameter (m)
D tube outer diameter (m)
L tube length (m).

When examining the figures and formulas above, note that the value for kA is the inverse value of the sum of individual thermal resistance values, i.e., $kA = 1/\Sigma R$. The thermal resistance of the wall in heat exchangers is usually small compared with the other thermal resistance factors. Convection heat transfer is therefore the most important factor that decreases the heat capacity of a heat exchanger. Convection heat transfer can improve with acceleration of the flow. Pressure loss limits the flow velocity because the energy consumption of a pump or fan must not become uneconomically high.

Use of nondimensional parameters

Solutions of heat transfer problems are often quicker with nondimensional heat transfer parameters, ε, R, and z. They are easy to use when solving unknown temperatures for a known area and total heat transfer coefficient. One first defines the heat flow to the steam using average heat capacity:

$$\Phi_s = m_s c_{ps}(T_{so} - T_s) \tag{71}$$

In this case, the heat flow to the steam equals the heat flow from the flue gas:

$$\Phi_g = m_g c_{pg}(T_{gi} - T_{go}) = \Phi_s \tag{72}$$

The parameters describing the heat transfer can be defined as

$$G = k_t A_{eff} \tag{73}$$

$$C_{max} = Max(m_g c_{pg}, m_s c_{ps}) \tag{74}$$

$$C_{min} = Min(m_g c_{pg}, m_s c_{ps}) \tag{75}$$

$$\Delta T_{max} = Max(T_{gi} - T_{go}, T_{so} - T_{si}) \tag{76}$$

$$\Delta T_{min} = Min(T_{gi} - T_{go}, T_{so} - T_{si}) \tag{77}$$

$$\theta = Max(T_{gi}, T_{go}, T_{so}, T_{si}) - Min(T_{gi}, T_{go}, T_{so}, T_{si}) \tag{78}$$

The non dimensional parameters are defined as

$$\varepsilon = \frac{\Delta T_{max}}{\theta} \qquad R = \frac{\Delta T_{min}}{\Delta T_{max}} \qquad z = \frac{G}{C_{min}} \tag{79}$$

The dimensionless parameters can be used to express the heat flow to steam/water.

$$\Phi_s = \Phi_s(\varepsilon, R, z) = m_s c_{ps}(T_{so} - T_{si}) \tag{80}$$

For surfaces with constant temperature on the other side ($C_{max} = \infty$ and $R = 0$), i.e., boiling or condensing water as the other medium, the parameters are the following:

$$\varepsilon = \frac{\Delta t}{\theta} = 1 - e^{-z}$$
$$R = 0 \tag{81}$$
$$z = \frac{G}{C_{min}} = -ln(1 - \varepsilon)$$

In a countercurrent flow situation, the heat receiving media outlet temperature is on the same side as the heat source media inlet temperature. The heat receiving media typically has maximum heat capacity. The parameters are the following:

$$\varepsilon = \frac{\Delta t}{\theta} = \frac{1 - e^{-z(1-R)}}{1 - e^{-z(1-R)}R}$$
$$R = \frac{\Delta t_{max}}{\theta} \tag{82}$$
$$z = \frac{G}{C_{min}}$$

In parallel flow situations, the heat receiving media inlet temperature is on the same side as the heat source media inlet temperature. The heat receiving media typically has maximum heat capacity. The parameters are the following:

$$\varepsilon = \frac{\Delta t}{\theta} = \frac{1 - e^{-z(1-R)}}{1 + R}$$
$$R = \frac{\Delta t_{max}}{\theta} \tag{83}$$
$$z = \frac{G}{C_{min}}$$

Note that the above are valid only for the case when flue gas heats the steam and water.

7 Boiler efficiency and losses

7.1 Determining efficiency with the direct method

The ratio of the heat output to the energy input is boiler efficiency:

$$\eta = \frac{Q_{abs}}{Q_{in}} \tag{84}$$

where η is boiler efficiency
Q_{abs} heat output (absorbed by the working fluid)
Q_{in} energy input into the boiler.

This method of determining the efficiency by measuring the heat output directly from the boiler and the energy input into the boiler is called the direct method. To determine the efficiency of a boiler, a system must be defined and the energy flows that cross the system boundaries must be examined. Defining the system boundaries must consider that it is possible to define the energy flows in and out with sufficient accuracy. Figure 64 illustrates typical system boundaries. According to the definitions in Fig. 64 the system includes for excample the following components:

- Equipment for crushing and pulverizing the fuel
- Air preheaters
- Forced circulation pumps
- Fan for flue gas recirculation.

In this case flue gas fans (ID fans) and combustion air fans (FD fans) are outside the system. However, the system can be defined differently for different cases, in which the flue gas cleaning equipment for excample can be included.

The energy input has two components. One is proportional to the fuel flow, and the other does not depend on the fuel flow. Energy flows that depend on the fuel flow are as follows:

- Chemical energy included in the fuel, H_u
- Energy included in the fuel preheating, Q_f
- Energy included in the air preheating, Q_a.

When mass flows, specific heat values, and temperatures are known, the heat input with preheated air and fuel can be calculated using the following formulas:

$$\begin{aligned} Q_f &= m_f \cdot c_{p(fuel)} \cdot (t - t_0) \\ Q_a &= m_a \cdot c_{p(air)} \cdot (t - t_0) \end{aligned} \tag{85}$$

where
Q_f is heat input with preheated fuel (kW)
Q_a heat input with preheated air (kW)
m_f mass flow for fuel (kg/s)
m_a mass flow for combustion air (kg/s)
$c_{p(fuel)}$ specific heat for fuel (kJ/kgK)
$c_{p(air)}$ specific heat for air (kJ/kgK)
t temperature of air or fuel when entering the boiler (°C)
t_0 reference temperature (usually 25°C).

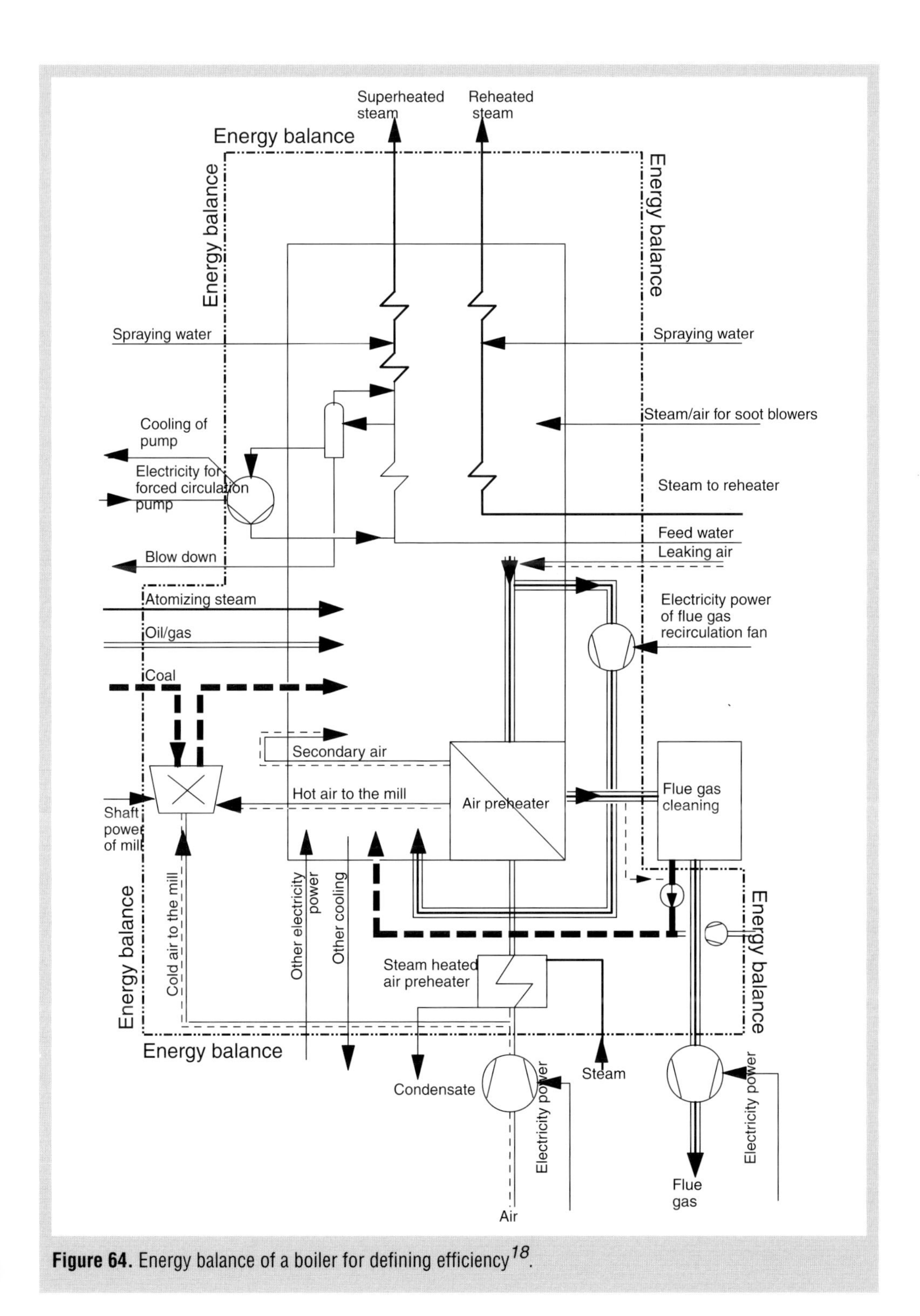

Figure 64. Energy balance of a boiler for defining efficiency[18].

If fuel and air are at reference temperature (t_0), the imported energy equals the chemical energy of the fuel multiplied by the fuel mass flow. Even at its highest, the energy from the preheating of the fuel and combustion air are only a few percentage points of the total energy input. Energy flows that do not depend on the amount of fuel entering the boiler are the following:

- Heat input with the steam heated air preheater, Q_{shap}
- Heat input with the fuel atomizing steam , Q_{atom}
- Use of the electric motors of mills, pumps and fans, P
- Spray water for adjusting the superheating, Q_{spray}.

Normally boiler plant is equipped with such measurements that these energy flows can be determined.

The following formulas provide calculation of heat output (large boilers may have a reheater) :

$$\begin{aligned} Q_s &= m_{ls} \cdot (h_{ls} - h_{fw}) \\ Q_{RH} &= m_{RH} \cdot (h_{RHa} - h_{RHb}) \end{aligned} \tag{86}$$

where Q_S is absorped heat for heating, vaporizing, and superheating (kW)
Q_{RH} absorped heat for reheating of steam (kW)
m_{ls} live steam flow (kg/s)
h_{ls} enthalpy of live steam (kJ/kg)
h_{fw} enthalpy of feed water (kJ/kg)
m_{RH} mass flow in reheater (kg/s)
h_{RTa} steam enthalpy after the reheater (kJ/kg)
h_{RTb} steam enthalpy before the reheater (kJ/kg).

In natural circulation boilers some absorbed heat exits the boiler with blow down. In practice, the blow down line is normally closed when measuring boiler efficiencies. Absorbed energy for heating the blow down can be calculated from the following equation:

$$Q_{BD} = m_{BD} \cdot (h_{BD} - h_{fw}) \tag{87}$$

where Q_{BD} is heat output with blow down (kW)
m_{BD} blow down mass flow (kg/s)
h_{BD} enthalpy of blow down (kJ/kg)
h_{fw} feed water enthalpy (kJ/kg).

The formula for calculating boiler efficiency is therefore the following:

$$\eta = \frac{Q_S + Q_{RH} + Q_{BD}}{m_f H_U + Q_F + Q_a + (Q_{shap} + Q_{atom} + P + Q_{spray})} \tag{88}$$

The following text denotes the energy flows in parentheses by Q_{oth} . Boiler efficiency depends on fuel, combustion method, and other items. Table 17 shows some typical boiler efficiencies.

Table 17. Typical boiler efficiencies calculated according to DIN 1942.

Fuel	Efficiency,%
Natural gas	94
Oil	93
Coal	92
Peat	90 – 91
Bark	90 – 91

7.2 Determining efficiency with the indirect method

The energy input is divided into heat output (absorbed heat) and losses:

$$Q_{abs} = Q_{in} - Q_{losses} \tag{89}$$

The formula for efficiency given earlier leads to the following:

$$\eta = \frac{Q_{in} - Q_{losses}}{Q_{in}} \tag{90}$$

or

$$\eta = 1 - \frac{Q_{losess}}{Q_{in}} = 1 - \Sigma q \tag{91}$$

where q denotes the ratio of single losses to the total input energy flow. The indirect method therefore gives a better idea of the factors influencing boiler efficiency and what measures are necessary to improve efficiency.

The losses consist of the following:

- Losses of unburned combustible fuel in ash and in flue gas
- Sensible heat in the ashes
- Sensible heat in the flue gases
- Radiation and conduction losses.

Losses related to the discontinuous use of the boiler such as losses in startups and losses in shutdowns lower the boiler efficiency. These losses are not topics for consideration here.

Losses of the unburned combustibles

Some fuel fed into the boiler might not combust. This is described by combustion efficiency. The combustion efficiency of modern combustion equipments is nearly 100%. The unburned fuel may leave the boiler in the gaseous phase with the flue gases or in the solid phase with the ashes. Determining the amount of unburned gases in the flue gases and the amount of unburned fuel in the ashes determines the combustion efficiency.

Losses of the unburned combustible gases

The unburned combustible gases are carbon monoxide and various types of hydrocarbons. The unburned gas found in the flue gases from large power plant boilers is primarily carbon monoxide. Hydrocarbons that occur in the flue gases of small boilers and combustion engines usually are not present. The following formula allows calculation of the losses caused by the unburned carbon monoxide:

$$q_{CO} = \frac{x_{CO} \cdot H_{CO} \cdot V_{fg}}{m_f H_U + Q_f + Q_a + Q_{oth}} \tag{92}$$

where q_{CO} is the ratio of loss due to unburned CO to the total input energy flow
x_{CO} concentration of unburned CO in dry flue gases (kg CO/m^{3}fg)
H_{CO} heat value for unburned CO (10.7 MJ/kg CO)
V_{fg} dry flue gas flow (m^3/s)
m_f fuel flow (kg/s)
H_u fuel heat value (MJ/kg)
Q imported energy flow.

The losses of unburned CO can also be determined according to the diagram of Fig. 65. In the diagram, the moisture in the fuel is represented by a moisture coefficient determined with dry fuel heat value and vaporization energy for water as follows:

$$k = \frac{H_{u(dry)}}{H_{u(dry)} - u \cdot l} \tag{93}$$

where $H_{u(dry)}$ is dry fuel heating value
u fuel moisture ratio (ratio of water to dry fuel)
l heat of vaporization for water (2.443 MJ/kg when t = 25°C).

One can see from the Fig. 65 that the CO concentration in flue gases of 500 ppm represents a deterioration of 0.24% in boiler efficiency. In this example it was supposed that fuel moisture was 50%, dry fuel heating value was 19 MJ/kg and excess air ratio was 1.2. So the fuel moisture ratio $u = 1$ kg$_{H_2O}$/kg$_{dry\ fuel}$ and moisture coefficient $k = 1.15$. CO emissions from power plant boilers are usually very small (approximately 50 ppm). The losses they cause are therefore very small i.e. ten times less than in the example. If the flue gases contain other unburned gases than CO, the losses they cause can be calculated with the same method as the CO losses by replacing the CO concentration and heat value with the values for the gas in question.

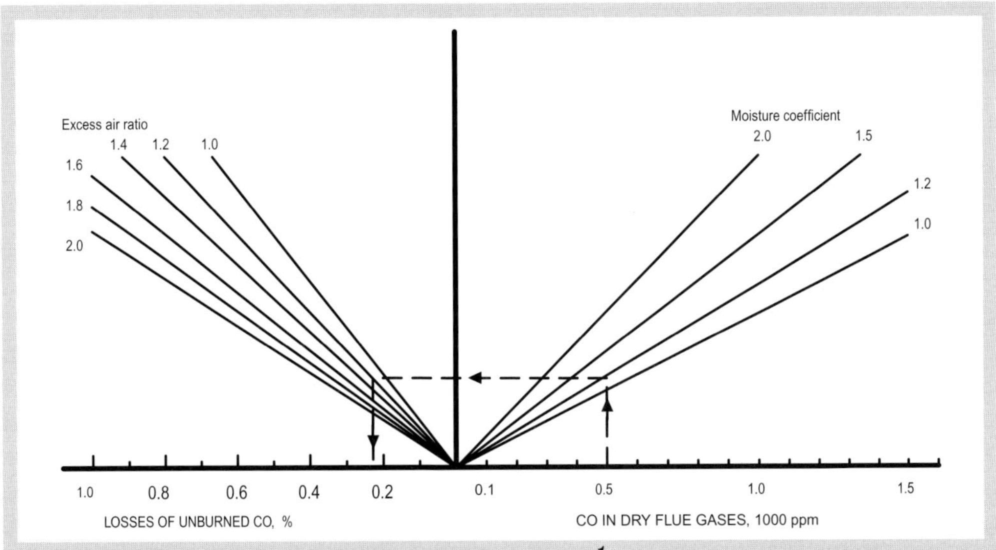

Figure 65. Diagram for determining the losses of unburned CO[1].

Losses of the unburned solids

Unburned solid fuel can leave the boiler with the fly ash or the bottom ash. Combusting a sample of dried ash in a laboratory oven in the temperature of 815°C and weighing the change of weight of the sample can determine the ratio of unburned fuel if there is no carbonate in the ash. If there are carbonate in the ash then the temperature should not be higher than 550°C because of decomposition of carbonate. The other possibility is to dissolve carbonate before burning. The following formula calculates the loss caused by the fuel that leaves the boiler with the ashes:

$$q_{unbs} = \frac{m_{unbs} \cdot H_{unbs}}{m_f H_U + Q_f + Q_a + Q_{oth}} \tag{94}$$

where q_{unbs} is the ratio of loss due to unburned solid fuel to the total input energy flow
H_{unbs} heating value for unburned solid fuel (MJ/kg)
H_u heating value for fuel (MJ/kg)
Q imported energy flow.

The ratio of unburned combustible solids to the amount of fuel fed into the boiler needed in the formula above is obtained from the following equation:

$$\frac{m_{unbs}}{m_f} = p \cdot \frac{x_t}{1-p} \cdot (1 - x_{H_2O}) \tag{95}$$

where m_{unbs} is mass outflow of unburnt fuel (kg/s)
m_f fuel mass flow (kg/s)
p weight loss of burned ash sample, (%/100)

x_t ash content in fuel, (%/100)
x_{H_2O} moisture content in fuel,(%/100).

Table 18 shows the heating value for unburned fuel.

Table 18. Heating value for unburned combustible fuel.

Unburned fuel	Heating value, H_{unbs}, MJ/kg
Coke from coal combustion	33
Coke from brown coal combustion	27.2

Ash leaves the boiler primarily as fly ash. If the amounts of fly ash and bottom ash are known and the amount of unburned fuel they contain is known, applying the method described above to each component (fly ash and bottom ash) separately can determine losses of the unburned solid fuel.

The losses caused by the unburned solid fuel can also be determined using the diagram of Fig. 66. In the example

- Weight loss of burned ash sample is 30%
- Moisture content in fuel is 50%
- Heat value for fuel is 9 MJ/kg
- Heat value for unburned fuel is 27 MJ/kg
- Ash content of dry fuel is 0.5%

With above values according to the Fig. 66 the losses of unburned solids is 0.32%.

The loss of unburned combustible fuel depends on the fuel, combustion method, and the condition and adjustment of the equipment. Low volatile content and high ash content raise the losses of unburned combustible fuel by several percentage points. Liquid and gaseous fuels and properly adjusted burners allow a loss of less than 0,1%. In modern fluidized bed boilers that burn bark or wood the loss of unburnt combustible is less than 1%.

Losses of sensible heat of ash and flue gases

The imported energy to the boiler leaves with the produced steam and other mass flows such as flue gas and ash that represent heat losses.

Sensible heat loss from flue gases

The biggest heat loss from a boiler is usually the sensible heat that exits with the hot flue gases. The flue gas loss depends on the final flue gas temperature and the amount of flue gases.

The flue gases should leave the boiler at a temperature as low as possible to minimize the flue gas losses. The temperature of the flue gases that contain sulfur should be higher than the temperature of the acid dew point (140–150°C) to avoid corrosion of the heat delivery surfaces.

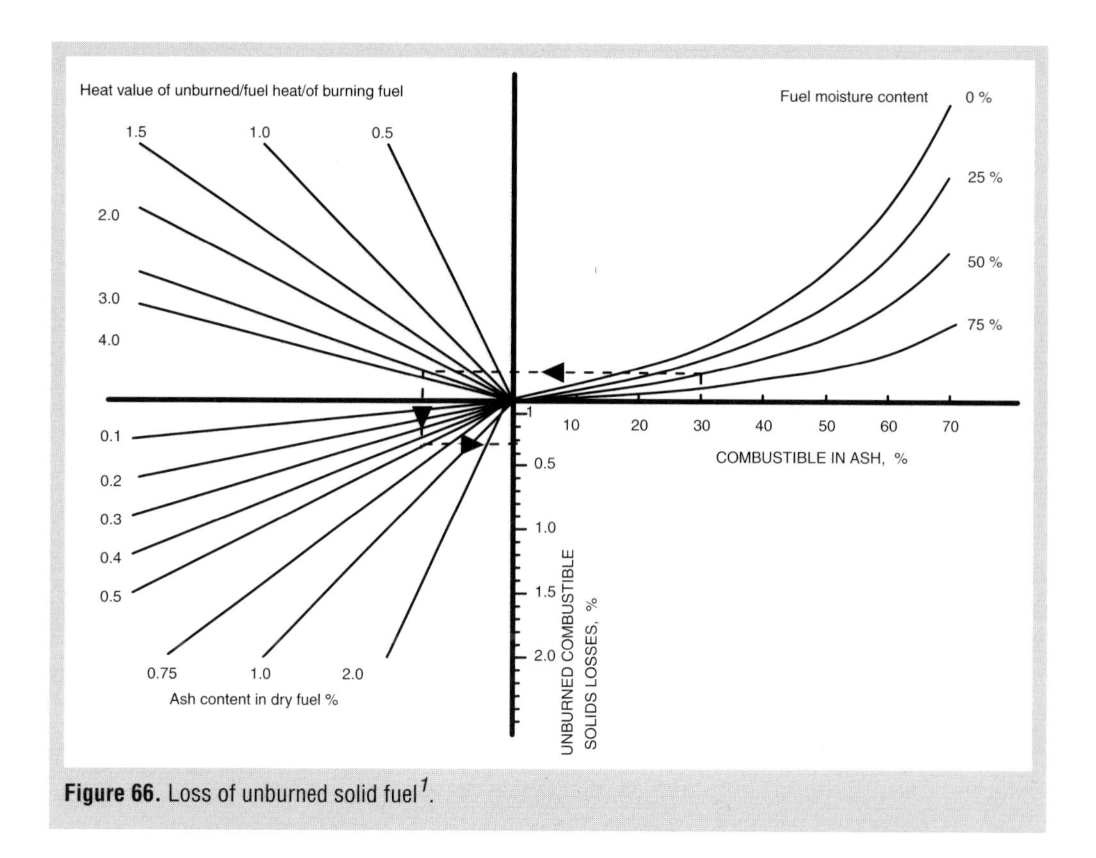

Figure 66. Loss of unburned solid fuel[1].

The heat transfer in the boiler deteriorates as the heat delivery surfaces become dirty. This raises the final temperature of the flue gases and the flue gas losses. The amount of air should also be as near the theoretical minimum value as possible because excess air increases the flue gas flow and the flue gas loss also increases. The amount of air however should not decrease too much or incomplete combustion will occur. Then the loss due to the unburned fuel may increase more than that gained by the lower flue gas loss (see Fig. 7). The following equation gives the flue gas losses:

$$q_{fg} = \frac{m_{fg} \cdot c_{pfg}(t_{fg} - t_0)}{m_f H_U + Q_f + Q_a + Q_{oth}} \tag{96}$$

where q_{fg} is ratio of flue gas loss (loss due to sensible heat of flue gases) to the total input energy flow

- m_{fg} flue gas mass flow
- m_f fuel mass flow
- c_{pfg} specific heat of flue gas
- t_{fg} flue gas end temperature
- t_0 reference temperature
- H_U heating value of the fuel
- Q imported energy flow.

If the amount of excess air and final flue gas temperature are known, the flue gas loss for wood fuels can be determined according to the diagram of Fig. 67. For example if

- Moisture content in fuel is 50%
- Flue gas temperature is 200°C
- Excess air ratio is 1.2
- Reference temperature is 25°C then flue gas losses are 11,4%.

Losses of sensible heat of ash

The ratio of the loss due to thermal energy that leaves the boiler with the ash to the total input energy can be calculated from the following:

$$q_{ash} = \frac{m_{ash} \cdot c_{pash}(t_{ash} - t_0)}{m_f H_U + Q_f + Q_a + Q_{oth}} \qquad (97)$$

where $c_{p(ash)}$ is specific heat of ash (fly ash 0.84 kJ/kg and bottom ash 1.00 kJ/kg)
t_{ash} ash temperature, °C
t_0 reference temperature, °C.

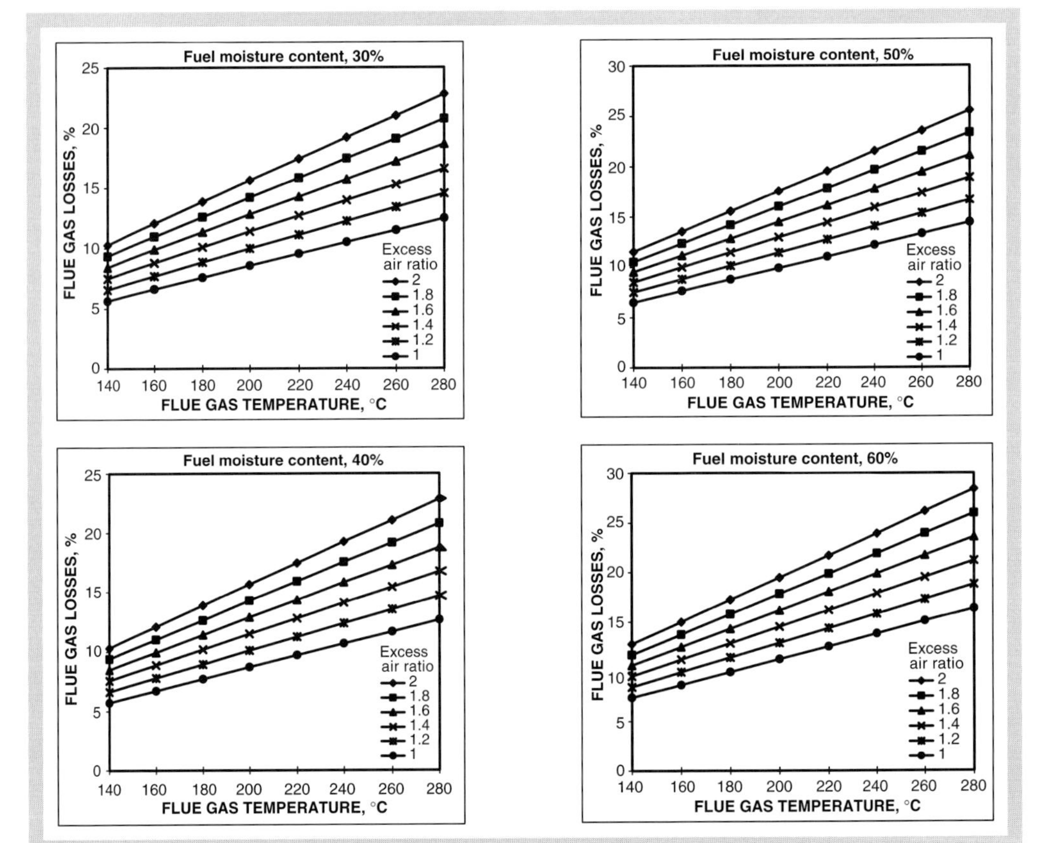

Figure 67. Diagram for determining the flue gas loss (reference temperature 25°C)[1].

The ratio of ash mass flow to fuel mass flow, m_{ash}/m_f, is obtained from the next equation. The ash flow includes the incombustible ash and the unburned fuel in the ash.

$$\frac{m_{ash}}{m_f} = \frac{x_{ash}}{1-p} \cdot (1 - x_{H_2O}) \qquad (98)$$

where x_{ash} is ash content in dry fuel (%/100)
x_{H_2O} moisture content in fuel (%/100)
p weight loss of burned ash sample (%/100).

With the method described above, the losses caused by fly ash and bottom ash can also be calculated separately. For calculation, the relative ratios of fly ash and bottom ash must be known.

If the ash content in fuel has its normally low value, the effect of loss of sensible heat of ash on the boiler efficiency is negligible.

Radiation and conduction losses

Since the temperature in the boiler is considerably higher than the temperature of the surroundings, heat transfers to the surroundings, although boilers are usually well insulated. In a well insulated boiler with a wall temperature below 55°C, the heat losses are 200–300 W/m^{2}K. Relative heat losses for large boilers are smaller than for small boilers because the exterior wall area does not grow in relation to the boiler capacity.

If the heat transfer surfaces, temperatures, and coefficients of heat transfer are known, the heat losses could be calculated but heat losses of power plant boilers are normally estimated based on experience . See the values in Table 19. Note from these values that the heat losses in power plant boilers are 1.0%–0.2% of the imported energy. At partial loads, the relative ratio of boiler heat losses increases since the absolute heat losses remain almost constant.

Table 19. Radiation losses of boilers,%[18].

Boiler capacity, MW	Oil	Fuel Bituminous coal	Brown coal
1	1.1		
10	0.6	1.1	1.6
100	0.3	0.55	0.8
500		0.35	

The heat that transfers through the boiler walls to the surroundings is partly used, since it heats the air of the boiler room. Combustion air usually comes from the upper part of the boiler room.

Figure 68 shows how boiler efficiency depends on boiler % capacity. As the % capacity drops from 100% to 60%, the efficiency improves slightly because heat exchangers work better with smaller temperature differences at partial load cooling the flue gases to a lower temperature and hence decreasing the flue gas loss. At very small partial load, the efficiency deteriorates due to heat losses unrelated to load. Also the losses of unburned fuel will increase because of poor mixing of fuel and combustion air at small partial load.

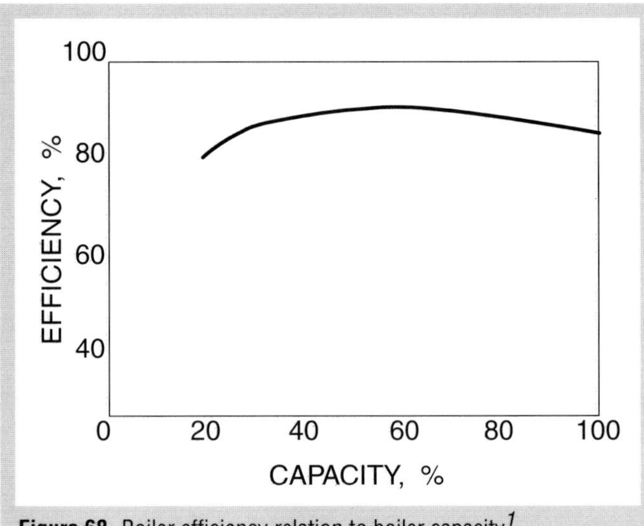

Figure 68. Boiler efficiency relation to boiler capacity[1].

References

1. *Huhtinen M., Kettunen A., Nurminen P.,Pakkanen H.: Höyrykattilatekniikka (in Finnish), Oy Edita Ab, Helsinki 1996*
2. *Neste Oy: Product data sheet for heavy and light oil, Espoo, 1996*
3. *Suomen Bioenergia Yhdistys (FINBIO): Puupolttoaineiden laatuohje (in Finnish), julkaisu 5 Jyväskylä , 1998*
4. *Singer J.G (ed.): Combustion Fossil Power, Fossil Fuels, Combustion Engineering , 1991, pp. 2–24/25*
5. *Kablitz: Marketing brochure of Kablitz-grate*
6. *Kunii D., Levenspiel A.: Fludization Engineering. John Wiley & Son Inc. New York/London/Sydney/Toronto 1969*
7. *Grewall N.S., Saxena S.C.: Maximum heat transfer coefficient between a horisontal tube and a gas-solid fluidized bed. Ind.Eng.Chem.Process Des.Dev.,20/1981 pp 108–118*
8. *Foster Wheeler Energia Oy: Training material of fluidized bed technology*
9. *Pellikka J., Savuharju K.: Kuoren ja puujätteen poltto,(in Finnish) Puumassan valmistus II osa 2 luku 29, Suomen Paperi-insinööri Yhdistys ry. 1983 pp 1519–1582*
10. *Iisa K.: Rikin oksidien muodostuminen ja poistaminen. Poltto ja palaminen.(in Finnish) International Flame Research Foudation Suomen osasto.1995 pp. 277–297*
11. *Foster Wheeler Energia Oy: P&I-diagrams of fluidized bed boilers*
12. *ABB Fläkt Oy: Puhallintekniikan käsikirja (in Finnish)*
13. *Kaijansaari S., Lumme E.: Ilman epäpuhtauksia päästävien pienten laitosten piipun korkeuden määritys (in Finnish). Ilmatieteen laitos, Tutkimus 79, 1986*
14. *Rothemuhle: Supply programme leaflet; Heat exchangers, dust collectors, fans*
15. *KSB: Type leaflet of high pressure centrifugal pumps*
16. *By-Cast: Leaflet of cast iron air preheaters*
17. *Nirafon: Technical quideline of acoustic cleaning*
18. *DIN 1942, Abnahme versuche an Dampferzeugerregelns (in German)*
19. *ABB Fläkt Oy: Marketing quidelines of axial and centrifugal fans*

CHAPTER 16

Heat and power co-generation

Carl-Johan Fogelholm and Jukka Suutela

Heat and power co-generation

1 Introduction

This chapter illustrates heat and power balances and co-generation alternatives in pulp and paper manufacturing. A modern pulp mill produces more heat and power than necessary for the process. This is due to low heat and power consumption, high dry solids content of the black liquor from the evaporation plant, and effective burning in the recovery and bio-fuel boilers. Power generation has increased after considering higher steam values. Power consumption has decreased but less than heat consumption because of environmental protection. New bleaching methods also require more power. Figure 1 shows the heat and power balance of a modern pulp mill.

2 Heat and power demand in pulp and paper manufacturing

Table 1 shows the heat and power demand for various pulp grades. A negative value indicates that the process generates heat.

Table 1. Heat and power demand of pulp.

		GJ/t	kWh/t
Mechanical pulp			
Groundwood (GW)	100 CSF	0–0.6	1 400–1 500
Groundwood	60 CSF	0–0.7	1 650–1 700
Pressure groundwood (PGW)	100 CSF	0–0.5	1 450–1 550
Pressure groundwood	60 CSF	0–0.6	1 700–1 800
Thermomechanical pulp (TMP)	100 CSF	-3.1	2 100–2 300
TMP	60 CSF	-4.3	2 600–2 900
Chemithermomechanical pulp (CTMP) 0		0	2 000
Neutral sulfite semichemical (NSSC)		3.8	400
Chemical pulp			
Softwood (pine)			
Unbleached		7.5–10.6	400–490
Unbleached dried		10–13.9	520–630
Bleached		9–12	540–680
Bleached dried		11.5–15	660–800
Hardwood (birch)			
Unbleached		7–10.0	390–500
Unbleached dried		9.5–13.2	500–650

Table 1. Heat and power demand of pulp.

		GJ/t	kWh/t
Bleached		8.5–11.0	510–620
Bleached dried		10.9–14.0	630–760
Recycled pulp			
Deinked		1.0	310
Not deinked		0	50

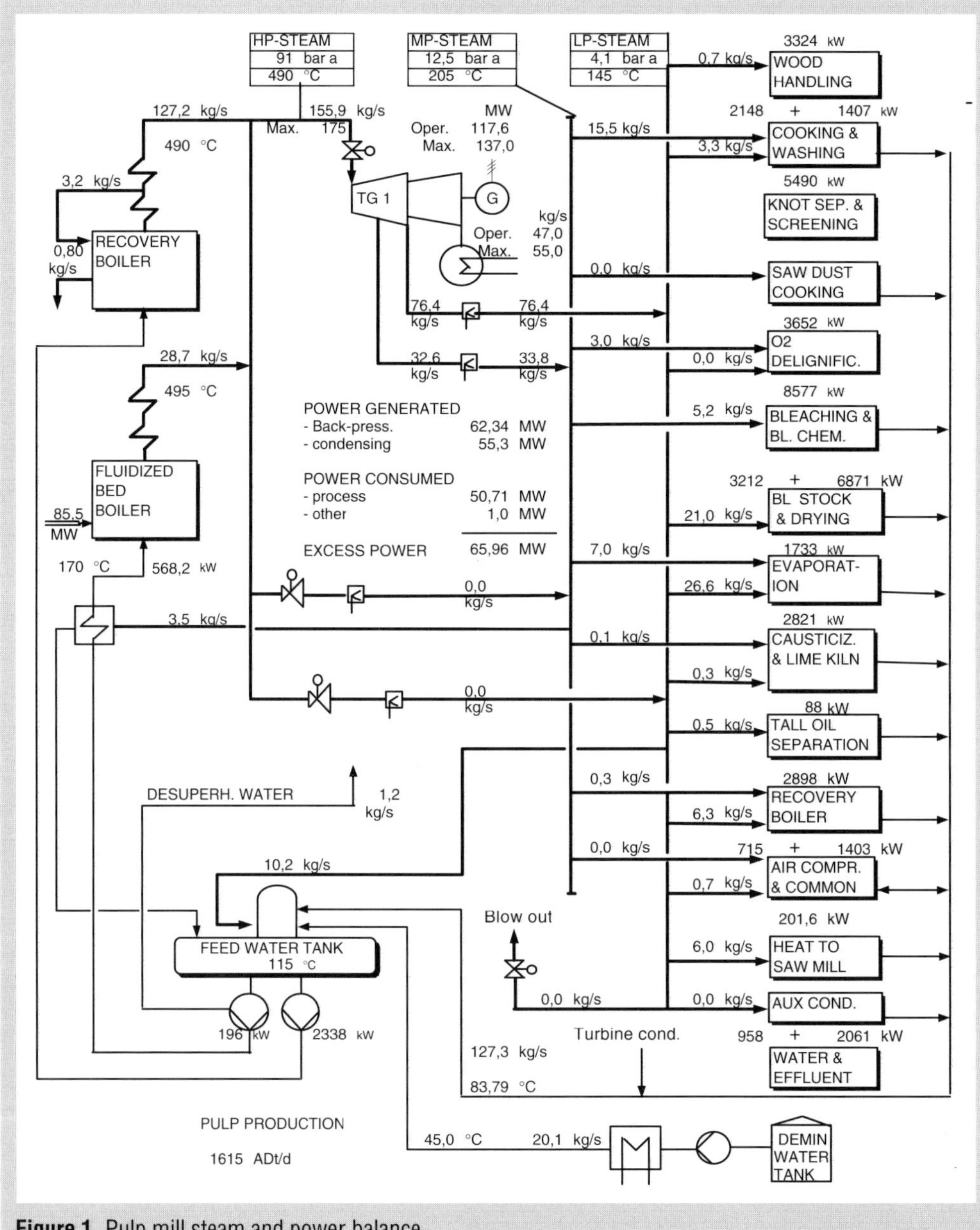

Figure 1. Pulp mill steam and power balance.

In the table, the first figure represents the potential of the late 1990s, and the last figure is the average of existing mills in the 1990s. A single figure represents the potential of the late 1990s.

Table 2 shows the heat and power demand for various paper grades.

Table 2. Heat and power demand of paper.

	GJ/t	kWh/t
Newsprint	4.5–5.3	550–585
Special newsprint	4.5–5.3	570–590
Uncoated wood containing	4.5–5.4	610–720
Coated wood containing	4.6–5.3	700–770
Uncoated wood free	6.6–7.1	535–670
Coated wood free	5.5–7.8	720–850
Tissue paper	6.9	1 010
Kraft liner	5.8	530
Fluting medium	5.6	510
Folding box board	6–7	560–740
Bleached kraft board	7.1–7.7	600–870
Sack paper	6.9	1 000
Other kraft paper and kraft board	7.3	850
Other paper and board	7.5	700

Table 3 shows the heat and power demand including the product drying for various mechanical wood industries.

Table 3. Heat and power demand of mechanical wood.

	GJ/unit	kWh/unit
Sawed timber	1.2	85
Plywood, veneer, and block board	5.7	300
Chipboard	2.7	190
Insulating board	10	1200
Hardboard	7.0	610

3 Heat and power demand of chemical pulp mills

This section covers the heat and power demand of various pulp mill departments. Middle pressure (MP) steam refers to 1 000–1 200 kPa gauge steam at the user point, and low pressure (LP) steam refers to 250–350 kPa gauge steam at the user point. The heat demand is the difference of heat in steam consumed and condensate returned.

3.1 Wood yard

The power demand of the wood yard including log handling, conveying, slashing, debarking, chipping, chip screening, chip storing, bark handling, etc., is about 10 kWh/m^3 sob (wood as solid over bark). Heat demand depends on the need for log deicing. In the Nordic countries, this is an annual average of 0.3–0.5 GJ/adt (air dry ton). During the winter, the value is about 1.0 GJ/m^3sob, but it is practically nil in summer.

3.2 Cooking

Tables 4 and 5 show the heat and power consumption of cooking methods.

Table 4. Heat and power demand of displacement batch cooking.

	Heat demand, GJ/adt	Condensate return, %	Power demand, kWh/adt
Softwood			
MP steam	1.6–1.9	85	35–50
LP steam	0.3–0.5	0	
Hardwood			
MP steam	1.4–1.6	80	30–45
LP steam	0.3–0.4	0	

In older cooking plants, the LP steam consumption can be over 2 GJ/adt when heat recovery is not proper.

Table 5. Heat and power demand of continuous flow cooking.

	Heat demand, GJ/adt	Condensate return, %	Power demand, kWh/adt
Softwood			
MP steam	1.7– 2.0	30–70	55–70
LP steam	0.2– 0.4	0	
Hardwood			
MP steam	1.5–1.7	30–70	40–50
LP steam	0.2–0.4	0	

The dry solids content in liquor to the evaporation plant is higher for continuous cooking than for batch cooking. This decreases the total steam demand for continuous cooking. The above power consumption includes diffusor washing.

3.3 Washing and screening

Power consumption of washing is 20–30 kWh/adt (50 maximum), and power consumption of softwood brown stock knot separation and screening is 70–100 kWh/adt. For hardwood pulp, the power consumption of knot separation and screening is approximately 20% lower. The washing and screening departments do not consume heat.

3.4 Bleaching

LP steam consumption of oxygen delignification is about 0.4–0.5 GJ/adt, and power consumption is about 40–50 kWh/adt. The LP steam consumption of elementary chlorine free (ECF) bleaching is about 0.4–0.6 GJ/adt, and power consumption is 80–120 kWh/adt depending on the system. Heat and power consumption of total chlorine free (TCF) bleaching is higher.

3.5 Bleach stock cleaning

The power consumption for bleach stock cleaning strongly depends on the method selected. Normal consumption is 25–85 kWh/adt, but power consumption can be as high as 150 kWh/adt. When using pulp in the same paper mill, bleach stock cleaning in the pulp mill is unnecessary.

Centricleaning is the most popular method for cleaning bleached pulp. The following are the most common types of centricleaning plants:

- 4–6 stage units removing only heavy rejects equipped with a stock thickener such as a decker or vacuum filter consuming 25–35 kWh/adt total
- 4–6 stage plant equipped with one thickening stage consuming 50–60 kWh/adt total
- 4–6 stage heavy reject and 3–4 stage light reject centricleaning plant consuming 75–85 kWh/adt total.

The first two plants can be before a drying machine head box because the outlet consistency of the stock is about 1.2%.

Bleach stock cleaning can also use slotted pressure screens. The width of the slot is normally 0.15–0.3 mm. Power consumption depends on the slot and the consistency of the stock. As an example, the power consumption is typically 25–30 kWh/adt with slotted 0.2 mm screens.

Bleach stock is normally screened in a consistency of about 2.5% or 1.2%. The latter value is valid when the stock feeds straight to the headbox.

3.6 Drying machine

The LP steam consumption of an airborne drying machine is about 2.7 GJ/t of water evaporated (about 2.9 GJ/t of water for cylinder driers). The web dryness after the press section is approximately 45%–48%. With improved press section configurations such as shoe presses, dryness values over 52% are possible. The approximate specific heat consumption of the drying is the following:

$$Q = hW \quad (1)$$

$$W = X_2(1/X_1 - 1/X_2)A \quad (2)$$

where Q is heat consumption, GJ
h heat demand of water evaporated, GJ/t water
W water evaporated
X_1 dryness after the press section
X_2 dryness after drying (normally 90%)
A production, adt.

Water evaporated with a dryness of 45% after the press section is therefore 1.0 t water/adt, and with a dryness of 52% it is only 0.73 t water/adt. The heat consumption for the pulp drier is approximately 2.0–2.7 GJ/adt pulp. Figure 2 shows the heat consumption for pulp drying.

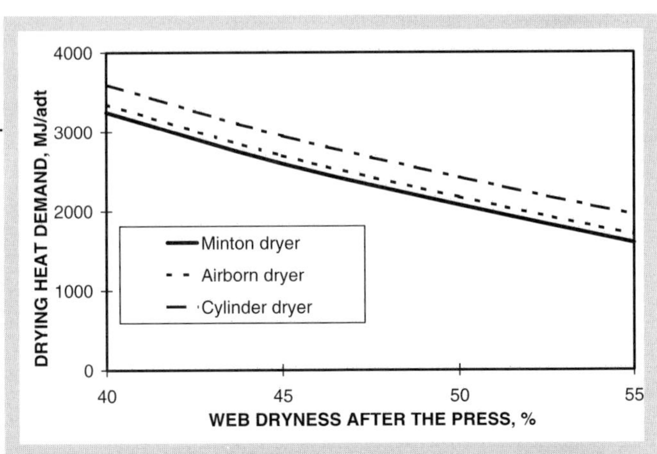

Figure 2. Heat consumption of pulp drying.

In any case, hot water results when condensate from the drier flashes to a temperature of about 105°C. This hot water generation is approximately 0.1 GJ/adt. In some cases, this hot water generation is sufficient. The total water heating demand of the drying machine is often about 0.5 GJ/adt. The condensate return rate is over 90% with a temperature of about 105°C. Power demand for the drying unit is approximately 80–130 kWh/adt.

3.7 Evaporation

The heat consumption of the evaporation plant depends on the number of effects, temperature of the incoming weak black liquor (normally about 90°C), and dry solids content of the incoming and exiting strong black liquor. Heat and power consumption of the evaporation plant is the following:

Number of heat effects	Power demand, MJ/t water	Power demand, kWh/t water
5	530+570	2.1+3.0
6	450+470	2.2+3.0
7	380+420	2.3+3.0

The black liquor dry solids content from the fiber line including spills is approximately 15%–18%. With low pressure steam of about 250–300 kPa, the liquor can be evaporated to 72%–75% dry solids content. If the black liquor is evaporated to more than 80% dry solids, the first body of effect 1 must use steam with a higher pressure over 600 kPa. In this way, the heat consumption of a 7-effect evaporation plant can be 80–100 MJ/t evaporated water of MP steam and 300–320 MJ/t evaporated water of LP steam.

Calculation of the heat consumption of a softwood pulp black liquor evaporation uses the following:

$$Q = a(1/X_1 - 1/X_2)h \qquad (3)$$

where Q is heat demand, GJ/adt
a dry solids per adt
X_1 dryness from the fiber line, ton dry solids/t water
X_2 dryness after evaporation plant, ton dry solids/t water
h heat demand of water evaporated, GJ/t water.

Consider the following example. For a seven-stage evaporation plant where the dry solids is 1.85 t/adt, the dryness of the incoming weak black liquor is 15.5%, and the dryness of the strong black liquor is 80%, the heat demand is as follows:

$Q = 1.85*(1/0.155–1/0.8)*(0.4) = 3.85$ GJ/adt

Fresh steam condensate return is about 95% at a temperature of approximately 105°C. Calculation of the power consumption for the same evaporation plant is the following:

$P = 1.85*(1/0.155–1/0.8)*(2.5) = 24$ kWh/adt

If the feed liquor temperature is 10°C lower or higher than 90°C, the heat consumption increases or decreases, respectively, by 30 MJ/t evaporated water.

The evaporation plant heat consumption can be higher if mill hot water or district heat is generated with evaporation plant extraction steam. The use of evaporation plant extraction steam is always more economical than low pressure steam use. Its use depends on application when turbine plant condensing steam is available. Chapter 12 provides further discussion.

3.8 Recovery boiler

The recovery boiler primarily uses steam for soot blowing, primary and secondary air preheating, liquor preheating (unnecessary in modern plants), and in a smaller amount for smelt shattering.

Soot blowing steam can be extracted from HP steam leaving the first super heater and then reduced to a suitable pressure of 2–3 MPa or extracted from the steam turbine. Soot blowing steam demand is 2.5%–4% of HP steam generated. In older and smaller overloaded boilers, soot blowing can be as much as 10% of generated steam.

Primary and secondary air is about 75%–85% of total air heated from 20°C–40°C to 110°C–120°C using LP steam and then to 150°C with MP steam. Total air demand for softwood black liquor firing is about 3.5–3.9 m^3N/kg dry solids when black liquor higher heat value (HHV) is 14–15 MJ/kg dry solids and excess air is 15%. Total air demand for hardwood black liquor firing is about 3.3–3.6 m^3N/kg dry solids when black liquor higher heat value is 13.5–14.5 MJ/kg dry solids and excess air is 15%. Calculation of heat demand for air heating is then the following:

$$Q = abc_p d(T_2 - T_1) \tag{4}$$

where Q is heat demand, kJ/kg dry solids
a amount of total air to be heated
b air demand per kg liquor dry solids, m^3N/kg dry solids
c_p air specific heat, kJ/kg °C (1.015 kJ/kg °C at 0°C–150°C)
d air density, kg/m^3N
T_1 air temperature before the heating battery, °C
T_2 air temperature after the heating battery, °C.

The following is a sample calculation. If the amount of total air to be heated is 0.85, the air demand per kg liquor dry solids is 3.6 m^3N/kg dry solids, air density is 1.29 kg/m^3N, air specific heat for LP steam is 1.015 kJ/kg °C and for MP steam is 1.02 kJ/kg °C, the air temperature after the heating battery is 110°C for LP steam and 150°C for MP steam, and the air temperature before the heating battery for LP steam is 25°C and for MP steam is 110°C, then the heat demand of air heating calculated with the formula given above is the following:

LP steam demand = 0.85*(3.6)*(1.015)*(1.29)*(110–25) = 340 kJ/kg dry solids

MP steam demand = 0.85*(3.6)*(1.02)*(1.29)*(150–110) =161 kJ/kg dry solids

When the dry solids amount for softwood black liquor is 1.85 t dry solids/adt, the above heat demands are 630 MJ/adt as LP steam and 300 MJ/adt as MP steam. All condensate from air heaters returns at a temperature of approximately 135°C because MP steam condensate usually flashes to LP steam.

Some MP steam is necessary for smelt shattering. The amount is approximately 0.01 kg/t dry solids (20–30 MJ/t dry solids).

Power demand for recovery boiler forced draft (FD) and induced draft (ID) fans is about 15–20 kWh/t dry solids. An additional 4–5 kWh/t dry solids is necessary for pumps, agitators, and electrostatic precipitator (ESP). The power demand for recovery boiler feed water pumps is an additional 4.3–5.0 kWh/t steam generated (8–10 MPa superheated steam).

Total power demand for the recovery boiler including fans, ESP, pumps and agitators, and feed water pumps is therefore about 35–40 kWh/t dry solids (about 55 kWh/adt for hardwood pulp and about 70 kWh/adt for softwood pulp).

3.9 Causticizing and lime kiln
Causticizing

LP steam demand of causticizing is about 4–5 MJ/m^3 white liquor (WL) produced (20 MJ/adt pulp), and power demand is 2–6 kWh/m^3 WL produced depending on the process selected (8–22 kWh/adt pulp). Power demands of different types of processes calculated per m^3 WL produced are the following:

- Green liquor clarifier with tube filter for WL, 2–4 kWh/m^3WL
- Cross flow filter for green liquor with tube filter for WL, 6–7 kWh/m^3WL
- Cross flow filter for green liquor with disc filter for WL, 6–7 kWh/m^3WL.

Lime kiln

MP steam demand for the lime kiln oil burners (atomizing steam) is about 50 MJ/t CaO produced (20 MJ/adt). Power demand of a lime kiln including ESP is about 25–40 kWh/t CaO (8–13 kWh/adt pulp). Kiln fuel demand (heavy fuel oil or natural gas) is about 5.5–7 GJ/t lime.

Bio-gas produced in a gasification plant using waste wood and bark can replace the fossil fuel for the lime kiln. The pulp mill then does not use fossil fuel or electricity from the grid.

3.10 Bleaching chemicals preparation

Table 6 gives the power and steam demand for production of various bleaching chemicals.If a vacuum pressure swing absorption (VPSA) process produces oxygen, the power consumption is about 750 kWh/t oxygen, and steam consumption is zero.

Normally, only ozone, oxygen, and ClO_2 plants are inside the mill. Other chemicals such as liquid chlorine, NaOH, peroxide, H_2SO_4, SO_2, etc., are normally purchased from external sources.

Table 6. Power and steam demand for production of various bleaching chemicals.

Chemical	Power, kWh/t (100%)	Steam, GJ/t
Peroxide	380	2.0 (MP)
Ozone	12 000	0
NaOH	300–2300	0.3
Oxygen	800	0.6
Integrated ClO_2	8 450–230	9 (MP)+32 (LP)
ClO_2	200	12
SO_2	27	0.7

Table 7 shows the bleaching chemical use for some pulp mills.

Table 7. Bleaching chemical use.

Mill Wood Bleaching type	A Pine TCF	B Pine TCF	B Birch TCF	C Pine TCF	C Pine ECF	D Pine ECF	D Birch ECF
Chemicals used, kg/adt							
Peroxide	42	35	30	59	0	3	3
Ozone	5	2	2	0	0	0	0
NaOH	35	33	30	42	21	26	26
Oxygen	13	13	12	12	10.5	5	4
H_2SO_4	23	18	16	19	0	0	0
ClO_2	0	0	0	0	20	26	22
$MgSO_4$	1	0	0	1	1	0	0
EDTA	4	3	2	4	2	0	0
SO_2	1	1.5	1.5	0	0	0	0
Power consumption, kWh/adt	70	35	34	10	13	8	8

If the mill actually produced all the chemicals that it theoretically could, the power demand in the case of mill C with ECF pulp would be about 240 kWh/adt. For mill D, the value would be about 260 kWh/adt. For pulp mills producing all chemicals inside the mill, power consumption for chemical production is about 260–300 kWh/adt of pulp, and steam demand is about 0.7–0.9 GJ/adt.

3.11 Water and effluent

Power demand for raw water pumping and distribution is 10–20 kWh/adt depending on water consumption and the distance and height difference of the raw water intake to the mill. Additionally, about 2–3 kWh/adt is necessary for chemical treatment of the water.

Power demand of the effluent treatment strongly depends on the effluent amounts for treatment of the selected process. It is normally about 25–45 kWh/adt.

3.12 Power boiler

Power demand of a fluidized bed (FB) boiler is about 8–12 kWh/t steam generated. The demand of the ESP is about 1.5 kWh/t steam generated, and the consumption of the feed water pumps is 4–5 kWh/t steam generated (8–10 MPa superheated steam). Power demand of an oil or gas fired boiler is only 3 kWh/t steam generated. The power in the FB boiler is mostly consumed in the primary (fluidizing air) air fan with pressure of about 15 kPa, in the secondary air fan (about 4 kPa), and in the flue gas fans.

3.13 Steam turbines and water cooling

Power demand of steam turbines is negligible, since the turbine oil pump is shaft driven. Most power demand is necessary for condensate pumping and cooling water pumping.

The following formula gives the condenser cooling water demand:

$$m_{cw} = m_c(h_s - h_c)/((T_{out} - T_{in})c_p) \tag{5}$$

where m_{cw} is cooling water demand, kg/s
m_c maximum condensate flow, kg/s
h_s enthalpy of steam to condenser, kJ/kg
h_c condensate enthalpy, kJ/kg
T_{out}-T_{in} outlet minus inlet water temperature, °C
c_p specific heat of water, kJ/kg°C.

The following is an example. When maximum condensate flow is 1 kg/s, steam enthalpy to condenser is 2 300kJ/kg, condensate enthalpy is 126 kJ/kg, condensate temperature is 30°C, the temperature difference is 10°C, and the specific heat of water is 4.19 kJ/kg, the cooling water demand is the following:

$m_{cw} = 1*(2\ 300 - 126)/(10 \times 4.19) = 52$ kg/s

Cooling water flow is normally constant regardless of condenser load.

For an open cooling water system, the cooling water pumping power calculation is as follows:

$$P = m_{cw}g((dH)/\eta_p) \tag{6}$$

where P is power requirement, kW
m_{cw} cooling water demand, kg/s
g specific gravity, m/s^2
dH pumping head, m
η_p pump efficiency.

An example of calculation for power demand for the condenser, in the previous example, is the following:

$P = 52 \times 9.81 \times 20/\ 0.73 = 14$ kW/ kg/s steam to condenser

when the cooling water demand is 52 kg/s, specific gravity is 9.81 m/s^2, and pumping head is 20 m. With a forced draft cooling tower, the power demand is higher because the tower cooling air fans can double the necessary power demand in warmer countries.

Alternator cooling demand is about 1.5%–2%, and turbine oil cooling demand is about 2%–2.5% of the alternator maximum effect for a nongeared turbine and about 3.5%–4% for a geared turbine. This cooling water demand is as follows:

$$m_{cw} = (1 - \eta_g - \eta_{oc}) \times P_g/(T_{out} - T_{in})c_p \tag{7}$$

where η_{oc} is relative oil cooling
η_g alternator efficiency
m_{cw} cooling water demand, kg/s
P_g maximum alternator effect, kW.

Cooling water demand for a 50 MW turbo alternator is then as follows when the cooling water temperature rises by 7°C:

$$m_{cw} = 0.04*50\,000/(7 \times 4.19) = 68 \text{ kg/s}$$

3.14 Other

Additional power is necessary for the following:

- Mill and instrument air, 10–15 kWh/adt pulp
- HVLC and LVHC gases, 5–10 kWh/adt pulp
- Lighting, workshops, and offices, 15–20 kWh/adt pulp.

In Far Eastern countries, power is commonly supplied for outside infrastructures such as villages for the mill workers and their families. This may increase the total power demand by 1–3 MW. The electrical room, control room, and office air conditioning also require additional power of 1–2 MW compared with Nordic countries.

4 Heat and power balance of chemical pulp mill

4.1 Installed power vs. demand

Table 8 shows power demand vs. installed power and department power demand in a mill that started in the beginning of the 1990s.

Table 8. Power demand.

Mill department	Installed power normal AC, MW	Installed power speed controlled, MW	Power demand installed without spares, %	Power demand per t produced, kWh/adt
Wood handling	0.6	3.3	74	45
Cooking	4.9	-	66	48
Washing & screening	5.6	0.6	78	71
Bleaching	12.1	0.4	63	112
Cleaning & drying	13.8	2.4	68	160
Total fiber line	37.0	6.7	68	436
Evaporation	2.7	0.4	53	24
Recovery boiler	2.3	4.4	40	39
Condensate, demineralization, and feed water	0.5	3.3*2	57	33
Causticizing	1.1	4.4	58	13
Lime kiln	0.2	0.5	69	8
Total recovery	6.8	9.0	50	117
River & chemical water	2.0	0.3	58	22
Waste water treatment	2.8	0.3	73	33
Mill air	1.1	-	85	14
Ventilation and general	2.3	-	59	20
Total auxiliaries	8.2	0.6	68	89

Table 8. Power demand.

Mill department	Installed power normal AC, MW	Installed power speed controlled, MW	Power demand installed without spares, %	Power demand per t produced, kWh/adt
Total pulp mill				
Without spare equip.	52.1	16.2	64	642
With spare equip.	52.7	20.5	60	-
Bark boiler and feed water pump	8	38	56	31
Total mill	536	243	595	673

In the table above, the mill has been operating at an annual production rate of 1.12 times the designed production rate. At the design production, the power consumption is about 715 kWh/adt. The power consumption has two components: a fixed part that does not depend on the production rate and a part that depends on the production rate. This means the fixed part of the power consumption is about 55% calculated from the power consumed at the design production rate. In this mill, speed control has been fully used only in the recovery and power boilers and their feed water pumps.

Assuming that the power demand is a linear function of the relative production at a production rate of 0.8–1.2 times the design production, the function for a mill is the following:

$$SP_a = SP_d(a(R_d/R_a) + 1 - a) \quad (8)$$

where SP_a is specific power demand for actual production, kWh/adt
SP_d specific power demand for design production, kWh/adt
R_a actual production rate, adt
R_d design production rate, adt
a fixed part of the power consumption.

In a mill that started in the late 1970s, the total power consumption is about 730 kWh/adt with the same proportional production rate. The power consumption is about 59% of the installed effect without considering spare equipment. In this mill, the fixed part of the consumption is about 48%.

When a pulp mill operates with a production rate of 0.8–1.2 of the design production, about 45%–55% of the power consumed is fixed and the remainder depends on the production rate. Normally, the fixed portion should be lower in newer mills than in older ones. In mills with more speed controlled equipment (frequency controlled motors or hydraulic couplings between the motor and equipment), the fixed consumption can be as low as 45%. Variations do exist in different departments of the mill.

4.2 Heat generation

Heat results primarily from firing heavy black liquor in the recovery boiler and bark and wood waste in the power boiler. Additionally, some heat comes from firing low volume high concentration (LVHC) gases. Firing places are recovery boiler, lime kiln, or a separate MP steam boiler.

Recovery boiler

When cooking to lower kappa number with oxygen delignification before the bleaching plant (ECF type bleaching), black liquor with softwood will be about 1.8–1.9 t dry solids/adt bleached. With hardwood, the value is about 1.6–1.7 t dry solids/adt. Black liquor heat value depends strongly on the pulp process. Higher dry solids per ton pulp (lower pulp yield) gives a lower dry solids calorific heat value when pulping the same wood species. This is the case with TCF pulp compared with ECF pulp.

From the above higher heat value, about 72%–75% (hardwood about 71%–74%) will convert to steam when black liquor dry solids content is 72%–85%. About 15% of the HHV will go with the smelt, and the rest will be lost with the flue gases. Table 9 shows specific values, and Fig. 3 illustrates black liquor dry solids information.

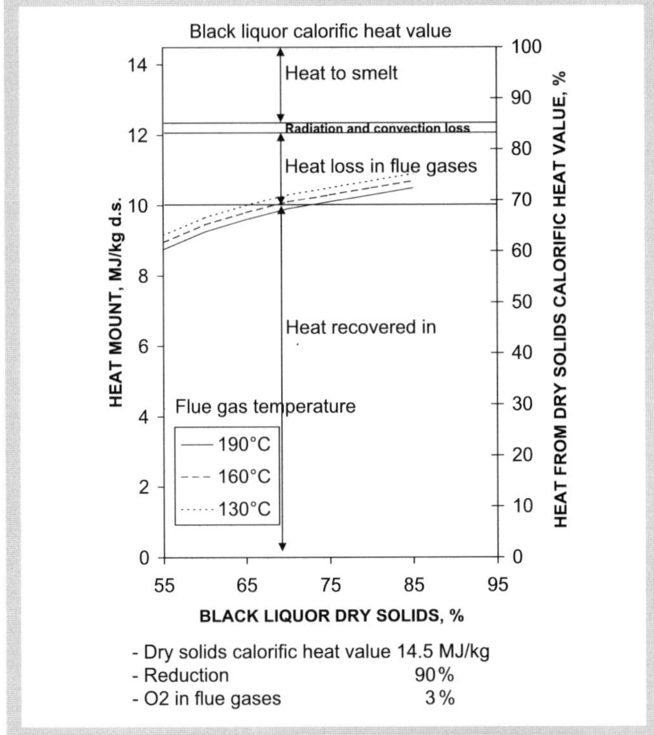

Figure 3. Black liquor dry solids heat to smelt, flue gases, and steam.

Table 9. Recovery boiler balance.

Wood species		Hardwood	Softwood
Black liquor dry solids	tds/ADt	1.6–1.7	1.8– 1.9
- dry solids HHV	GJ/tds	14.5–13.5	15–14
- black liquor ds	%	70–85	70–85
- heat to steam	%	69–73.5	70–74.5
- heat generated	GJ/ADt	15–17	18–20
Steam generation			
- HP steam 8.4 MPa/480 °C	GJ/t	3.345	3.345
- feed water 115 °C	GJ/t	0.485	0.485
- steam generation	ts/tds	3.3–3.7	3.4–3.9
- steam generation	ts/ADt	5.2–6.0	6.3– 7.0

Power boiler

Bark and wood waste are normally burned in a power boiler designed for such fuel. Table 10 shows the calculation of a bark and wood waste boiler balance using only internal fuel firing.

Table 10. Bark and wood waste boiler balance.

Wood species	Units	Hardwood	Softwood
Wood to cooking	m^3sub/adt	3.67	5.11
Chip screening loss	%	1.5	2
Wood as roundwood	%	70–100	50–100
Debarking loss	%	1.5	1.5
Wood to firing	m^3sub/adt	0.096–0.113	0.144–0.184
Wood dry substance	kg/m^3s	520	400
Wood dry substance	%	50	50
Wood to firing	kg/adt	100–117	115–147
LHV in dry solids	MJ/kg	19.1	19.2
LHV as fired	MJ/kg	8.3	8.35
- wood heat content	GJ/ADt	0.83–0.97	0.96–1.23
Wood to bark drum	m^3sub/adt	3.79	5.30
Wood as roundwood	%	70–100	50–100
Wood bark content	%/m^3sob	13	12
Bark to firing	m^3sub/adt	0.40–0.57	0.36–0.72
Bark dry substance	kg/m^3s	500	305
Bark dry substance	%	50	45
Bark to firing	kg/adt	400–570	245–490
LHV in dry solids	MJ/kg	22.5	20
LHV as fired	MJ/kg	10	7.6
Bark heat content	GJ/adt	4.0–5.7	1.85–3.7
Total wood & bark fired	GJ/adt	4.8–6.7	2.8–4.95
Boiler efficiency	%	87	87
Heat to steam	GJ/adt	4.2–5.8	2.44–4.3
Steam generation			
HP steam 8.4 MPa and 480°C	GJ/t	3.345	3.345
Feedwater 115°C	GJ/t	0.485	0.485
Steam generation	ts/adt	1.47–2.0	0.85–1.5

As Table 10 shows, steam generation with bark and wood fines depends on the amount of wood received as nonbarked roundwood and saw mill chips. Bark and wood fines from the pulp mill may also be transported to a paper mill when the distance is reasonable. In a paper mill, this bio-fuel can then provide back pressure power generation.

4.3 Heat and power balance

Heat and power balance of the mill depend on many parameters:

- Pulp yield (depends on wood type and pulp type)
- Amount of wood received as chips (barked or nonbarked roundwood)
- Process heat and power consumption
- Steam parameters in the power plant and pulp mill
- Pulp mill integration into paper mill.

Table 11 gives the heat and power balance of a modern pulp mill producing hardwood and softwood pulp for three cases.

Table 11. Heat and power balance.

Mill type	Units	Nonintegrated and bark sold	Nonintegrated and bark fired	Fully integrated and bark fired
Heat generation				
Black liquor	GJ/adt	18.0	18.0	18.0
Bark & wood waste	GJ/adt	-	4.2	4.2
Heat consumption				
Pulp mill process	GJ/adt	11.7	11.7	9.0
Paper mill process	GJ/t paper	–	–	6.5
Back pressure power	GJ/adt	3.2	3.2	4.4
Condensing power	GJ/adt	3.1	7.3	2.3
Power generation				
Back pressure power	kWh/adt	870	870	1200
Condensing power	kWh/adt	300	710	225
Total	kWh/adt	1170	1580	1425
Power consumption				
Pulp mill process	kWh/adt	660	700	550
Paper process	kWh/t paper	–	–	650
Power sold	kWh/adt	510	880	225

4.4 Steam turbines

In an integrated pulp and paper mill, the steam turbine can be the extraction back pressure type or the multiextraction condensing type. In a nonintegrated pulp mill, the turbine is normally a multiextraction condensing type. High pressure steam values are from 6 MPa and 450°C to 10 MPa and 500°C. Medium pressure steam is extracted at 1.1–1.3 MPa, and low pressure steam is extracted at 300–400 kPa. Steam not necessary in the process goes to a condensing tail or a separate condensing turbine. Figure 5 shows the turbine and steam distribution line diagram for a nonintegrated pulp mill with a capacity of 600 000 t/year of bleached pine and birch pulp.

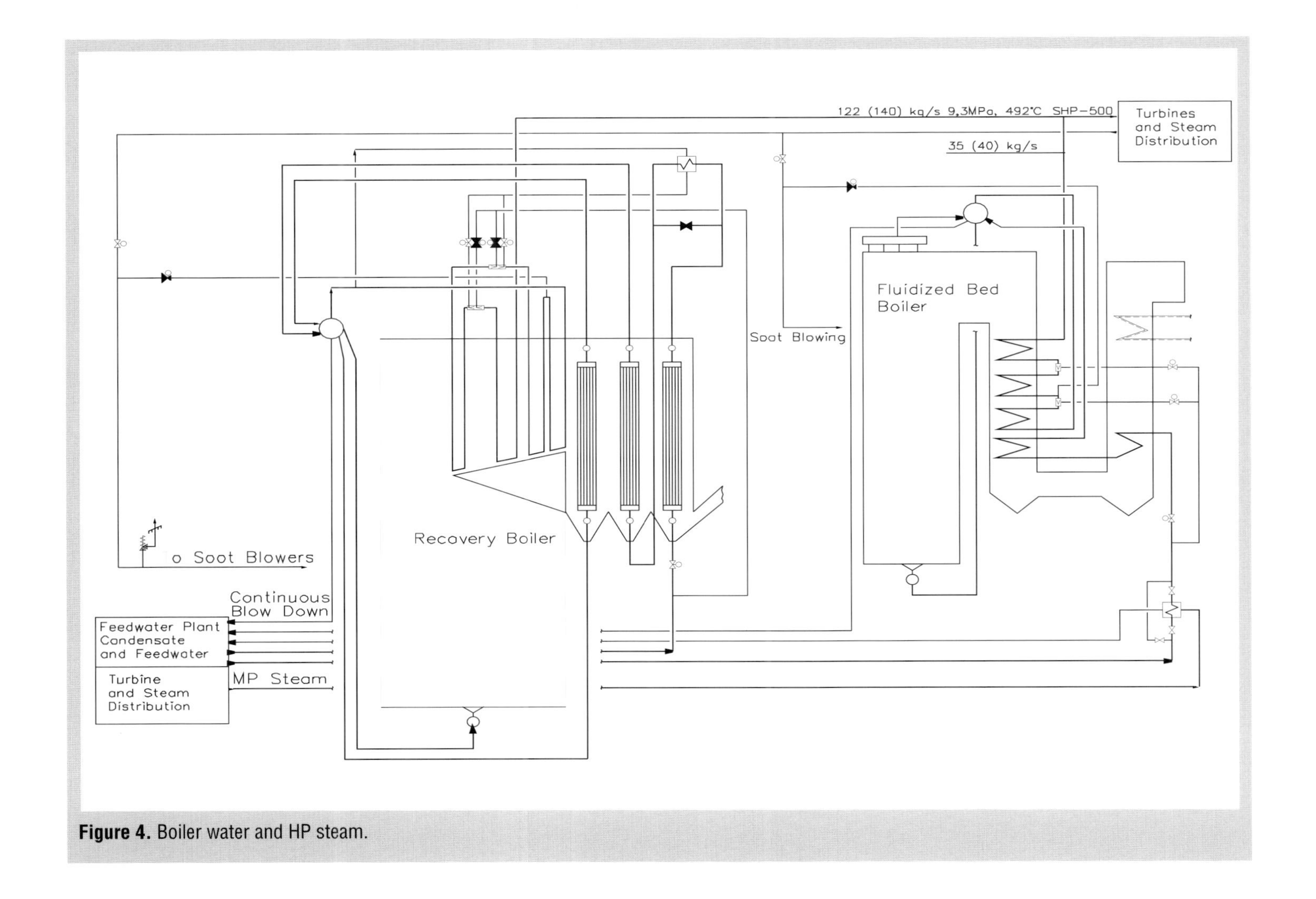

Figure 4. Boiler water and HP steam.

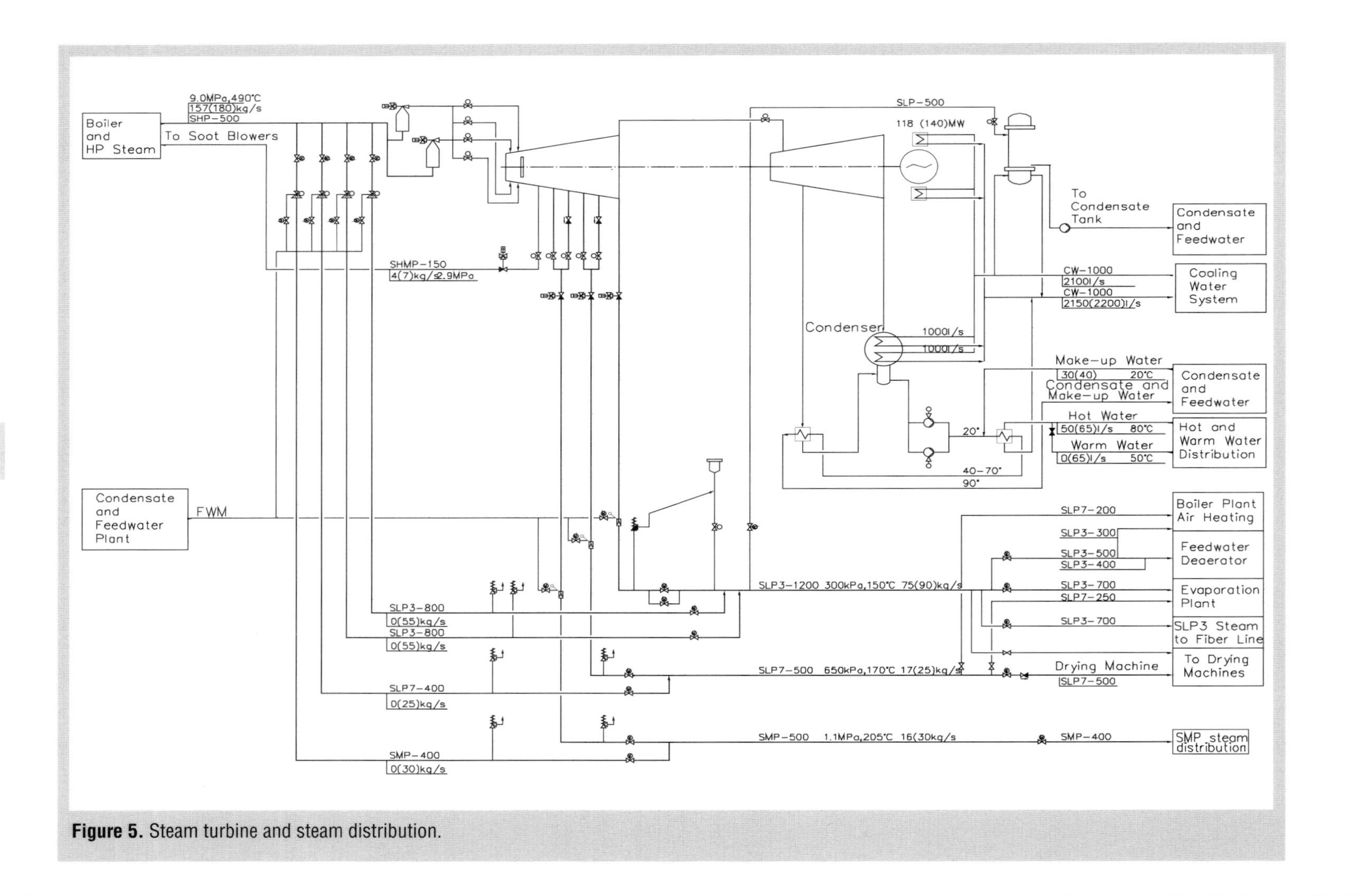

Figure 5. Steam turbine and steam distribution.

The following can increase power generation:

- Using higher inlet steam values while considering that the restrictions of the recovery boiler corrosion will increase with higher steam values
- Extracting soot blowing steam from the turbine at about 2.5–3 MPa instead of removing it after the recovery boiler primary super heater
- Using controlled (2–3 bleeds in the turbine) extractions for higher MP and lower MP steam such as 1.2 and 1.0 MPa
- Using controlled extraction (LP steam at 600–800 kPa) for evaporation plant concentrators and recovery boiler air heaters following 300 kPa steam air heaters
- Using make-up (demineralized) water preheating first with mill excess warm or hot water
- Using make-up water and turbine condensate heating with turbine condensing part bleed steam
- Making certain that the turbine guarantee points match the process needs in different operating cases such as lower and higher pulp production rate and steam demand variation between winter and summer months
- Considering that turbine and alternator efficiency is normally best in the range of 60%–100%
- Remembering that pulp design production can later be increased 15%–30% by cost effective projects to eliminate bottlenecks, but the turbine throughput cannot normally be increased from the original dimensioning
- Judging the efficiency of the turbine and alternator and not only the equipment price.

Table 12 shows an example of pressure and temperature selection influencing the steam turbine power generation with a 600 000 t/year nonintegrated pulp mill.

Table 12. Steam turbine power generation.

High pressure steam values	MPa ^{0}C	6 455	9 492
Power to heat ratio MP steam (1.2 MPa) LP steam (300 kPa)		 0.123 0.271	 0.154* 0.307
Condensing power efficiency with 87% boiler efficiency	 %	 29.3	 30.6
Power generation Back pressure power Condensing power Total	 MW MW MW	 55.8 54.0 109.8	 63.8 54.3 118.1
Power demand Power sold	MW MW	48.0 61.8	48.9 69.2

* *Normally, about 20%–25% of the heat consumption is medium pressure steam. The remainder is low pressure steam. (The MP steam share can be higher if the drying machine operates with MP steam.)*

4.5 Back pressure power process

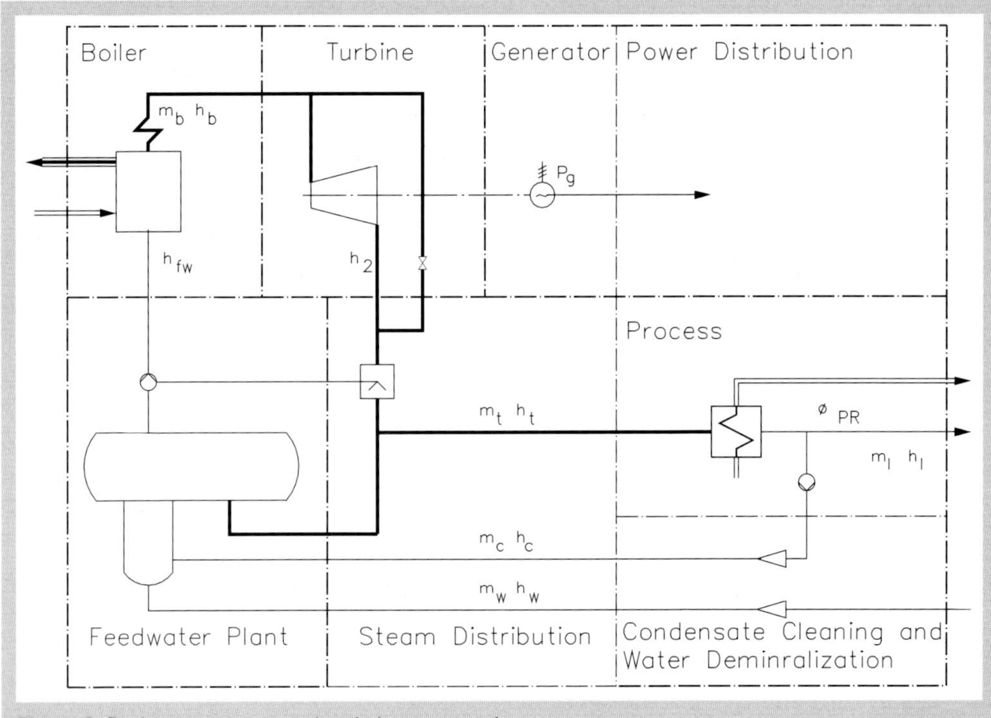

Figure 6. Back pressure power plant balance network.

When calculating back pressure power processes, using well chosen energy balances is important. Figure 6 shows a simplified network of a back pressure power plant and its heat consumers. The total balance of the figure has eight partial balances.

The most important characteristic in back pressure power generation is the power to process heat ratio:

$$\alpha = P_g / Q_t \tag{9}$$

where P_g is power generation at the alternator terminals
Q_t process heat demand

Heat balance over the turbine and the alternator is the following:

$$m_b h_b - m_b h_2 - P_g - Q_{tl} = 0 \tag{10}$$

or

$$P_g = m_b(h_b - h_2) - Q_{tl} \tag{11}$$

where Q_{tl} is turboalternator mechanical and electrical losses (about 3%–5%)
m_b boiler steam flow, kg/s
h_b boiler steam enthalpy, kJ/kg
h_2 steam enthalpy after turbine, kJ/kg.

The heat balance over boiler feed water and steam distribution is the following:

$$m_b h_2 - m_t h_t + m_c h_c + m_w h_w - m_b h_{fw} - Q_{sl} = 0 \tag{12}$$

where Q_{sl} is loss

and

$$m_t h_t - m_c h_c = m_b(h_2 - h_{fw}) + m_w h_w - Q_{sl} = Q_t \tag{13}$$

where Q_t is process heat demand, kW
Q_{sl} feed water and steam distribution loss
m_t steam to process, kg/s
h_t process steam enthalpy, kJ/kg
m_c process condensate return, kg/s
h_c process condensate enthalpy, kJ/kg
h_{fw} feed water enthalpy, kJ/kg
m_w makeup water, kg/s
h_w makeup water enthalpy, kJ/kg.

The above gives appropriate meaning for the process heat demand, Q_t. Since the terms m_w and h_w (makeup water heat over 0°C) and Q_{sl} are very small and have opposite signs, the process heat demand can in practice be the following:

$$Q_t = m_b(h_2 - h_{fw}) \tag{14}$$

The turboalternator power is as follows:

$$P_g = \eta_{gm}(m_b)(h_b - h_2) \tag{15}$$

Combining these formulas, the back pressure power characteristic is then:

$$P_g = \eta_{gm}(h_b - h_2)/(h_2 - h_{fw})Q_t = \eta_{gm}(h_b - h_2)/(h_b - h_{fw})Q_b \tag{16}$$

and

$$Q_b = Q_t + P_g/n_{gm} \tag{17}$$

and

$$\alpha = \eta_{gm}(h_b - h_2)/(h_2 - h_{fw}) \tag{18}$$

where Q_b is boiler heat generation, kW
η_{gm} alternator and turbine mechanical efficiency.

These formulas show that increasing the boiler steam enthalpy and feed water enthalpy will increase the power to heat ratio.

The heat demand of the power plant requires handling like other mill departments. Note that only real heat consumers like air and oil heating, soot blowing steam, etc., are considered. Feed water heating in the deaerator and preheaters is not heat consumption but internal circulation where heat is returning to the boiler. Figure 7 shows the power to heat ratio for co-generation plants, and Fig. 8 gives the heat rate of additional power in steam.

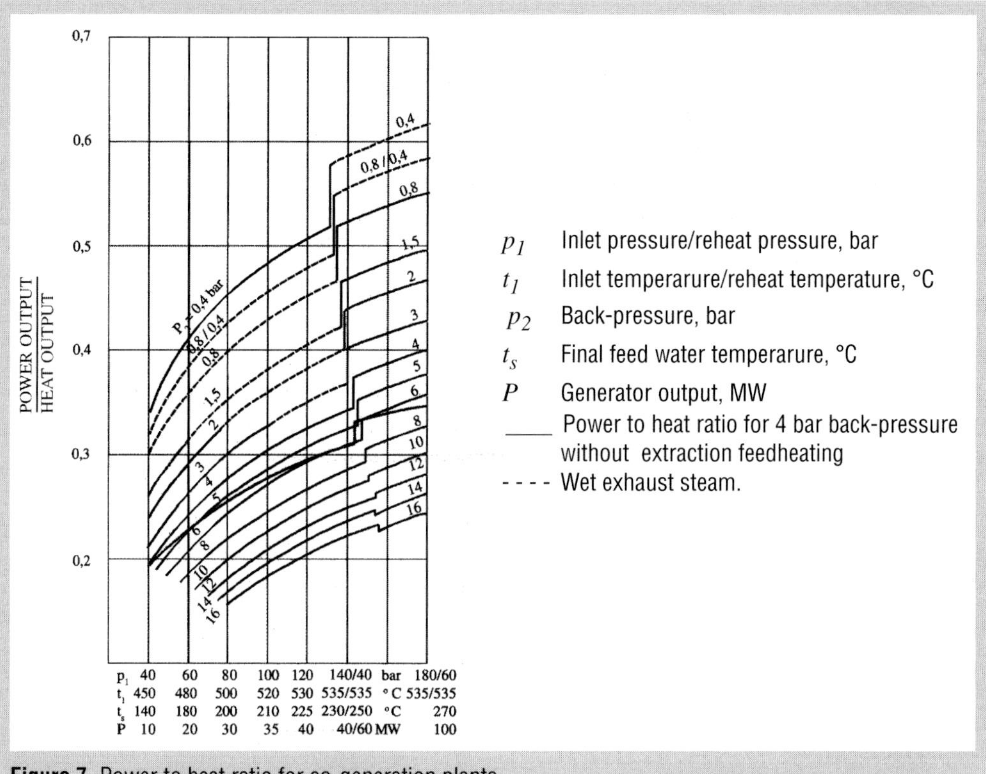

Figure 7. Power to heat ratio for co-generation plants.

Steam turbine characteristics

Steam turbine efficiency depends on the size of the turbine. Figure 9 gives the optimum load efficiency of axial flow back pressure turbines depending on the size of the turbine

The x-axis of the curve describes the turbine size = $P_{is}/(p_1-p_2)$ [kW/bar]

where

P_{is} is optimum load isentropic power, kW

p_1 inlet pressure, bar

p_2 outlet pressure, bar.

Calculation of the power given by the alternator then uses the following:

$$P[kW] = \eta_{is} m dh_s \qquad (19)$$

where

η_{is} is optimum load efficiency

m steam flow, kg/s

dh_s isentropic enthalpy difference, kJ/kg

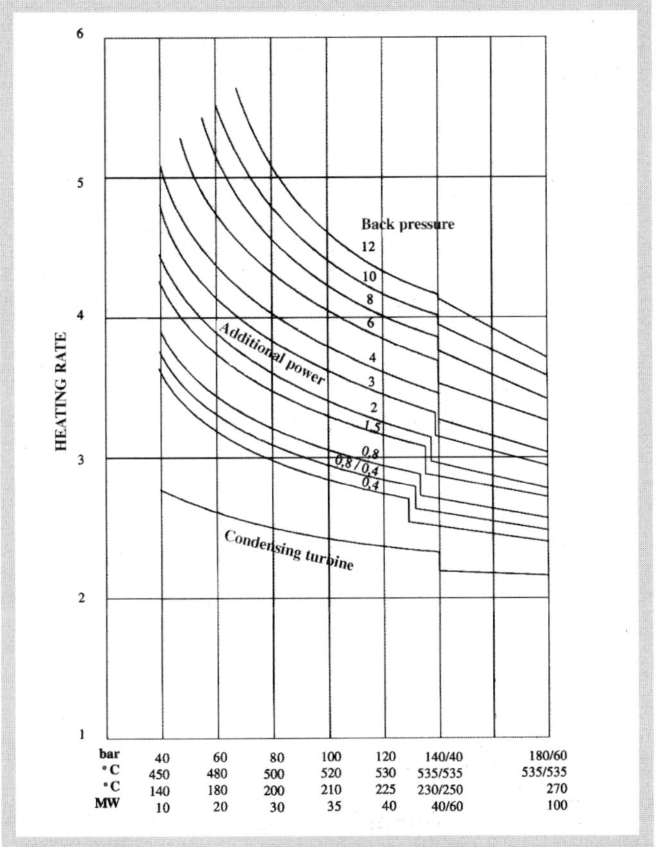

Figure 8. Heat rate of additional power in steam.

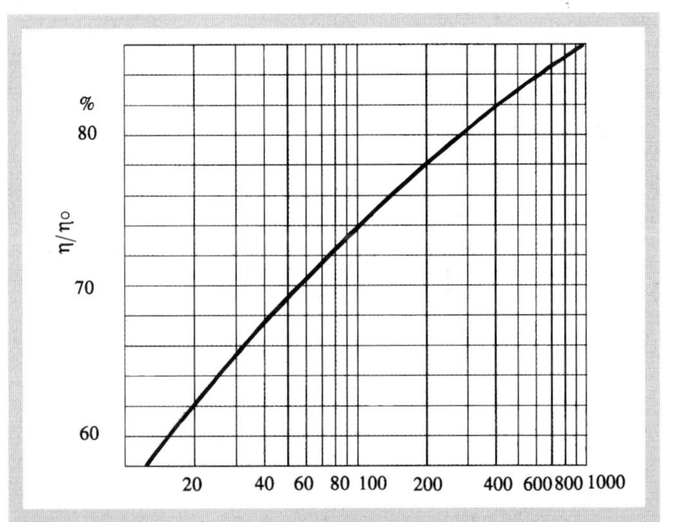

Figure 9. Optimum load efficiency of axial flow back pressure turbines with partial admission at generator terminals.

The steam consumption of a back pressure turbine depends almost linearly on the relative load of the turbine. The steam consumption at zero load can be calculated from Fig. 10.

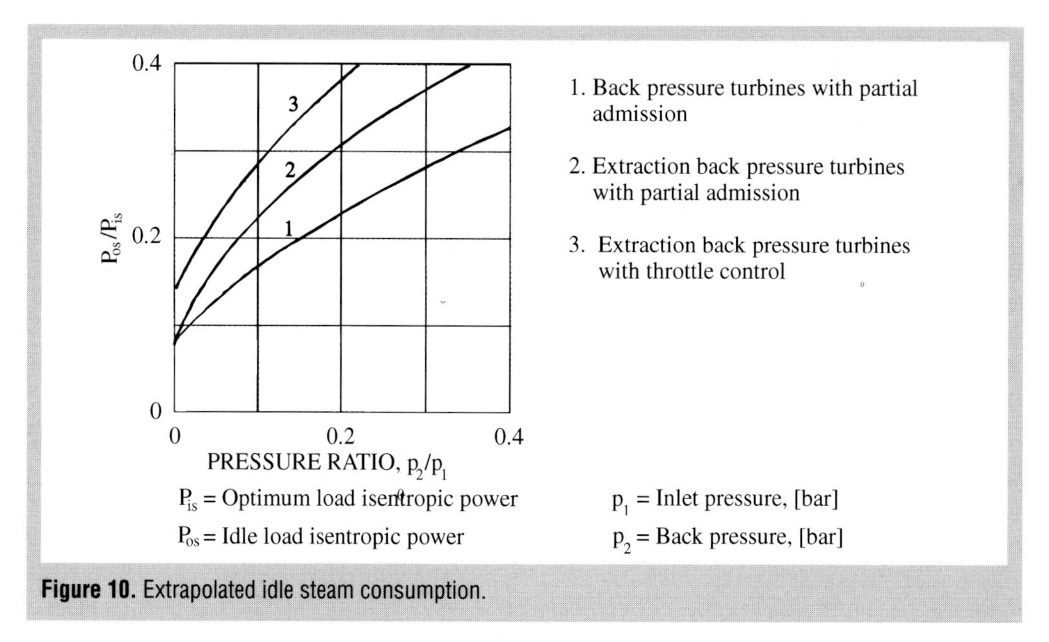

Figure 10. Extrapolated idle steam consumption.

4.6 Co-generation alternatives

The following discusses the most common recovery and power plant configurations.

Alternate A: Recovery boiler operates with steam parameters of 9–10 MPa and 490°C–500°C, and bark is sold outside the mill. Figure 11 shows basic power plant configuration.

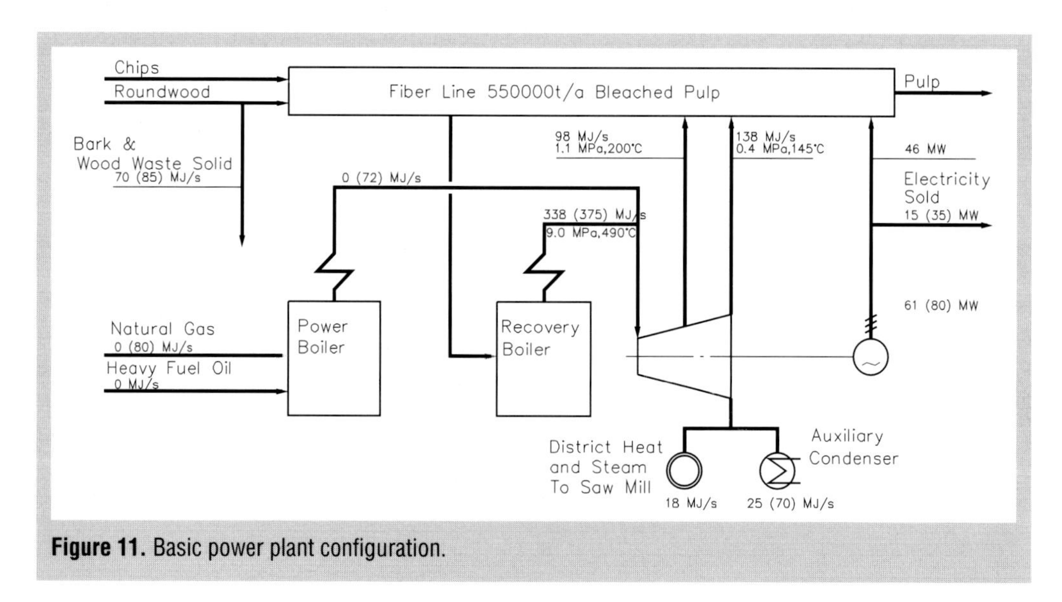

Figure 11. Basic power plant configuration.

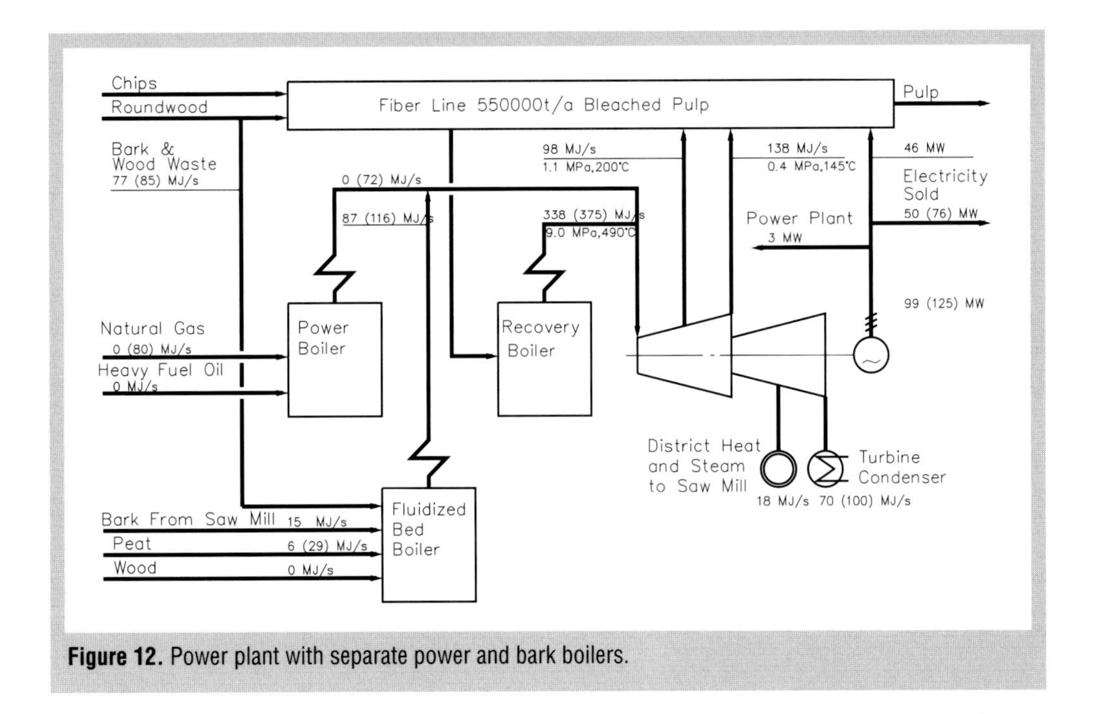

Figure 12. Power plant with separate power and bark boilers.

Alternate B: Recovery and bark boilers operate with steam parameters of 9–10 MPa and 490°C–500°C. Bark and wood waste from pulp wood is fired with some bio-fuel from outside the mill when available at an economical price.

Both these conventional processes have wide use in pulp mills but many times with lower steam parameters. Figure 12 shows a power plant with separate power and bark boilers.

Alternate C: Reheating bio-fuel power plant.

Recovery boiler operates with the same steam values as above, and steam is led to an extraction back pressure turbine. Bark, wood waste, and purchased bio-fuel from outside the mill are fired in a high pressure reheat boiler, and steam is led to a condensing turbine.

This concept is not economical in integrated pulp and paper mills because the benefit of reheating disappears with the process MP and LP steam desuperheating need.

With nonintegrated pulp mills, this concept offers some benefits such as common fuel supply organization, common operating and maintenance staff, common mill site and departments with the other mill, and access to outside grid. Because of these, a better economy may exist than a conventional coal fired power plant located in a new site. Economy becomes better if the taxation of fossil fuels becomes more strict. Figure 13 shows a power plant with separate reheat boiler and condensing turbine.

Chips
Roundwood
Fiber Line 550000t/a Bleached Pulp
Pulp
Bark &
Wood Waste
77 (85) MJ/s
0 (72) MJ/s
98 MJ/s
1.1 MPa,200°C
138 MJ/s
0.4 MPa,145°C
46 MW
Electricity
Sold
96 (138) MW
338 (375) MJ/s
9.0 MPa,490°C
Power Plant
5 (6) MW
60 (80) MW
Natural Gas
0 (80) MJ/s
Heavy Fuel Oil
0 MJ/s
Power
Boiler
Recovery
Boiler
40 MJ/s
186 (254) MJ/s
87 (117) MW
Bark From Saw Mill 15 MJ/s
Peat
56 (102) MJ/s
Wood
59 (80) MJ/s
Fluidized
Bed
Boiler
District
Heat and
Steam to
Saw Mill 18 MJ/s
Turbine
Condenser
119 (156) MJ/s

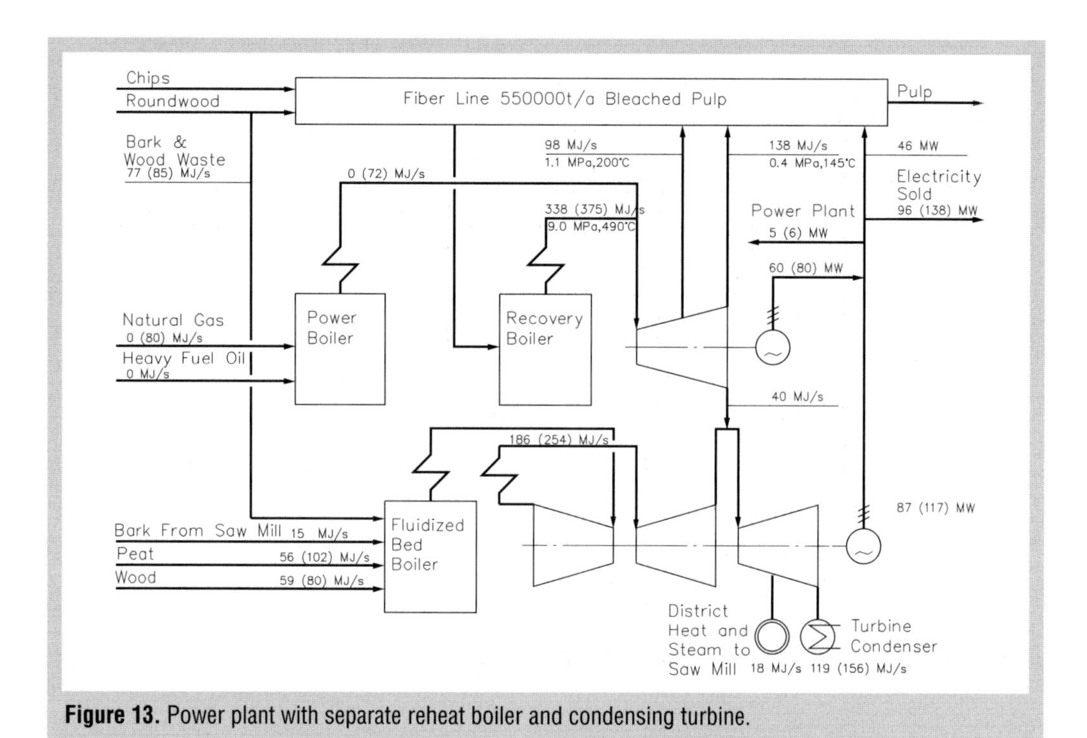

Figure 13. Power plant with separate reheat boiler and condensing turbine.

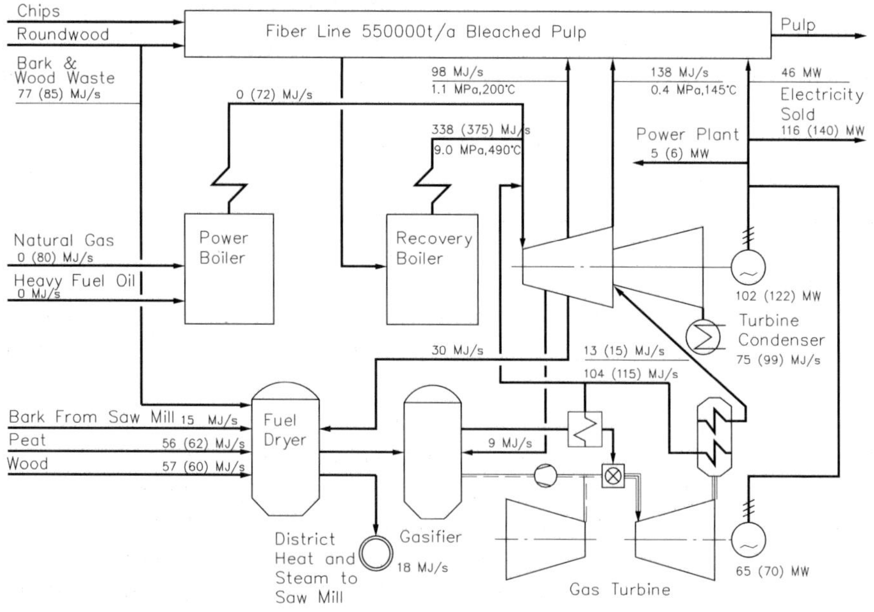

Chips
Roundwood
Fiber Line 550000t/a Bleached Pulp
Pulp
Bark &
Wood Waste
77 (85) MJ/s
0 (72) MJ/s
98 MJ/s
1.1 MPa,200°C
138 MJ/s
0.4 MPa,145°C
46 MW
Electricity
Sold
116 (140) MW
338 (375) MJ/s
9.0 MPa,490°C
Power Plant
5 (6) MW
Natural Gas
0 (80) MJ/s
Heavy Fuel Oil
0 MJ/s
Power
Boiler
Recovery
Boiler
102 (122) MW
Turbine
Condenser
75 (99) MJ/s
30 MJ/s
13 (15) MJ/s
104 (115) MJ/s
Bark From Saw Mill 15 MJ/s
Peat
56 (62) MJ/s
Wood
57 (60) MJ/s
Fuel
Dryer
9 MJ/s
District
Heat and
Steam to
Saw Mill
Gasifier
18 MJ/s
65 (70) MW
Gas Turbine

Figure 14. Combined cycle power plant with pressurized bio-fuel gasification.

Alternate D: Integrated gasification combined cycle (IGCC)

Recovery boiler operates as above. Bark and wood waste are purchased outside the mill. Bark and wood waste are dried and gasified. Gas is purified and fired in a gas turbine. Steam is generated in a gas turbine heat recovery steam generator (HRSG). Steam from the recovery boiler and the HRSG are led to a common steam turbine

The most difficult parts in this process are fuel drying and product gas purification. This concept is ready for commercialization but requires financial support at least for the initial plants. An integrated gasification combined cycle process offers a high power to process heat ratio of about 0.65. A value of 0.75 is possible with a demand for low temperature heat. The condensing power generation efficiency is also reasonably high at about 43%–45% or even higher. It offers the same synergies as alternate C.

Black liquor gasification has also been studied. Because of the complicated process and problems with materials, the process needs further development. Figure 14 shows a combined cycle power plant with pressurized bio-fuel gasification.

Table 13 shows the power to heat ratio and efficiencies of different power plant processes.

Table 13. Power to heat ratio and efficiencies for power plant processes.

	Fuel	Power to heat ratio	Efficiency with fuel LHV,%
Pulp and paper			
- Conventional	Bark and peat	0.35	85
- GTCC	Natural gas	0.90	85
- IGCC	Bark and peat	0.65–0.75	85
District heat			
- Conventional	Coal	0.48–0.58	85–90
- GTCC	Natural gas	1.0–1.1	85–92
- IGCC	Coal and bio-fuel	0.75–0.9	85–90
- Pressurized firing (PFBC)	Coal	0.65	85–90
Condensing power			
- Conventional	Coal and peat		40
- Conventional large units	Coal		41–43–(47)
- GTCC	Natural gas		50–55–(58)
- IGCC	Coal and peat		43–45–(50)
- Pressurized firing (PFBC)	Coal		42–44–(50)

In pulp and paper mills, the gas turbine combined cycle process (GTCC) normally involves a gas turbine, a heat recovery steam generator with a high pressure cycle (7–9 Mpa), a low pressure cycle (300–600 kPa), a hot water cycle, and a steam turbine.

5 Paper mills

This section discusses the heat and power balance of different paper mills.

5.1 Heat and power balance of newsprint production

Table 14 is a study of newsprint production of 280 000 t/year with different furnish and possible power plant configuration:

Alternate 1 100% TMP+TMP heat recovery and saturated steam bark boiler
Alternate 2 100% PGW with high pressure bark and oil fired boiler and steam turbine
Alternate 3 50% deinked and 50% TMP with saturated steam waste fuel boiler
Alternate 4 90% deinked and 10% purchased fiber with HP waste fuel and oil fired boiler and back pressure steam turbine
Alternate 5 90% deinked and 10% purchased fiber with HP waste fuel boiler, gas turbine with HRSG boiler, and back pressure steam turbine, Fig. 15.

Table 14. Newsprint production with different alternatives.

		Alt. 1	Alt. 2	Alt. 3	Alt. 4	Alt. 5
Furnish						
TMP	%	100		50		
PGW	%		100			
Deinked	%			50	90	90
Purchased pulp	%				10	10
Heat demand						
Process steam	MJ/s	43	43	48	51	51
Hot water	MJ/s	5	5	9	11	11
Excess heat	MJ/s	4	-	-	-	-
Power generation						
Steam turbine	MJ/s		12	-	14	14.5
Gas turbine	MJ/s		-	-	-	26
Total heat demand	MJ/s	52	60	57	76	102.5
Fuel consumption						
Bark and waste	MJ/s	13	13	6	-	-
Sludge and rejects	MJ/s			12	24	24
Gas or oil		1	51	21	62	89.5
Power plant losses	MJ/s	-2	- 9	- 5	- 10	-11
Heat recovery	MJ/s	40	5	23	-	-
Total generation	MJ/s	52	60	57	76	102.5
Power demand						
Paper machine	MW	19	19	19	19	19
Pulp manufacturing	MW	74	52	51	16	16
Other	MW	4	4	4	4.5	4
Total demand	MW	97	75	74	39.5	39
Power generation						
Gas turbine	MW	-	-	-	-	25
Steam turbine	MW	-	11.5	-	13.5	14
Total generation	MW	-	11.5	-	13.5	39
Power purchase	MW	97	63.5	74	26	0

A gas turbine combined cycle power plant with deinking sludge and reject firing in a fluidized bed boiler is a suitable power plant configuration for a newsprint mill with recycled paper as raw material.

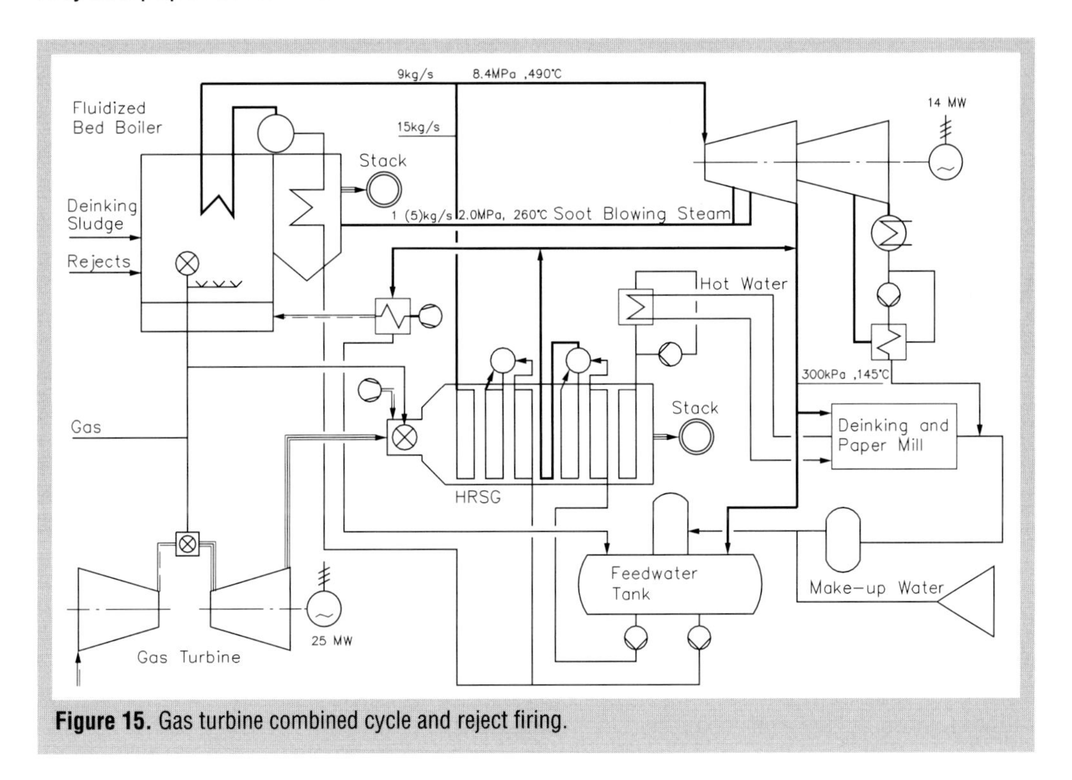

Figure 15. Gas turbine combined cycle and reject firing.

5.2 Heat and power generation in SC and LWC production

Table 15 shows the furnish for supercalendered (SC) and light weight coated (LWC) paper grades.

Table 15. Furnish.

	SC, %	LWC, %
PGW or TMP	30–55	30–42
Chemical pulp	15–20	28–30
Recycled fiber	0–20	0–10
Filler	30	3
Coating	0	27

TMP alternative

The following are possible power plant configurations when most mechanical fiber is TMP:

- TMP heat recovery, bark and wood waste fired boiler generating 1.5–2.0 MPa saturated steam reduced to low pressure
- TMP heat recovery, HP bark and wood waste fired boiler and a steam turbine
- TMP heat recovery, HP bark and wood waste fired boiler, gas turbine with HRSG, and an extraction condensing turbine.

When the amount of TMP pulp is high, only the two initial power plant alternatives are possible because TMP recovery steam does not leave space for gas turbine combined cycle back pressure power generation.

Table 16 shows an example of heat recovery from a TMP plant for steam and hot water.

Table 16. Heat recovery.

Refiners	Power demand, kWh/t	Heat recovery	
		Steam, %	Water, %
First stage refiner	1400	55	25
Second stage refiner	1100	65	0
Reject refiner	400	40	40
Total	2900	57	18

Steam is normally recovered from the refiners at a pressure of 300–320 kPa, and fresh steam from the reboiler goes to the paper machine at 250 kPa.

PGW alternative

The following are possible power plant configurations when most mechanical fiber is

- HP bark and wood waste fired boiler, HP gas or oil fired boiler, and a back pressure steam turbine
- HP bark and wood waste fired boiler, gas turbine with HRSG, and a back pressure turbine
- HP bark and wood waste fired boiler, gas turbine with HRSG, and an extraction condensing turbine.

All three power plant alternatives are possible because secondary heat from the PGW process is normally recovered only as hot water.

6 Time dependent variation in power demand

When making energy and economical decisions, the treatment becomes more simple and clear if time dependent variation of power demand is transformed into variation in size. Time dependent variation can only be used when the occurring times of different size values have no relevance to the end result. Figure 16 shows the method of creating a time dependent duration curve.

The following is the definition of a time dependent duration curve. The abscissa value, w_x ($0 \leq w \leq 1$), of each point of the time dependant duration curve expresses how large the amount of time for the varying quantity, *P*, is equal to or greater than the corresponding quantity, P_x.

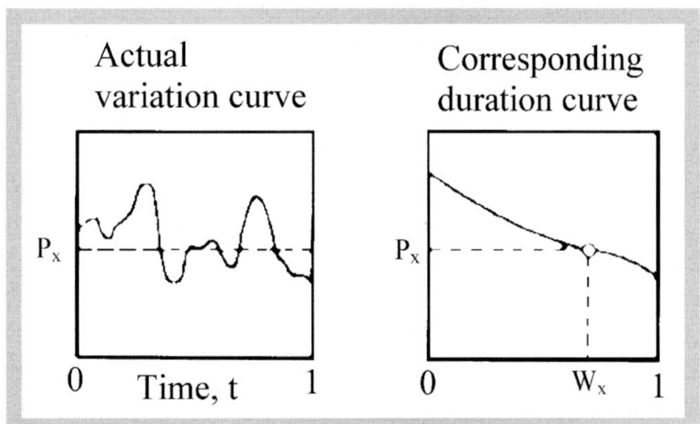

Figure 16. Power demand variation and corresponding duration curve.

The following basic concepts are components of time dependent duration curves:

- Maximum value, P_{max}
- Minimum value, P_{min}
- Mean value, P_m
- Load factor, $f_m = P_m / P_{max}$

$$W = \int_0^1 P dw \tag{20}$$

- Energy or efficiency average, W

$$s^2 = \int_0^1 (P - P_m)^2 dw \tag{21}$$

The concept of dispersion, s, is the measure of variation in power demand.

The following is the definition of dispersion:

Dispersion has the same dimension as the quantity in question: kW.

Integrating variation curves is relevant when the power demand of two factory compartments or two factories are known and their joint duration curve is necessary.

A more common case is calculating the difference between two separate duration curves such as when defining the purchased power when creating power.

Considering that loads do not appear on duration curves in actual time order, duration curves cannot be added as such. The maximum values of power demand will probably not appear at the same time for both cases.

The duration curves for two power demands can be integrated according to probability calculations if they vary completely independently. This means that if the value of one of these curves is taken at any given time, the value of the other curve at the same time can be anything between its maximum and minimum values.

Integrating duration curves according to probability calculations are only valid with this condition. One must always therefore carefully consider whether this condition is fulfilled before integrating the curves. The methods use the law of probability calculations. This states that the probability of the sum or the difference between two quantities is the product of the factors.

In practice, the calculations use computer programs. This allows analysis of the observation material. The variations will fall into systematic and random factors. The most advanced programs use the mean hour values of daily curves and their random variations.

6.1 Variations in energy production

The production of back pressure power varies depending on the heat demand. If the variation in power demand can be solved and expressed in the form of a duration curve or otherwise, the variation of back pressure power is also clear. To achieve this, the duration curve of power demand and the ϕ_T -, P_g - characteristics simply need to be multiplied with one another. This can be done graphically by creating applicable formulas or by dividing the power consumption into idle motion, normal load, and overload. The principle is the same in all cases. Figure 17 shows co-generation power as a function of heat demand and turbine characteristics.

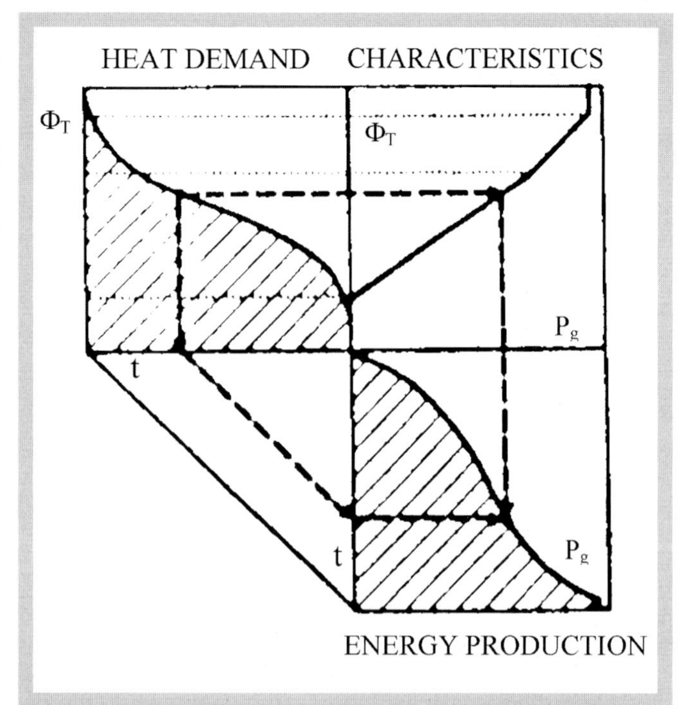

Figure 17. Co-generation power as a function of heat demand and turbine characteristics.

6.1.1 Variations in power demand in industry

The main factors that influence the demand for process heat in industry are the quantity and quality of production, outside temperature, process water temperature, random breakdowns, and planned shutdowns.

Examining variations in heat demand usually divides them into three main categories:

- Production dependent variation
- Variation depending on outside temperature
- Residual variation.

Careful examination of heat demand requires separation for different departments since the dependancy functions differ considerably from each other. To obtain unambiguous results, a large amount of observation material is necessary. The best way to do the work is with computer programs that use statistical mathematics. The knowledge on variations on loads primarily uses yearly duration curves gathered from different types of factories.

6.1.2 Power balance drawings

The variations in energy demand and power production are best presented as an integrated power balance drawing. In Fig. 18, the operating time is divided into two phases: night and day. For both, a duration curve for consumption, back pressure power production, and the difference that signifies the buying and selling curve or the result of a balance calculation in the shape of a duration curve is available.

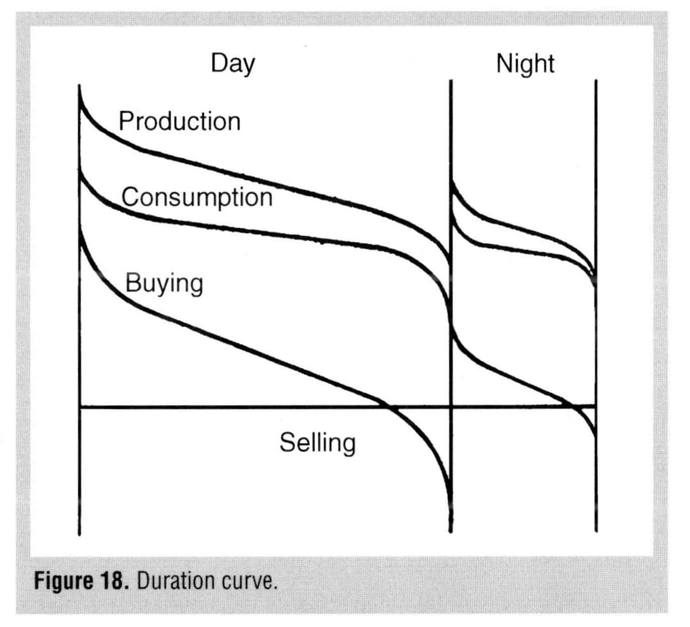

Figure 18. Duration curve.

The planning of power maintenance in an industrial plant uses examination of the situation and judgement of the difference curve of the power balance. The variations in energy prices due to the alterations made in the process or in power production should be calculated according to the variation curve of shortage. This is because the power prices are very dependent on the type of load and the amount. In addition, purchased energy and sold excess energy have different bases for pricing.

CHAPTER 17

Secondary heat systems and pinch technology

CHAPTER 17

Kari Parviainen, Carl-Johan Fogelholm, Jussi Manninen

Secondary heat systems and pinch technology

1 Introduction

In pulp and paper mills, chemical energy is not bound into the products. Therefore, all the energy that goes into the process will also come out at a lower temperature level. Pulp and paper processes does not consume energy, they consume temperature. A process that operates at a low temperature can use the waste heat from other processes operating at a higher temperature level. Heat from a process at a temperature of 45°C–110°C is secondary heat.

Secondary heat from processes goes into the secondary heat system that is a water-based heat supply system. The secondary heat system transports waste heat from a process to another process that can use this heat. A modern pulp mill has an excess of secondary heat. Which then can be used, as a heat source, in district heating or district cooling systems.

Secondary heat system can be very complicated. Process integration techniques provide an efficient way to analyze the efficiency of the secondary heat systems.

2 Secondary heat

2.1 Definition of secondary heat

In general, secondary heat means the heat which is expelled from a process in own streams without the heat of the product. The secondary heat is defined as %:

Secondary heat as %

$$\frac{Q_3}{Q_1} \times 100 = \frac{Q_3}{Q_2 + Q_3} \times 100 \tag{1}$$

where Q_1 is the heat in feed
Q_2 the heat in product flow
Q_3 the heat in side streams = secondary heat.

Part of the secondary heat is transferred into warm water or hot water in the pulp mill. The warm water temperature varies between 35°C–50°C and hot water 60°C–80°C. The transferred secondary heat calculated from the primary heat in the department is normally:

Mill department	Transferred secondary heat from primary heat in warm or hot water %
Cooking plant	70–80
Evaporation plant	100
Drying machine	50

In a pulp mill with a production of 1 000 adt/d, the transferred secondary heat amounts to 8 000 MJ/adt, corresponding to a fuel oil consumption of 230 t per day. The utilization of secondary heat is, however, difficult owing to the low temperature pinch available. For example, in a modern closed pulp mill the utilization of warm water from the evaporation can only 20%–50%, and the rest of the heat is dissipated in the cooling tower or into sewer.

This chapter discusses the secondary heat sources and the consumption points in the pulp mill and suggests solutions to optimize the secondary heat system.

2.2 Secondary heat production in the pulp mill

2.2.1 Continuous Cooking

Hot water is produced in continuous cooking from the following sources, as shown in Fig. 1:

- primary and secondary condensers
- black liquor cooler
- turpentine cooler
- condensate flash condenser

The primary and secondary condensers are connected in series. Vapors from the chip pre-steaming vessel and from the second flash cyclone are condensed partially in the primary condenser, and the rest of turpentine-rich vapor is condensed in the secondary condenser. The condensed turpentine/water mixture is led through the turpentine cooler into the turpentine decanter. Outcoming hot water from the condensers and the turpentine cooler is led to the hot water storage tank normally located at the cooking plant.

Hot water is produced also in the black liquor cooler where black liquor from the second flash cyclone is cooled down from 115°C–110°C to 90°C, suitable for storage in the weak black liquor tank at the evaporation plant.

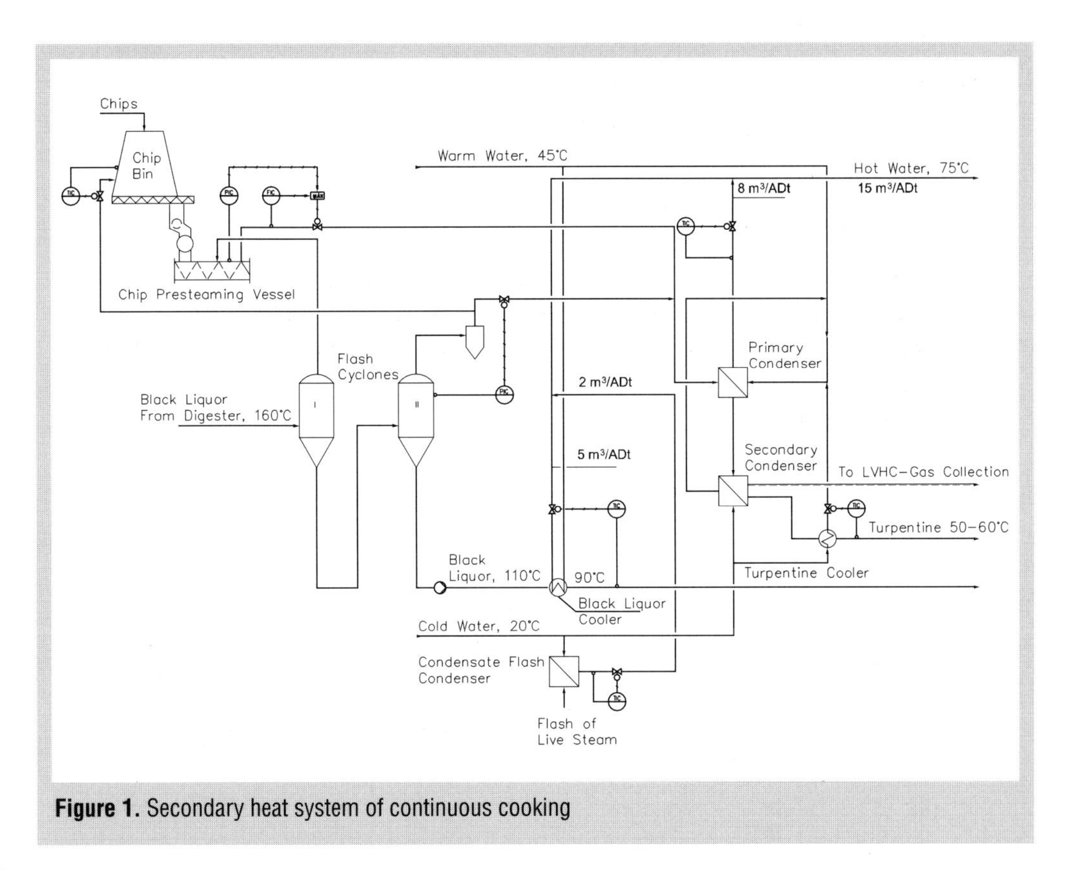

Figure 1. Secondary heat system of continuous cooking

It is possible to prepare hot water with flash steam of live steam condensate if the power plant requires cooled condensate.

Typical hot water amounts are given in Table 1

Table 1. Hot water (75°C) amounts from continuous cooking (Kamyr).

	Hot water, t/adt
- from primary and secondary condensers and turpentine cooler	8.0
- from black liquor cooler	5.0
- condensate flash condenser	2.0
Total	**15.0**

The continuous cooking may have three flash cyclones and flash steam from the first cyclone is directed into the heater of the digester wash circulation. In this case the total produced hot water from the primary, secondary, etc. condensers amounts to about 10.0 t hot water/adt.

2.2.2 Displacement batch cooking

The following cooking methods are available today:

- Rapid displacement heating (RDH)
- Super batch cooking (S-B)
- White liquor impregnation (WLI)

Secondary heat is recovered in a slightly different way in the above processes, but the common features are that the liquor heat of the preceding cook is transferred to the preheating of chips and incoming white liquor of the succeeding cook and in this way the utilization of heat is effective, and as a consequence less secondary heat is liberated into hot water.

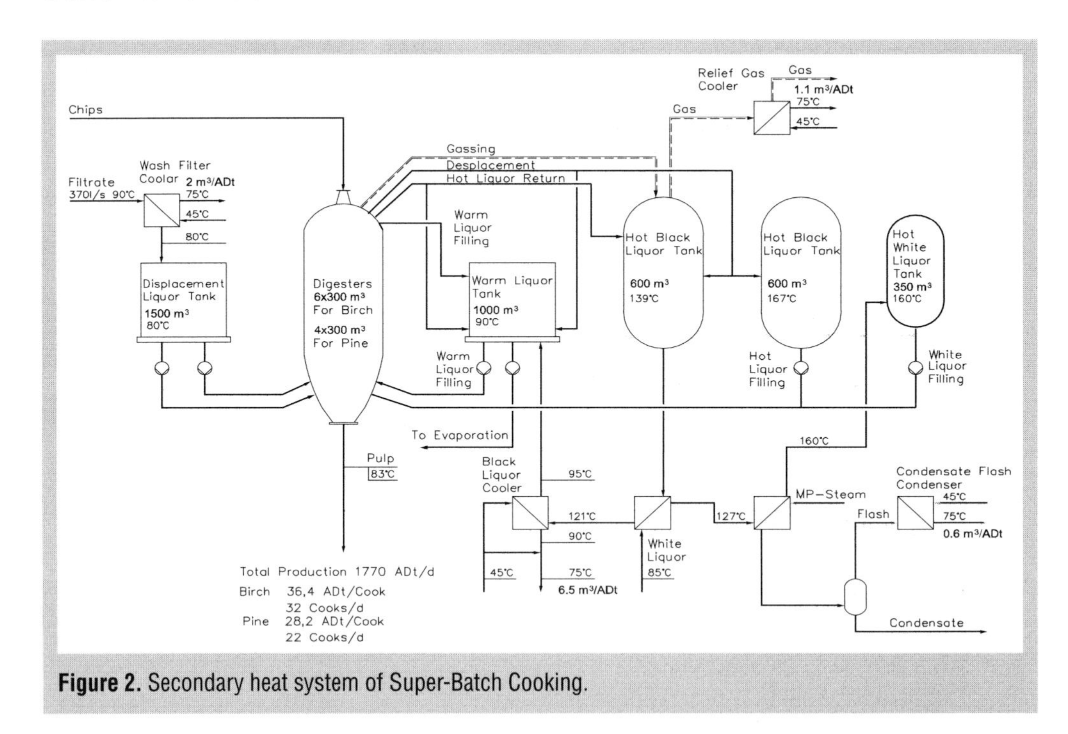

Figure 2. Secondary heat system of Super-Batch Cooking.

In Super Batch Cooking the sequence of the process steps is the following:

1) Chip filling
2) Impregnation: The chips are pre-impregnated and preheated with warm black liquor by filling the digester by that liquor.
3) Displacement with hot black liquor
4) Displacement and cooking with white liquor.
5) Displacement with wash liquor at the end of cooking.
6) Digester discharge by the cold blow.

During the above process steps hot water is produced from the sources:

- black liquor cooler
- wash filtrate cooler
- relief gas cooler
- condensate flash condenser

There is a continuous pumping from the hot black liquor tank, through the white liquor heater and the black liquor cooler into the warm liquor tank, and during that time the black liquor cooler makes hot water.

Also the wash filtrate cooler generates continuously hot water. The filtrate shall be cooled down so that the digester content can reach the correct temperature after the wash liquor displacement. Then pulp is discharged from the digester using a cold blow.

The relief gas cooler gives hot water during de-gassing of a digester and bigger amounts when displacement liquor enters the hot black liquor tank.

The typical hot water amounts are given in Table 2.

Table 2. Hot water amounts from super batch cooking (Sunds).

	Hot water, t/adt
- from black liquor cooler	6.5
- from wash filtrate cooler	2.0
- from relief gas cooler	1.0
- from condensate flash condenser	0.5
Total	**10.0**

2.2.3 Conventional batch cooking

The batch cooking includes the following secondary heat recovery facilities:

- primary and secondary condenser
- blow heat accumulator
- relief gas condenser
- trim condenser
- heat exchangers to make hot water

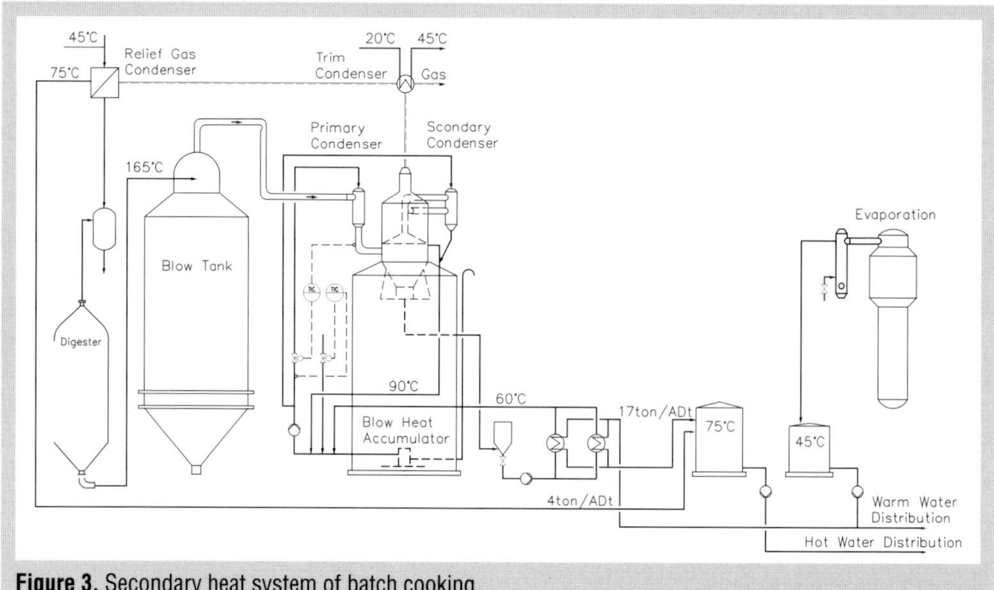

Figure 3. Secondary heat system of batch cooking.

Hot water is produced during the blow of a digester. From the digester is released a big amount of water vapor which is condensed first in the primary condenser and then in the secondary condenser. Cooling water into the condensers is pumped from the bottom of the accumulator tank and the vapor condensate/cooling water mixture flows to top of the accumulator. The accumulator works so, that hot condensate is collected in the top and cooled condensate in the bottom part of the accumulator. Hot condensate from the top is circulated through the chip strainer and the heat exchangers back in to the bottom of the accumulator. In the heat exchangers hot water is prepared and then stored in the hot water tank.

Temperature controllers (TIC-controllers) are activated in the beginning of the blow admitting cooling condensate into the primary and secondary condensers.

The condensers are normally of spray type. Some mills have indirect surface condensers where black liquor or water is used as coolant.

A trim condenser is added in some mills to complete the vapor condensing before malodorous gases are led in to the concentrated gas collection system.

The relief gas condenser receives turpentine-containing vapor from the digester during the cooking and makes hot water simultaneously.

Table 3. Hot water from batch cooking.

	Hot water, t/adt
- from blow heat recovery	17.0
- degassing	4.0
- from condensate flash condenser	4.0
Total	**25.0**

2.2.4 Warm water production from surface condenser

Warm water is produced from the surface condenser of the multiple stage evaporation, see Fig. 6.

The amount of warm water Q can be calculated by the formula

$$Q = qfi/[n(T_1 - T_2)c_p] \quad (2)$$

where Q is cooling water amount from condenser, kg/s (t/adt)
q evaporation capacity, kg/s (t/adt)
n number of stages
f flash factor, normally 1.1–1.15
T_1 warm water temperature out, °C
T_2 cooling water temperature in, °C
i latent heat of vapor from the last stage, kJ/kg vapor
c_p specific heat, kJ/kg °C.

The flash factor *(f)* means that vapor amount to surface condenser is bigger than the average evaporation per effect.

The specific warm water amounts t/adt from the surface condenser when the evaporation rate is 8 t water/adt, are expressed in the Table 4.:

Table 4. Warm water amounts from surface condenser, t/adt.

Number of stage	Water in at 30 °C	Water in at 20 °C	Water in at 4 °C
5	70	42	26
6	58	35	21
7	50	30	18

Calculation parameters		
cooling water in	°C	30, 20 and 4
warm water out	°C	45
flash factor	-	1.15

Example:		
number of stages	-	5
water in	°C	30
water out	°C	45
flash factor	-	1.15
latent heat of steam	MJ/kg	2361.36

When the Eq. 2 is applied the warm water amount is:

$$Q = q \times f \times i/n \times (T_1 - T_2) \times c_p = (8 \times 1.15 \times 2361.36)/(5 \times (45-30) \times 4.19) = 70\text{t/adt}$$

2.2.5 Secondary condensates from evaporation

The amounts of vapor condensate pumped from the evaporators are presented in the chapter 12.5.5 in Table 5.

2.2.6 Secondary heat production from recovery boiler flue gas scrubber

Flue gas from the recovery boiler is led through the electrostatic precipitators into the flue gas scrubber, where hot water of 50°C–65°C can be produced. The hot water temperature is defined by the flue gas wet bulb temperature and the number of heat transfer steps that are constructed in the scrubber. The heat of the flue gas is transferred into water directly or indirectly. Plate heat exchangers are used in case clean hot water is needed. The scrubber for the hot water production shall comprise of two sections:

1) The lower section for the flue gas washing
2) The upper section for the heat recovery with the heat transfer internals

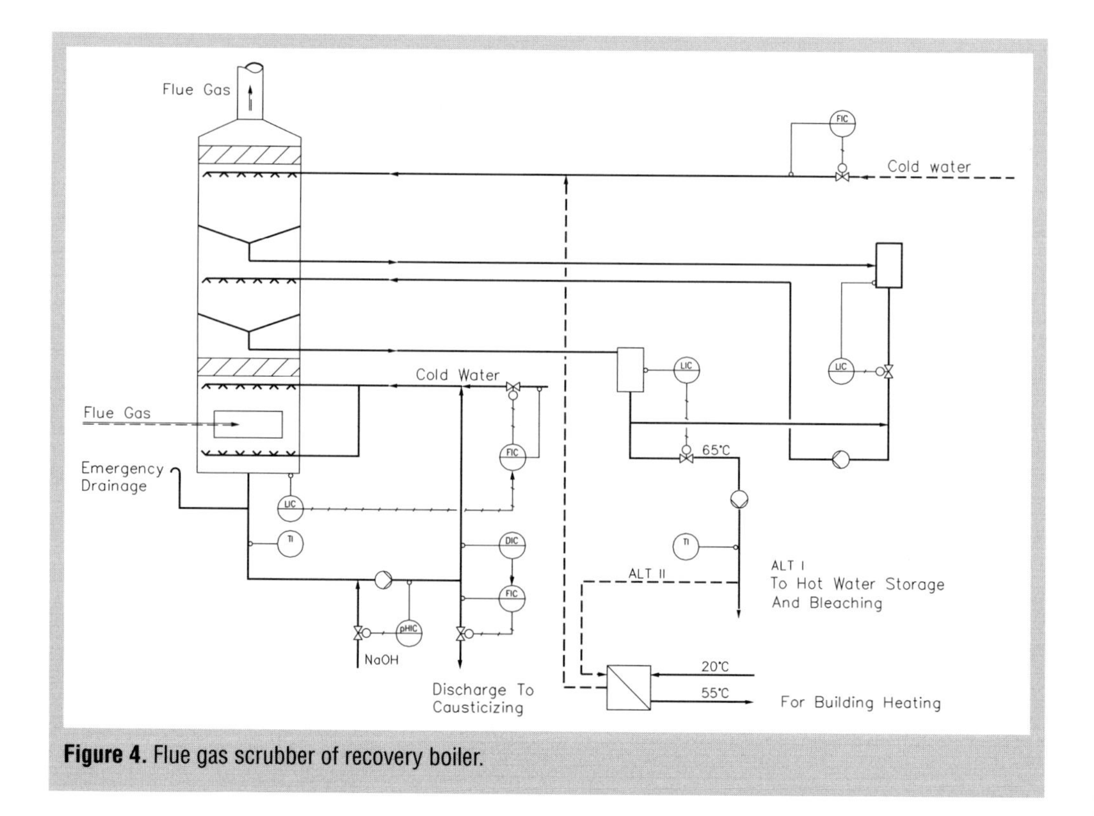

Figure 4. Flue gas scrubber of recovery boiler.

The hot water amount Q can be estimated by the formula

$$Q = Q_1 + Q_2 \tag{3}$$

where Q is total hot water produced, kg/s
Q_1 cooling water into scrubber, kg/s
Q_2 water condensed from flue gas inside the scrubber, kg/s.

The cooling water amount Q_1 into the scrubber is defined by the formula

$$Q_1 = G_1 \times (i_1 - i_2)/(T_2 - T_1) \times c_p \tag{4}$$

where Q_1 is cooling water into scrubber, kg/s
G_1 dry flue gas amount, kg/s
i_1 flue gas enthalpy in the scrubber inlet, kJ/kg dry gas at the inlet gas moisture x_1, kg water/kg dry gas and the temperature, °C
i_2 flue gas enthalpy in the scrubber outlet, kJ/kg dry gas at the outlet gas moisture x_2, kg water/kg dry gas and the temperature, °C
T_1 cooling water temperature, °C
T_2 hot water temperature, °C.
c_p specifice heat, kJ/kg °C

The water amount Q_2 condensed in the scrubber is defined by the formula

$$Q_2 = (x_1 - x_2) \times G_1 \tag{5}$$

where Q_2 is condensed water inside the scrubber
x_1 the moisture of inlet gas, kg water/kg dry gas
x_2 the moisture of outlet gas, kg water/kg dry gas
G_1 dry flue gas amount, kg/s.

By applying the above Eqs. 3–5 the hot water amount from the scrubber would be as expressed in Table 5.

Production of hot water of 65°C from the recovery boiler flue gas scrubber when:

- Flue gas amount, 3 200 Nm^3 dry gas/t DS
- Density, 1.35 Kg/Nm^3
- Dry solids amount, 2 t DS/adt
- Moisture content of flue gas before scrubber, 25 %
- Flue gas temperature before the scrubber, 170 °C
- Flue gas temperature after the scrubber, 40 °C

- Scrubber has the sections:

1) the lower part for flue gas washing

2) the upper part for hot water production (at least 3-stages)

Example:

By applying the above process data and a water temperature of 2°C into the scrubber, the amount of water produced can be calculated:

Flue gas amount:

$G_1 = 3\,200 \times 2 \times 1.35 = 8\,650$ kg/adt = 8.65 t/adt

The moisture before the scrubber:

$x_1 = (0.25 \times 18)/(22.4 \times 0.75 \times 1.35) = 0.20$ kg H_2O/kg dry gas

From Mollier`s diagram by 0.20 kg H_2O/kg dry gas and by 170°C:

$i_1 = 732.69$ kJ/kg dry gas

From Mollier´s diagram on the saturation curve by 40°C:

$i_2 = 167.42$ kJ/kg dry gas

and the moisture:

$x_2 = 0.05$ kg H_2O/kg dry gas

The formula (4) gives:

$Q_1 = 8.65 \times (732.69 - 167.42) / (65 - 2) \times 4.19 = 18.5$ t/adt

The formula (5) gives :

$Q_2 = (0.20 - 0.05) \times 8.65 = 1.5$ t/adt

The formula (3):

$Q = Q_1 + Q_2 = 18.5 + 1.5 = 20$ t/adt

Table 5. Production of hot water from recovery boiler scrubber.

Water temperature before scrubber, °C	2	20	30
Water amount into scrubber, t/adt	18.5	25.5	32.5
Water condensed from flue gas, t/adt	1.5	1.5	1.5
Hot water from scrubber at 65°C, t/adt	20.0	27.0	34.0

2.2.7 Lime kiln flue gas scrubber

The lime kiln scrubber is placed after the electrostatic precipitator. Normally a venturi-type scrubber is used because this type is not as sensitive to plugging by the $CaCO_3$ particles entrained in the flue gas as most of the other types. To use a lime kiln scrubber for production of hot water can be justified only if there is a hot water demand in the causticizing plant. Normally, the amount of hot water produced in the cooking plant and the vapor condensate from the black liquor evaporation are sufficient for the causticizing process.

The calculation method for the production of hot water from the lime kiln scrubber is similar to the calculation of the recovery boiler scrubber, see Eqs. 3–5.

Typical hot water production from the lime kiln scrubber would be as given in the Table 6.

Table 6. Hot water from lime kiln scrubber.

Production of hot water from lime kiln flue gas scrubber:				
flue gas amount	700 kg dry gas/adt			
flue gas moisture content before the scrubber	30 vol.%			
flue gas temperatures:				
- before the scrubber	200°C			
- after the scrubber	60°C			
hot water produced	60°C			
Cooling water temperature	°C	2	20	30
Cooling water in	t/adt	1.4	2.0	2.7
Water condensed from flue gas	t/adt	0.1	0.1	0.1
Hot water produced	t/adt	1.5	2.1	2.8

2.2.8 Production of hot water from drying machine

Secondary heat recovery of the drying machine can be made in different ways. It is possible to use a simplified system where the dryer hood heat is transferred into the air and is introduced to the pulp drying.

A more complete system is presented in Fig. 5 "Secondary heat recovery in drying machine". In this system the vapor/gas mixture from the hood is led at 100°C–120°C into air/air heat exchangers where the air of 25°C is heated and sent into the pulp drying. After the heat exchangers the gas/air mixture goes:

a) into the scrubber producing water at 60°C from white water and then into the air conditioning heat exchanger. Scrubber water is led through a strainer into the scrubber water tank and further through the flash steam condenser at 80°C into the wire pit or into the hot water boxes of the wire pit.

b) into the process water heat exchanger and then into the air conditioning heat exchanger. Process water goes into the warm water tank from where it is pumped to the low and high pressure sprays of the wire and felt cleaning.

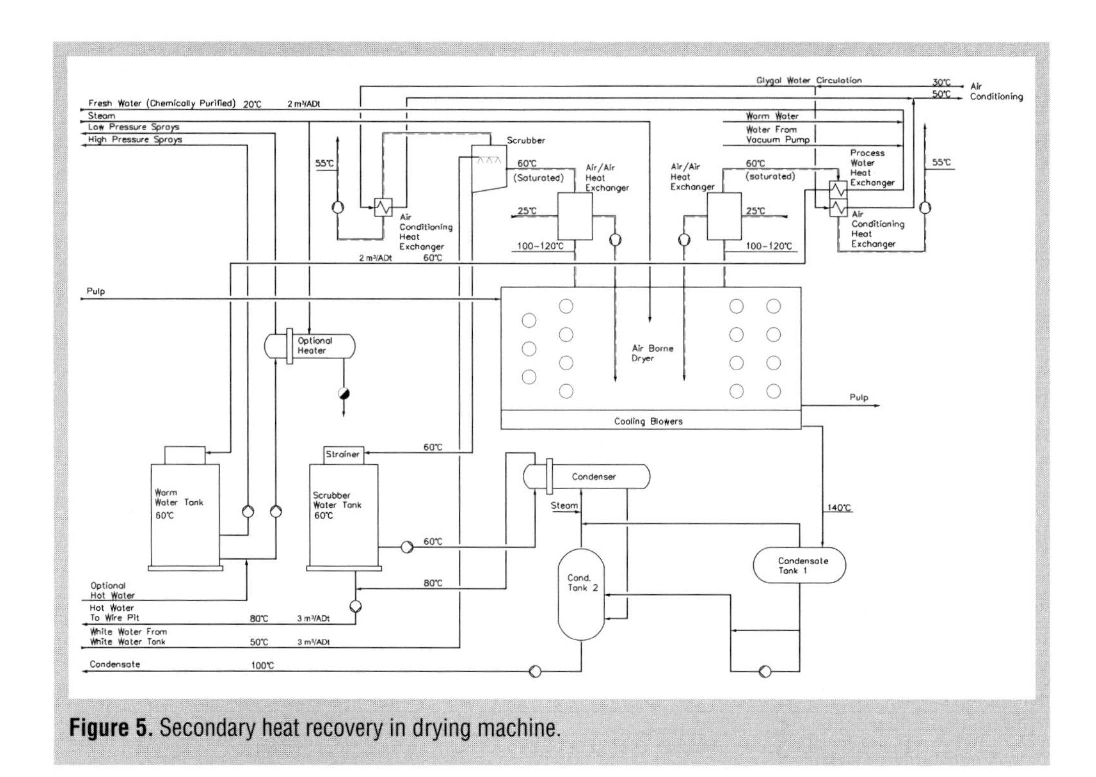

Figure 5. Secondary heat recovery in drying machine.

Live steam condensate at 140°C is taken first into the condensate tank 1 to remove air from condensate during the start up, then the condensate is flashed to the temperature of 100°C in the condensate tank 2 and pumped into the power plant condensate tank and hot water is prepared by flash steam.

The hot water production of the above system is summarized in the Table 7.

Table 7. Hot water production at drying machine.

Hot water	m^3/adt
- scrubber and the live steam flash condenser at 80°C	3.0 white water/fresh water
- process water heat exchanger at 60°C	2.0 fresh water
Total	**5.0**
Heat from	**kWh/adt**
- air conditioning heat exchangers	200

2.2.9 Other secondary heat sources in pulp mill

The other secondary heat sources can be found from the Table 8 where the specific flows, t/bleached adt are given for the different streams.

Table 8. Secondary heat production in pulp mill (pulp production = 333 bleached adt/d).

Department	Warm water, t/bl adt	Hot water, t/bl adt	Condensate of liquor vapors, t/ bl adt	Cooling water to cooling towers, t/bl adt
Cooking				
- From chemically purified water		3.3		
- From warm water		14		
Oxygen delignification				
- From chemically purified water		5.4		
Evaporation plant				
- Warm water from chemically purified water	17.2			
- Warm water from cooling water				75.9
- Vapor condensate			8.4	
- Stripper			1.6	
- Trim condenser			0.5	
- Methanol distillation				2.1
Turbine				
- Condenser				175 (201)
- Generator cooler				7.2
Causticising plant				
- Green liquor cooling				1.6
Tall oil plant				1.2
ClO_2 plant				23.5
Hot water tank overflow (as WCC make up)				7
Control room and electrical space cooling				35
Oxygen plant				2.2
Compressor station				5.4
Tank vents in evaporation				2.6
Total	**17.2**	**22.7**	**10.5**	**339 (365)**

2.3 Secondary heat consumption

The specific secondary heat consumptions as ton water/bleached adt is presented in the Table 9 as well.

Table 9. Secondary heat consumption in pulp mill.

Department	Warm water, t/bl adt	Hot water, t/bl adt	Condensate of liquor vapors, t/bl adt	Cooling water, t/bl adt	Chemically purified water, t/bl adt	White water from drying machine, t/bl adt
Wood handling				2.6		
Cooking						
- Warm water	14					
- Chemically purified water					3.2	
Oxygen delignification						
- To filter			6.5			
- Filtrate cooling					5.5	
- Pulp cooling				2.8		
Bleaching						
- To filter		11.7				
- To D2 filter						8.3
Drying						
- To hot water production	2.7	3.8				
- To vacuum system				0.8		
Evaporation plant						
- Surface condenser				64.7	30.9	
- Vapor condensate tank level control		0.1			(incl. WCC make-up)	
- Stripping and methanol distillation	0.6			2.1		
Turbine condenser				176 (202)		
- Generator and oilcooling				12.1		
Recovery boiler				3.9		
Hot water tank overflow		7				
Power generation make up water					2	

Table 9. Secondary heat consumption in pulp mill (continued).

Department	Warm water, t/bl adt	Hot water, t/bl adt	Condensate of liquor vapors, t/bl adt	Cooling water, t/bl adt	Chemically purified water, t/bl adt	White water from drying machine, t/bl adt
Causticising and lime kiln			4.2			
- Jet condensers						
- Green liquor cooler				4.6		
- Other						
Tall oil plant				1.2		
ClO_2 plant		0.3		23.6	4.7	
Control room and electrical space cooling				35.2		
Sealing water					~ 4	
Tank vents in evaporation				2.6		
Compressor station				2.6		
Oxygen plant				2.2		
Total	**17.3**	**22**	**10.7**	**340 (364)**	**50**	**8.3**
Surplus (+) Shortage (–)	0	0	0	0	- 50	- 8.2

2.4 Optimizing of secondary heat system

2.4.1 Optimizing of warm water production

There is normally an excess of warm water from the evaporation surface condenser and from the other sources in a modern pulp mill. Principally the warm water from the surface condenser can be decreased by increasing the number of evaporation stages, maximally to 7-stages. Anyhow the number of the evaporation stages is defined by the total mill steam and energy balance. In the case of a modern market pulp mill, there is an excess of bark/wood residues from the debarking, and these fuels must often be burned for environmental reasons, generating a lot of heat energy for the pulp mill. In this case a 5-stage evaporation is normally selected.

On the other hand, mills with an integrated paper production require a lot of electrical power, and in this case the power generation is maximised with condensing power generation and by selecting a 7-stage evaporation plant.

The excess warm water from the evaporation surface condenser is led into a cooling tower and back to the condenser. Because the cooling tower water might contain impurities, it is not often used for the production of hot water which will be used in the bleaching plant. Therefore the surface condenser must be divided into two sections, the part producing clean warm water from raw water, and the part using cooling tower water as the coolant.

In some cases the hot water demand is so small that hot water can be directly prepared from raw water (4°C–25°C) instead of warm water (45°C–50°C) and no division is required for the surface condenser at the evaporation plant.

2.4.2 Optimizing hot water production

In a modern closed pulp mill hot water from the cooking plant can normally satisfy the mill´s hot water demand. If additional hot water must be produced, there are two ways of doing it as is shown in the following calculation example:

Example

Hot water at 75°C and 100 l/s will be prepared from warm water of 45°C. Calculate which alternative is the cheapest to produce hot water:

a) by fresh steam of 0.45 MPa(a)

b) by vapor of the last three evaporation effects

Parameters:

purchased power price	$/MWh	38
fuel price	$ / GJ	6
boiler thermal efficiency	%	85

Solution

Cost of high pressure steam

= fuel price/boiler thermal efficiency

= 6/0.85

= 7.1 $/GJ

Heat transferred in heat exchangers = 100 x (75 - 45) x 4.2 = 12 600 kW

When process steam at 0.45 MPa(a) is led into the evaporation effect I and then vapor from the effect I is utilised in the next effect II etc., the so called equivalent vapor price or heat consumption decreases correspondingly, as presented in Table 10.

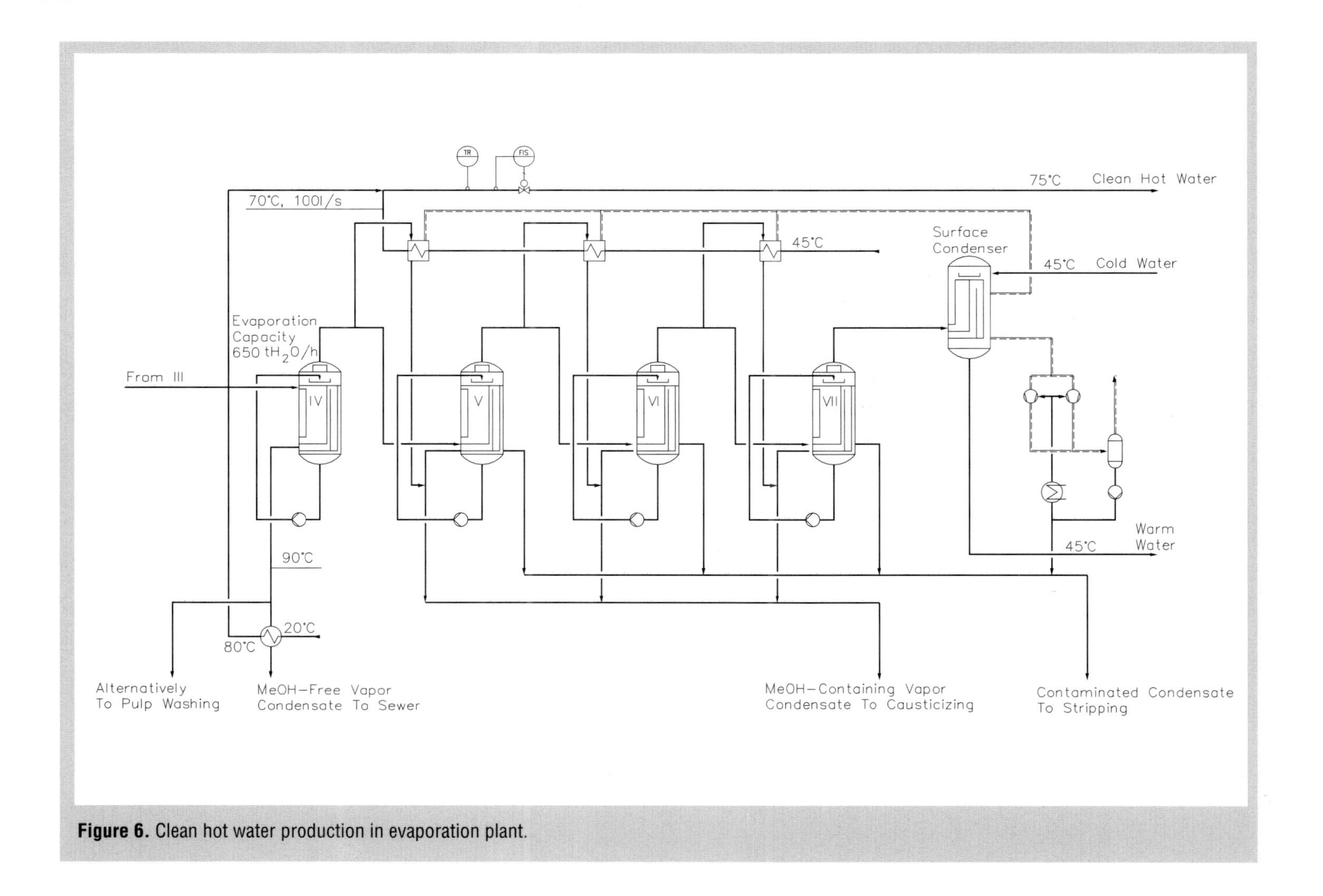

Figure 6. Clean hot water production in evaporation plant.

Table 10. Equivalent vapor price or equivalent heat consumption in the different effects compared to fresh steam.

Vapor source		Equivalent vapor price at 6-stage plant	Equivalent heat consumption at 7-stage plant
Fresh steam to effect I		1	1
Vapor from effect	I	5/6	6/7
Vapor from effect	II	4/6	5/6
Vapor from effect	III	3/6	4/7
Vapor from effect	IV	2/6	3/7
Vapor from effect	V	1/6	2/7
Vapor from effect	VI	0	1/7
Vapor from effect	VII		0

The equivalent vapor price from the last effects (from the effect VI or VII) can be valued to zero if there is an excess of warm water in the pulp mill.

If hot water is prepared with vapor from the effects IV, V and VI in a 7-stage plant, compare with the Fig. 6, and assuming that the vapor amounts are equal, the equivalent heat consumption can be written:

= 1/3 x 3/7 + 1/3 x 2/7 + 1/3 x 1/7

= 0.29

Heat in fresh steam to evaporation in the alternative b.

0.29 x 12.6 = 3.3 MW

Heat consumption and power generation in the alternatives a) and b)

- 0.25 = heat consumption in back-pressure power generation per process heat
- 0.97 = mechanical efficiency of turbo generator

	Alt a MW	Alt b MW
Heat for hot water preparation	12.6	3.3
Heat in power generation	0.25 x 12.6	0.25 x 3.3
	= 3.2	= 0.8
Total heat consumption	15.8	4.1
Back pressure power generation	0.97 x 3.2	0.97 x 0.8
	= 3.1	= 0.77
Heat cost, milj. USD/a	8 500 x 7.1 x 15.8 x 3.6	
	= 3.4	= 0.9
Bonus from power, milj. USD/a	8 500 x 38 x3.1	
	= 1.0	= 0.3
Total operation cost, milj. USD/a	2.4	0.6

Investment costs

Alternative b) 0.50 milj USD

Alternative a) 0.10 milj USD

The pay back time is: (0.5 - 0.1) / (2.4 - 0.6) = 0.22 years, if alternative b is used instead of alternative a.

In actual projects the pay back times are normally below one year when hot water is produced by the condensers connected to evaporation vapors.

2.5 Mill wide secondary system

2.5.1 Design considerations

When the secondary heat system of a new pulp mill is designed or the system of an old mill is modified, it is essential to compile the balance sheet for the water producers and consumers. A typical water balance is presented in Table 8.

When this balance is known the warm/hot water production can be optimized.

The mill wide water distribution system shall be secured so that all water streams are available to the process all the time. In case some hot or warm water producer is occasionally shut down, it is accepted that the water temperatures are lower. A good practice to minimize the temperature drop is to use level control management for the tanks:

1) Hot water tank level is safeguarded by warm water
2) Warm water tank is safeguarded by raw water
3) Evaporation vapor condensate tank level is safeguarded primarily by hot water and secondly by warm water.

The main consumers should be provided with integrating flowmeters so that the overall water balance can be compiled and checked afterwards. The main water headers are provided with temperature and pressure measurements to be used for process adjustments.

2.5.2 Water quality control

The quality of mill water shall be followed by analysis. Especially the quality of raw water, cooling tower water and evaporation vapor condensate should be determined frequently. The recommended analysis and analysis intervals could be:

Chemically purified raw water analysis:

- to be done daily:
  - conductivity
  - pH
  - $KMnO_4$ consumption
  - color
- to be done twice a week:
  - suspended solids

- to be done weekly:
  - hardness and Langelier saturation index
  - turbidity
  - P-value and M-value
- to be done monthly:
  - SiO_2
  - O_2-content
  - NH_3-content
  - Fe-content
  - Al-content
  - Cu-content
  - Chloride-content

The determination of Langelier saturation index will unveil how corrosive (if Langier index <0) or scaling (if Langier index >0) water will be when heated in a heat exchanger. The iron (Fe) content should be 0.1 mg/L when water used in pulp bleaching.

Cooling Tower Analysis:

- to be done weekly:
  - conductivity
  - non-volatile matter
  - pH
  - chloride
  - suspended solids
  - micro-organisms (total)
- to be done monthly:
  - $KMnO_4$-consumption

To prevent the growth of the micro-organisms cooling tower water shall be treated by hypochlorite or ozone at regular interval, normally twice per month or according to experience. If hypochlorite is used, a concentration of 2–5 g active Cl/m^3 water shall be used.

Vapor condensates analyses (evaporation plant):

- conductivity weekly

Vapor condensates especially to causticizing and pulp washing must be clean enough, the conductivity below 100 mS/m and 50 mS/m respectively.

2.5.3 Secondary heat system applications

Case I

The system has been divided into

1) clean raw, warm, and hot water distribution
2) cooling tower circulation water containing impurities

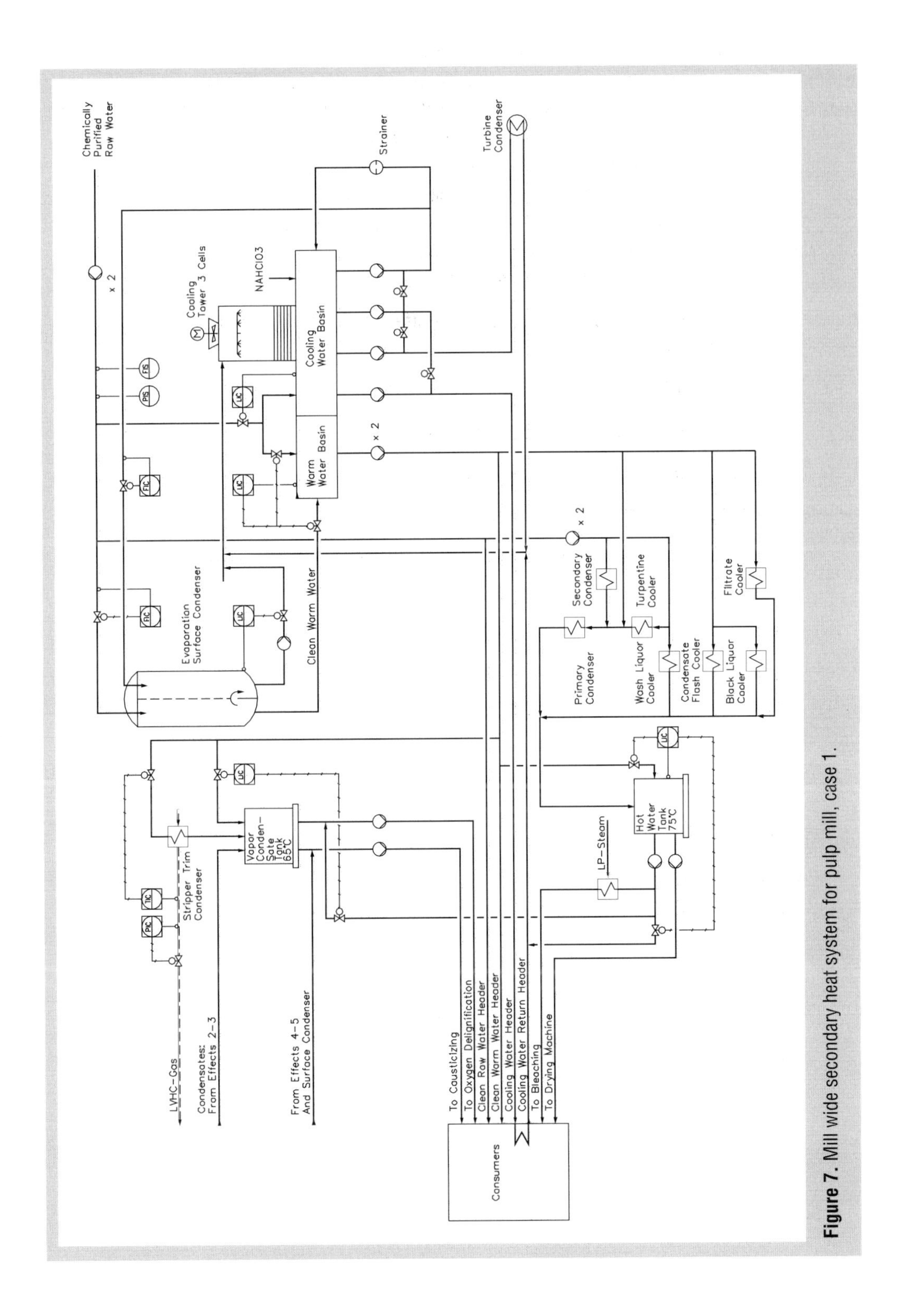

Figure 7. Mill wide secondary heat system for pulp mill, case 1.

Chemically purified raw water is pumped into the raw water distribution and into the "clean" section of the evaporation surface condenser. From the condenser the clean warm water flows to the warm water basin and is further pumped to the hot water generation in the stripper trim condenser and in the condensers and coolers of the cooking plant. The produced clean hot water is used in the bleaching plant.

Cooling tower water is pumped into the "dirty" section of the surface condenser, into the turbine condenser and into the various smaller consumers. About 90% of the cooling tower water is returned back to the cooling tower.

Evaporation vapor condensates are pumped

- methanol-rich condensate from the effect 4–5 and surface condenser to the causticizing
- methanol-free condensate to the fiber line

The tank levels are safeguarded as follows:

- cooling tower basin by raw water
- warm water basin by raw water
- vapor condensate tank by hot water and warm water
- hot water tank by warm water

Case II

The system has the water distribution headers with prevailing pressures

Header	Prevailing pressure MPa(g)
1. cooling tower water header	0.40
2. warm water header	0.32
3. hot water header	0.24
4. cooling water return header	0.16

The water distribution headers are safeguarded as follows:

1. Cooling tower water by raw water into the cooling tower basin and by the pump pressure/speed controller
2. Warm water by cooling tower water into the suction side of warm water distribution pump.

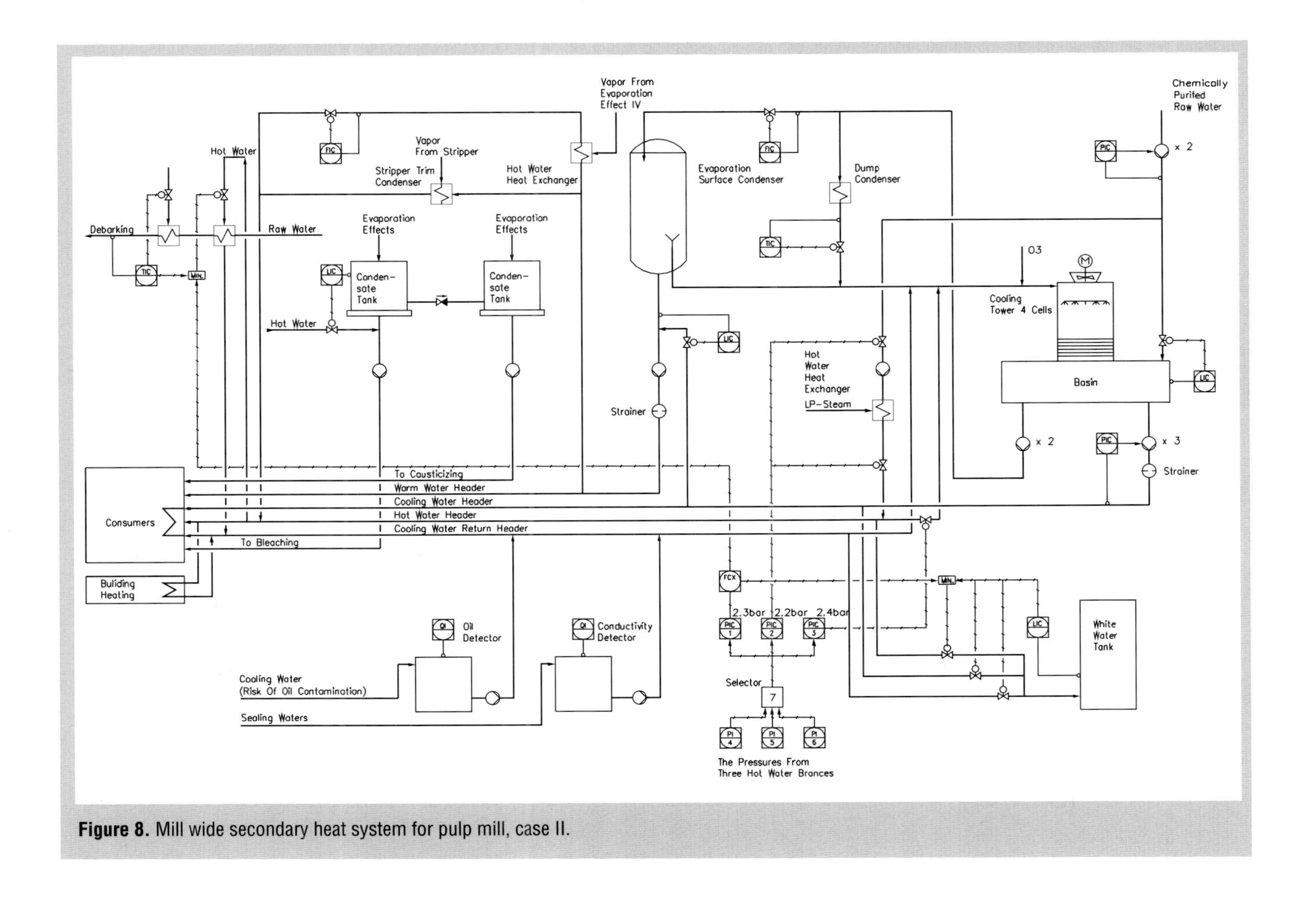

Figure 8. Mill wide secondary heat system for pulp mill, case II.

3. The hot water main branches are controlled by the pressures PI-4, PI-5 and PI-6. The selector 7 takes the pressure which is the intermediate value and sends the set point to one of the controllers:
   - to PIC-3, if the pressure is under 0.24 MPa(g) closing the hot water delivery into the cooling tower
   - to PIC-1, if the pressure is under 0.23 MPa(g) closing hot water supply into the raw water heating for the debarking and into the white water tank. The MIN selectors will assure that hot water is taken into debarking with a very low temperature and into the white water tank with a very low level.
   - to PIC-2 , if the pressure is under 0.22 MPa(g) actuating the hot water preparation by LP-steam to produce hot water into the hot water distribution header.

The mill is going to recover the possible oil containing cooling waters and pump sealing waters into the cooling water return header. The possible contamination is monitored by oil detectors and conductivity meters.

3 Process integration techniques

The purpose of this section is to provide the basic concepts for the most commonly used process integration techniques for heat integration. These are pinch technology, mathematical programming, and exergy analysis. For detailed information on any of these techniques, a list of relevant references appears at the end of this section.

3.1 Basic principles of pinch technology

Pinch technology started with heat recovery and heat exchanger networks. It finds application today in heat and power separation systems and waste water minimization.

The main philosophy of pinch technology is to obtain targets before design. Targeting makes extensive use of various graphical representations of the process. Rules are also available for designing heat and water networks that will satisfy the necessary targets. This also gives the designer full control of the design task. Design by pinch rules can be very time-consuming for big problems. A growing tendency exists for using pinch technology in the analysis and screening stage and then allowing mathematical programming to handle the complex interactions involved in the final design stage.

This section introduces the basic concepts of pinch analysis. The main emphasis is on the heat recovery targets for individual processes and for total sites. The pinch design methodology of heat exchanger networks is not part of this discussion. It is very well documented[2]. Another important application namely water pinch is also not part of this discussion. It is available elsewhere[4].

3.1.1 Composite curves

The process streams in a heat exchanging process can have two categories: heat sources (hot streams) and heat sinks (cold streams). The hot streams require cooling. While cooling, they can heat cold streams.

Consider a simple example using only one hot and one cold stream. Table 11 contains the stream data.

Table 11. Stream data for example 1.

Stream	Type	Supply temperature, °C	Target temperature, °C	CP, kW/°C	Enthalpy change, kW
1	Cold	40	100	2	120
2	Hot	150	40	1	110

The term supply temperature is the initial temperature of the stream. Target temperature is the final temperature. CP is the heat capacity obtained by multiplying the mass flow by the specific heat capacity, c_p.

The streams can be represented on a temperature and enthalpy (T-H) diagram. For feasible heat exchange between the streams, the hot stream must obviously be hotter than the cold stream at all points. Figure 9 shows the T-H diagram of these two streams assuming a minimum temperature difference, ΔT_{min}, of 10°C. The overlap of the curves indicates the heat exchange potential between these streams at this specified ΔT_{min}. In this case, it is 100 kW. The part of the cold stream that extends beyond the hot stream requires warming by a hot utility such as steam. Correspondingly, the hot stream must cool to its target temperature using a cold utility such as cooling water. The amounts of utility usage shown by the T-H diagram are the theoretically lowest achievable values. They are therefore utility targets.

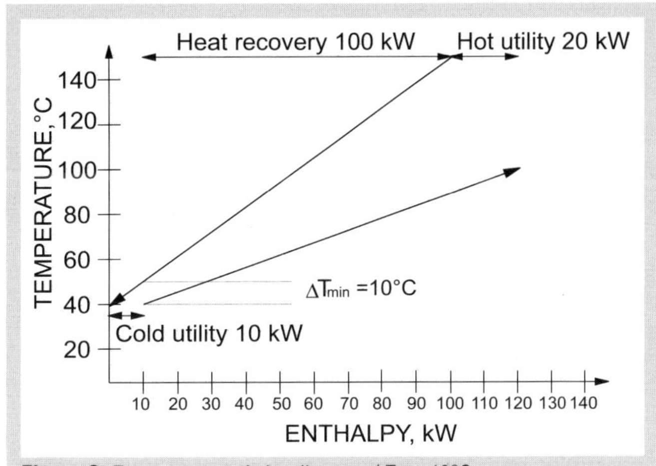

Figure 9. Temperature-enthalpy diagram, ΔT_{min} 10°C.

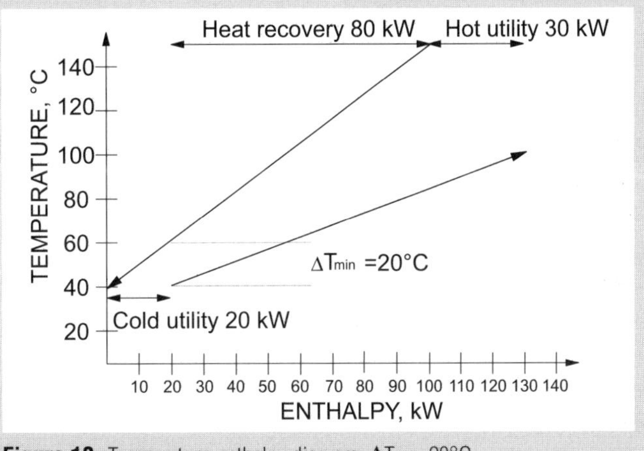

Figure 10. Temperature-enthalpy diagram, ΔT_{min} 20°C.

The T-H diagram can screen the energy targets at various values of ΔT_{min}. To examine the system with ΔT_{min} of 20°C, one simply moves the curve(s) horizontally as Fig. 10 shows. The slopes of streams or the temperatures cannot be altered because they are absolute values. Since the enthalpy change is a relative measure, the curves can therefore be moved horizontally. As Fig. 10 shows, larger ΔT_{min} results in a decrease in the heat exchange potential and an increase in the utility consumption. Since the area required for heat exchange decreases, the selection of the optimum ΔT_{min} is a balance between capital costs (area) and the operating costs (cold and hot utilities).

In a case with many hot and cold streams, we can join all the hot streams into a hot composite curve and all the cold streams into a cold composite curve. The composites result from adding all the CP values of the streams within temperature ranges. These temperature ranges result from changes in overall CP. Changes in overall CP occur when streams start or finish. In the case of nonconstant CP, they occur when the segment changes. An example can illustrate this. Table 12 contains data for four streams: two hot streams and two cold ones.

Table 12. Stream data for example 2.

Stream	Type	Supply temperature, °C	Target temperature, °C	CP, kW/°C	Enthalpy change, MW
1	Cold	20	180	200	32
2	Hot	250	40	150	-31.5
3	Cold	140	230	300	27
4	Hot	200	80	250	-30

Because all the streams have a constant CP, the supply and target temperatures of the streams determine the temperature ranges. Consider the hot composite curve of Fig. 11. The temperature ranges are plotted on the T-H diagram. The streams within the temperature ranges can be summed like vectors. The enthalpy change is the sum of all the individual enthalpy changes. The same process works for the cold streams of Fig. 12.

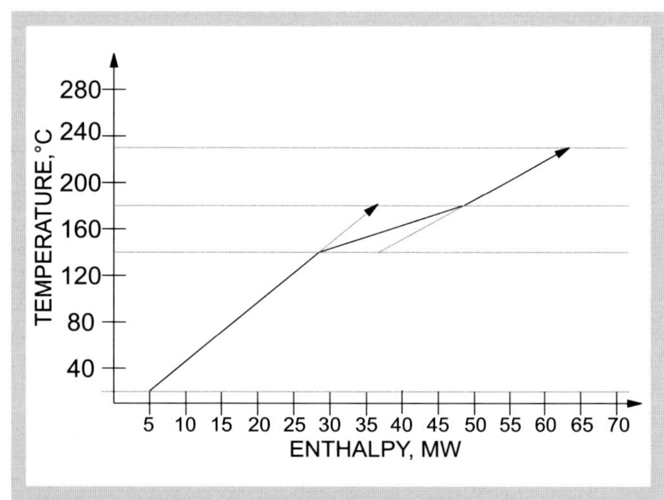

Figure 11. Construction of hot composite curve.

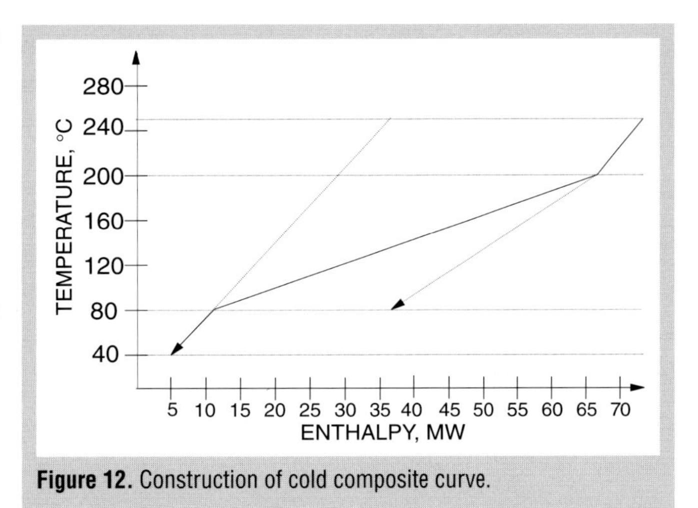

Figure 12. Construction of cold composite curve.

Figure 13 shows the curves placed on the T-H diagram with ΔT_{min} of 10°C. As with single streams, the overlap represents the heat exchange from hot streams to the cold streams. The temperature where the temperature difference between the curves is at its minimum value, ΔT_{min}, is the pinch temperature or pinch point. This point has a special significance because it divides the process into two separate processes.

The process above the pinch is a heat sink that receives heat from the hot utility. The process below the pinch is a heat source that rejects heat to the cold utility. The utility targets require that no heat transfer across the pinch. Now consider a case of transferring X MW from the hot composite above the pinch to the cold composite below the pinch. This means increasing the hot utility usage by X MW to compensate for the deficit in the upper part of the curve. Using X MW more cold utility will also be necessary to cool the excess heat in the lower part of the curve. The same situation will arise when using hot utilities below the pinch or cold utilities above it. To achieve the energy targets, the following must not occur:

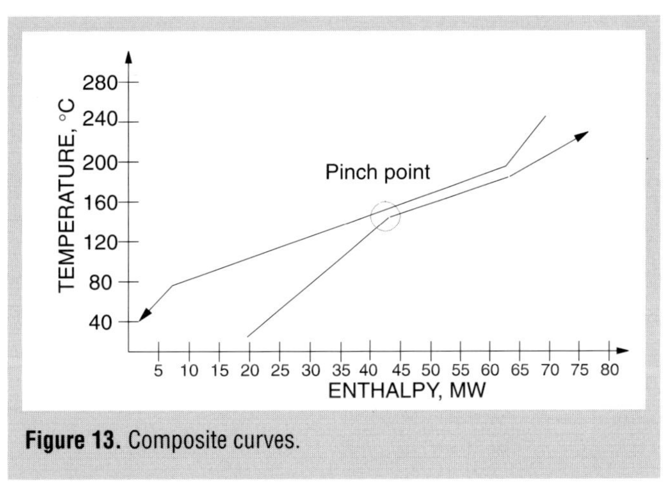

Figure 13. Composite curves.

- Transfer of heat across the pinch
- Use of hot utility below the pinch
- Use of cold utility above the pinch.

Some cases require neither cold nor hot utility. When ΔT_{min} decreases between the curves as Fig. 14 shows, the requirement for cold utility becomes zero at a certain ΔT_{min}. With further decrease of ΔT_{min}, the utility consumption remains constant. This type of problem is a threshold problem.

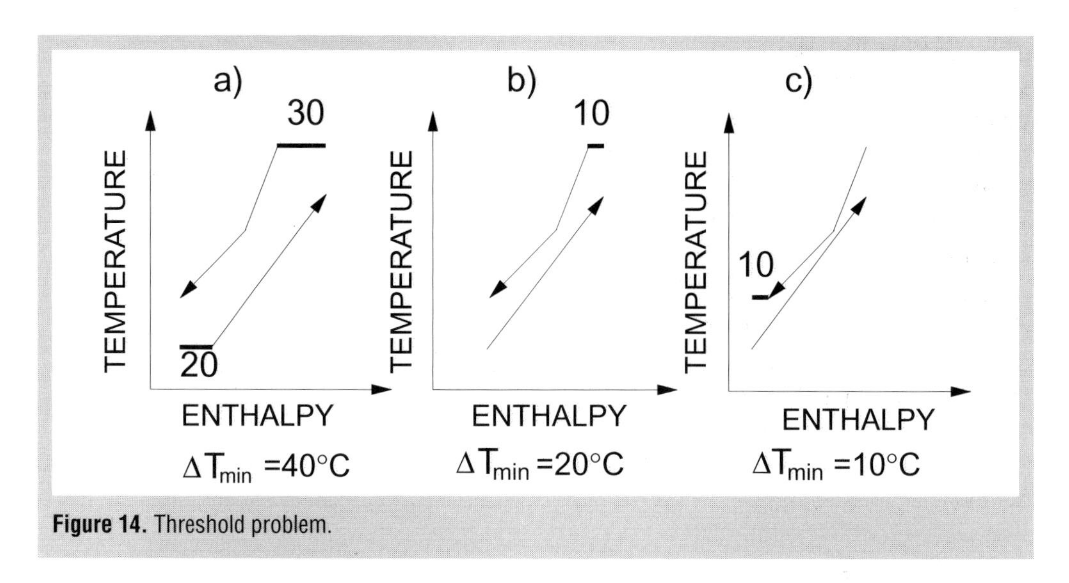

Figure 14. Threshold problem.

The optimal ΔT_{min} for threshold problems is at or above the threshold, since the utility cost remains constant below the threshold. The area does increase.

3.1.2 Grand composite curve

Composite curves are very suitable for energy targeting. For utility selection, another graphical representation called the grand composite curve (GCC) is necessary. The easiest way to construct the GCC is with the problem table algorithm. This is a nongraphical way of calculating the energy targets. The problem table algorithm is easy to use in spreadsheets and other software. The construction of a problem table for the previous example is as follows:

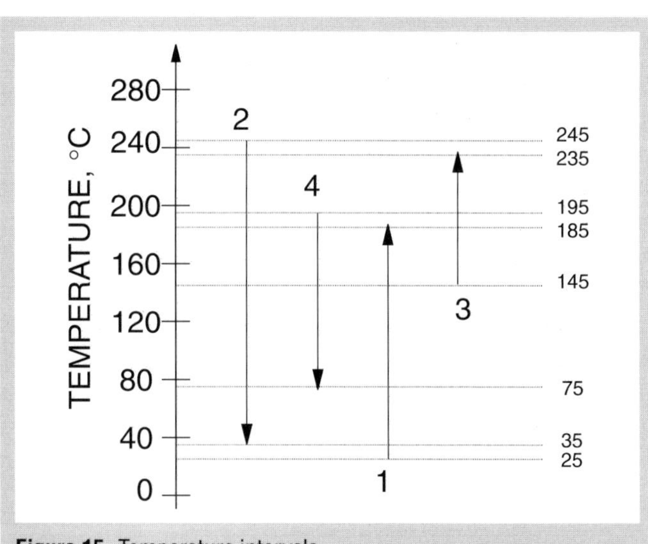

Figure 15. Temperature intervals.

- Shift down the temperatures of hot streams by ΔT_{min} /2, shift up the temperatures of cold streams by ΔT_{min} /2, and identify temperature intervals as Fig. 15 shows.
- Perform a heat balance around each temperature interval by calculating the following:

$$\Delta H_i = (\Sigma CP_{cold,i} - \Sigma CP_{hot,i}) \bullet \Delta T_i \tag{6}$$

where i is the temperature interval.

If ΔH is positive, then a net deficit of heat in the temperature interval exists. If DH is negative, then a net surplus of heat exists as Table 13 shows.

Table 13. Heat balances for temperature intervals.

Temperature interval, °C	ΔT	$\Sigma CP_{cold}-CP_{hot}$	ΔH
245–235	10	-150	-1 500
235–195	40	-150+300 = 150	6 000
195–185	10	-150 - 250+300 = -100	-1 000
185–145	40	-150 - 250+200+300 = 100	4 000
145–75	70	-150 - 250+200 = -200	-14 000
75–35	40	-150+200 = 50	2 000
35–25	10	200	2 000

- Cascade the heat down the temperature intervals by first assuming that the hot utility consumption is zero as Fig. 16 shows.
- Locate the interval with the largest negative heat flow. This temperature is the pinch point, and the absolute value of the largest negative flow is the minimum hot utility requirement.
- Add the hot utility to the heat flows as Fig. 17 shows.

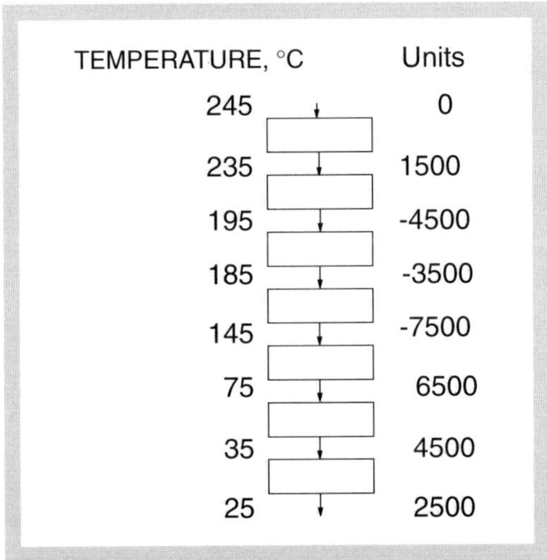

Figure 16. Heat cascade for zero hot utility consumption.

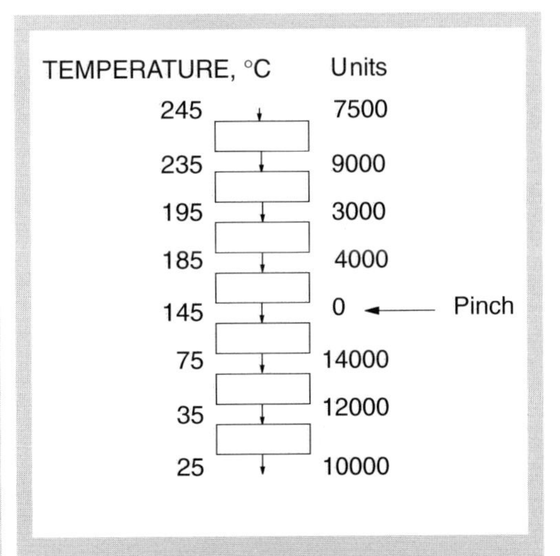

Figure 17. Final heat cascade.

Converting the information of the problem table into graphical form is now possible. Figure 18 shows the grand composite curve.

The point with zero heat flow is the pinch dividing the curve into heat sink and heat source. Both ends of the curve give the utility requirements. The pockets in the curve indicate process-to-process heat transfer.

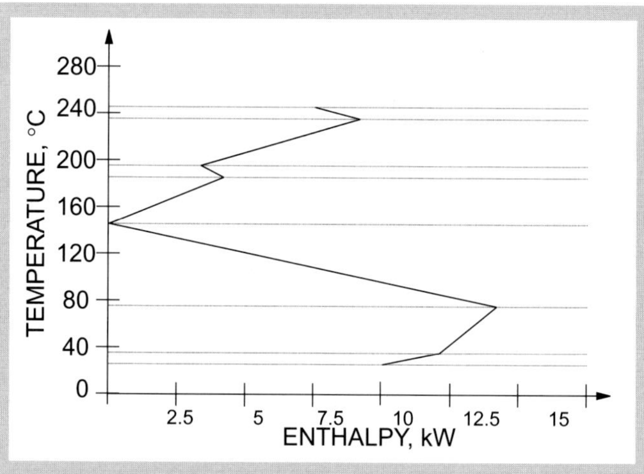

Figure 18. Grand composite curve.

Matching different utilities against the curve is now possible. In the example shown in Fig. 19, steam is used originally at high pressure to satisfy the hot utility demand. Cooling water is used as a cold utility. As the curve indicates, using medium pressure and low pressure steam will partially satisfy the hot utility requirement. The utilities touch the curve, but the ΔT_{min} is still preserved because of the use of shifted rather than actual temperatures.

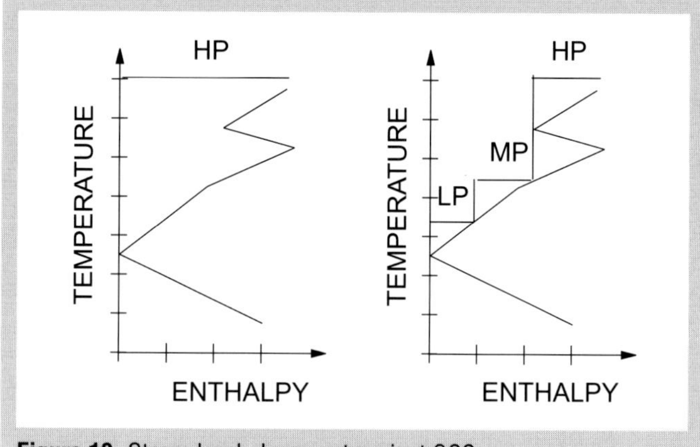

Figure 19. Steam level placement against GCC.

Other hot utilities such as flue gas, gas turbine exhaust, and heat pump cycle condensation can be placed on the curve. Cold utility options can consist of feed water preheating, low pressure steam generation, or heat pump cycle evaporation. Figure 20 shows some options.

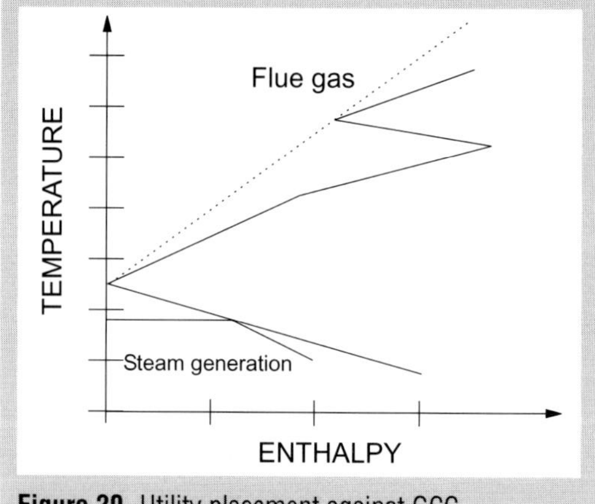

Figure 20. Utility placement against GCC.

3.1.3 Total site analysis

Composite and grand composite curves are useful in analyzing single processes. Interest often exists in the possible interactions of various processes on the site. The heat integration of processes independently of each other may lead to a less than optimum total energy use. Total site analysis provides the tools to optimize the processes and utility system in the context of the overall site.

Site composite curves

The construction of site composite curve starts with the construction of total site profiles. Total site sink and source profiles result from adding all sink and source elements from the grand composite curves of all individual processes on site after making the following modifications:

- The pockets indicating process-to-process heat transfer are sealed off.
- The temperatures of all source elements are reduced by $\Delta T_{min}/2$, and the temperatures of all sink elements are increased by $\Delta T_{min}/2$.

Figure 21 shows the construction.

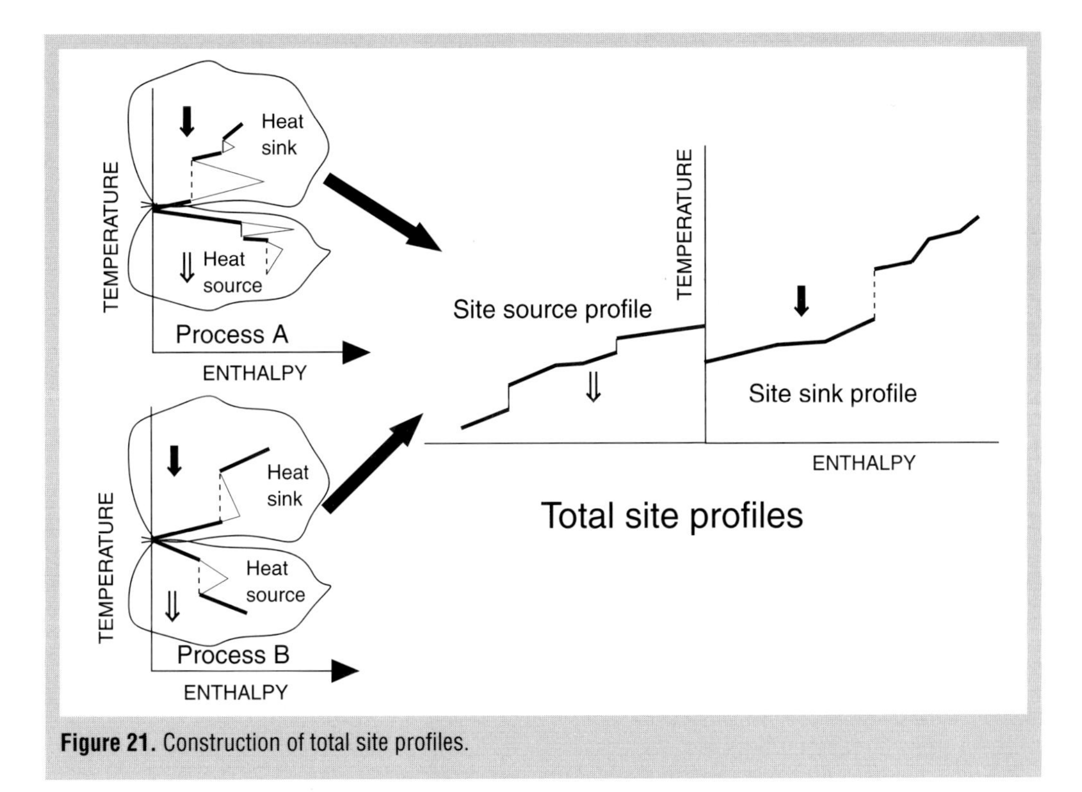

Figure 21. Construction of total site profiles.

Next, one introduces the steam levels and their corresponding heat loads to the total site profiles. This curve is the site composite curve of Fig. 22. As the curves shift closer together, the vertical overlap indicates the heat recovery through the utility system. The curves can be shifted toward each other until there is a region of zero approach between the utility loads. This area of zero approach is the site pinch. Now the heat recovery on site is at its maximum value.

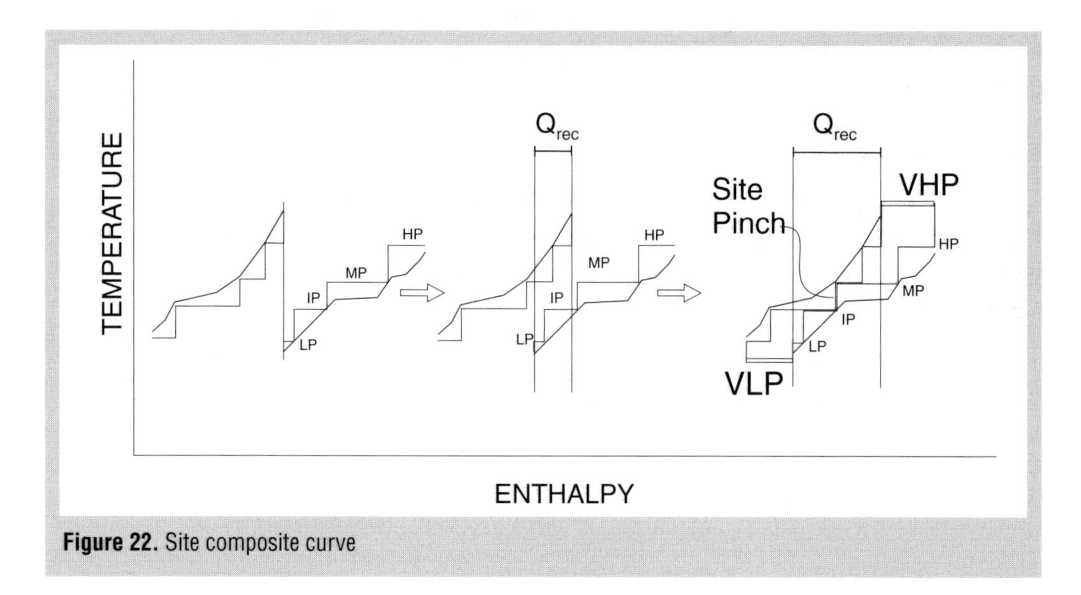

Figure 22. Site composite curve

Targeting for co-generation

The overlap between steam levels indicates a potential for shaft work production by a steam turbine. Correlations can relate this area into actual shaft work by using rigorous models or simple correlation coefficients. Obviously, increased co-generation will pull the site profiles apart and result in increased fuel consumption. Again a balance exists between the increased fuel consumption and co-generation.

3.1.4 Advantages and limitations

The main advantages of pinch technology are the following:

- Targets are available before design.
- The design engineer controls the design procedure.
- Pinch analysis provides insights.

The main disadvantages are as follows:

- Design according to pinch rules can be time consuming.
- Methodology cannot address all the possible trade-offs.

3.2 Process integration using mathematical programming techniques

Mathematical programming techniques are a set of mathematical formulations to handle optimization problems. Due to the rapidly increasing capacity of computers and the development of better and more reliable mathematical solvers, mathematical programming can apply to many industrial design problems. These include heat exchanger network design, utility system design, and scheduling of batch processes.

This section presents the basic concepts of mathematical optimization. The focus is on general formulation of the problems and the main programming techniques. More detailed information on optimization principles and techniques are available[5–6]. Information is also available on applications of mathematical programming in the chemical and process industries[7-9].

Every practical optimization problem usually contains the following characteristics:

- Objective function for optimization, e.g., minimize f(x)
- Equality constraints, e.g., h(x) = 0
- Inequality constraints, e.g., g(x) = 0.

Here x is a vector of n variables, x_1, x_2, x_3,. . ., x_n. An optimum solution is the set of variables that satisfies the constraints and provides an optimum value for the objective function.

In engineering design problems, the constraint functions are often known as design equations because they model the behavior and the specifications of the equipment. They also define the region of feasible solutions and where the optimal solution occurs.

The variables are often called design variables. They can consist of two types. The first type of design variable is a continuous variable. This is a variable such as temperature and pressure described in a continuous manner in the equations. The second type of variable is a discontinuous integer variable. It has frequent use in binary form to describe the existence of process units or for logical decisions.

3.2.1 Overview of techniques

The programming techniques can have three main categories depending on the formulation of the problem. The first two categories are linear and nonlinear programming. They are continuous formulations. Mixed integer programming (MIP) uses discontinuous integer variables.

Linear programming (LP)

The problems falling in this category involve linear formulation of the objective function and all the constraints. The main advantages of linear formulation are that the objective function is convex and the constraints form a convex set. In practice, this means that a local optimum is always also the global optimum. Another unique characteristic of linear formulation is that the optimum solution always lies on a constraint or at an intersection of constraints. This property speeds the search for the optimum.

Although many engineering problems are nonlinear by nature, some will transform into a linear form. The simplest way to accomplish this is to linearize the functions over their entire range. This may lead to large errors between the original and linearized function. Alternatively, one may try piecewise linearization. This divides the functions into various linear sections over their range. It improves the accuracy, but the formulation requires integer variables to indicate which linearized sections are being used. It then becomes a mixed integer linear programming (MILP) problem. One may also try successive linear programming where the linearization occurs at an estimate of the solution.

Numerous LP algorithms are available. They can handle very large numbers of variables and constraints, and they achieve a rapid convergence toward the global optimum.

Nonlinear programming (NLP)

If the formulation involves nonlinear terms in the objective or in any constraint, then the problem becomes nonlinear. For NLP problems, the global optimum solution usually cannot be guaranteed unless the objective and constraints form a convex set. In practice, this means the solver finds a local optimum point and cannot find the way toward the global optimum. Changing the initial starting point of the optimization may therefore change the solution.

Mixed integer programming (MIP)

Many problems in design, operation, and scheduling involve noncontinuous variables that take integer values. These decision variables that often are binary (0–1) variables denote the existence or nonexistence of process units. In other words, one can use MIP formulation for simultaneous optimization of structure and parameters. LP and NLP primarily find use for parameter optimization.

Structural optimization of process design requires construction of a superstructure. The superstructure contains all the feasible process operations and their connections. The existence of the units and their connections are modeled by using binary variables. A unit or connection exists in the final solution if its integer variable is set to one and does not exist if its integer variable is set to zero.

If the formulation of the objective and constraints is linear, then it is MILP. Similarly, nonlinear formulations with integers become mixed integer nonlinear programming (MINLP) problems. Combinations may cause problems when the number of integer variables is large. For a formulation with N binary variables, the number of possible combinations becomes 2^N. With 10 binary variables, we have 2^{10} (1024) combinations, and with 20 binary variables 1 048 576 combinations exist!

3.2.2 Advantages and limitations

Advantages of mathematical programming include the following:

- Many design options and their trade-offs can be handled simultaneously.
- The entire design procedure can be handled by software with quick results.

Following are the disadvantages:

- The design engineer is not part of the decision making.
- In structural optimization, the optimum structure must be embedded in the superstructure.

- Combinations and nonconvexities cause numerical problems .
- Global optimum is not guaranteed for nonlinear problems.

3.3 Exergy analysis

The most common analysis method of thermal performance is the energy analysis. It uses the first law of thermodynamics. This states the preservation of energy and treats all forms of energy equally. The energy analysis is easy when one knows the enthalpy or the specific heat of the process flows. The major drawback is that the energy analysis does not consider the quality of the energy. In energy analysis, it is equally acceptable to use high quality energy like steam or electricity or energy of lower quality like hot water.

Exergy analysis uses the second law of thermodynamics. This states that entropy increases in real processes. The increase of entropy in a process is a sign of internal losses within the process. A greater increase of entropy means the quality of the energy is lower.

Exergy is the ability to do work in relation to the environment. Szargut[10] defined exergy as "the amount of work obtainable when some matter is brought to a state of thermodynamic equilibrium with the common components of the natural surroundings by means of reversible process, involving interaction only with the components of nature."

Exergy of a flow is the following:

$$E = E_{kin} + E_{pot} + E_{ph} + E_{ch} \tag{7}$$

where E_{kin} is kinetic exergy
E_{pot} potential exergy
E_{ph} physical exergy
E_{ch} chemical exergy.

The two first terms are fully convertible into work because no entropy is produced. The sum of the final two terms that include entropy generation is called thermal exergy, E_{th} .

$$E_{th} = E_{ph} + E_{ch} \tag{8}$$

The physical exergy is the following:

$$E_{ph} = (H - H_0) - T_0 \bullet (S - S_0) \tag{9}$$

where the subscript 0 denotes the reference state. The physical exergy of a substance can be calculated provided the thermodynamic data for enthalpy and entropy are available.

In heat exchangers, the exergy loss calculations can be simplified by using the Carnot-factor:

$$\eta_c = \left(1 - \frac{T_0}{T}\right) \tag{10}$$

where T_0 is the ambient temperature
T the average temperature of the medium.

The exergy loss is then

$$\Delta E_{loss} = Q \cdot (\eta_{c,\,hot} - \eta_{c,\,cold}) \tag{11}$$

Due to the nonlinear scale of the Carnot-factor, the same ΔT_{min} at lower temperatures results in a larger loss than at high temperatures. The Carnot-factor can be used in connection with pinch technology curves for shaft work targeting, co-generation, and refrigeration systems[1].

The calculation of chemical exergy is not as straightforward because the chemical exergy depends on the composition of the substance. Tabulated values for chemical exergy exist for pure substances and binary mixtures. For some complex substances such as industrial fuels, approximative formulas are available in the literature[11].

When applying exergy analysis to pulp and paper mills, major problems will develop when trying to determine the chemical exergy of black liquor, smelt, and pulp. Their composition is unknown, and chemical reactions occur during the process.

Since exergy calculation refers to the environmental state, the exergy efficiency varies with the environmental temperature.

References

1. Dhole, V. R., Smith, R., and Linnhoff, B., Computer Applications for Energy-Efficient Systems, Encyclopedia of Energy Technology and the Environment, John Wiley & Sons, 1995.
2. Linnhoff, B., et al, A User Guide on Process Integration for the Efficient Use of Energy, IchemE UK, 1982.
3. Smith, R., Chemical Process Design, McGraw-Hill, 1995.
4. Wang, Y. P., and Smith, R., Wastewater minimisation, Chemical Engineering Science Vol.49, No. 7, pp. 981–1006, 1994.
5. Edgar, F., and Himmelblau, D. M., Optimization of Chemical Processes, McGraw-Hill, 1988.
6. Floudas, C. A., Nonlinear and Mixed-Integer Optimisation Fundamentals and Applications, Oxford University Press, 1995.
7. Grossmann, I.E., Kravanza, Z., Mixed-Integer Non-linear Programming Techniques for Process System Engineers, Computers and Chemical Engineering, Vol. 19 (suppl.), pp. 189–204, 1995.
8. Grossmann, I.E., Santibanez, J., Applications of Mixed-Integer Linear Programming in Process Synthesis, Computers and Chemical Engineering, Vol. 4, pp. 205–214, 1980.
9. Papoulias, S.A., Grossmann, I.E., A Structural Optimization Approach in Process Synthesis – Utility Systems, Computers and Chemical Engineering, Vol. 7, No. 6, pp. 695–706, 1983.
10. Szargut, J., Exergy analysis of thermal, chemical and metallurgigal processes, Springer – Verlag, Berlin, 1988.
11. Kotas, T.J., The exergy method of thermal plant analysis, Butterworths, 1985.

CHAPTER 18

Byproducts of chemical pulping

Johan Gullichsen and Harry Lindeberg

Byproducts of chemical pulping

Pulping spent liquors from acid and alkaline pulping contain hundreds of different compounds originating from lignin, carbohydrates, and extractives in the original wood. Species and processing conditions strongly influence the composition of spent liquors. The global pulping industry produces more than 150 million t of these organic substances each year; most of the material serves as a renewable energy source for the paper industry. Those compounds that have a higher market value than fuel are separated from pulping wastes. This chapter describes some techniques used in manufacturing such pulping side products.

1 Turpentine

Turpentine is produced from volatile organic compounds in wood. Softwoods contain mono-, sesqui-, and di-terpenes with the general formula $(C_5H_8)_n$. Hardwoods primarily contain higher terpenes and their derivatives, or triterpenes, sterols, and polyprenoles. The most important terpenes are monoterpenes. Of these, α-pinene has prime importance. It is the main constituent of commercial turpentine. Other minor components are β-pinene, Δ^3-karene, and limonene (Fig. 1). Table 1 shows the typical composition of a Nordic softwood kraft turpentine.

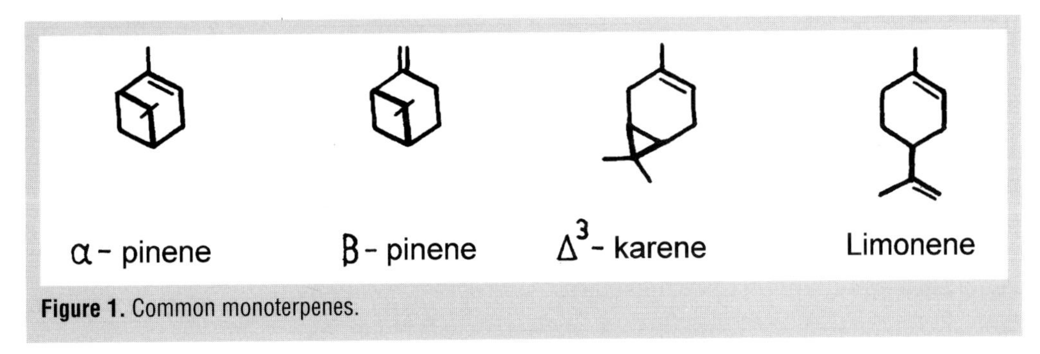

Figure 1. Common monoterpenes.

Table 1. Composition of Nordic softwood kraft turpentine.

Compound	Weight fraction, %
α-pinene	60–83
β-pinene	2–7
Δ^3- karene	11–28
Others	2–6

The terpenes are volatilized from wood during the steaming of chips and in digester degassing in the early phases of a kraft cook. In processes that do not apply steaming, such as displacement batch cooking, turpentine collection is from off-gases from the warm and hot liquor accumulators.

The turpentine-rich gas stream contains other volatiles, steam, perhaps black liquor droplets, and possibly fibers. Droplets and fibers are separated in a cyclone before condensation. Condensation in a surface cooler separates turpentine and water. The noncondensibles go to the mill low volume, high concentration (LVHC) malododrous gas collection system. Since turpentine is lighter than water and only poorly soluble, it will float on top of the water layer in the decanter that collects condensate. The raw turpentine is removed with the decanter overflow. The water phase goes to the mill condensate stripping system.

The purity of turpentine varies, depending on the phase of steaming or cooking from which it comes. Steaming produces the purest fractions. Turpentine from cooking contains considerable amounts of reduced organic sulfur compounds and other impurities. Some turpentine dissolves in the black liquor and cannot be recovered.

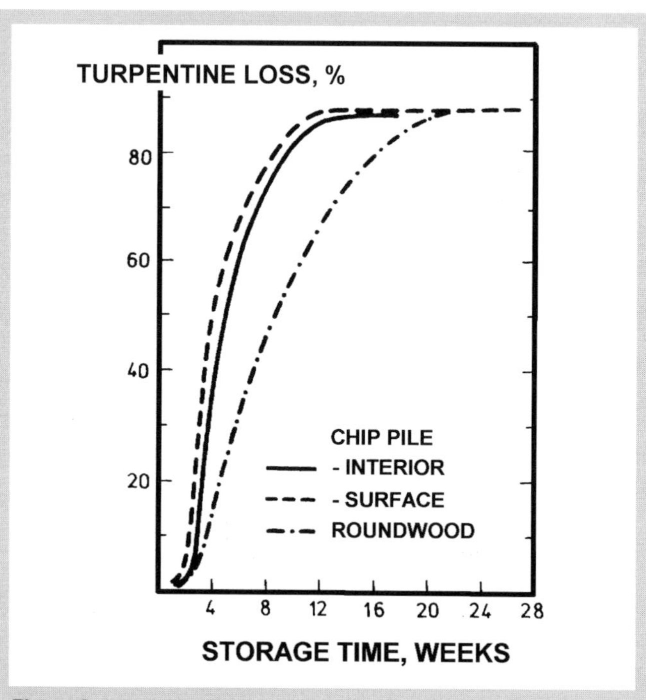

Figure 2. Loss of turpentine on storing pine chips vs. time.

Pine kraft pulping may yield 2–15 kg of turpentine per ton of pulp, depending on the growing conditions of the tree, wood storage time, and conditions of steaming time and temperature. The yield of turpentine from trees growing in arctic conditions is generally higher than that of trees growing in warmer ones. Spruce produces substantially less turpentine (2–3 kg/t pulp). The loss of turpentine during wood storage is dramatic if storage time exceeds four weeks. Processing of fresh wood will result in higher turpentine yields as Fig. 2 shows.

Figure 3 shows a typical turpentine collection system for conventional batch cooking. The raw turpentine is distilled into fractions at a centralized distillation facility to separate the main fractions as Fig. 4 shows.

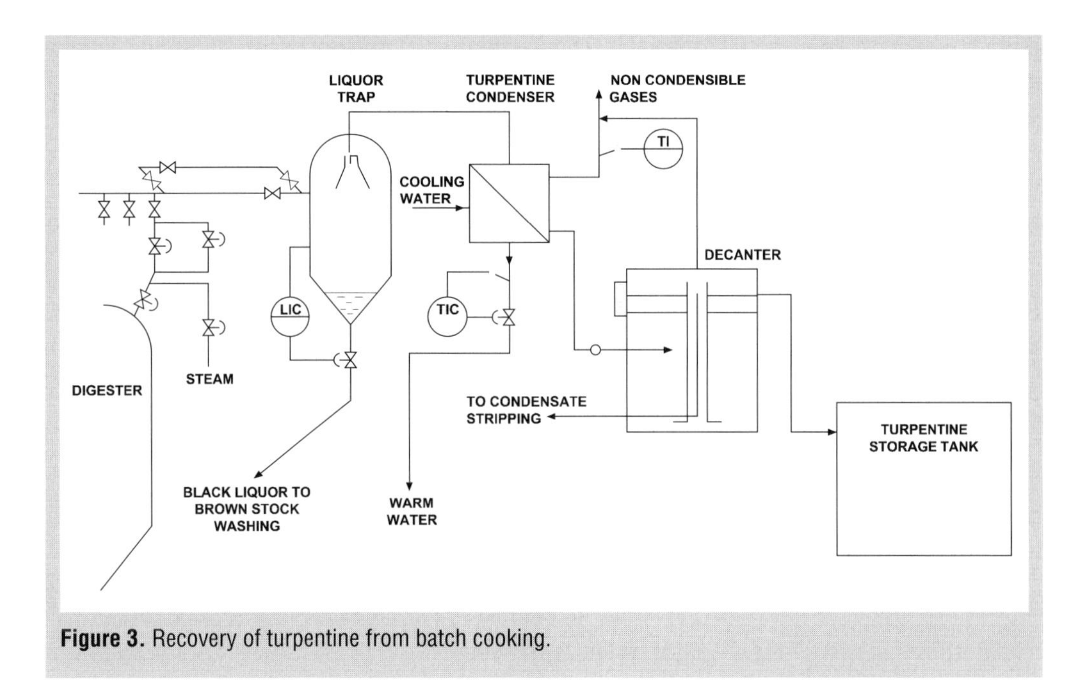

Figure 3. Recovery of turpentine from batch cooking.

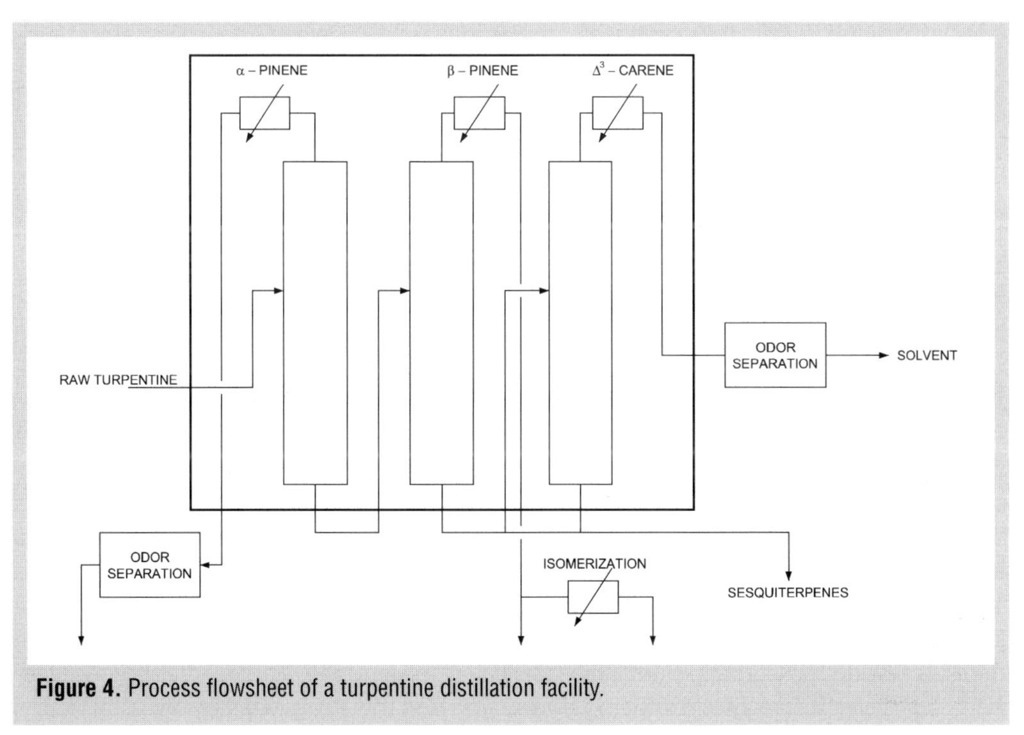

Figure 4. Process flowsheet of a turpentine distillation facility.

The different fractions can undergo further processing by polymerization, hydration, or oxidation into various products. The main uses of turpentine products are as solvents and diluents for oil-based paints and varnishes, pharmaceutical industry products, and perfume additives.

2 Soap and tall oil

2.1 Introduction

The major chemical by-product derived from the kraft pulp industry is crude tall oil (CTO). The name comes from the Swedish word "tall" (English, pine). This indicates that a primary source is pine. In the kraft pulping of coniferous wood, some of the extractives in the wood react with the alkaline pulping liquor to form sodium soaps. These dissolve in the pulping liquor, but float to the surface upon concentration of the liquor. These "soap skimmings" are separated from the black liquor early in the recovery process, and are further treated with acid to form CTO. This CTO is a saleable product for a pulp mill, but it represents only about 1%–1.5% of the total revenue for the mill. Globally, about 2 million t/year of CTO are refined.

CTO can also serve as fuel in the lime kiln. This alternative is attractive when producing low quality CTO. Soap can also be sold as such. This practice is uncommon in North America, but is occasionally practised in other parts of the world. Another alternative is to burn soap in the recovery boiler. This can be expensive if the recovery boiler is a bottleneck in the pulp mill.

Soap CTO is not a very important commercial product from the pulp mill point of view, but removing most of the soap formed during the pulping process is simply necessary.

With the closing of kraft pulp mill circulations, the role of the CTO plant has become more important. It is a major source of sulfur addition and a part of the pulp mill waste handling and removal system.

2.2 Chemistry

2.2.1 General

The precursors of soap CTO are the extractives found especially in coniferous trees. The composition of extractives in hardwoods such as birch can have detrimental effects on the quality of the CTO end products. Pine typically contains extractives as free resin acids, fatty acids in the form of glycerides, and terpenes. The extractives also contain "neutrals" or "unsaponifiables" that are primarily different alcohols. The most important alcohol is sitosterol.

During kraft pulping, the acids form sodium salts that are soluble in the cooking liquor. By concentrating the resulting black liquor, the sodium salts of the acids and the neutrals float to the surface, from which they can be skimmed. The resultant product is the crude soap. It has strong foaming properties, and therefore requires removal from the recovery cycle.

2.2.2 Composition of tall oil

The composition of CTO varies considerably, depending on the location of the mill and the wood furnish used. A commonly used measure of quality is acid number (AN) expressed as grams of KOH per gram. Mills that use only pine have the highest quality CTO with AN values of 160–165. Mills that use 50% or more hardwood in their furnish have a CTO AN of 125–135. Table 2 gives an analysis of CTO samples from different parts of the world[1].

Table 2. Composition of CTO.

	Southeastern USA	Northern USA and Canada	Scandinavia
Acid number	165	135	132
Saponification number	172	166	142
Resin acids,%	40	30	23
Fatty acids,%	52	55	57
Unsaponifiables,%	8	15	20

The CTO from the southeastern United States is an example of a very high quality product with pine as the wood furnish. The Scandinavian example is from a mill using a mixture of pine, spruce, and birch.

2.3 Soap recovery system design

2.3.1 Skimming and storage of soap

Soap is removed from black liquor in the tanks connected to the pulp washing plant and the evaporators. Due to the strong tendency of soap to create foam, removing as much soap as possible before black liquor enters the evaporators is important. The systems to do this vary from mill to mill, and no standard solution is practised.

A thorough investigation that included mill and laboratory trials was performed in Sweden in the early 1970s. The results, published in a compendium[2],

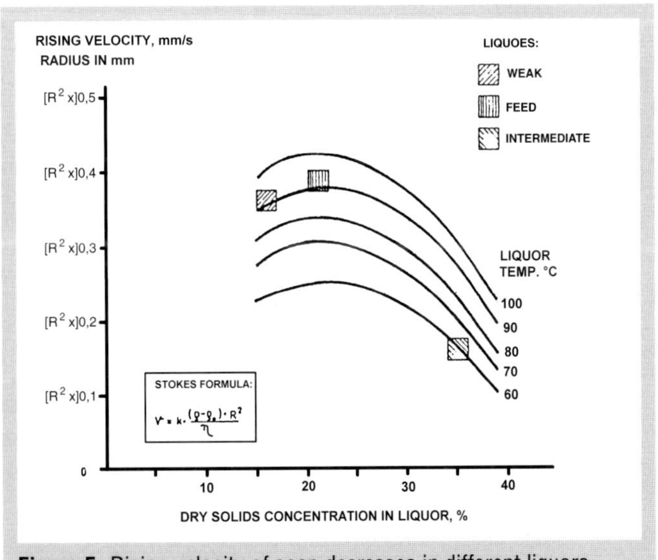

Figure 5. Rising velocity of soap decreases in different liquors.

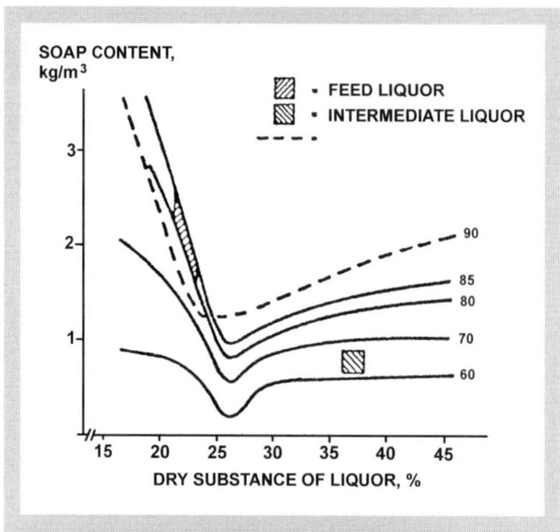

Figure 6. Solubility of soap in black liquor per volume of liquor.

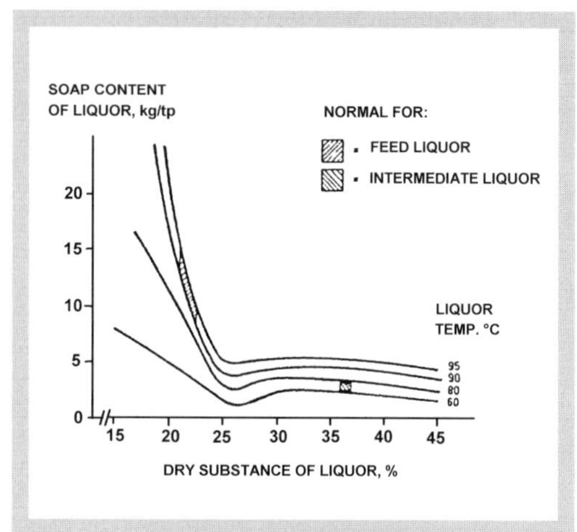

Figure 7. Solubility of soap in black liquor per t of pulp.

are probably the best source of practical information on sulfate soap handling. Here are some of the results:

- Effective separation of soap from black liquor requires that the soap particles rise rapidly and that their solubility in black liquor is low. The soap particles apparently follow Stoke's law within reasonable limits as Fig. 5 shows.
- Soap solubility reaches a minimum at 26% dry substance in the black liquor, and does not increase (especially as calculated per t of pulp). Figures 6 and 7 show solubility curves.

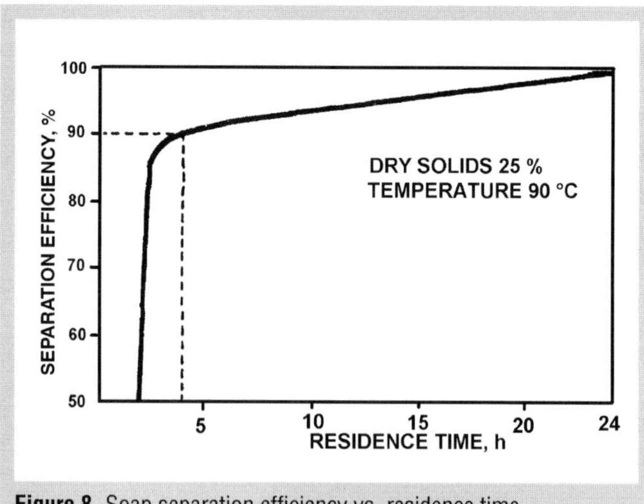

Figure 8. Soap separation efficiency vs. residence time.

- Residence time in soap separation tanks is at least 4 h for good separation as Fig. 8 shows. With allowance for fluctuations in operation and other factors, a residence time of 8 h is recommended in feed liquor tanks and intermediate liquor tanks. The flow of liquor in the tanks should avoid "short circuiting" to guarantee the best possible separation of soap.

These factors determine the arrangement for separation of soap in the pulp washing department and the evaporators. No package systems are available for soap removal. Although every mill has its own tailor-made tank farm, the flowsheets are similar to the system of Fig. 9.

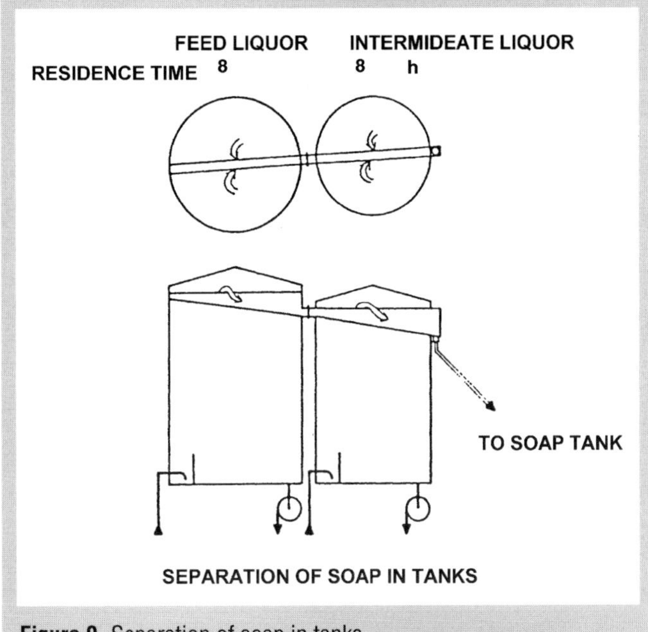

Figure 9. Separation of soap in tanks.

2.3.2 Soap handling

For further processing to CTO, soap must undergo cleaning and homogenization. When burned in the recovery boiler, soap requires no special treatment. The primary impurities in soap are:

- Lignin (as black liquor)
- Calcium
- Fibers.

All these components create operational disturbances in the acidification process, and they can also severely degrade CTO yield and quality. The simplest and most common procedures allow the soap to settle in the tank. Black liquor is pumped from the bottom of the soap tank intermittently. Using this method, only some of the free black liquor is removed.

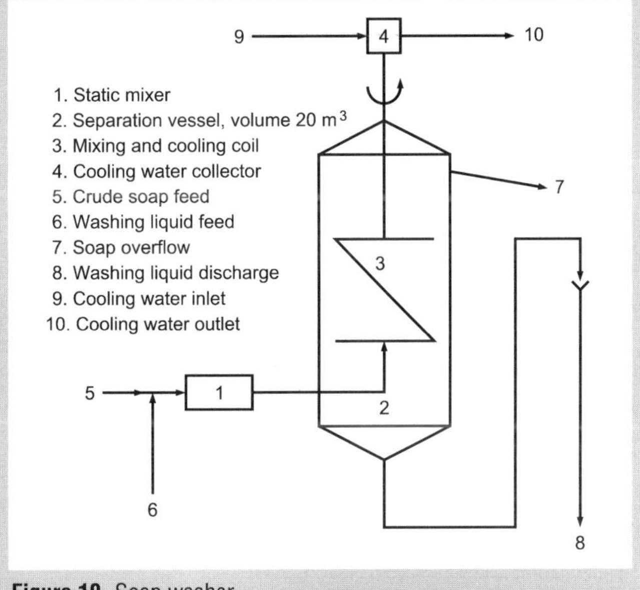

Figure 10. Soap washer.

Installing a revolving rake, the so-called Lundberg rake, in the soap tanks[3] can improve the separation method. A soap washing system developed in 1979 uses an electrolyte (sodium sulfate = neutralized chlorine dioxide generator spent acid) as a washing medium with cooling of the soap to 40°C–45°C by a revolving cooler and rake[4]. This soap washer is in use in several mills around the world. About 75% of the black liquor is removed, and some fibers are apparently also removed by this system. Figure 10 gives a simplified drawing of the system. The decanter centrifuge in Fig. 11 serves the same purpose.

No method to clean soap for further processing has been effective in removing calcium and fibers.

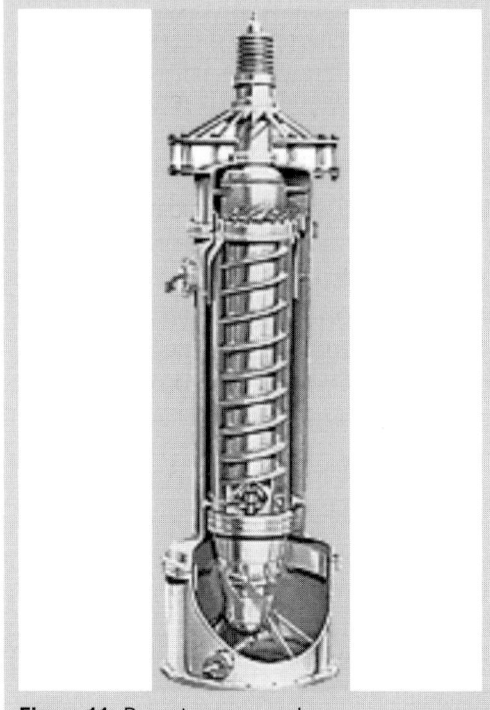

Figure 11. Decanter soap washer.

2.4 Tall oil plant design

2.4.1 Tall oil yield

In designing CTO plants, the expected yield of CTO obviously has prime importance. But yield varies over a wide range, depending on the following factors:

- Type of wood
- Time of year of harvest
- Log or chip storage
- Storage time and temperature
- Soap skimming system
- Type of CTO plant.

All these factors can influence the yield considerably. The classic textbook on CTO[5] gives a few examples quoted here. "In the southern United States where among others slash pine is used, the following yields are reported as kg CTO/t of pulp: 22.5–40.5 (Casey) and 18–45 (Herrlinger). Wegelius gives as a normal yield 45 for southern Finland and 60 for northern Finland. As a maximum for Finland, Juvonen gives 70 and for Sweden, 90."

One must assume that the high values are for 100% pine as the wood furnish. When using spruce in the wood furnish, the yield is considerably lower. The increased use of hardwood (mainly birch) in Scandinavia since the 1970s has lowered the yield and quality of CTO even more.

Thus, one can conclude that the calculation of CTO yield in a new plant being considered involves many factors. Consider the following example:

Pulp production: 500 000 t pulp/year

Wood furnish: 40% pine, 30% spruce, and 30% birch

Location: Middle of Sweden

Yield calculation:

Pine: 0.4 x (500 000) x (0.045) = 9 000 t CTO/year

Spruce: 0.3 x (500 000) x (0.025) = 3 750 t CTO/year

Birch: 0.3 x (500 000) x (0.015) = 2 250 t CTO/year

Total:15 000 t CTO/year

2.4.2 Design parameters

Systems approach, design capacity, and soap quality are the main criteria to consider in CTO plant investments. This section will consider the effect of soap and some other parameters; later sections will cover the selection of process and equipment.

Soap contains three major impurities: lignin in the form of black liquor, fibers, and calcium. Lower amounts of these substances make the operation of a CTO plant easier. CTO content in soap should be above 50%. This indicates the total amount of impurities. Here a typical list of target values for soap to the CTO plant with analytical methods in brackets:

- Tall oil content in soap > 50% (PCA-7)
- Calcium content in soap < 0.3% (TAPPI T635)
- Lignin content in soap < 2% (PCA-8)
- Fiber content in soap < 1% (PCA-9).

Considering the example in the previous section, a suitable design capacity in a situation where the pulp mill runs with 100% pine and a peak yield of pine would specify a CTO plant with a capacity of 3.5–4 t CTO/h.

Hydrogen sulfide gas is emitted from a CTO plant. A scrubber is normally necessary for effective removal of this toxic and malodorous gas.

Normally, CTO fractionators do not accept CTO with a water content above 1.5%. This means that a CTO drier (vacuum drier) is necessary to ensure water content below 1%. CTO picks up water from the atmosphere during storage. The CTO drier can combine with CTO washing. This improves CTO quality by lowering the ash content.

2.4.3 Selection of process

The standard process for converting soap to CTO is the addition of sulfuric acid to soap in a batch reactor or continuous reactor system. The sulfuric acid is sufficiently strong to convert the fatty acid and rosin acid sodium salts into acids. An advantage of sulfuric acid is its low cost.

Pulp mills that have a chlorine dioxide plant already have sulfuric acid available as by-product. The disadvantage of generator waste acid is its high corrosivity, requiring special care in materials of construction.

In recent years, environmental restrictions on low sulfur emissions prompted finding alternative sulfur-free acids for use in the acidification process. The sulfur input in the standard acidification process is of about the same order of magnitude as the total sulfur losses in the pulp mill.

The best candidate is carbon dioxide. Under atmospheric pressure, it can convert about one-third of the sodium salts to acids, primarily the rosin acids[6]. A few commercial installations exist but operating results are not available. Another alternative is hydrochloric acid[7], but this has had no commercial use to date.

2.4.4 Selection of equipment

The following basic alternatives are available:

- Batch processing
- Continuous processing and separation by centrifuge
- Continuous processing and separation by decanting.

Batch processing is still a common practice in North America. The batch plant is a combination of reactor and decanter; soap and sulfuric acid are mixed using an internal mixer or external mixing with a circulation pump. After mixing, the product has about 3–4 h to settle. CTO separates at the top, and brine is pumped out through the bottom valve. The reactor tank is usually brick lined, and the mixer must be a high alloy steel to withstand the extremely corrosive mixture at the beginning of the process. In Scandinavia, installation of batch processing plants is rare.

A batch plant is suitable for small production units of less than 5000 t of CTO/year. Such a plant is easy to run, and can be designed as a 1–3 shift operation. Automation is difficult, but a few instruments, litmus paper, and common sense are themain requirements. Disadvantages of the process are low yield, high sulfuric acid consumption (250–300 kg sulfuric acid/t CTO), and high gas emissions during a short period.

Continuous processing, using a centrifuge to separate CTO and brine, became available in the 1960s. Figure 12 shows a diagram of the process.

Using a continuous process eliminates most of the disadvantages of the batch process. The corrosive mixture of soap and sulfuric acid created new problems, primarily due to high maintenance costs. The need for better soap treatment before processing (soap washing) is also greater. Improved materials of construction and better pH control have improved matters to the point where most plants in the Nordic countries run with full automation. All the continuous centrifuge plants in the Nordic countries operate in three shifts. In North America, running continuous plants on the day shift five days per week is common.

Continuous processing with decanting became available in the early 1970s using two decanters coupled in series[8]. A newer version of continuous decanting is the unit in Fig. 13 developed in the early 1980s[9].This unit eliminates the problem of accumulation of a lignin layer during separation. It can be built in one unit for large capacities. The space requirement is also considerably less than for decanting in tanks.

When building a CTO plant, great care is necessary in selecting materials of construction[10]. Floor material selection is also crucial, since CTO is extremely sticky and difficult to remove unless handled immediately after a spill.

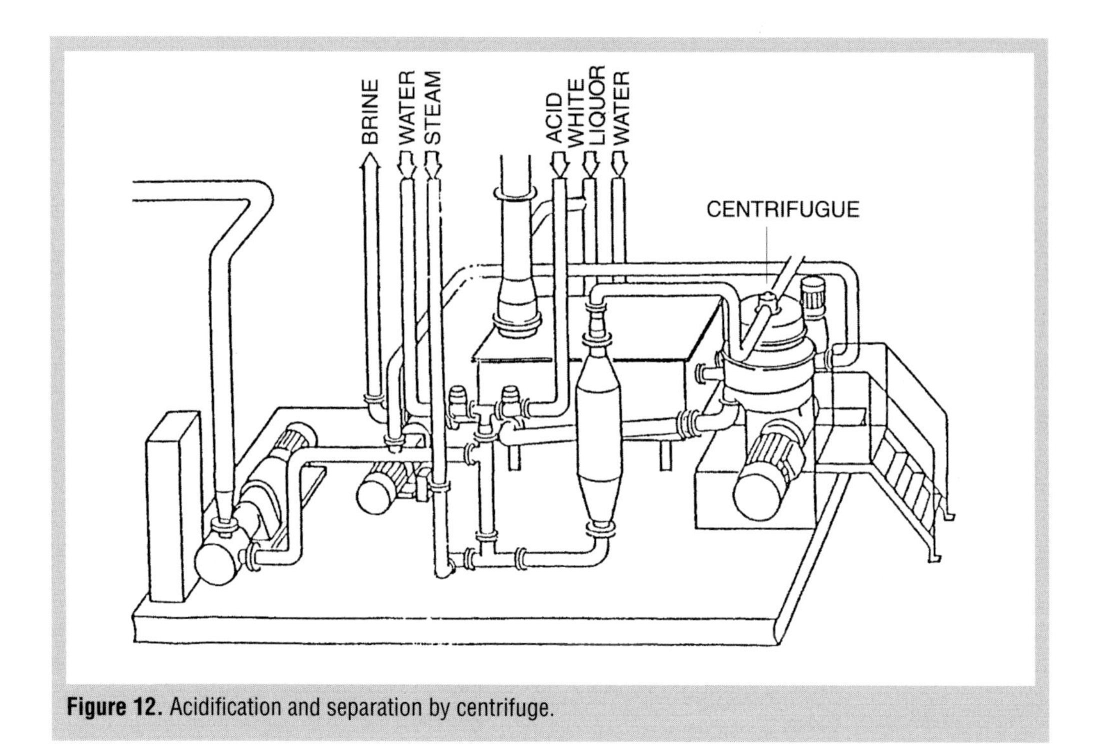

Figure 12. Acidification and separation by centrifuge.

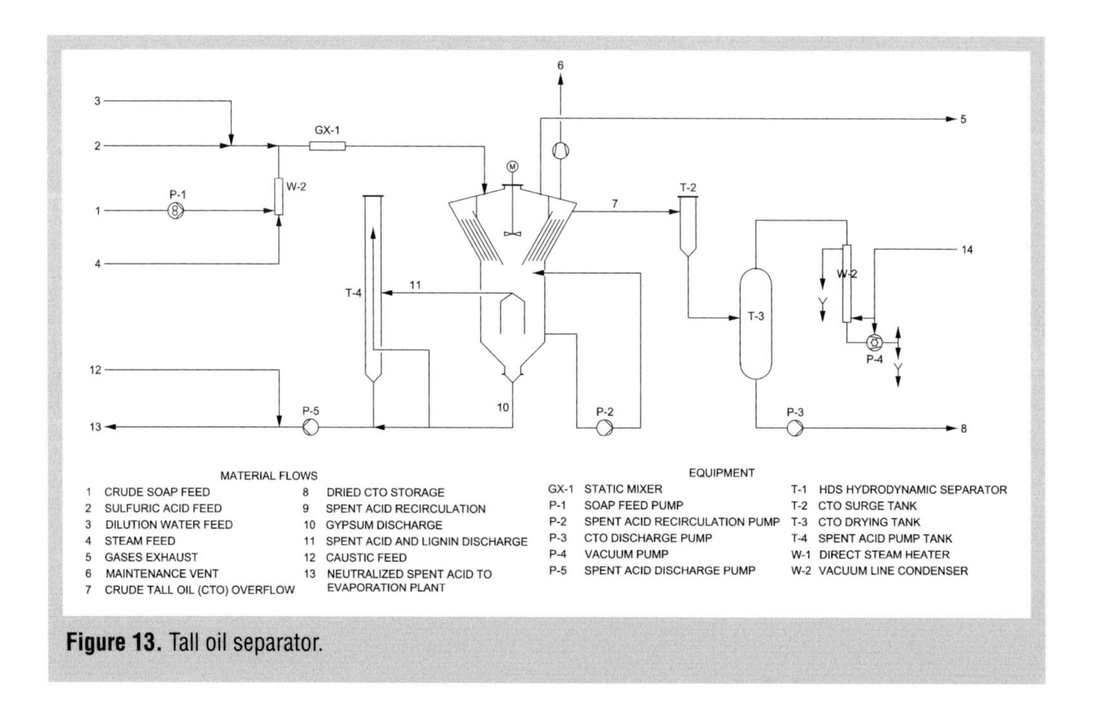

Figure 13. Tall oil separator.

2.4.5 Tall oil handling and storage

CTO is stored at the plant site in tanks that have heating coils. CTO shoud be stored below 65°C, since the occurrence of unwanted chemical reactions at higher temperatures reduces the sales value. Lining the bottom and walls of the tanks to the height of 1–2 m with acid-resistant steel is a good idea, since the water separating at the bottom of the tanks is corrosive.

For loading into tank cars (that should also have heating coils), CTO requires heating to about 60°C–70°C. CTO shipments today use railway cars, special trucks, and small tankers[11].

2.5 End use of tall oil products

2.5.1 Typical tall oil products

CTO or tall oil has little use in its raw state today. It is distilled to the following base products:

- Tall oil fatty acids
- Tall oil rosin
- Distilled tall oil
- Tall oil heads
- Tall oil pitch.

Tall oil products have a wide range of applications. The different base products can be further refined into a wide variety of products and formulations, some of which are listed here[12]:

Tall oil fatty acids are used in:

- Alkyd resins
- Detergents
- Drilling fluids
- Fungicides
- Lubricants
- Paint resins.

Tall oil rosin has traditionally been used as a paper size, but this use is declining. Other applications:

- Ink resins
- Adhesive tackifiers
- Disproportionated rosins
- Phenolic resins.

Tall oil pitch has some special applications, but its primary use is still fuel (sulfur free).

2.5.2 Special tall oil products

The wood processing industry has sometimes produced chemicals that have attracted interest beyond the industry. One is furfural. This is the base chemical in furan chemistry, and is a starting chemical for nylon. Another is xylitol can be an important ingredient in chewing gum to prevent tooth decay. Sitosterol in tall oil soap is the chief ingredient in a margarine.

A process invented[13] in the beginning of the 1970s improved the quality of tall oil soap from kraft mills that were using pine and birch. The invention underwent further development[14]. The soap extraction process produced a by-product of neutrals with a high content of sitosterol. A patented process[16] produced sitosterol for the pharmaceutical and cosmetics industry. In the 1980s, development work[17,18] resulted in a sitostanol product that clinical studies proved was a cholesterol lowering chemical. This led to the development of margarine using this chemical. The development has now attracted global attention.

3 Other byproducts

3.1 Byproducts from acidic pulping

The spent liquors from sulfite pulping contain lignosulfonates, sugars, and organic acids. Table 3 shows an example.

Table 3. Gross composition of pulping spent liquor from acid sulfite pulping of Norway spruce.

Lignosulfonates	50%–65%
Sugars	15%–29%
Volatile acids	2%–6%
Total sulfur	2%–5%
Calcium	4%–10%
Sulfate ash	16%–19%
Other ashes	10%

This spent liquor can be used (after concentration and drying) as a binder in several applications of brickette and pellet manufacturing. It also finds use as a colloidal additive in oil drilling, flotation aid, emulsifier, and simple glue.

A variety of phenolic compounds can be separated from the lignin fraction after desulfurization. Vanillin is one example; it can be made from lignosulfonate decomposition products after alkaline oxidation.

The most widely applied use of sufite spent liquor from spruce is the manufacture of ethanol from the carbohydrate portion of the liquor. Table 4 shows the relative composition of monosaccharides in spent liquors from acid cooking for some wood species.

Table 4. Relative proportion of monosacharides in the organic fraction of sulfite spent liquors.

	Wood species		
Monosacharide	**Spruce,%**	**Birch, %**	**Aspen, %**
Galactose	2.6	0.6	0
Glucose	2.6	1.1	0.5
Mannose	11	6.4	3.1
Xylose	4.6	21.1	24.3
Arabinose	0.9	0	15
Hexoses/pentoses	16.5/5.5	8.1/21.1	3.6/25.8

The table shows that hydrolysis products from softwoods such as spruce are excellent raw materials for products using hexoses. The hardwoods are more suitable for making products from pentoses. The hexoses are easily fermented into ethanol. The pentoses can be converted to make furfurol or xylose. The xylose can be further converted into xylitol, a widely used artificial sweetener.

Figure 14 shows a simple flowsheet for a typical sulfite mill ethanol process.

The spent liquor is processed in a stripper column to eliminate free sulfur dioxide, after which it is concentrated in an evaporator, cooled, and neutralized. The liquor is then fermented by adding nutrients. The ethanol is distilled and rectified to ~95% ethanol solution. The remaining spent liquor is further evaporated to a concentration suitable for incineration. A typical yield is 60–75 L of 95% ethanol per t of sulfite spruce pulp.

Other fermentation products made from acid pulping spent liquors are torula yeast and pekilo protein that can be used as livestock fodder.

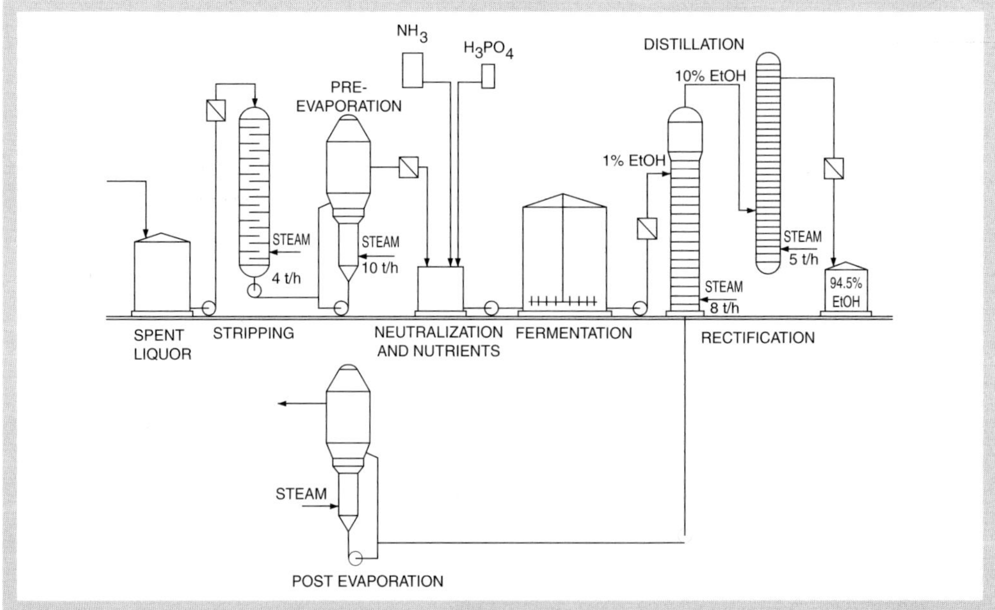

Figure 14. Flowsheet for making ethanol from sulfite spent liquor.

References

1. *McSweeney, E. E., Arlt, H. G. Jr, and Russell, J., Tall Oil and its Uses II; Pulp Chemical Association, New York, 1987, p. 13.*
2. *Jönsson, S.-E. et al, Såpkompendium; Ångpanneföreningen 1972 (in Swedish).*
3. *McSweeney; Naval Stores; Pulp Chemical Association, New York 1989, p. 168.*
4. *A. Johansson; Finnish Patent 55519 (1979).*
5. *Sandermann, W.; Naturharze, Terpentinöl, Tallöl; Berlin 1960 p. 322 (in German).*
6. *Bills, A. M., U. S. Patent 3,901,869 (1975).*
7. *Ukkonen, K. A. et al. Finnish patent 9006451.*
8. *Linotek Oy Personal communications.*
9. *Linotek Oy Personal communications.*
10. *Pelttari, M., Cutting down maintenance costs in CTO acidulation and fractionation plants; International Tall Oil Symposium, Imatra, Finland, 1983.*
11. *McSweeney, E. E., Arlt, H. G. Jr, and Russell, J., Tall Oil and its Uses II, Pulp Chemicals Association, New York, 1987, pp. 8–9.*
12. *Ukkonen, K. A., Personal communications.*
13. *Holmbom, B; Avela, E., U.S. Patent 3,965,085 (1976).*
14. *Linotek Oy Personal communications.*
15. *Ukkonen, K. A. The Crude Soap Refining Process and the Production of Sitosterols from the Unsaponifiables of Crude Sulphate Soap, Pulp Chemicals Association Meetings, Washington (1980).*
16. *Johansson, A., U.S. Patent 4,044,031 (1977).*
17. *Hamunen, A., U.S. Patent 4,420,427 (1983).*
18. *Hamunen, A., U.S. Patent 4,422,974 (1983).*

CHAPTER 19

Preparation and handling of bleaching chemicals

CHAPTER 19

Johan Gullichsen

Preparation and handling of bleaching chemicals

Some bleaching chemicals are manufactured and prepared in the pulp mill, others are shipped to it. Several factors decide whether bleaching chemicals are manufactured on-site or elsewhere:

- chemical stability of the product
- safety of handling, transport, and storage
- economy of scale
- availability of raw materials and energy
- geographic location and infrastructure
- the pulp mills chemical balance
- environmental factors

Bleaching chemicals such as chlorine dioxide and ozone are always made on-site due to their instability. Hypochlorite is also made on-site from waste streams (tail gases) from chlorine handling and chlorine dioxide manufacturing. Sulfur dioxide may be a by-product from the mill's chemical regeneration system.

Some remote, sufficiently large mills or mills with availability of suitably priced electric energy may produce their own chlorine, caustic soda, chlorate and oxygen. Most mills import and store these chemicals for use.

Mills using large quantities of peroxide might choose to manufacture that also on-site. Others may need to make a part of their sodium hydroxide for bleaching and oxygen delignification by white liquor oxidation in their chemicals regeneration departments.

1 Manufacture and handling of chlorine

Molecular chlorine (Cl_2) is as a rule made in special chlor-alkali plants that may be at the pulp mill. One reason favoring off-site manufacturing is that the economy of scale is such that one chlor-alkali plant can supply several pulp mills. Another is that both liquid chlorine and caustic soda (NaOH) can be transported and stored safely without difficulty.

1.1 Preparation of chlorine and caustic

Chlorine is made by electrolysis of sodium chloride according to the general formula:

$$2NaCl + 2H_2O \Rightarrow 2NaOH + H_2 + Cl_2 \quad (1)$$

The products are sodium hydroxide, hydrogen, and chlorine gas. The hydrogen is burned to water and the sodium hydroxide and the chlorine recovered.

The salt brine (NaCl) must be well purified before use to avoid cation contamination. This is usually done by reprecipitation. The purified brine is pumped to either mercury, diaphragm, or membrane electrolysis cells.

Mercury cells

The salt brine is electrolysed between a graphite anode and a mercury cathode. Sodium ions are discharged at the cathode and are dissolved in mercury as amalgam. Chlorine is produced at the anode according to the following reaction:

$$2Cl^- + 2e \Rightarrow Cl_2 \quad (2)$$

Sodium metal forms at the cathode and amalgamates with mercury:

$$2Na + 2Hg + 2e \Rightarrow 2NaHg \quad (3)$$

The sodium amalgamate is taken to a decomposer or caustic cell where it decomposes:

$$2HgNa + 2H_2O \Rightarrow 2NaOH + 2Hg + H_2 \quad (4)$$

The sodium hydroxide and mercury are pumped back to the electrolysis cell (Fig. 1).

The main advantage of the mercury cell is that it produces a very pure sodium hydroxide solution of high concentration (60%). This cell has been the most commonly used by the chlor-alkali industry, but it is now in decline because of health and environmental hazards related to the use of large quantities of mercury.

This process would theoretically need 1.65 kg of sodium chloride to produce 1 kg of chlorine gas, 1.13 kg of sodium hydroxide and 28 g of hydrogen. In practice 1.8 kg of NaCl, 2.2 kg of ion exchanged water, 7–8 kg of cooling water and 3. 5 kWh of electric energy are required to produce 1 kg of chlorine.

Diaphragm cell

The diaphragm cell eliminates the need for mercury (Fig. 1), but uses asbestos, another hazardous compound. The electrodes are graphite and steel. An asbestos diaphragm separates the anode chamber from the cathode chamber and does not allow sodium to pass to the cathode chamber.

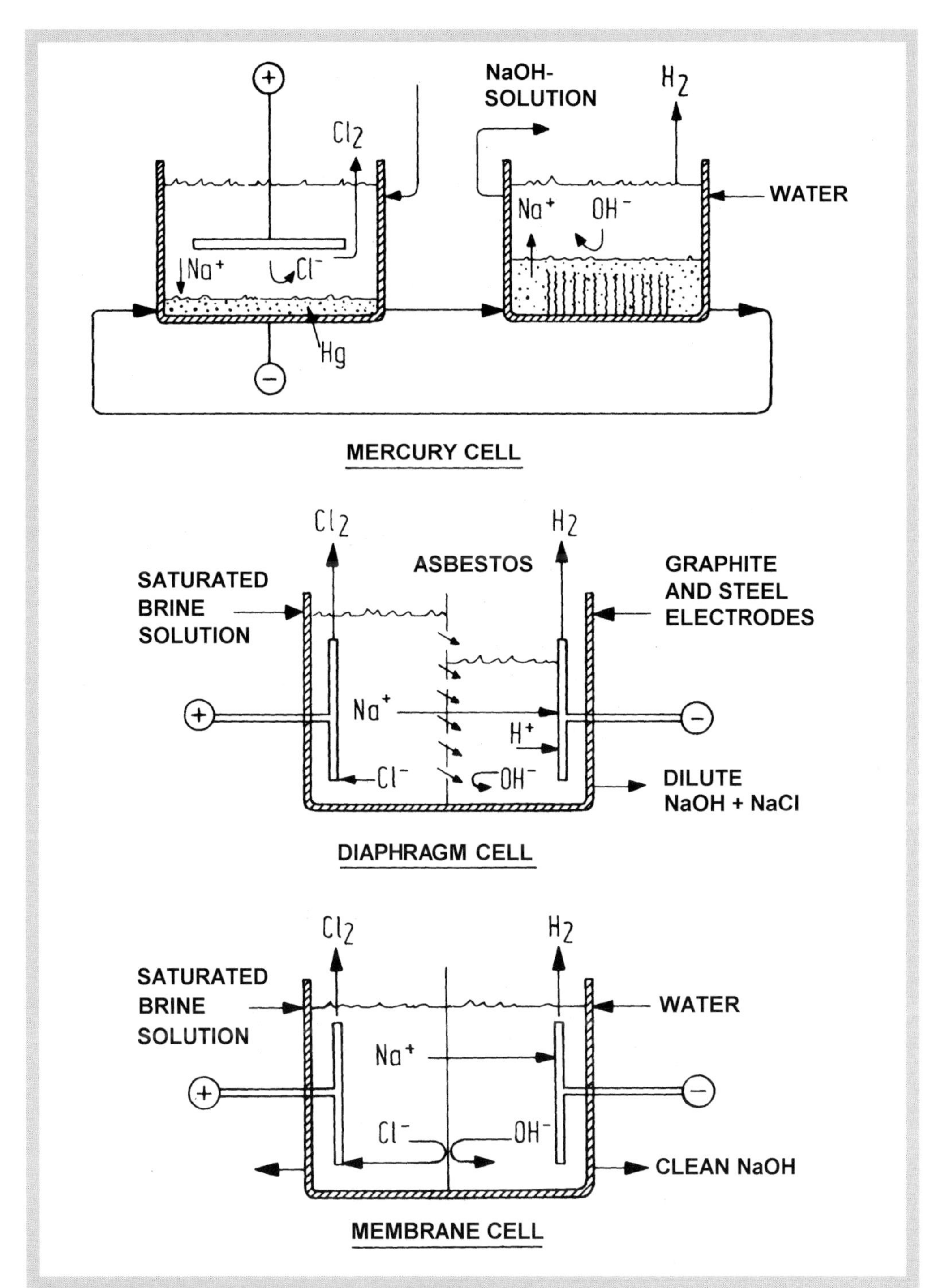

Figure 1. Principles of some commonly used chlorine manufacturing systems.

This method produces a dilute caustic soda solution which is contaminated with a substantial quantity of chloride. Chloride must be crystallized from the caustic, and the solution concentrated by evaporation before the caustic soda is usable.

Membrane cell

This method (Fig. 1)uses a cation exchange membrane containing perfluorosulfonic acid – and perfluorocarboxylic acid groups. The membrane allows $H^+$and Na^+ ions to pass through the membrane, while Cl^- ions remain in the anode chamber. The method eliminates any need to use asbestos or mercury, and produces a virtually chloride-free sodium hydroxide solution of 10%–20% concentration. This sodium hydroxide solution can be used directly in on-site bleaching. If it requires transportation for long distances or bulk storage, it is usually concentrated by evaporation to 50% dry solids concentration. All new chlorine manufacturing plants use the membrane technique.

1.2 Chlorine gas handling

Chlorine gas is evacuated from the electrolysis cell under a slight vacuum to prevent leakage into the surroundings. The chlorine gas is then liquefied. It is first cooled to remove most of the water, dried with >98% sulfuric acid in packed towers, and finally compressed to liquid. Liquid chlorine has a considerable vapor pressure as Table 1 shows. It requires storage under pressure in well insulated tanks to protect it against excessive heat.

Chlorine gas can be safely stored and transported in pressurized containers by sea, road or rail. Regulations require that these cars are pressure coded for 22-bar pressure and that they are never filled over 1.25 t Cl_2/m^3 of container volume. This avoids excessive expansion at elevated temperatures. The density of chlorine drops significantly at increased temperature (Table 1). A typical railway car is 40 m^3 in volume and carries 50 t of dry chlorine gas. Cars with almost 100 t capacity are available.

Table 1. Temperature, pressure and density of liquid chlorine

Temperature, °C	Pressure, bar	Density, t/m^3
-34.6	0.98	1.561
-30	1.20	1.550
-20	1.80	1.524
-10	2.56	1.496
0	3.58	1.468
+10	4.86	1.438
+20	6.44	1.408
+30	8.42	1.377
+40	10.9	1.344
+50	13.8	1.310

Chlorine containers are commonly made of steel and have safety devices for excessive pressure. They are inspected before refilling and are pressure-tested regularly. Most chlorine cars have the design shown in Fig. 2. They are loaded and unloaded from the center of the car through a valve arrangement covered by a protective steel housing. One of the unloading line goes to the bottom of the tank for liquid unloading and another only to the top for gas unloading.

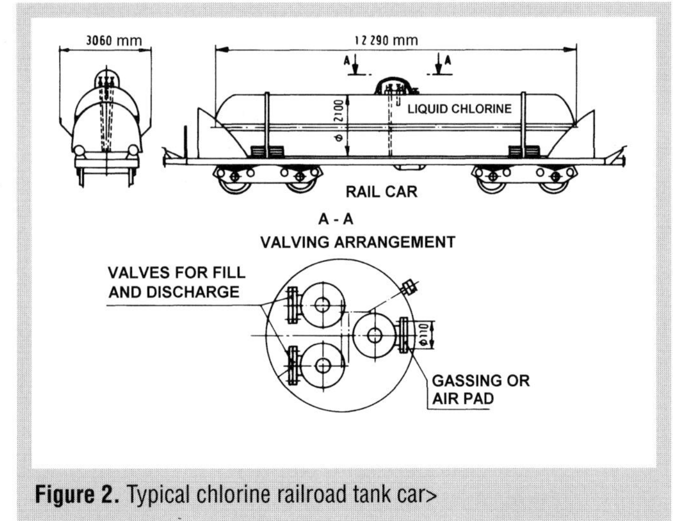

Figure 2. Typical chlorine railroad tank car>

Chlorine can be taken to the process directly from the car or via a reserve tank (Fig. 3). The reserve tank system is preferable to secure continuous chlorine flow during car changes. Chlorine containers are unloaded with pressurized dry air connected to the gas phase of the container. The air should be dry and oil-free. Its pressure must always override the chlorine pressure in the container. The unloading system must have dedicated compressors. Calcium chloride is used to dry the air.

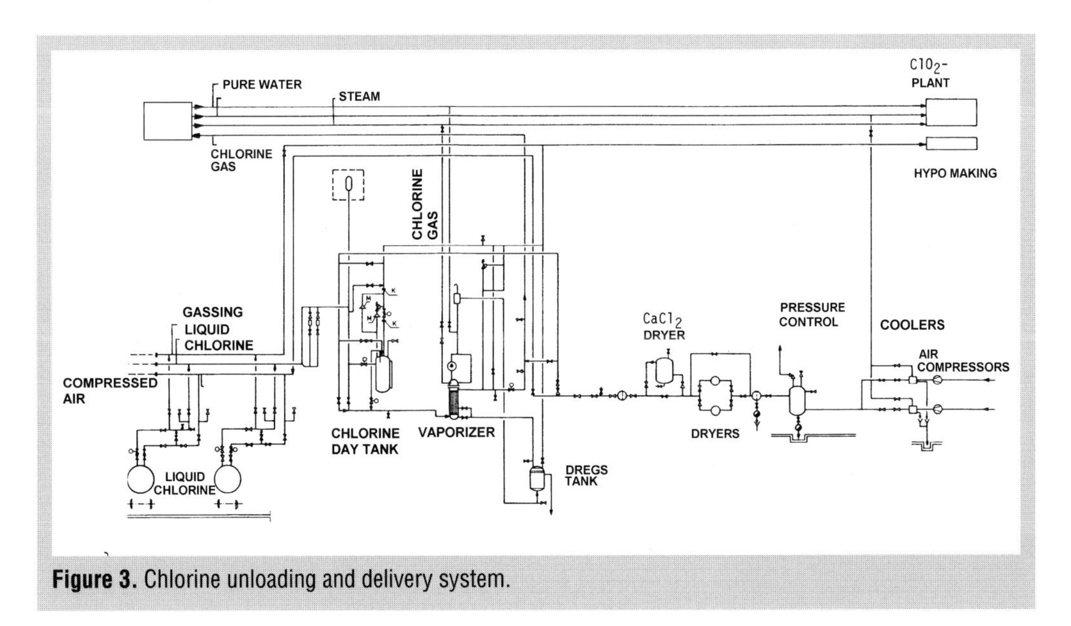

Figure 3. Chlorine unloading and delivery system.

Liquid chlorine is vaporized in steam or hot-water-heated vaporizers before introduction to the process. The chlorine charge system must be built so that water and fibers cannot get into the gas system.

Dry chlorine gas or liquid chlorine does not corrode steel. Chlorine becomes extremely reactive to steel if water or humidity is present. Titanium resists wet chlorine gas but is explosively reactive with dry gas. The choice of materials of construction in chlorine handling systems is thus critical.

1.3 Handling of caustic soda

Chlorine electrolysis produces an equivalent amount of sodium hydroxide. The caustic is diluted to 50% concentration and shipped as such to pulp mills when using the mercury cell process. A dilute (10%–20%) sodium hydroxide solution is produced by the membrane process. The caustic can be transported also as such to the pulp mill if transportation distances allow. Otherwise, it is concentrated by evaporation to 50%.

Transportation is by rail or road in 10–20 m^3 containers. The transportation containers, storage tanks, and pipelines must have good insulation, since a 50% NaOH solution crystallizes below 14°C. The caustic soda is diluted to a 5%–10% concentration before use in the pulp mill.

2 Manufacture of hypochlorite

Hypochlorite is produced by reacting an alkaline solution (slaked lime or caustic soda) with chlorine as follows:

$$2NaOH + Cl_2 \Rightarrow NaOCl + NaCl + H_2O + heat \quad (5)$$

or

$$2Ca(OH)_2 + 2Cl_2 \Rightarrow Ca(OCl)_2 + CaCl_2 + 2H_2O + heat \quad (6)$$

3 Generation and handling of chlorine dioxide

Because chlorine dioxide is an unstable gas and can only be stored in solution, it must be generated on-site and immediately dissolved in cold water. In solution, chlorine dioxide is stable and can be stored in containers at low temperature (~5°C) and without a free gas phase for long periods.

3.1 Short chemistry

Chlorine dioxide can be produced in either of two ways; by reduction of chlorate ion (ClO_3^-) in acidic medium, or by oxidation of chlorite ion (ClO_2^-).

Reduction:

$$ClO_3^- + 2H^+ + e^- \Rightarrow ClO_2 + H_2O \quad (7)$$

Oxidation:

$$ClO_2^- \Rightarrow ClO_2 + e^- \tag{8}$$

Generation by oxidation is not economical, but may occur as side reaction in chlorine dioxide bleaching. The choice of reducing agent in chlorine dioxide generation is important for optimal reaction conditions, for by-product formation and the mills chemical balance.

The most common commercially used reducing agents are sulfur dioxide (SO_2), methanol (CH_3OH), chloride ion (Cl^-) and hydrogen peroxide (H_2O_2). They react with chlorate as follows:

$$2ClO_2^- + SO_2 \Rightarrow 2ClO_2 + SO_2^{2-} \tag{9}$$

$$4ClO_3^- + CH_3OH + 4H^+ \Rightarrow 4ClO_2 + HCOOH + 3H_2O \tag{10}$$

$$ClO_3^- + 2Cl^- + 2H^+ \Rightarrow ClO_2 + Cl_2 + H_2O \tag{11}$$

$$2ClO_3^- + H_2O_2 + 2H^+ \Rightarrow 2ClO_2 + O_2 + 2H_2O \tag{12}$$

The side products and demand for protons vary. The use of sulfur dioxide gives sulfate that can be a sulfur make-up source in the kraft recovery cycle if required. Methanol produces acetic acid that easily decomposes in recovery boilers. The use of chloride ion inevitably leads to generation of chlorine that is active in bleaching, but this may cause disposal problems when chlorine-free bleaching is necessary. Peroxide forms no harmful by-products; its disadvantage is the high cost.

All systems will involve side reactions like decomposition of chlorate to chloride:

$$ClO_3^- + 6H^+ + 6e^- \Rightarrow Cl^- + 3H_2O \tag{13}$$

This side reaction may be crucial for the efficiency of chlorine dioxide generation. It appears that chloride must be present to promote formation of chlorine dioxide.

3.2 Generation systems

All ClO_2 generation systems start from solid chlorate made elsewhere and shipped to the pulp mill in crystalline form. Chlorate is dissolved in hot clean water to form a 40%–50% solution and pumped to the reactor. Reactants are added, and the released ClO_2 gas is evacuated and absorbed in chilled clean water. Controlling the gas partial pressure sufficiently low is important to avoid spontaneous decomposition. This is done by steam or air dilution combined with vacuum. The concentration of the chlorine dioxide solution formed is about 10 g/L.

The Mathieson process

The Mathieson process, the first large commercial-scale generator system developed in the 1950s, uses sulfur dioxide as the reducing agent. Strong sulfuric acid is the acid source to ensure that chlorate will not excessively decompose to chloride. The acid concentration in the reaction system is as high as 450–500 g H_2SO_4/L or 2–2.5 t H_2SO_4/t ClO_2 generated. The excess sulfuric acid overflowing from the reactor system must be used elsewhere in the mill. Figure 4 shows a simplified flowsheet of the Mathieson process.

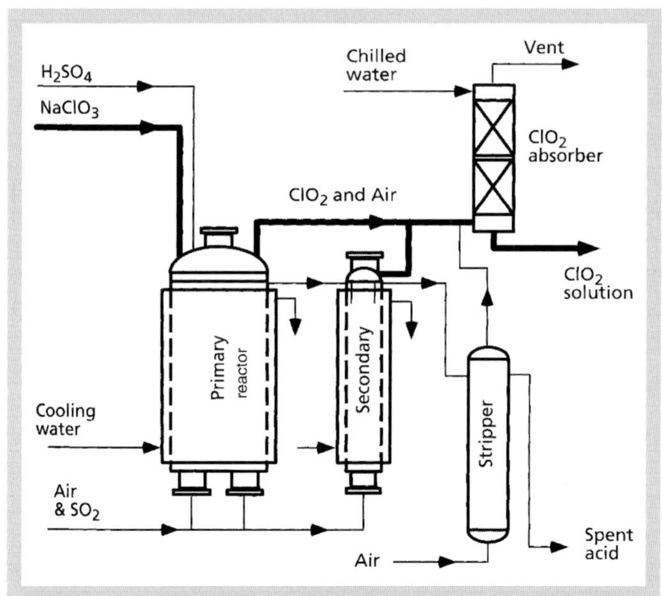

Figure 4. Simplified flowsheet of chlorine dioxide generation using the Mathieson process.

Concentrated sulfuric acid and sodium chlorate solutions are fed to the primary reactor. A mixture of sulfur dioxide and air is blown into the bottom of the reactor. The up-flowing gas mixes gas and liquid. The generator content is cooled to ~40°C by a water jacket surrounding the vessel. Chlorine dioxide is stripped from the solution with air and fed to an absorption tower where it is absorbed into chilled water. The resultant chlorine dioxide solution exits the bottom section of the absorber. Liquor from the primary reactor overflows to a secondary reactor where residual of chlorate is depleted with sulfur dioxide and air. Spent liquor from this secondary reactor is stripped with pure air to separate any remaining dissolved chlorine dioxide, and the overflow is pumped to a spent acid tank.

Later versions of the Mathieson process scrub the gas coming out from the primary reactor with feed acid and chlorate to reduce acid droplet carryover to the absorption system. Sodium chloride was also mixed into the feed to improve conversion efficiency.

The conversion efficiency of the Mathieson system rarely exceeds 90%. The quantity of spent acid generated is very large, as Table 2 shows. The original Mathieson process did not produce by-product chlorine, meaning that the acidity of the product chlorine dioxide solution was low. Acid had to be used in some bleaching applications for pH control. Newer applications that use chloride makeup generate some by-product chlorine in the gas.

The R3/SVP generator

The R3/SVP generator combines chlorine dioxide generation, sodium sulfate crystallization, and evaporation in a single vessel. This single vessel process (SVP) is essentially the same as that shown in Fig. 5. The main body of the generator is a forced circulation evaporator. A solution of sodium chlorate and sodium chloride is fed into the recirculating loop. Sulfuric acid is added later, usually after the heat exchanger. Chlorine dioxide forms immediately on contact with sulfuric acid. The slurry returned to the generator body contains a considerable amount of chlorine dioxide gas. The gas is separated in the generator, along with released steam, and is fed into a indirect-contact cooler which condenses water from the gas stream. The remaining chlorine dioxide and chlorine gas mixture then passes to an absorption system where the chlorine dioxide goes into solution. Some chlorine may also be dissolved here, but most will continue to a second adsorption system where it its absorbed into water or alkali to generate sodium hypochlorite.

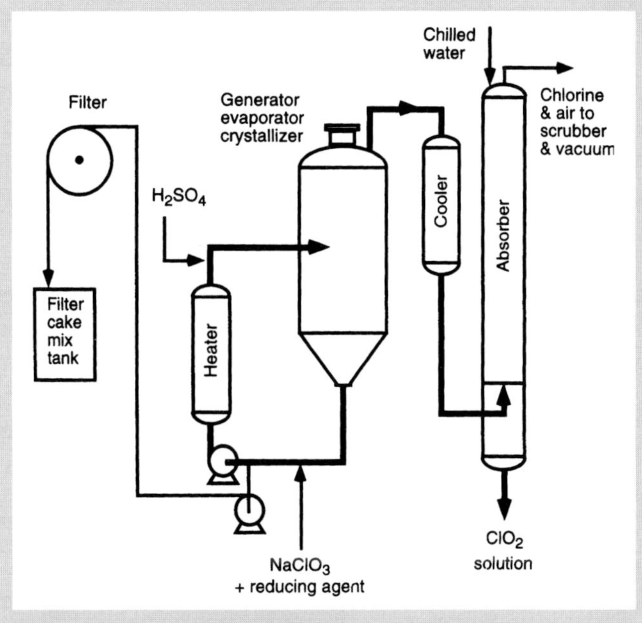

Figure 5. Flowsheet for R3/SVP chlorine dioxide generation.

The reaction residues are primarily a mixture of sulfuric acid and saltcake. They go via a cyclone to a filter where crystallized saltcake (Na_2SO_4) is separated by dewatering and is dissolved in black liquor as makeup. The acid returns to the generator loop. Acid is efficiently reused and the amount of sulfur compounds decreases, as Table 2 shows. Residual chlorine from the first absorber can also be reacted with sulfur dioxide to form a mixture of sulfuric and hydrochloric acid to be reused to replace make-up acid and chloride:

$$\frac{1}{2}Cl_2 + \frac{1}{2}SO_2 + H_2O \Rightarrow HCl + \frac{1}{2}H_2SO_4 \qquad (14)$$

Processes based on hydrochloric acid

As pulp mill chemical recovery systems improved and smaller quantities of makeup chemicals were necessary; mills increased their use of chlorine dioxide and, the waste streams from chlorine dioxide generation exceeded the makeup needs of the chemical recovery loop. This stimulated development of processes which replaced sulfuric acid by hydrochloric acid as acid source:

$$NaClO_3 + HCl + \frac{1}{2}H_2SO_4 \Rightarrow ClO_2 + \frac{1}{2}Na_2SO_4 + \frac{1}{2}Cl_2 \tag{15}$$

The required mixture of hydrochloric acid and sulfuric acid may be produced according to Eq. 14.

Hydrochloric acid processes can operate independently or can be integrated with chlorate manufacturing (Fig. 6). This does not use sulfuric acid:

$$NaClO_3 + 2HCl \Rightarrow \frac{1}{2}ClO_2 + Cl_2 + H_2O + NaCl \tag{16}$$

The chlorate is made from by-product salt according to reaction (17) by electrolysis:

$$NaCl + 3H_2O \Rightarrow NaClO_3 + 3H_2 \tag{17}$$

and the by-product chlorine is used to make hydrochloric acid:

$$Cl_2 + H_2 \Rightarrow 2HCl \tag{18}$$

This process does not produce any side product chemicals. Integrated plants require however, essentially more capital, on site power and real estate than other chlorine dioxide generating methods.

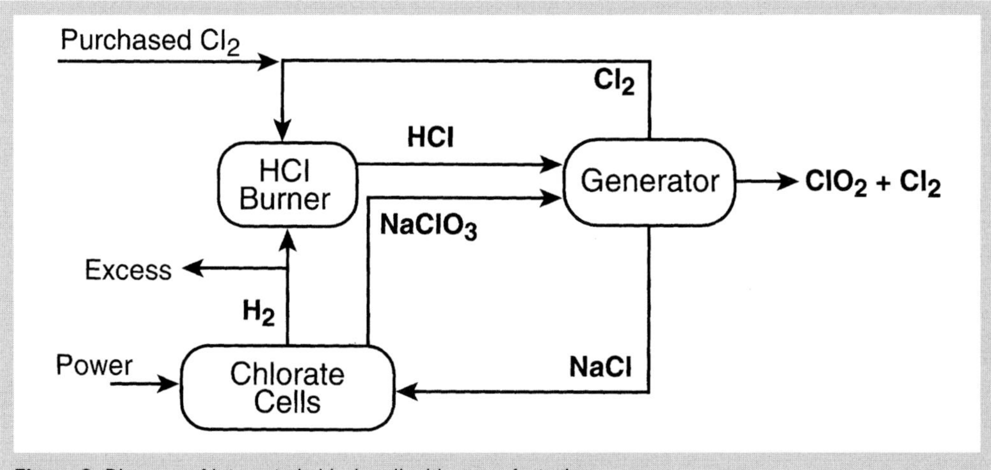

Figure 6. Diagram of integrated chlorine dioxide manufacturing.

Methanol processes

The need to reduce by-product chlorine and sodium sulfate generation stimulated development of the R8/SVP process. The overall reaction can be written as:

$$3NaClO_3 + 2H_2SO_4 + 0.8CH_3OH \Rightarrow 3ClO_3 + Na_3H(SO_4)_2 + 2.3H_2O + 0.8HCOOH \qquad (19)$$

More methanol is needed in practice than the Equation 19 indicates because some methanol escapes with generator gas. Some of the formic acid reacts further to form carbon dioxide and hydrogen ions. The vapour pressure of formic acid is similar to that of water. Most of it is therefore stripped from the generator and is reabsorbed into water to be found in the product solution.

Methanol processes virtually eliminate by-product chlorine and reduce the salt-cake production considerably over those based on purchased chlorate.

Peroxide process

When using hydrogen peroxide as reducing agent, no by-product chlorine results:

$$2NaClO_3 + H_2O_2 + H_2SO_4 \Rightarrow 2ClO_2 + O_2 + Na_2SO_4 \qquad (20)$$

The high cost of sodium peroxide limits its use in ClO_2 generation.

Table 2. Typical feed chemical demands and by-products of some chlorine dioxide generation systems (t/t ClO_2).

Feed chemical demand	Mathieson	R3	R3H	R5	R6	R7	R8
$NaClO_3$	1.75	1.68	1.68	1.75	-	1.68	1.68
NaCl	-	1.15	-	-	-	0.35	0.025
H_2SO_4	1.3	1.73	0.8	-	-	0.4	1.2
HCl	-	-	0.7	1.4	-	-	-
SO_2	0.65	-	-	-	-	0.4	-
CH_3OH	-	-	-	-	-	-	0.14
Cl_2	-	-	-	-	0.8	-	-
Power, MWh	-	-	-	-	8.5	-	-
Byproducts generation							
Na_2SO_4	1.17	2.3	1.2	-	-	1.6	-
$Na_3H(SO_4)_2$	-	-	-	-	-	-	1.3
H_2SO_4	1.5						
Cl_2	-	0.7	0.7	0.85	0.32	0.25	0.04
H_2	-	-	-	-	0.05	-	-
NaCl	-	-	-	0.96	-	-	-
As total Na	0.38	0.75	0.38	0.4	-	0.51	0.34
As total S	0.75	0.52	0.19	-	-	0.36	0.32

The R# headers designae different commercial processes for generating chlorine dioxide

3.3 Safe handling of chlorine dioxide

Because chlorine dioxide gas is unstable, it requires careful handling to avoid spontaneous decomposition according to the formula below:

$$2ClO_2 \Rightarrow Cl_2 + O_2 + heat \tag{21}$$

Excessive chlorine dioxide in the gas phase, 20%–30% by volume (corresponding to a chlorine dioxide atmospheric partial gas pressure of 190–230 mm Hg) causes this explosion-like decomposition. Another cause may be organically contaminated feed chemicals. Maintaining clinical purity of raw materials and avoiding free oxygen-containing gas and elevated temperatures in reactors and tanks containing chlorine dioxide can prevent reactor puffing. Systems must have explosion hatches and rupture discs to avoid equipment damage.

4 Hydrogen peroxide

Hydrogen peroxide is safely stored and transported as a concentrated solution. It is usually manufactured outside pulp mills.

4.1 Manufacture of hydrogen peroxide

Most industrially used hydrogen peroxide is manufactured from hydrogen and oxygen by the anthraquinone process of Fig. 7. A solution of alkyl anthraquinone reacts with hydrogen in the presence of a catalyst in a hydrogenation reactor. The resulting hydrogenated anthraquinone solution is filtered to remove the catalyst and contacted with oxygen in an oxidizer where hydrogen peroxide and alkyl anthraquinone are formed. The hydrogen peroxide is extracted from the solution with demineralized, leaving the anthraquinone solution for recycling to hydrogenation. The crude hydrogen peroxide solution is purified and concentrated.

Hydogen peroxide is unstable, and must be stabilized before shipment. Stabilizers are mainly tin and phosphate compounds. They may also contain nitrates to inhibit chloride corrosion of aluminum storage equipment.

Storage and handling of hydrogen peroxide

Design of peroxide storage and handling facilities minimizes product contamination that may lead to spontaneous decomposition. The decomposition is very exothermic and can cause rapid pressure increases as oxygen gas forms. Storage tanks and pipeline systems must therefore have venting to avoid system damage in case of decomposition. Hydrogen peroxide is stored as a 50% aqueous solution and diluted down to 5% before use.

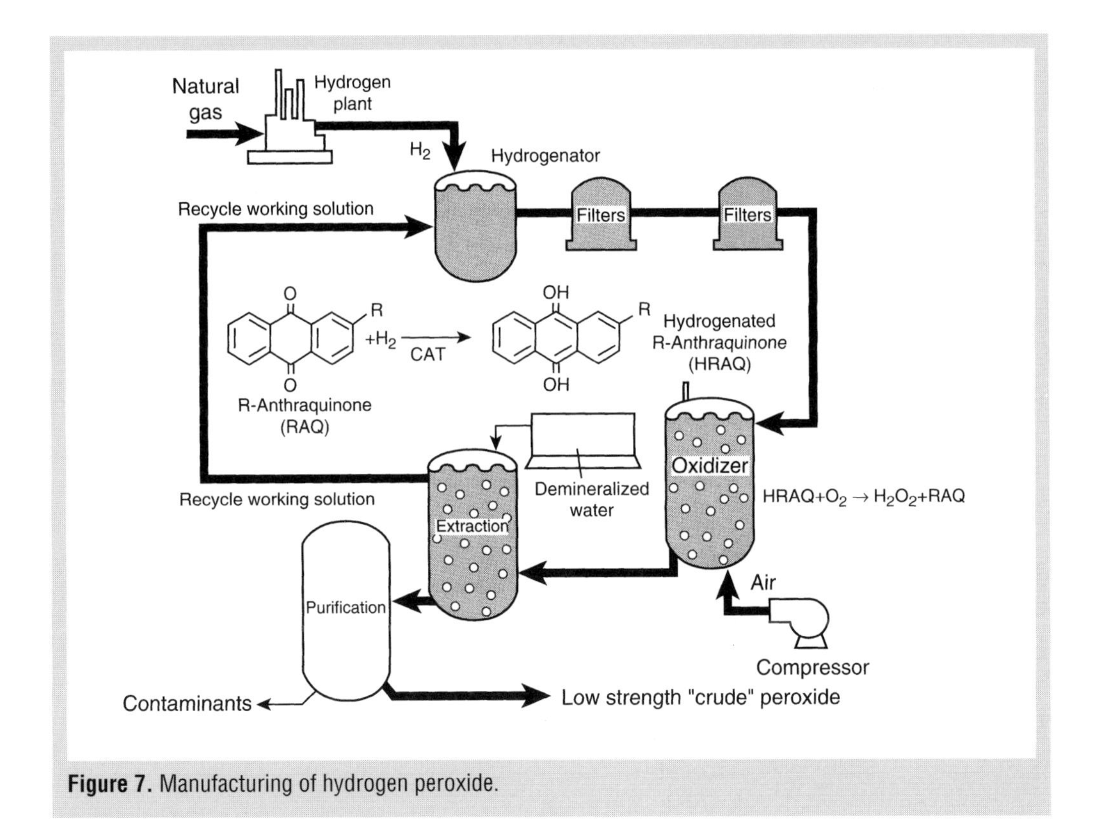

Figure 7. Manufacturing of hydrogen peroxide.

4.2 Peroxy acids

Peroxy acids or peracids are manufactured by combining hydrogen peroxide with a suitable acid to form perhydroxyl (OOH^-) groups. The acid can be an organic (e.g. acetic or formic acid) or an inorganic acid (e.g. sulfuric acid). Peroxy acids are generally weaker acids than their parent acids, but still very corrosive. For example peracetic acid has a boiling point of 103°C and a vapor pressure of 20 mm Hg at 25°C. The vapors are toxic and causes lung damage on exposure. The material is very agressive to skin tissues. Personal protective gear is necessary when handling it.

The sulfuric acid based peracid, H_2SO_5, is also called Caro's acid. In mixtures with peracetic acid it can detonate at certain molar ratio's of the mixture. Peroxy acids decompose on heating to form oxygen gas. Metal ions of iron, copper chromium and cobalt will accelerate the decomposition. Containers, pumps and pipelines are therefore lined with synthetic plastics.

Peroxy acids are not sufficiently stable for shipping or storing for long periods, so they are manufactured at the pulp mill. Highly concentrated reactants are used to drive the reactions toward formation of peroxy acids. High conversion also requires use of a significant excess of parent acid.

5 Oxygen

Oxygen is a colorless, tasteless gas that is slightly heavier than air. It is the essential constituent of air, and accounts for roughly 21% of air by volume. Oxygen is a transparent bluish liquifid at -183°C that is slightly heavier than water. Its main properties are listed Table 3.

Table 3. Chemical and physical properties of oxygen.

Specific gravity of gas at 21.1°C and atmospheric presure	1.1053
Density of gas at 21.1°C and 1 atm	31.33 kg/m^3
Density of gas at bailing point	4.47 kg/m^3
Boiling point at 1 atm.	-183^{0}C
Density of liquid at boiling point	1141.0 kg/m^3
Latent heat of vaporization at boiling point	213 kJ/kg
Specific heat at 21.1°C and 1 atm.	0.917 kJ/kg
Solubility of gas in water at 21.1°C as volume ratio	0.031

5.1 Manufacture of oxygen

Oxygen has been manufactured by liquefaction of air and distillation since the early 1900s. Developments from the 1960s and 1970s included oxygen manufacturing by absorption and membrane separation. The cryogenic process is still the favorite process but absorption processes are gradually gaining ground.

In the cryogenic process, air is cooled until it liquefies, followed by fractional distillation of the constituents. The resulting oxygen and nitrogen are then liquefied by further cooling. The absorption process air passes over a suitable absorbent such as a molecular sieve that selectively absorbs most of the constituent gases and leaves an oxygen-enriched product.

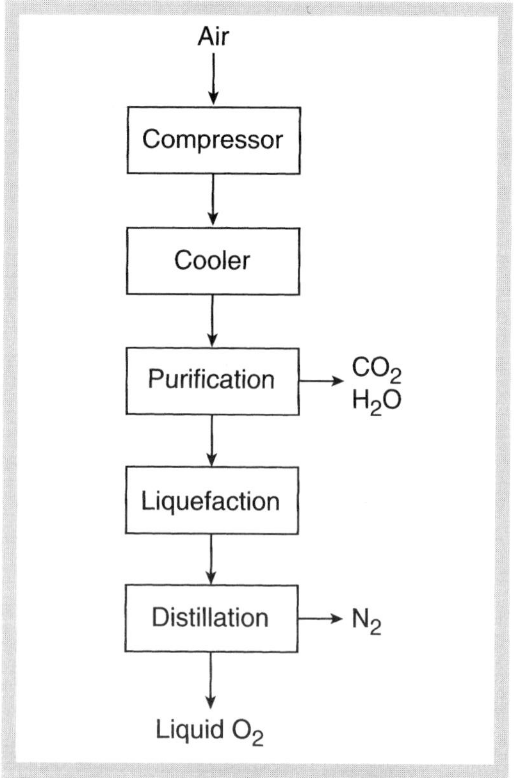

Figure 8. Block diagram for cryogenic separation of oxygen.

Cryogenic separation

The primary product of cryogenic air separation is high-purity gaseous oxygen. Air is first purified to remove all moisture and carbon dioxide that might freeze during liquefaction. Filtering removes particles.

The air is then compressed and cooled to condense the water and carbon dioxide. Then the cold clean air is further cooled in an expansion turbine to liquefaction temperatures. The liquid is distiiled to produce liquid oxygen. Figure 8 shows a simple block diagram.

Oxygen separation by absorption

Absorption processes produce a gas stream of 95% purity in on-site installations. The process consists of a cyclic sequence in which nitrogen in air is absorbed onto a molecular sieve at high partial pressure. Nitrogen released at low pressure regenerates the sieve. Several approaches are as follows:

- PSA (pressure swing absorption) where absorption is at high pressure and regeneration is at atmospheric pressure
- VSA (vacuum swing absorption) where absorption is performed at atmospheric pressure and regeneration under vacuum
- VPSA(vacuum pressure swing absorption) is a combination of the previous approaches.

The flowsheet of Fig. 9 shows a three-bed system where unit A produces oxygen, unit B is being drained of free nitrogen to the atmosphere via a vacuum pump, and unit C is in a stand-by position.

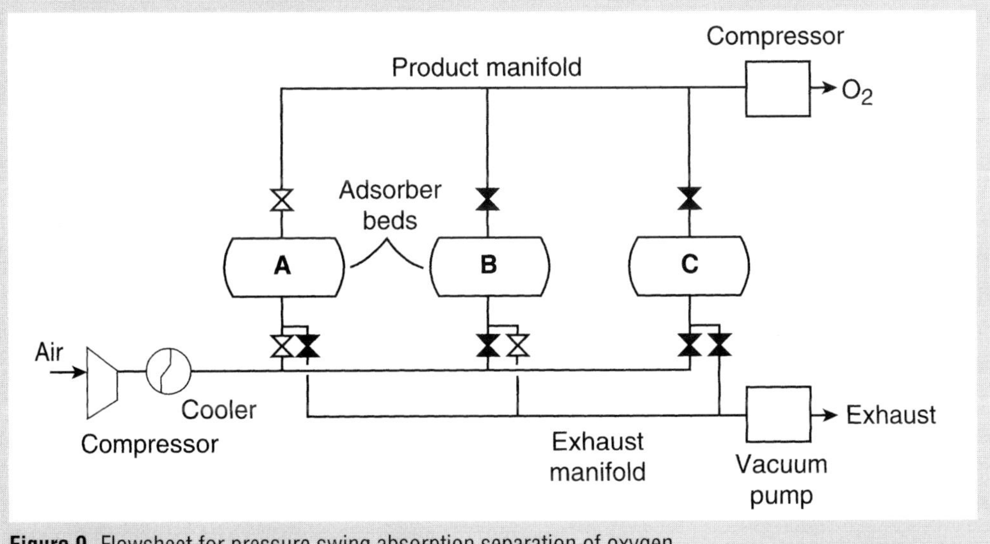

Figure 9. Flowsheet for pressure swing absorption separation of oxygen.

5.2 Transportation and handling of oxygen

Special attention is required in systems when handling highly concentrated oxygen gas or oxygen as liquid. All systems must be dry and free from easily oxidizable material like grease. The low handling temperature of the liquid (-180°C or -190°C) puts special requirements on metallurgy, and also requires that operators use proper gloves and clothing when manually handling oxygen. Proper relief valve arrangements must avoid pipeline trapping between valves. Table 4 lists the properties of some common oxygen supply alternatives.

Table 4. Oxygen supply alternatives

System	Adsorption separation	Cryogenic separation	Purchased liquid
Operating temperature	Ambient	–190°C	–190°C
Maximum purity, %	95	99.9	99.9
Usual purity,%	90–93	98-99.5	98–99.5
Pressure kPa	near atmospheric	70–1 400	up to 1 600
Flexibility in operation	Start-up in 10 min	Cold start 45 min	Instantaneous
Turndown flow,%	from 100 to 0	from 100 to 50	from 100 to 0
Liquid production	no	yes	–
Specific power at site, kWh/t O_2	265	300	50 for vaporization only

6 Ozone

Ozone is a bluish gas under normal atmospheric conditions. It is a powerful oxidising agent and is toxic. Ozone is always present in the atmosphere at 0.025–0.045 ppm by volume. It can be stored without decomposition at -50°C but at room temperature it decomposes slowly into oxygen. Its solubility in water is poor.

Ozone has a very strong electronegative oxidation potential and is thus also a very powerful oxidant in pulp bleaching and delignification. All handling and safety measures that apply for oxygen are also valid for ozone.

6.1 Ozone manufacture

Ozone is always produced on-site because of its instability. It cannot be produced as a pure gas from oxygen, but only in a mixture of ozone in oxygen at 8%–12% by volume. This means that off-gas oxygen, after reactions with pulp, must be recirculated via purification to the ozonator, and makeup oxygen is taken into the system as required for conversion into ozone. Another use of off-gas oxygen could be oxygen delignification. Since ozone is very toxic, any residual must be completely destroyed.

Ozone is commonly made in a corona discharge process in which molecular oxygen (O_2) is partly converted into ozone (O_3). The ozone production is proportional to the applied voltage, electrical frequency, dielectric conditions, temperature, pressure and the geometry of the discharge gap. Generated heat must be efficiently removed to avoid ozone decomposition. The feed gas can be either air or oxygen. A lower ozone concentration results when air is used as feed gas. Therefore, oxygen is the preferred feed gas.

Ozone forms in the oxygen gas stream when it passes a high-voltage high-density electrical field. The corona discharge forms in the oxygen-filled gap as the voltage applied exceeds the ionization potential of the dielectric material and electrons travel across the gap. Collisions between electrons and oxygen molecules produce oxygen atoms that recombine with oxygen to form ozone. The dielectric material prevents arcing and provides a uniform discharge surface. The selection of materials is an important efficiency issue in the process.

Most commercial installations use horizontal dielectric glass tubes (Fig. 10).

Current ozone generators are not very energy efficient, and need 8–12 kWh/ kg ozone generated. They require an efficient cooling system to maintain sufficiently low temperature for maximum gas concentration and minimum decomposition. The ozone and oxygen gas mixture must be compressed to high pressure to minimize the absolute gas volume mixed into the fiber suspension.

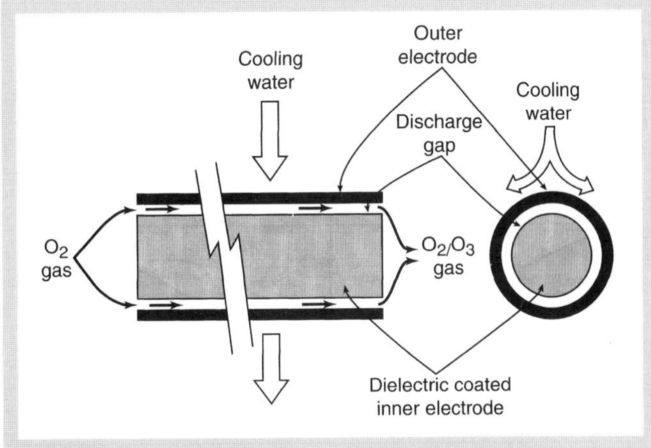

Figure 10. Schematic diagram of a ozone generation tube>

Medium-consistency ozone reactors operate frequently at 10 bar pressure or higher for this purpose. High-consistency gas phase reactors do not require operation at high pressure.

Residual ozone must be destroyed after use, and carrier oxygen must be reused. Thermal destruction of ozone is generally carried out at 300°C. Catalytic manganese dioxide or palladium coated alumina catalysts destroy ozone at 150°C.

The ozone consumption is about 95% of the charge in commercial pulp reactors. This leaves 5% of the ozone for destruction. Destruction systems must be built for full ozone capacity to ensure good operation also during upset conditions. The vent gas, free of ozone is still very rich in oxygen. It may contain small quantities of carbon monoxide but carbon dioxide is the prime contaminant. Once-through systems such as in Fig. 11 are easy to handle since off-gas is not reintroduced to the generator, but the oxygen-rich fraction must be used elsewhere for economic reasons.

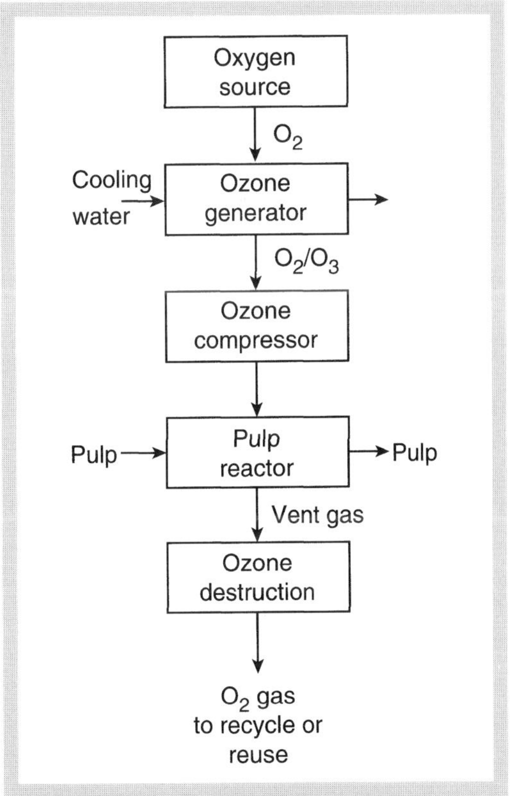

Figure 11. Block diagram of ozone bleaching system.

Off-gas oxygen can also be recirculated to the ozone generator before it is contacted with pulp. This eliminates extensive gas clean-up before recirculation, but requires that carrier oxygen is replaced with nitrogen in a sorption and desorption system. It is possible to absorb ozone onto silica gel or moleqular sieves and pass the oxygen directly back to the generator. Adsorbed ozone is then released by nitrogen gas, and the nitrogen and ozone mixture is used as bleaching gas. The off-gas nitrogen can then be passed to atmosphere after ozone bleaching and destruction of any residual ozone. Such systems are called short-loop oxygen recovery systems (Fig. 12).

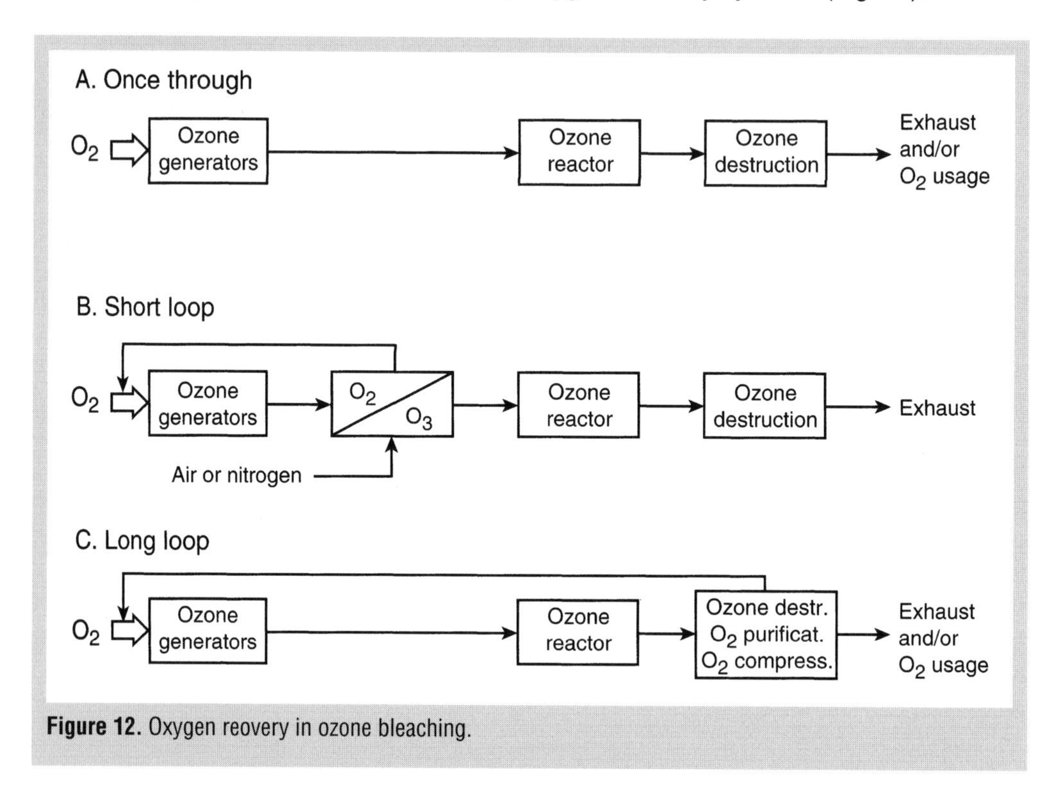

Figure 12. Oxygen reovery in ozone bleaching.

The long-loop system recovers off-gas oxygen only after ozone bleaching. The contaminated oxygen requires efficient purification before reuse in ozone generation. These purification methods involve drying, chemical conversion, and separation.

Current ozone systems are the once-through or long-loop type. They are built for 5-10 kg/adt ozone charges in bleaching. This means that the use of ozone bleaching or delignification will increase pulp mill's in-house power consumption by 10%–15%.

References

1. *Nils-Erik Virkola (editor). Puumassan valmistus. Suomen Paperi-insinöörien yhdistyksen oppi- ja käsikirja. Toinen uudistettu painos. 1983 osa II p. 909–926. (In Finnish)*
2. *Carlsson W. Dence and Douglas W. Reeve (editors) "Pulp Bleaching, Principles and Practise" TAPPI PRESS, Atlanta 1996 p. 226.*

CHAPTER 20

Organosolv pulping

CHAPTER 20

Jorma Sundquist

Organosolv pulping

1 Introduction

Organosolv (solvent-based or solvolysis) pulping is a chemical pulping method in which delignification of the biomass (usually wood) is done in an organic solvent or solvent plus water system. For delignification, the important process variables are the ability of the solvent to dissolve lignin and the reaction products, pH, temperature, and the additives used to promote lignin degradation and prevent its condensation.

Although attempts have been made since the late 1800s to dissolve wood components in a variety of organic solvents such as alcohols and organic acids, the real pioneers of solvent-based pulping are Kleinert and Tayenthal. In 1931, they published results from experiments using mixtures of alcohol and water[1]. During the decades that followed, relatively little research involved organosolv cooking. A revival in the late 1970s was largely the result of a growing concern for the environment. The oil crisis of the 1970s and the idea of using biomass instead of fossil raw materials for the production of chemicals and energy also stimulated interest in solvent-based fractionation methods for plant raw materials and in further processing of the nonfiber products obtained from them.

More recent organosolv research has sought a pulping method that would not only replace conventional inorganic processes but also be economically feasible on a much smaller scale than the kraft process. Another aim is to develop a method to allow total use of the biomass so that greater economic benefit can result from all its different chemical components: cellulose, other polysaccharides, lignin, and extractives. Essential to both these aims is that the products should have the highest possible quality. Any new process must also be more environmentally friendly than existing processes.

2 Organosolv pulping methods

In an excellent review article, Hergert[2] divides organosolv methods into six categories based on the cooking chemistry involved:

- Methods involving thermal autohydrolysis that use the hydrolyzing effect of organic acids cleaved from the wood during cooking
- Acid catalyzed methods using acidic materials to cause hydrolysis
- Methods using phenols and acid catalysts (This could also be part of the previous category.)
- Alkaline organosolv cooking methods

- Sulfite and sulfide cooking in organic solvents
- Cooking using oxidation of lignin in an organic solvent.

The primary function of the organic solvent in organosolv cooking is to render the lignin more soluble in the cooking liquor. In many cases, the solvent actually takes part in the delignification reactions in one way or another. The most widely used solvents are methanol, ethanol, and acetic and formic acids. Others include various phenols, amines, glycols, nitrobenzene, dioxane, dimethylsulfoxide, sulfolane, and liquid carbon dioxide.

3 Delignification chemistry in organosolv processes

In conventional cooking and organosolv cooking, the lignin structure must be broken into smaller parts before it will dissolve in the cooking liquor. No solvent or combination of solvents has yet been found that will dissolve lignin directly from the wood matrix. Without chemical reactions, lignin will not dissolve. One reason is that it forms chemical bonds with wood polysaccharides. The delignification reactions are different in acidic and alkaline conditions and in the presence of oxidizing and reducing agents. Process variables such as cooking time and temperature also influence the reactions in the same way as in conventional cooking.

The cleavage of ether bonds in the lignin structure is a key delignification reaction[3]. At acid pH, alpha aryl ether bonds (α-O-4 in Fig. 1) break easier than beta bonds (β-O-4). The significance of beta bond cleavage grows with decreasing pH and is more important in hardwood than softwood cooking. In acidic organosolv cooking, the ether bonds between lignin and carbohydrates are also easy to break.

The reactions taking place during alkaline organosolv cooking are similar to those in the corresponding conventional processes (kraft, soda, and alkaline sulfite). At alkaline pH, alpha aryl ether bonds break only in those phenyl propane units containing free phenolic hydroxyl groups. Etherified phenol groups cannot undergo conversion to quinone methides. This means that the alpha bond cannot be broken. Beta aryl ether

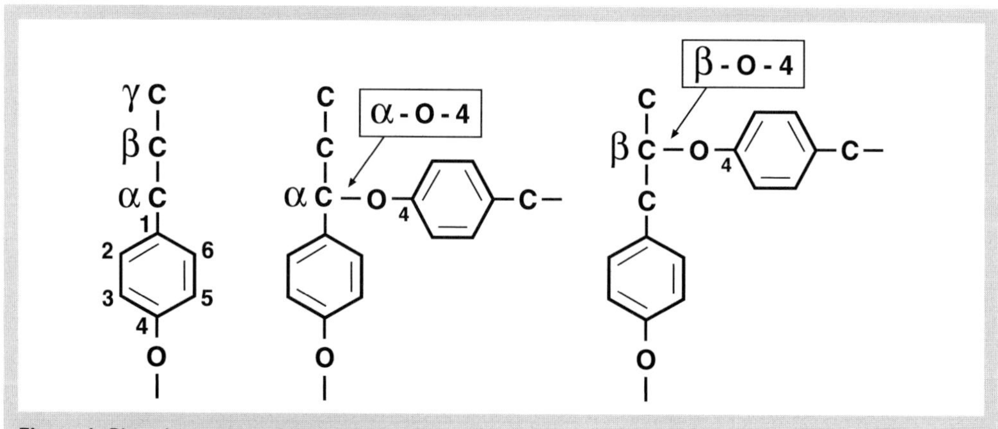

Figure 1. Phenyl propane unit bonds in lignin structure.

bonds can be broken under alkaline conditions whether the phenyl hydroxyl groups are free or etherified. Nucleophilic additives such as hydrosulfide and anthraquinone promote the cleavage of beta bonds.

Lignin undergoes condensation reactions under acidic and alkaline conditions. Condensation reactions such as the formation of carbon-carbon bonds and intermolecular condensation particularly hinder delignification and subsequent bleaching.

As with conventional methods, organosolv cooking delignifies hardwoods more readily than softwoods. This is because hardwood lignin is more reactive and is present in smaller amounts than softwood lignin.

Besides these depolymerization reactions, delignification can be promoted by reactions such as sulfonation and oxidation that introduce into the lignin molecule substituents such as sulfonate, carbonyl, carboxyl, and hydroxyl groups that render the lignin more soluble.

Lignin and polysaccharides react differently in various delignification processes, affecting their selectivity. Hemicelluloses are attacked more readily than cellulose due to their amorphous state and lower degree of polymerization. For autohydrolysis, an important reaction is the cleavage of acetic acid from wood xylan at high temperature. This free acetic acid initiates the acid hydrolysis of lignin.

4 Technical and economic demands on new processes

Over 80% of all chemical pulp comes from the kraft process. The advantages of this process are its suitability for a wide range of different raw materials, the excellent papermaking pulp it produces, and its energy self-sufficiency that uses combustion of material dissolved from the wood during cooking. Its disadvantages include process emissions and the high cost of the equipment necessary to reduce them. Ways of reducing emissions include collecting and burning odorous gases, treating effluents, and various internal measures within the process itself. The investment cost for a modern kraft mill in Finland in 1997 is approximately US$ 500–600 million. The minimum capacity for economic feasibility is over 500 000 t/year.

To be competitive, any new process and its products must be much cheaper, better, or in some way more attractive than existing processes and products. If a new process primarily produces chemical pulp for papermaking, the minimum capacity for economic feasibility should be much smaller (100 000–250 000 t/year) than for the kraft process.

The pulp must also compete with corresponding kraft pulps in quality. In special cases, dissolving pulps intended for chemical processing might be preferable, and their quality requirements are naturally quite different.

Any new process would be more attractive if it were suitable for different types of fiber raw material: softwoods, hardwoods, and nonwood materials. Recovery and regeneration of the cooking solvent and other chemicals should be simpler than in the present kraft process. Only very small solvent losses (< 1%–2%) are acceptable. The new process must also be acceptable in its emissions. Rather than preventing harmful emissions from entering the environment, the aim should be a process that produces no such emissions. This means processes that function without sulfur and chlorine chemicals are preferable. The absence of chlorine would allay environmental concerns. The additional absence of sulfur would help lower the cost of regenerating the process chemicals.

For an organosolv process whose production is intended partly or entirely as raw material for the chemical industry, the requirements placed on the products differ from those for papermaking. The purity of the different fractions and their behavior in chemical processes (reactivity, solubility, filtration properties, swelling, etc.) are more important factors than the mechanical and optical properties required of papermaking pulps. The requirements regarding mill capacity, environmental impact, and solvent regeneration are the same as for a mill producing pulp for papermaking.

In 1999, no full-scale organosolv mill was in operation. The first full-scale Organocell mill has closed. This chapter will deal with the Organocell process and other organosolv processes that have progressed to, or almost to, the pilot plant stage. These are Alcell, Acetocell, Formacell, Milox, ASAM, impregnation-depolymerization-extraction (IDE), and alcohol-reinforced kraft cooking.

5 Alcell

This pulping method comes from research work by Kleinert. Earlier, it was known as the alcohol pulping and recovery (APR) process. A pilot plant with a capacity of 30 t/day was built in 1988. While working with this, plans were made for a 100 000 t/year full-scale mill. But the economic situation deteriorated to the extent that no trials or plans for organosolv cooking currently exist.

The Alcell process uses the acid autohydrolysis of wood[4]. Chips are cooked at high temperature and pressure in an ethanol and water solution resulting in cleavage of acetyl groups to form free acetic acid and delignification of the wood. Little lignin condensation occurs. The average molecular mass of the lignin passing into solution is low: 1000 Dalton compared with that of kraft lignin or lignosulfonates. Since acid hydrolysis is not selective, some hydrolysis of wood hemicelluloses also occurs during cooking. Xylan is particularly affected. Some reacts further to produce furfural. Cellulose withstands cooking better. Pulp yield is lower than that for the corresponding kraft pulp. Another disadvantage of the process is that it is only suitable for hardwood delignification. Its advantages include the absence of sulfur and the fact that the cooking liquor contains no inorganic components. This makes recovery a comparatively simple matter.

The first stage of the process involves steaming the chips. They are then subjected to countercurrent extraction in three stages at 190°C–200°C with 50% aqueous ethanol. When this is complete, the temperature and pressure are lowered in stages, and the ethanol is recovered by condensation. The pulp is blown out and taken for washing, first with an ethanol solution and then with water. The aqueous ethanol solution is returned to the final cooking (extraction) stage, and the washed pulp is screened and bleached.

The black liquor from cooking is pumped to the alcohol recovery unit. The liquor, which contains dissolved lignin and carbohydrates, is diluted with water. This causes the lignin to precipitate. The lignin is filtered off, washed, and dried. The remaining liquor is distilled to provide ethanol for the next cook. Furfural is a by-product of this distillation. Some of the aqueous solution containing carbohydrates is taken from the base of the column to the lignin precipitation unit. The remainder is used for furfural production, fermentation, or burning[4,5]. Figure 2 shows the Alcell process.

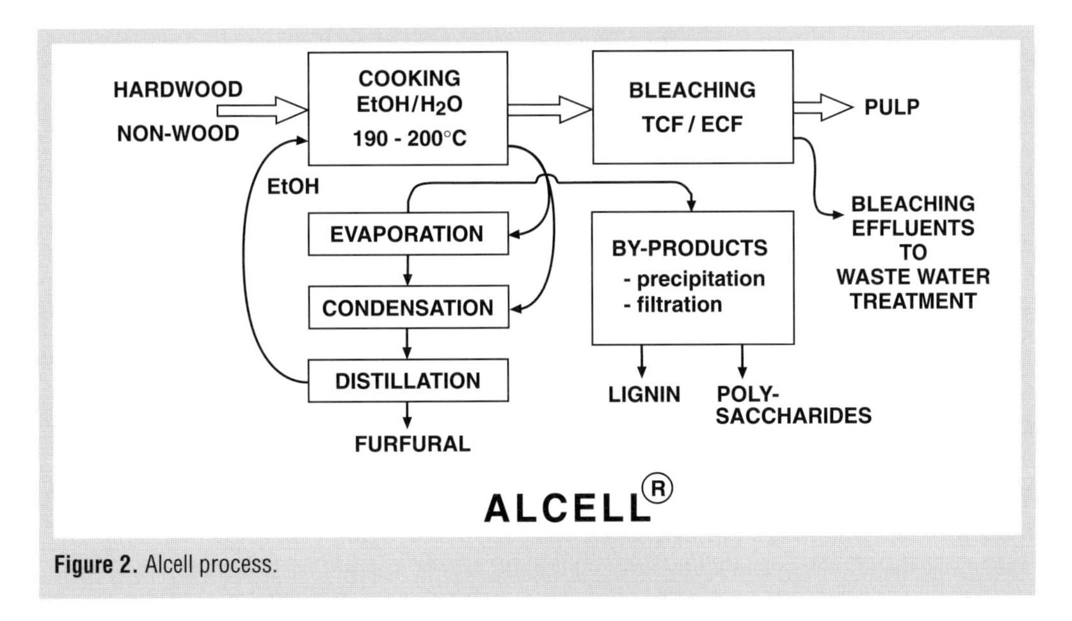

Figure 2. Alcell process.

Alcell hardwood pulp is readily bleached by both ECF and TCF processes. The kappa reduction in the oxygen delignification stage for a pulp with a starting kappa number of 25–30 is about 60% without any major drop in viscosity. Final bleaching with DED and ZQP sequences yields pulp of full brightness.

Although the pulp does not quite match the corresponding kraft pulp in strength, its optical characteristics are better. As with organosolv pulps in general, the extractives content is extremely low, and the brightness stability of the pulp is therefore excellent.

The Alcell process is also suitable for silica-containing nonwood materials. Silica does not dissolve in the acidic ethanol and water cooking liquor and therefore remains in the fibers. This means no fouling of heat exchangers in the recovery unit.

The development of the Alcell process paid special attention to the recovery and further processing of by-products. The primary by-products are sulfur-free, low molecular mass lignin, pentoses, acetic acid, and furfural. Alcell lignin can be an ingredient in the production of polyurethane plastics through its reactions with di- and polyfunctional isocyanates. The pentoses can be used for production of special sugars such as xylitol and for the manufacture of polymers and detergents. Furfural is a useful starting material for the production of a variety of chemicals and polymers such as nylon.

Advantages of the Alcell process are the following:

- Cooking liquor regeneration is simple
- Additional cooking chemicals besides ethanol are unnecessary

- Sulfur-free process
- Suitable for use with silica-containing nonwood raw materials
- Readily bleachable pulp
- Sulfur-free by-products for further processing.

Disadvantages of the Alcell process are as follows:

- Unsuitable for pulping softwoods
- Low pulp yield
- High pressure demands special cooking equipment and safety measures
- Papermaking pulp quality is inferior to that of kraft pulp.

6 Organocell

The Organocell process also has its origins in Kleinert's earliest research work. Studies were started at a mill site in 1978. Since the goal was to produce softwood pulp, the focus of the research soon switched from acid alcohol cooking to a two-stage cook in which the acid stage was followed by an alkaline stage with anthraquinone as catalyst. Cooking trials resulted in good yields of pulp comparable to softwood kraft. Using these results in 1984, a newly formed development company began planning a demonstration-scale mill. A few years later they purchased an old sulfite mill at Kelheim in Germany as a new site for conversion to the new process. Although a pilot-scale two-stage process had already been commissioned, this was changed to a single-stage methanol-AQ-alkali process to save costs. The first full-scale trials were conducted in the winter of 1992/1993. Because of numerous technical difficulties, the start-up of full-scale production was deferred, and the mill never actually started regular production. Another problem was that pulp quality did not match that achieved at pilot scale. Financial difficulties forced the mill to close toward the end of 1993.

The Organocell process[2,6,7] or methanol-AQ-alkali process resembles the soda-AQ and kraft processes. As with soda cooking, the main delignifying species is hydroxide ion. Unlike the soda process, Organocell cooking is also suitable for softwood pulp production, because of AQ addition. This is an advantage it shares with the kraft process. Methanol improves the ability of the cooking liquor to penetrate into the wood chips and renders the lignin more soluble. Anthraquinone functions in the same way as in soda cooking by stabilizing polysaccharides and accelerating lignin dissolution. In cooking, anthraquinone (AQ) is reduced to anthrahydroquinone (AHQ) by the polysaccharide end groups. In turn, these are oxidized to alkali-stable aldonic acid groups. The reduced AQ now acts as an effective cleaving agent for the lignin β-aryl ether linkages and is simultaneously oxidized to anthraquinone (Figure 3). Organocell pulps produced at pilot scale are almost as good as the corresponding kraft pulps in yield and physical characteristics.

At the original mill that had a continuous pilot digester with a capacity of 5 t/day, chips were first impregnated and then cooked at about 200°C in aqueous methanol, followed by cooking at 160°C–180°C in aqueous methanol containing caustic soda and a catalytic amount of anthraquinone. At the new mill that had a rated capacity of about 330 t/day, impregnation was followed directly by an alkaline cooking stage. The cooking temperature was 160°C–170°C, and the cooking time (including the preceding chip impregnation) was 3 h. For bleachability and suitable bleaching methods, Organocell pulp does not differ from corresponding kraft pulps.

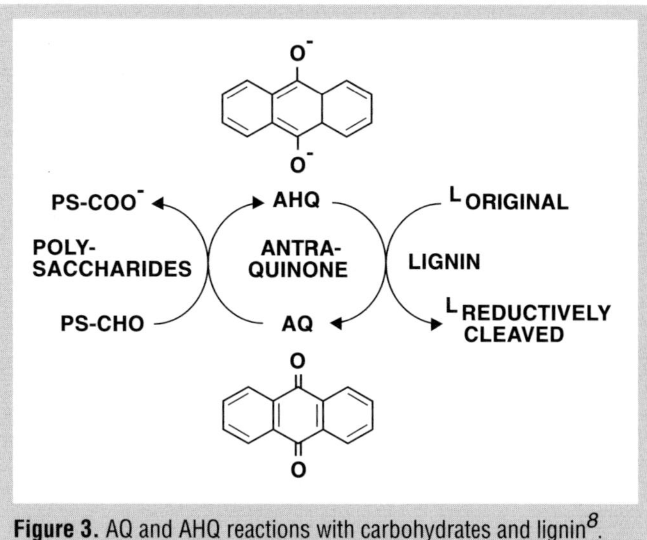

Figure 3. AQ and AHQ reactions with carbohydrates and lignin[8].

The chemical recovery system consists of two parts: methanol recovery and recycling, and caustic soda regeneration. Most of the methanol is recovered when the cooked pulp is blown from the digester and the remainder with hot steam after the first pulp washing stage. The condensate is distilled and the methanol returned to the digester. The black liquor is evaporated to about 60% dry solids and burned under oxidative conditions. The soda obtained is causticized with calcium hydroxide in the normal way and returned to the process. Figure 4 shows the Organocell process.

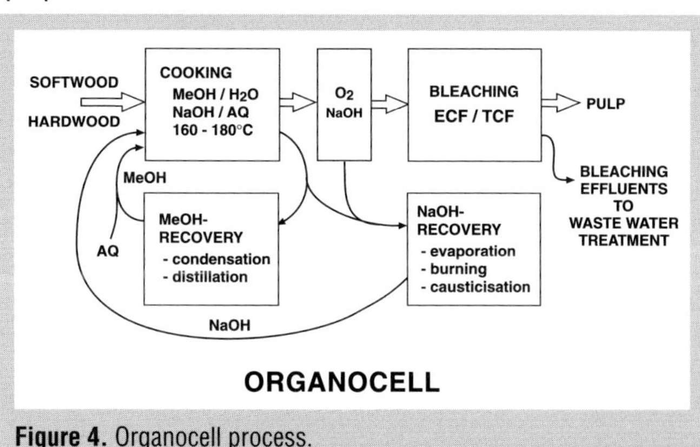

Figure 4. Organocell process.

The potential use of the by-products, especially lignin, from Organocell cooking was investigated when the process was under development. Sulfur-free lignin is a useful carrier for long-acting fertilizers, for one example. It biodegrades without forming any acidic or otherwise detrimental foreign inorganic substances.

Advantages of the Organocell process from pilot trials are the following:

- Suitable for hardwood and softwood pulping
- Strength comparable to that of kraft pulp
- Sulfur-free process
- Sulfur-free by-products for further processing.

Disadvantages of the Oganocell process are:

- Two parallel chemical recovery systems (methanol and alkali)
- High pressure requires special cooking equipment and safety measures
- Original scale-up attempts were unsuccessful.

7 IDE

The IDE process uses an alkaline ethanol solution for cooking[9]. Laboratory studies are still continuing. The process has three consecutive stages, each optimized to favor specific reactions and phenomena.

In the first stage, impregnation liquor is forced into the chip filled digester system and heated to 100°C. The liquor-to-wood ratio is 8:1. The cooking chemicals are sodium hydroxide and sodium carbonate. The temperature is held at 100°C for 1–3 h. After the impregnation, the system is drained and the impregnation liquor collected. The impregnation temperature is low to prevent excessive consumption of the cooking chemicals in this stage. The impregnation time is long to allow uniform penetration and diffusion of the cooking liquor chemicals.

The second stage of the IDE method is depolymerization, or more precisely, delignification. The impregnated chips are treated with an ethanol and water solution containing 50% ethanol by volume. The temperature is raised to 190°C over 30 min and kept there for another 30 min.

The third step is extraction. It starts at 190°C and continues as the temperature decreases slowly to 160°C. The sodium carbonate acts as a buffer by keeping the pH at 11–12 and preventing lignin from reforming large complexes that would be difficult to extract. It also prevents the precipitation of solubilized lignin on the surfaces of fibers.

The IDE method is suitable for softwood pulping, and the resulting pulp characteristics are similar to those of kraft pulps.

8 ASAM

In the early 1980s, R. Patt and O. Kordsachia working at the University of Hamburg developed an alkaline sulfite (AS) method reinforced with anthraquinone (A) and methanol (M). In 1985, the rights to the ASAM process were sold, and a 5 t/day pilot plant was built. The plant was completed and officially opened in spring 1990. The process has been developed for several years. The process is now probably ready to be tested at full-mill scale[10–12].

The inventors have stated that rather than being an actual organosolv process, the ASAM method is a modified sulfite process in which every individual stage has been tried on an industrial scale. The process has been modified through the inclusion of methanol and anthraquinone to make it suitable for softwood and hardwood delignification. It also can be used for nonwood materials.

The process consists of a single-stage batch cook. ASAM cooking liquor normally contains about 10% methanol by volume. The active cooking chemicals are sodium hydroxide, sodium carbonate, and sodium sulfite. The anthraquinone dose is 0.05%–0.1% by weight on wood. The liquor-to-wood ratio is 3–5:1, and the cooking temperature is 175°C. The cooking time is 60–150 min. By varying the ratio of sulfite to sodium bases, one can control the hemicellulose content and therefore the yield and optical properties.

In the ASAM process, the primary delignification reactions are alkaline hydrolysis and sulfonation of lignin brought about by the alkali and the sulfite, respectively. Both reactions promote lignin solubility. Methanol facilitates impregnation of the chips with cooking liquor and renders the anthraquinone more soluble in the cooking liquor. It also prevents lignin from entering into condensation reactions by methylation of reactive sites and promotes the solubility of the lignin decomposition products. Under the alkaline conditions of ASAM, anthraquinone acts according to the same mechanism as in other alkaline processes. It stabilizes polysaccharides and helps to cleave beta-aryl ether bonds in lignin.

Chemical recovery has three parts: methanol recovery and recycling, sodium hydroxide production by causticizing, and sodium sulfite production from sodium sulfide. Most of the methanol is recovered when the pulp is blown from the digester. Methanol remaining in the black liquor is recovered by steam stripping. The methanol fractions are condensed and distilled, with the methanol returning to the process.

Recycling the inorganic cooking chemicals is more complex. The black liquor is concentrated by evaporation and then burned. The inorganic content is dissolved to produce green liquor that contains sodium carbonate and sodium sulfide. Carbon dioxide from the flue gases is introduced into the green liquor. This causes the pH to fall and sulfide to be released as hydrogen sulfide gas. This leaves sodium carbonate and sodium bicarbonate in solution. The hydrogen sulfide is burned to produce sulfur dioxide that is then returned to the carbonate and bicarbonate solution to produce sulfite. The hydroxide necessary for cooking is produced in the normal way by causticizing the carbonate in the green liquor. The desired $NaOH/Na_2CO_3/Na_2SO_3$ ratio results from manipulating the component processes listed above. Figure 5 shows the ASAM process.

The disadvantages of the ASAM method concern the sulfur-containing chemicals and chemical recovery. Odor problems can arise during the stripping and burning of hydrogen sulfide. The alkaline pH used in the process means that the sulfite does not cause any more sulfur dioxide emissions than the sulfide emissions that are characteristic of the kraft process.

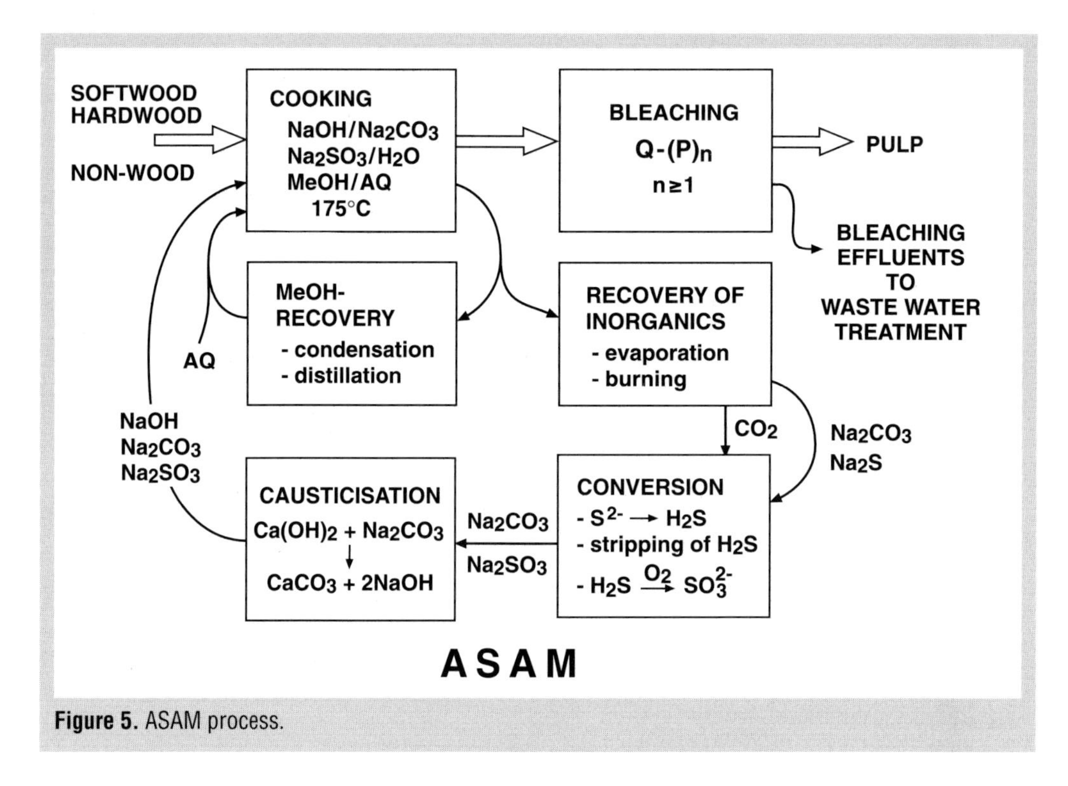

Figure 5. ASAM process.

The pulp yield from ASAM cooking is higher than that from kraft cooking. In one case starting with the same spruce chips, bleached, ASAM pulp was obtained at 48% yield, and that from modified kraft cooking at 42% yield[12].

ASAM pulps are easily bleached, even without chlorine chemicals. After chelation, a softwood pulp of kappa number 20–25 can be bleached using a peroxide/oxygen sequence without ozone to over 86% brightness. ASAM softwood pulps retain their viscosity during TCF bleaching much better than kraft pulps. Their papermaking characteristics are also good.

Advantages of the ASAM process are as follows:

- Suitable for all raw materials
- Higher yield than kraft cooking (48% vs. 42%)
- Pulp easy to bleach, even using only a Q-(O/P) sequence
- Strong pulp of high quality.

Disadvantages of the ASAM process are the following:

- Involves sulfur-containing chemicals
- Complex recovery system
- High cooking pressure requires special equipment and safety measures.

9 Methanol-reinforced kraft pulping

The good results achieved with methanol-reinforced alkaline sulfite pulping (ASAM process) and methanol-reinforced soda pulping (Organocell process) have naturally led to the idea of reinforcing the kraft process with methanol. Norman, Olm, and Teder have studied this[13]. They found that methanol has a positive effect on delignification even at a charge of only 2%. The lignin content of the pulp decreases as the methanol charge increases, but the decrease levels off at higher methanol charges. Carbohydrate degradation is also accelerated, but the effect on delignification is greater. This means that delignification can be extended to lower kappa numbers without major loss of pulp viscosity and yield. Adding methanol to the kraft process does not affect the bleachability or strength properties of the pulp.

The benefits of adding methanol to the kraft process have been relatively minor compared with the advantages gained with ASAM cooking relative to the alkaline sulfite process. Reinforcing the kraft process with methanol is therefore not particularly attractive.

10 Acetosolv, Acetocell, and Formacell

At the Institute of Wood Chemistry and Chemical Technology of Wood in Hamburg (BHF), Nimz *et al.*[14] have spent considerable time developing wood pulping using acetic acid as the main cooking solvent. The first such method to reach the pilot stage was the Acetosolv process. The cooking liquor was 90% acetic acid containing 1% hydrochloric acid. The cooking temperature was 110°C, and the cooking time was 3–5 h. In 1988, a rotating carousel digester divided into several segments was built. The carousel with segments filled with wood chips revolved counter-clockwise, and the cooking, washing, and bleaching solutions revolved clockwise. Every revolution, one segment of pulp was produced and the empty segment was then filled with new chips. Serious corrosion problems soon stopped these pilot trials.

Investigations returned to the laboratory. The process changed significantly in 1990. HCL-catalyzed cooking was replaced by cooking at high temperature with acetic acid alone. A new Acetocell pilot plant was started in June 1992.The digester capacity was 0.8 m^3, and the dry chip charge was about 200 kg. The acetic acid concentration in the cooking liquor was about 85%, the cooking temperature was 170°C–190°C, and the total cooking time was 5.5 h. Three-stage washing with acetic acid followed the cooking. The bleaching sequence was ZEPPaa. The ozone (Z) stage used acetic acid as solvent. This improved the selectivity of delignification since ozone is nine times more soluble in acetic acid than in water. The acetyl groups formed during cooking through the reaction of acetic acid with polysaccharides are hydrolyzed in the alkaline extraction stage (E) following ozonation. After the alkaline peroxide stage (P), the final stage in the bleaching sequence was a treatment with peracetic acid that gave pulp of full brightness. The bleached pulp yield in the case of softwood was 43%–44%. The spent cooking liquor and acetic acid washings were evaporated to 40%–50% dry solids, and the dry solids recovered by spray drying for burning or further processing.

The Acetocell knowledge and patent have been for sale, and the pilot plant was dismantled.

The third stage in this development work is the Formacell method, still at the laboratory stage. As with earlier versions, delignification in the Formacell process occurs primarily through acid hydrolysis. Because the acid hydrolysis is not selective, some wood polysaccharides also undergo hydrolysis and react further to form furfural. Under pulping conditions, acetic acid also reacts with the hydroxyl groups of lignin and polysaccharides to produce acetates.

The chips are dried to about 20% moisture content and cooked in acetic acid/water/formic acid (75/15/10). The liquor-to-wood ratio is 5:1, the cooking temperature is 160°C–180°C, and the cooking time at maximum temperature is 1–2 h. The cooked pulp is screened and washed with the acid mixture mentioned and bleached with ozone in one or two stages at high consistency (20°C–50°C, 10 min, 1% O_3 on pulp) still in the acid mixture. The ozone-bleached pulp is washed free from acid with butyl acetate and finally bleached with peracetic acid in butyl acetate for 2 h at 90°C. The final stage is to strip the butyl acetate with hot steam. Since the bleaching is entirely in an organic medium, no aqueous effluents result.

The spent cooking liquor is evaporated to 50% dry solids, and the concentrate is spray dried. The acetic acid/water/formic acid/butyl acetate/furfural mixture obtained from evaporation and spray drying is taken to a distillation column, and the azeotropic mixtures of furfural/water and butyl acetate/water are separated from the cooking acid. Figure 6 shows the Formacell process.

The highly corrosive nature of the cooking acid requires the use of special steel wherever the equipment is in contact with hot acid: the digester, evaporation unit, distillation column, spray dryer, and some piping.

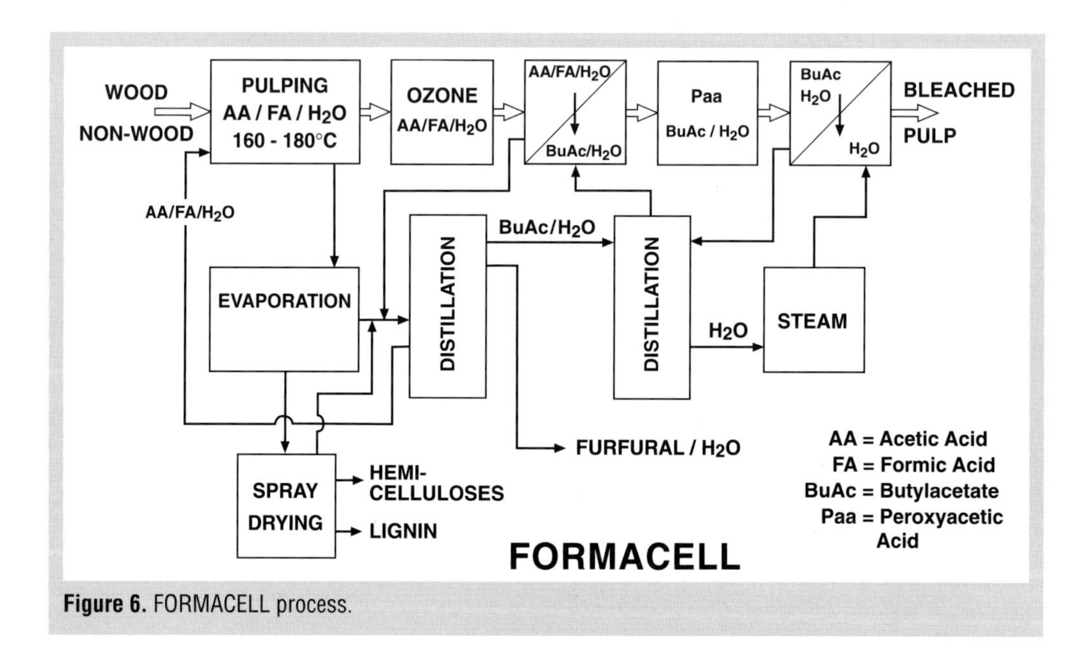

Figure 6. FORMACELL process.

According to the inventors, the Formacell process can be used to pulp hardwoods, softwoods, and grasses. Based on general experience with acid cooking methods, pulp quality will probably not rival that of kraft pulp at least with softwoods.

Advantages of the Formacell process are as follows:

- No inorganic cooking or bleaching chemicals
- Selective ozone delignification in an organic acid mixture
- Sulfur-free by-products
- No bleaching effluents.

Disadvantages of the Formacell are:

- Laboratory-scale use only
- Complex chemical recovery concept
- Corrosion problems
- Requirement for safety equipment.

11 Milox

The Finnish Pulp and Paper Research Institute (KCL) began work on developing a pulping process using formic acid and its derivative, peroxyformic acid, in 1984. The goal was to find a process free from sulfur and chlorine chemicals that would produce fully bleached pulp suitable for fine paper production[16–20].

Following laboratory experiments, the process progressed to pilot scale in 1990–1994.

The pilot plant is a batch operation. The capacity is approximately 100–150 kg pulp/batch. The name of the pulping process comes from milieu pure oxidative pulping.

Formic acid is an excellent solvent for the lignin and extractives in wood. Simultaneously, it reacts with wood polymers but at lower temperatures than the weaker acetic acid. The main reactions with lignin are beta-ether cleavages, acid hydrolysis, and intra- and intermolecular condensation reactions. Under Milox conditions, formic acid reacts with the free aliphatic and phenolic hydroxyl groups of lignin to produce formate esters. The polysaccharides also react with formic acid. The principal reaction is acid hydrolysis. Hemicelluloses are attacked more easily than cellulose. Polysaccharides also form formate esters.

Peroxyformic acid is simple to prepare by equilibrium reaction between formic acid and hydrogen peroxide. It is a highly selective chemical that does not react with cellulose or other wood polysaccharides in the same way as formic acid. Peroxyformic acid oxidizes lignin, renders it more hydrophilic, and therefore raises its solubility.

Milox pulping is a three-stage process. Hardwood chips are first dried to a moisture content of less than 20% and then impregnated with 80%–85% formic acid from the third stage of the previous cook to which has been added 1%–2% hydrogen peroxide based on the dry weight of the chips. The liquor-to-wood ratio is 4. In the first stage, the temperature increases from 60°C to 80°C. The peroxyformic acid that forms is allowed to react with the chips for 0.5–1 h. The temperature is raised to the boiling point of the formic acid (ca. 105°C) and cooking proceeds for 2–3 h. The softened chips are then blown into another reactor, and the pulp is washed with pure formic acid. The washed pulp is then reheated with peroxyformic acid at 60°C at about 10% consistency. Peroxide is charged to the liquor at 1%–2% of the original dry weight of the chips. After cooking, the pulp is washed with strong formic acid, pressed to 30%–40% consistency, and washed under pressure with hot water at 120°C. This removes the chemically bonded formic acid. After washing and screening, the pulp is ready for bleaching.

The procedure for softwood is similar. The only differences are the hydrogen peroxide charge and temperature in the second stage. The total peroxide is 5%, and the temperature is 120°C–140°C. Unlike the case with wood species, two-stage Milox pulping of agricultural plants is more effective than three-stage cooking. The two-stage process uses cooking with formic acid alone, followed by treatment with formic acid and hydrogen peroxide.

Formic acid flows countercurrently through the process. From the cook, the black liquor is evaporated to a dry solids content of 65% and spray dried. The solids are burned in the power boiler to produce steam and electric power. Some formic acid from evaporation is fed back into the digester. All other formic acid is fed to the distillation unit where it is concentrated and the acetic acid originating from hemicelluloses is removed. Figure 7 shows a three-stage Milox process.

Formic acid is highly corrosive especially at high temperatures. In the Milox pilot plant, the digester is carbon steel coated with zirconium. All other equipment, pipes, and valves in contact with hot formic acid are of duplex steel.

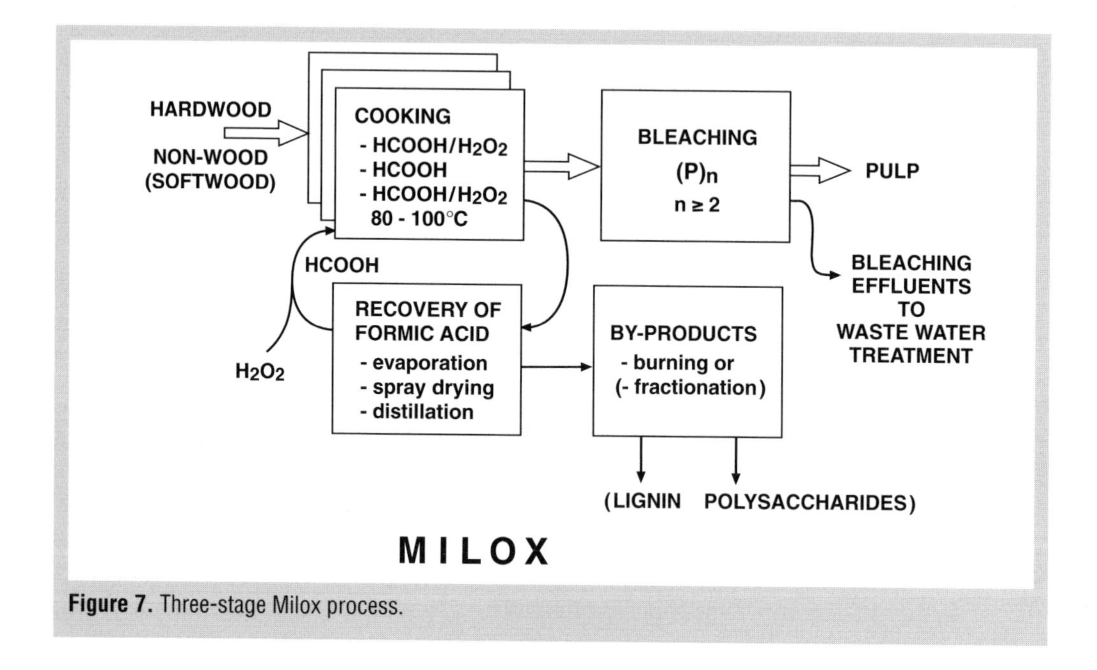

Figure 7. Three-stage Milox process.

The response of Milox pulps toward alkaline hydrogen peroxide bleaching illustrates their unusual character: they show a significantly greater brightness response than the less reactive conventional kraft pulps or oxygen-delignified kraft pulps under similar bleaching conditions. Hardwood and softwood pulps with low residual lignin contents reach full brightness in a two-stage alkaline bleaching sequence.

Birch and nonwood pulps prepared at pilot scale have been used for papermaking on an experimental paper machine. The properties of the resulting Milox birch and grass-containing papers ranked with those of high-quality wood free printing and writing papers. In contrast to hardwood pulps, the production of softwood Milox pulps is not profitable because of the high consumption of hydrogen peroxide and poor strength properties of the fibers.

The primary by-products of the Milox process (if not burned) are lignin (300–400 kg/adt) containing about 300 g sulfur/t, soluble carbohydrates (mainly xylose), and about 100 kg acetic acid/adt pulp when birch is the raw material.

According to a feasibility study, the investment cost for a Milox mill of 200 000 adt/year is the same as that of a kraft mill of the same size. Variable production costs are US$ 50–60/adt higher for the Milox mill. The role of the by-products will have major significance for the profitability of the Milox process as for other organosolv processes .

Advantages of the MILOX process are as follows:

- Low temperatures and unpressurized reactors
- Simple TCF bleaching: P-P-(P)
- Simple chemical recovery system
- Sulfur-free by-products
- Suitable for silica-containing nonwood materials.

Disadvantages of the MILOX process are:

- Poor quality of softwood pulps
- Corrosion problems
- Chip drying.

12 Future of Organosolv processes

At the present stage of development, imagining organosolv processes replacing the kraft process in the production of pulp for papermaking is difficult. Although perhaps more friendly to the environment, organosolv processes do not yet offer any economic or quality advantages over the kraft process that would provide the incentive for the necessary major investments. Closest to the goal in this sense is probably the ASAM process with regard to pulp quality and yield. The ASAM process uses sulfur-containing chemicals, and its recovery system is not simple. The Alcell process has the simplest and most straightforward fiber line and chemical recovery, but it is not suitable for cooking softwood. Milox, Formacell, and ASAM pulps are best for bleachability. None of the organosolv processes described in this chapter is likely to offer a smaller economic mill size than the kraft process.

In the last ten years, progress in organosolv pulping processes has been encouraging, and will undoubtedly continue. The application of catalysis in delignification and the processing of sulfur-free by-products is just beginning. The possibility still remains for developing chemical processing methods in the chemical fractionation of wood components and regeneration of cooking liquors. The next organosolv mill designed for full-scale production might not produce papermaking pulp but a variety of chemical products. Instead of, or in addition to, producing pulp and energy, it could convert renewable biomass into sulfur-free raw materials for the chemical industry, as Fig. 8 shows.

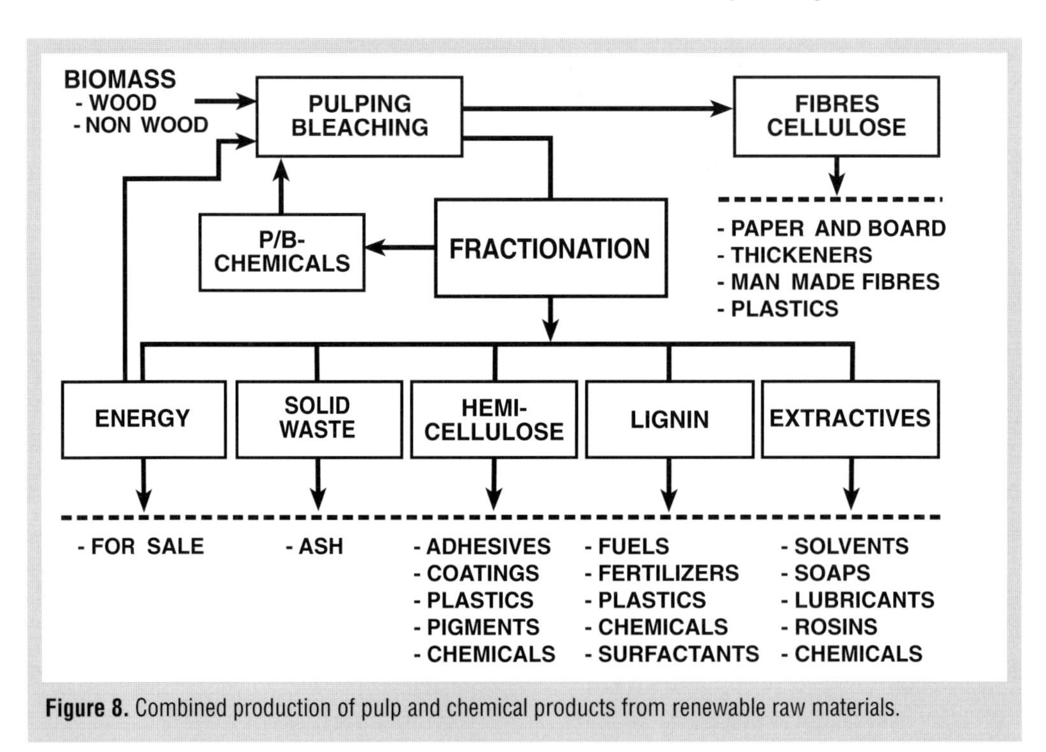

Figure 8. Combined production of pulp and chemical products from renewable raw materials.

References

1. Kleinert, T. and von Tayenthal, K., Angew. Chem. 44(39):788(1931).
2. Hergert, H. L., TAPPI 1992 Solvent Pulping Symposium Notes, TAPPI PRESS, Atlanta, p. 9.
3. McDonough, T. J., TAPPI 1992 Solvent Pulping Symposium Notes, TAPPI PRESS, Atlanta, p. 1.
4. Lora, J. H. and Pye, E.K., TAPPI 1992 Solvent Pulping Symposium Notes, TAPPI PRESS, Atlanta, p. 27.
5. Pye, E. K. and Lora, J.H., Tappi J. 74(3):113(1991).
6. Leopold, H., 25th EUCEPA Conference, 1993, p. 87.
7. Leopold, H., Das Papier 47(10A):V1(1993).
8. Sjöström, E., Wood Chemistry - Fundamentals and Applications, 2nd edn., Academic Press, San Diego, 1993, p. 157.
9. Backman, M., Lönnberg, B., Ebeling, K., et al., Paperi ja Puu 76(10):644(1994).
10. Schubert, H. -L., Fuchs, K., Patt, R., et al., Das Papier 47(10A):V6(1993).
11. Glasenapp, A., Kordsachia, O., Odermatt, J., et al., Das Papier 50(6):300(1996).
12. Patt, R. and Kordsachia, O., Das Papier 51(1):1(1997).
13. Norman, E., Olm, L., and Teder, A., Tappi J. 76(3):125(1993).
14. Nimz, H. H., Internal report., Federal Research Centre for Forestry and Forest Products., Institute of for Wood Chemistry and Chemical Technology of Wood., Hamburg (1994).
15. Neumann, N. and Balser, K., Das Papier 47(10A)V16(1993).
16. Poppius-Levlin, K., Hortling, B., and Sundquist, J., Proceedings 7th International Symposium on Wood and Pulping Chemistry, CTAPI, Beijing, 1993, vol. 1, p. 214.
17. Pohjanvesi, S., Saari, K., Poppius-Levlin, K., et al., Proceedings 8th International Symposium on Wood and Pulping Chemistry, 1995, vol. 2, p. 231.
18. Seisto, A., Poppius-Levlin, K., and Jousimaa, T., TAPPI 1995 Pulping Conference Proceedings, TAPPI PRESS, Atlanta, Book 2, p. 487.
19. Sundquist, J., Paperi ja Puu 78(3):92(1996).
20. Sundquist, J. and Poppius-Levlin, K., in Environmentally Friendly Technologies for the Pulp and Paper Industry (R.A. Young and M. Akhtar editors). John Wiley & Sons, Inc. New York 1998 pp. 157-190

CHAPTER 21

Process calculations and simulation

Johan Gullichsen

Process calculations and simulation

The chemical pulping process is a complex entity that requires consideration by examining the entire process simultaneously. A change in one department will radiate through the whole mill. Previous chapters gave the principles of detailed calculations. This chapter discusses methods and techniques available for whole mill projections necessary for accurate mill design and process modification.

1 Simulation principles

Several computer based simulation techniques are available. Steady state simulation gives an instantaneous look at an average process state in balance. Dynamic simulation considers the time dimension and describes transient conditions of a process in change. Both techniques require expressing accurate material and energy balances over each unit operation. Detailed process calculations and especially those that help equipment sizing may also require calculating detailed mass, force, and energy balances over elements within the process.

Simulation requires models to describe reactions and their kinetics and the transfer of heat and mass. These models preferably use continuity of mass and energy and the first principles of chemistry and physics. Processes unknown in detail use "black box" models that are statistical or simplified first principle models. Some models may be composites of both types. The use of any model requires verification by practical mill scale observation or valid laboratory testing. The range of validity is also necessary before using as an engineering tool.

Computer modeling can accurately describe and simulate complex processes. Complex process models are difficult to operate without user-friendly and simple user interfaces. Model developers are often specialists on modeling, programming, and mathematics. They may not have skills to express applied techniques to lay people. The process engineer knows his process but can seldom express it in a useful, systematic way for model building. People of many skills cooperate to create good simulation systems using a multi discipline exercise.

Modern simulators uses graphic interfaces with which the user draws a process flowsheet from already developed unit operation blocks or preprogrammed process modules. The blocks or modules interconnect by drawing connection lines (pipelines) from and to appropriate addresses. The user can define several process flow properties or components to describe the properties of each flow. Graphical interfaces automatically open dialog boxes for necessary specification of each flow.

User selected results of simulation calculations require presentation in a form the reader can understand. Printed data sheets can contain detailed information. Such an output easily contains thousands of pieces of data and is meaningful only to the specialist interested in a detailed data. Most people require only selected data from areas of specific interest. A good simulator should provide the user with tools to specify his own printout mode. The tools should also provide for summary data of pertinent details. This should include simplified process flow sheets with labeled process flows and boxes with selected flow properties. Good process simulation generates pertinent numbers and data and organizes the results in an easily accessible format.

Accurate process simulation requires substantial computational capacity. Contemplating what level of detail is necessary in solving simulation problems at hand is therefore important. Process simulation systems should have a structure of several layers providing more detailed information and complexity as one moves deeper into the soul of the process by proceeding from one level to another. For example, a gross mass and energy balance over a cooking plant is sufficient to provide basic data for detailed washing or evaporation projections. Detailed process internal data from each unit operation within the cooking plant is necessary for the design of digester processes.

Good process simulators provide several levels of information. Most current models use a network of blocks. Each describes one simple operation such as mixing, splitting, heating, reaction, and separation. One such network of interconnected blocks can describe an entire process entity such as a digester or washer. Simple algebraic expressions may describe the performance of many blocks used. Others may contain advanced heat and mass transfer models.

Many compounds of dissolved solid and gaseous matter participate independently or interdependently in a complex flow system. The user should always have the option to select compounds of particular interest. The simulation system should have a structure making this easy and providing sufficient room for additional compounds if necessary.

Calculations in most simulators are iterative. Meeting set balance accuracy criteria by iteration reaches a certain steady state. Iteration stops when required balance accuracy requirements occur over the entire process and each of the process internal simulator blocks.

The user may specify input data or process conditions by which balance will not occur. A good simulator should inform the user that convergence does not occur and in which module or block the error happens. The user may also attempt to make illegal process connections. This might include trying to feed fibers into a steam line or omitting an essential process without a proper address.The simulator should automatically inform the user that the erroneous connection is illegal simply by not accepting it or by advising that essential process lines have not been properly connected.

2 General structure of a simulator

The structure of a simulator may be described by some examples.

Example 1: Consider a simple pulp washer. This is a machine where dissolved substance separates from the fiber by a combination of dilution, rethickening, and displacement by a washing liquid.

The simplest description is the box of Fig. 1. A pulp suspension of a given consistency and level of contamination feeds into the device for washing, thickening to a given consistency, and discharging.

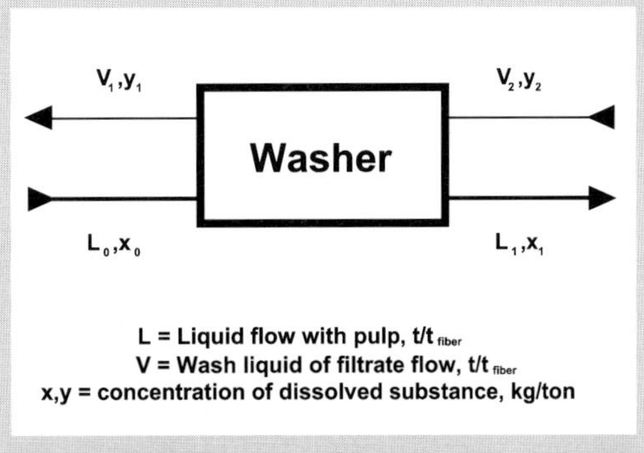

Figure 1. Black box model of a pulp washer operation.

A liquid with a given concentration of specified components and flux is the washing medium. A filtrate containing the material washed from the system goes to a preceding washer or to liquor evaporation. Equation 93 in Chapter 3 approximately describes the performance of this process indicated by the single black box model. The model tells how much dissolved material can separate from the washer knowing its specific efficiency parameter, E.

This same washer performs several operations. Breaking the machine into the single unit operations of Fig. 2 forms a process module. The feed suspension is diluted to feed vat consistency in a stirred dilution tank, D (diluter). The diluted stock is then pumped to the web former box where it is dewatered by filtration to a given consistency on a suction drum, T (thickener). Wash medium is then displaced through the formed pulp web, W (washing), and perhaps further thickened as the web continues on toward the washer discharge. Filtrates from thickening and displacement combine in this example in a mixing tank, M (mixer), from which a fraction of the flow is split, S (splitter). One stream goes to dilution and the remainder flows from the washer.

If steady state modeling without considering sorption and diffusion phenomena is in question, simple algebraic mass and heat balance equations can describe the performance of dilution (D), thickening (T), mixing (M), and splitting (S). The Norden efficiency function (Eq. 93 of Chapter 3) can describe the displacement (W).

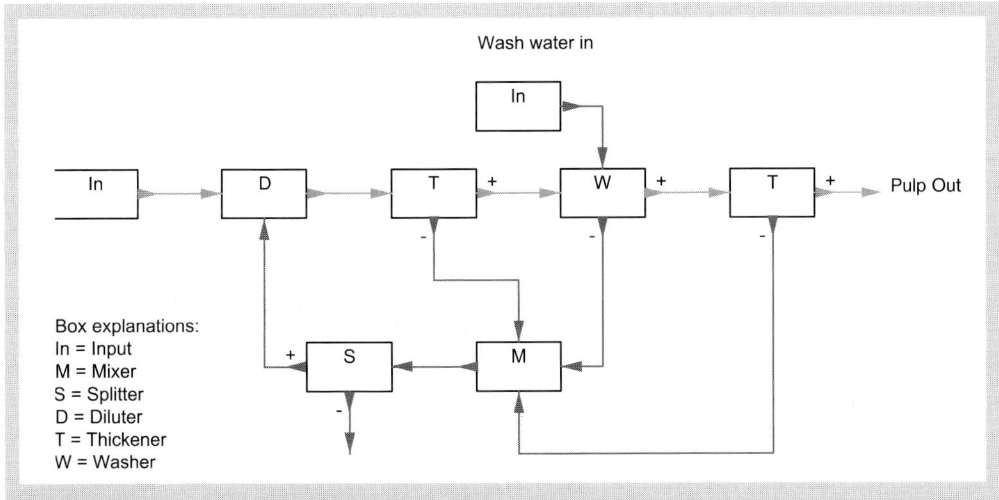

Figure 2. Washer module with unit operation blocks.

Including sorption and diffusion phenomena requires that sorption (Eq. 103 of Chapter 3) and diffusion (Eq. 102 of Chapter 3) functions become components of the models used to describe performance of the dilution (D) and displacement (W) blocks. Retention times in both these stages must also be known.

Dynamic simulation of the same operation also needs additional information about tank and operation zone volumes. Describing the dewatering and displacement functions hydrodynamically also requires filtration resistance functions (Eq. 30 of Chapter 3) and detailed physical dimensions and functional data of the machine.

The operator does not need to use or even be aware of all these details provided he has the means to enter and receive pertinent data. The interface may only show the machine as Fig. 3 shows.

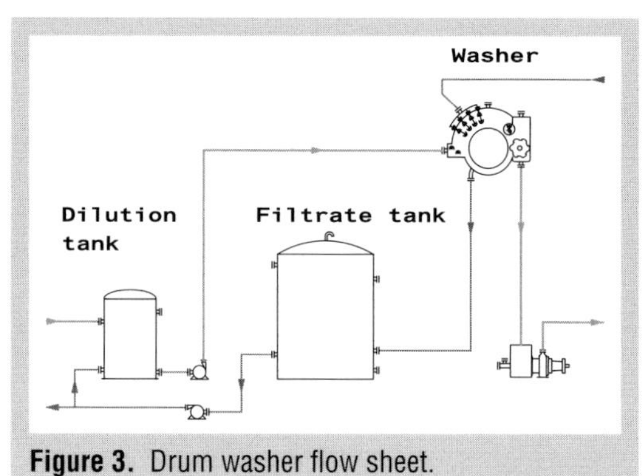

Figure 3. Drum washer flow sheet.

Example 2: A continuous hydraulic digester after chip steaming and impregnation is considerably more complex than the simple drum washer. Internal liquid circulation loops and countercurrent flow arrangements make a complex simulation model. Figure 4 shows the flow sheet and module matrix configurations.

Example 3: Figure 5 shows a simplified displacement batch cooking flow sheet and simulation module matrix.

Example 4: Multistage vacuum evaporation consists of several similar evaporation units coupled in series or cascade. The heat demand to reach a certain evaporation capacity is directly proportional to the number of stages used in an ideal case. With an optimal flow sheet, one can estimate the heat demand, the output of strong liquor, and condensates using the simplified model (Fig. 6) for a falling film evaporator application. Here the operator only gives data on the number of stages, steam data, and requested dry solids content exiting from evaporation. The simulation gives an estimate of steam required and the production of condensates.

Another structure is useful if more detailed evaporation data is necessary. Figure 7 shows the flow sheet designations and corresponding module block matrixes of the three typical evaporators frequently used in a system of falling film evaporators. The primary stage is a module driven by live fresh steam, the secondary stage evaporator receives its steam from another unit or process, and the last stage module includes condensers and vacuum systems. Each of the evaporator modules contains blocks to describe the heat exchanger, liquor flashing, stripping of volatile substances into departing vapor, and liquor recirculation rates and preheating. Models for flashing include the impact of dry solid content related boiling point elevation. Figure 8 shows a complete flow sheet and block matrix for a six-unit evaporator station.

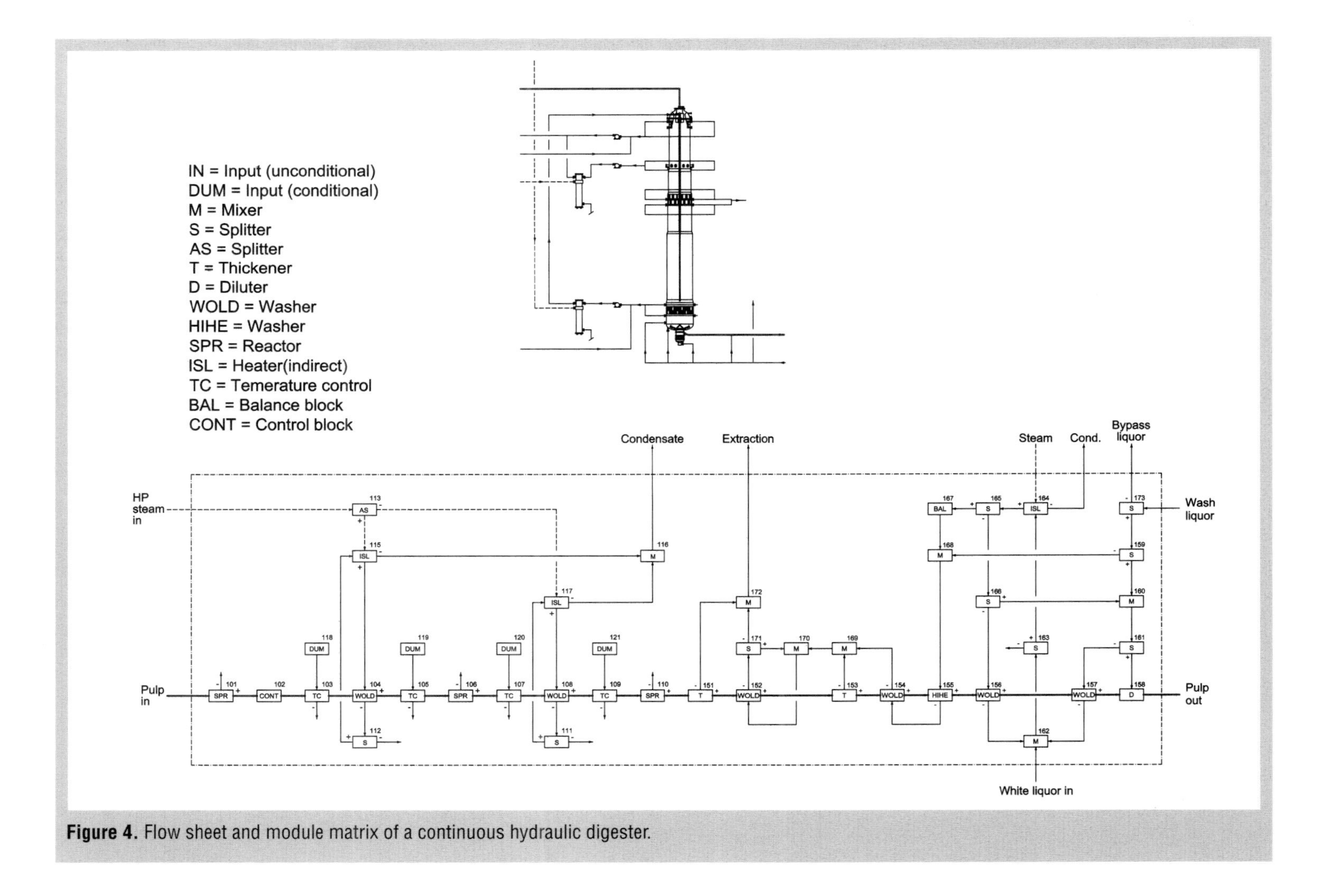

Figure 4. Flow sheet and module matrix of a continuous hydraulic digester.

3 Applications

The power of complex but flexible process simulation is considerable for selected system options, troubleshooting, and optimization of mill processes. Simulative comparisons of options may be accurate, although details of models used are incomplete. The strength is that comparisons use the same logic. The discussion on the impact of various cooking and bleaching options in Chapter 2.6 used projections derived by simulation. Another example is the following.

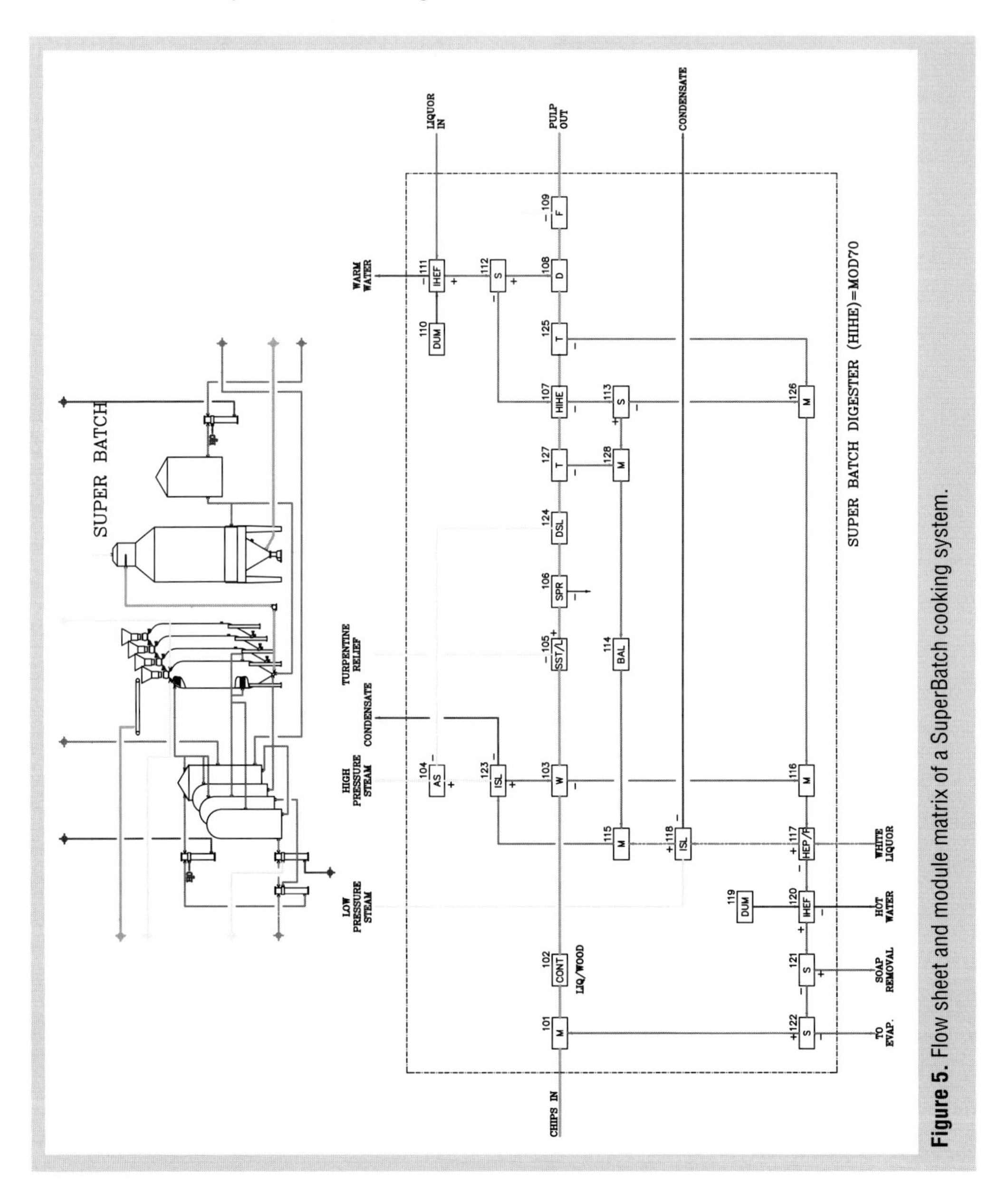

Figure 5. Flow sheet and module matrix of a SuperBatch cooking system.

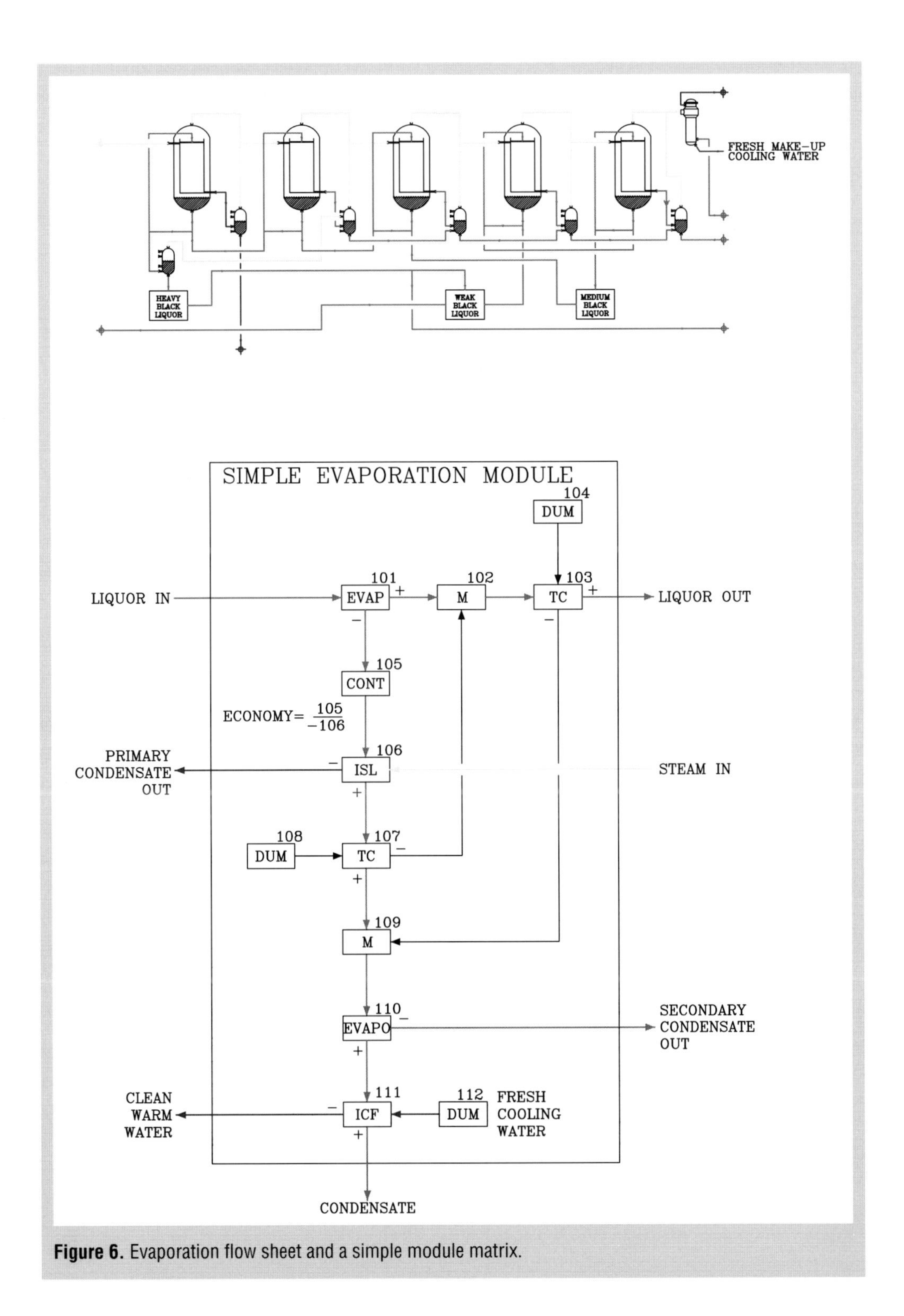

Figure 6. Evaporation flow sheet and a simple module matrix.

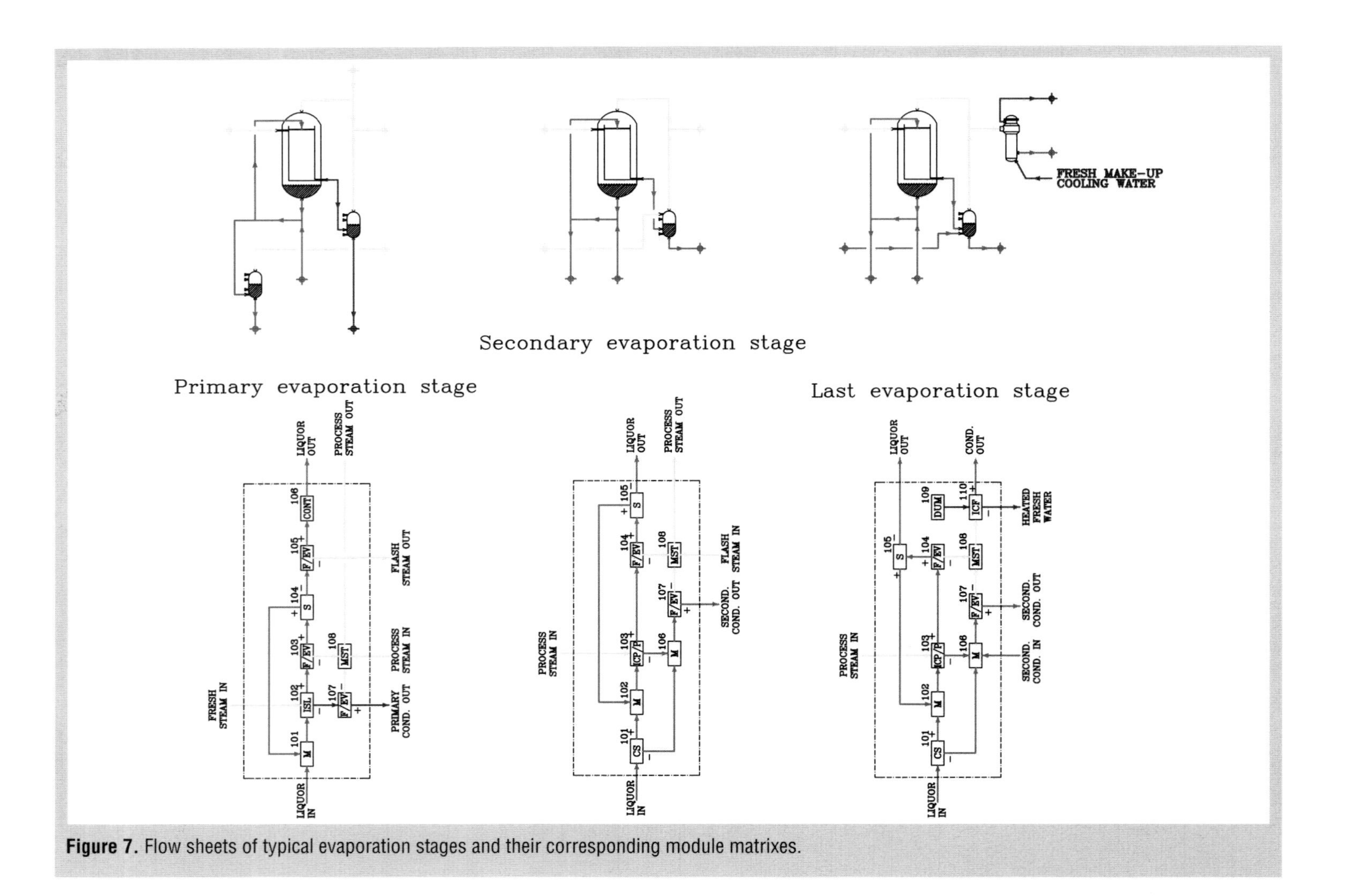

Figure 7. Flow sheets of typical evaporation stages and their corresponding module matrixes.

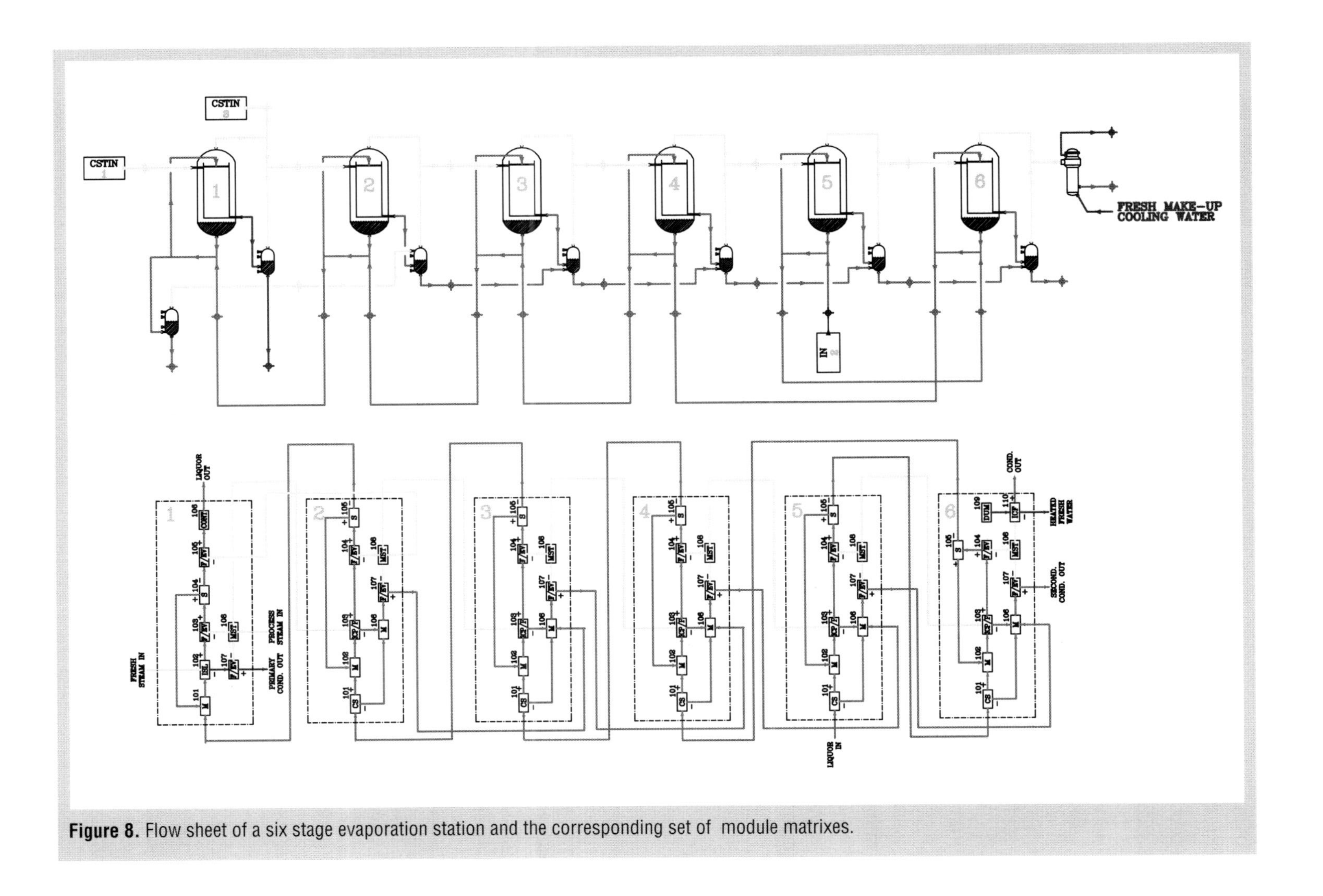

Figure 8. Flow sheet of a six stage evaporation station and the corresponding set of module matrixes.

Example 5: What is the difference between extended hydraulic continuous cooking and displacement batch cooking steam demand and black liquor dry solids when compared with systems where efficiencies and dilution factor in washing for both brown stock systems are equal? Figure 9 shows the process flow sheets of the systems simulated. Flow boxes on the flow sheet summarize the results. Table 1 provides more detailed data.

The conclusion is that a given primary steam demand benefit exists for continuous cooking. In addition, the displacement batch cooking process would require about 15% more evaporator capacity to reach the same black liquor dry solids content due to its lower weak liquor dry solids concentration. The reason is that continuous cooking systems recover heat as secondary steam from black liquor flashing from 157°C to atmospheric pressure. Displacement batch cooking recovers heat without flashing from a pressurized hot liquor.

The total heat energy demand for cooking and evaporation is 14% higher for displacement batch cooking, but it produces 2.4 times more secondary heat as 75°C water. This is because the efficiency of the final displacement is much lower in displacement batch cooking and that chips are preheated to a lower temperature by recovered heat. The demand to cool black liquors with water is considerably higher in the displacement batch cooking case.

Simulation can easily project the performance of the entire mill including chemical recovery, power and energy regeneration, and effluent treatment in systems having more than one-thousand blocks in tens of prescheduled modules. This task is impossible to perform by hand.

Several commercial steady state and dynamic simulation tools are commercially available. Most find primary use in the petrochemical industry and are difficult to adapt to chemical pulping or papermaking. Some systems have a special design or modification to meet the needs of the pulp and paper industry. All are in a state of rapid development as computer technology and mathematical applications evolve.

Table 1. A comparison by simulation of primary heat demand and secondary heat production from an extended continuous cooking plan and a SuperBatch plant as shown in Fig. 9.

Cooking method	Mass flow, t/t moisture-free pulp	Net energy flow, GJ/t moisturefree pulp
Extended modified continuous cooking		
LP steam demand to cook	-	-
HP steam demand to cook	0.91	1.84
LP steam to evaporation	1.2	2.62
Total steam	2.11	4.46
Hot water (75°C) produced	3.15	0.4
Displacement batch cooking		
LP steam demand	0.38	1.04
HP steam demand	0.5	1.03
LP steam to evaporation	1.38	3
Total steam	2.26	5.07
Hot water (75°C) produced	7.46	0.95

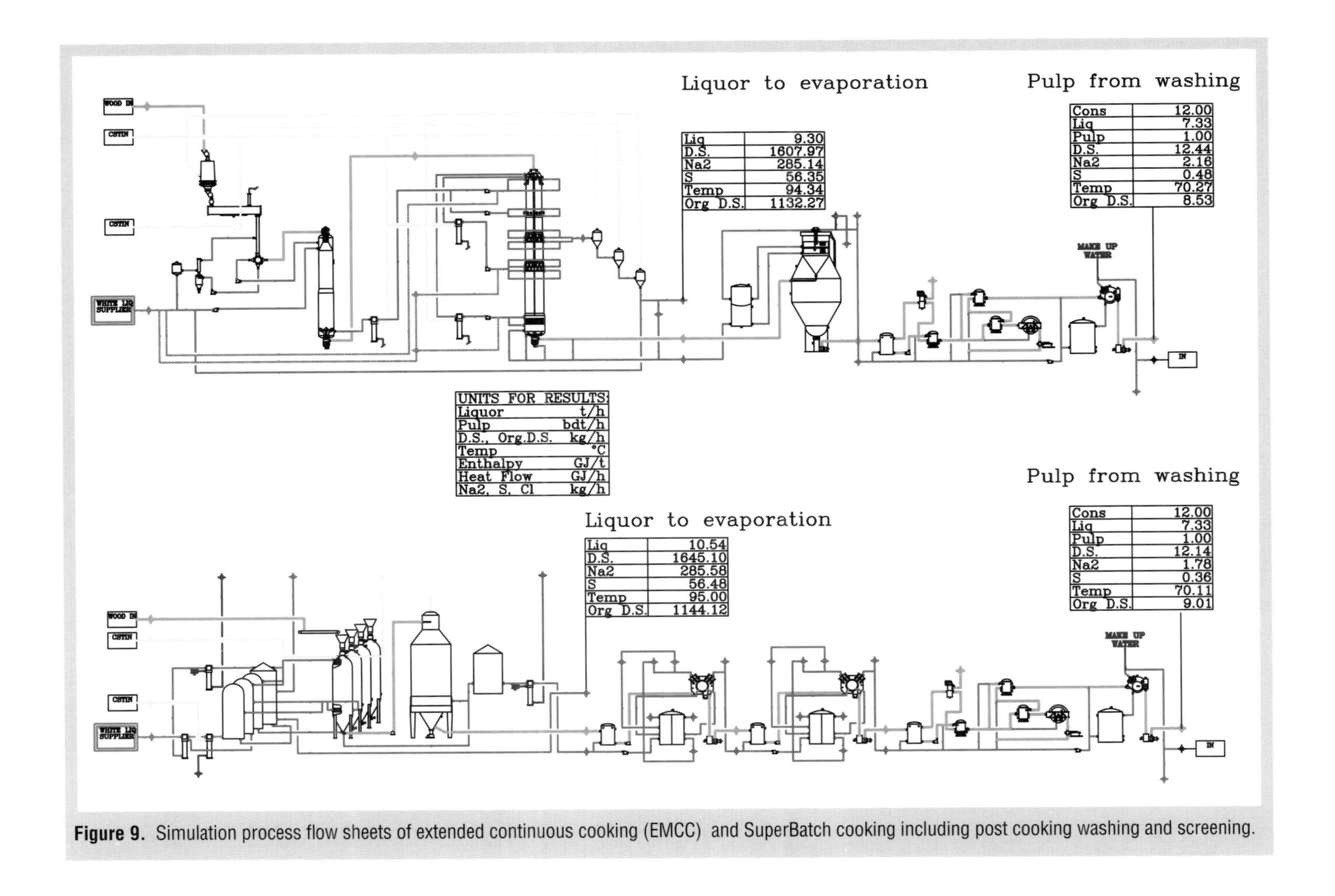

Figure 9. Simulation process flow sheets of extended continuous cooking (EMCC) and SuperBatch cooking including post cooking washing and screening.

CHAPTER 22

Closed cycle systems for manufacture of bleached chemical wood pulp

Douglas Reeve and Claudio Mudado Silva

Closed cycle systems for manufacture of bleached chemical wood pulp

1 Introduction

Closed cycle systems for manufacture of wood pulp of low lignin content, high brightness, strength, and cleanliness, i.e., bleached chemical pulp, use processes where water and other chemicals are recycled and reused. This minimizes waste disposal. Closed cycle systems are particularly intended to minimize aqueous effluent without increasing the burden of disposal elsewhere. The object of such systems is to protect the environment from the impact of disposal of effluents without jeopardizing processing cost or value of saleable products. Figure 1 shows the closed cycle concept[1]. Wood is a complex material, and any system design must account for each of its components and their transformation products. Any wood component entering must exit as product or solid waste or by emission to the air or aqueous environment.

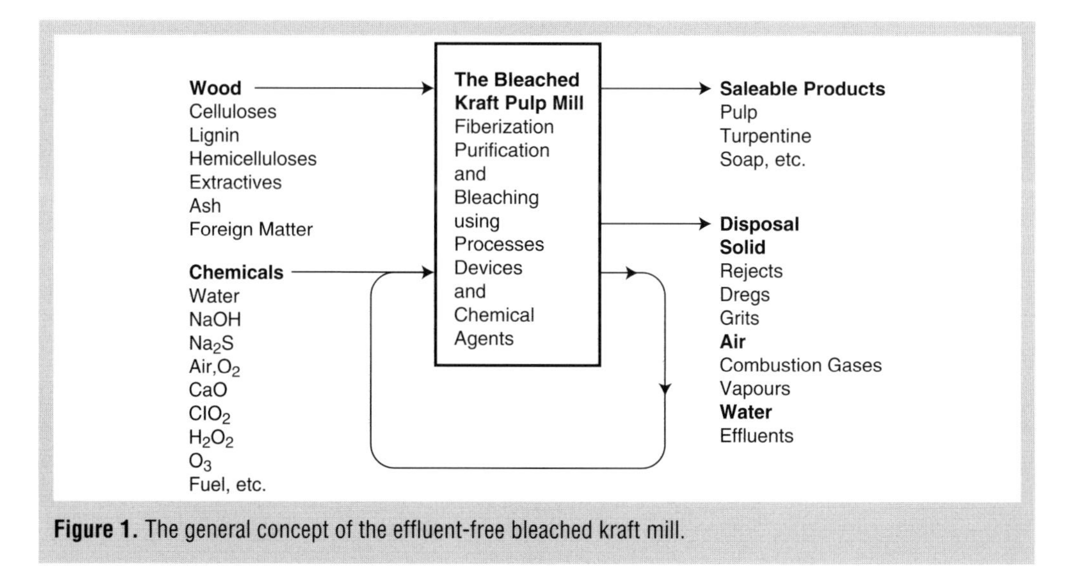

Figure 1. The general concept of the effluent-free bleached kraft mill.

Many distinct operations occur in the manufacture of bleached pulp: pulping, washing, screening, bleaching, cleaning. Chemical recovery operations include liquor evaporation, liquor burning, green liquor making, causticizing, and lime reburning. The foundation of any closed cycle system is the efficient and stable performance of each of these operations. Progress toward minimization of pollutant discharge often occurs with the improvement in performance of these basic operations without resorting to "closed cycle" operations. Much valuable information is in the other chapters of this book that deal separately with the basic operations. It is also important to recognize that there are many ways in which recycling is already widely practised in manufacture of kraft pulp; pulping chemicals are recovered and regenerated, as is lime for recausticizing. Water reuse is extensive in every mill. What distinguishes the "closed cycle" concept is the greater degree to which recycling is practised and, in manufacture of bleached kraft pulp specifically, the recycling of effluents from bleaching stages after the oxygen delignification.

This chapter describes a path-breaking development called the effluent-free mill proposed in 1967[2] and implemented in Thunder Bay, Ontario in 1977, and discusses the valuable lessons from that experience. Some system aspects are next presented including environmental performance, bleaching chemicals, water balance, sodium to sulfur balance, nonprocess elements, impact on the recovery system, and economics. Several recent mill-scale developments will be described and reviewed. Relevant technologies under development, such as membrane separation and ion exchange, are also reviewed. Finally, the chapter addresses the concept of a minimum impact mill and the debate concerning balance between environment and economy. Many overviews of closed cycle systems exist[1–9].

2 The effluent-free mill

2.1 Concept

Rapson first described the concept for recovery of bleached plant effluent to eliminate water pollution by kraft pulp mills in 1967[2]. Countercurrent washing used in the bleach plant minimized effluent volume through a (D+C)EDED bleaching sequence. Bleach plant effluent would then be used for brownstock washing. The organic matter dissolved during bleaching would be burned in the recovery furnace. The spent bleaching chemical, sodium chloride, would be recovered from the furnace flue gas fume captured in the electrostatic precipitator. Condensate from black liquor evaporation would be cleaned and reused for washing in the bleach plant. The benefits cited included elimination of water pollution, savings in heat, elimination of primary and secondary effluent treatment, associated capital and operating cost reductions, decrease in fresh water requirements, and the strategic opportunity of locating a pulp mill independent of a large water supply or receiving body of water. Full-scale testing of the various components of the system was suggested and justified by the potentially great advantages of an effluent-free kraft mill. In 1969, the concept was further advanced with the development of a new system for sodium chloride removal by white liquor evaporation. Figure 2 shows this[4]. A series of papers describes the salt recovery process development[10] and other aspects of the effluent free mill design[11].

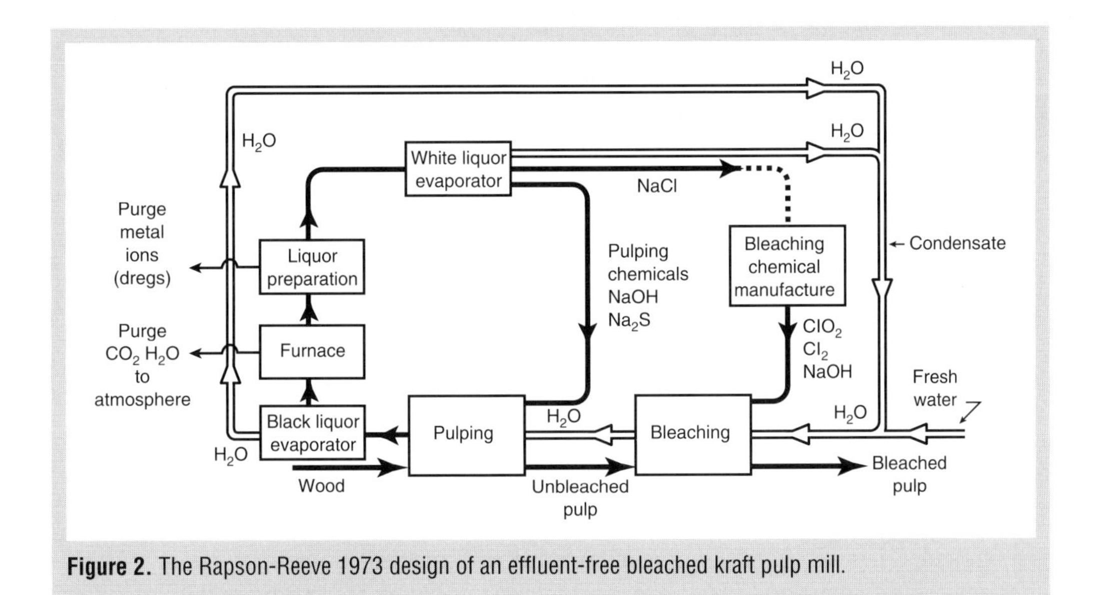

Figure 2. The Rapson-Reeve 1973 design of an effluent-free bleached kraft pulp mill.

2.2 The first commercial application

Design for the first commercial application of bleach plant effluent recovery and salt recovery began in 1974. The system started in 1977 at the Great Lakes Forest Products Company (Bowater Canada Inc.) in Thunder Bay, Ontario. This approach was an alternative to biological treatment that would have been very costly at the site. Information is available on the process design and operating performance of this system[4, 12-16]. Many technical difficulties and costs were significantly greater than expected. In 1985, all bleach plant effluent recovery and salt recovery ceased.

The process design of the fiber line used conventional equipment available then. A continuous digester produced kraft softwood pulp of 30 kappa number from an 80/20 mixture of jack pine and spruce. Aspen pulp was produced later. Washing in the digester was followed by a two-stage diffusion washer followed by screening and twin deckers. The mill design capacity was 700 o.d. t/day. The drum washer bleach plant used a (DC)EDED sequence. The original design was 70% chlorine dioxide substitution in the (DC) stage, but the unit usually operated at 50% substitution. Countercurrent washing in the bleach plant produced bleach plant effluent of 16 m^3/t of pulp when no effluent recovery was practised.

As Figure 3 shows, bleach plant effluent was incorporated into the unbleached part of the mill in several ways[14]. Brownstock washing used approximately 1.3 m^3/adt of extraction stage filtrate. This was introduced into the system where the filtrate concentration of dissolved organic matter approximated the concentration of dissolved organic matter from black liquor. In this scheme, introduction was in the filtrate going to the first brownstock diffuser. Approximately 2.7 m^3/adt of extraction stage filtrate went to the salt recovery plant for dilution of concentrated white liquor. Extraction stage filtrate concentration was closely monitored. Filtrate was purged to the sewer to maintain adequate net flow from the system. Acidic filtrate from the DC stage was neutralized with purchased sodium hydroxide and used as wash water on the last stage of brownstock washing, i.e., the decker after screening. Approximately 2.3 m^3/adt of acidic filtrate went to the lime kiln scrubber where it became part of the green liquor system. Oxidized white liquor was used to neutralize this filtrate. A deliberate purge of acidic stage filtrate maintained adequate net flow.

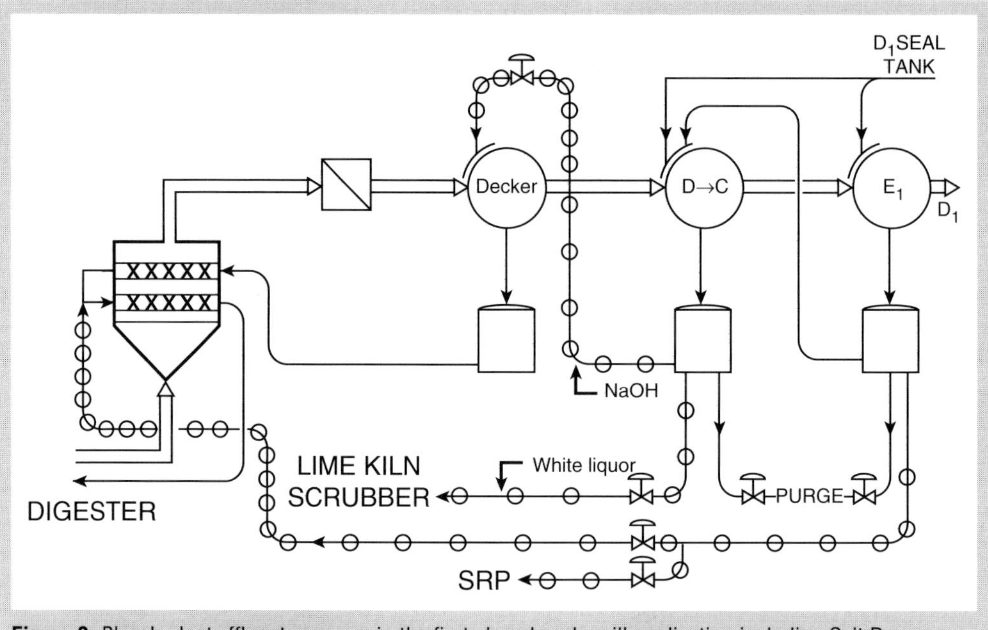

Figure 3. Bleach plant effluent recovery in the first closed cycle mill application including Salt Recovery Process (SRP).

Evaporation of white liquor provided salt recovery[15]. Triple-effect evaporators concentrated the white liquor 2.5 times. The sodium carbonate and sodium sulfate crystals thereby formed were removed and returned to the white liquor after salt removal. The clarified, partially concentrated white liquor then underwent further concentration and cooling to crystallize sodium chloride. The recovered sodium chloride was purified by leaching and then dissolved. The brine was used for chlorine dioxide manufacture.

Four spill collection systems were part of the process design. Fiber spills in the screening, beaching, and pulp machine area were collected and returned to the unbleached decker. Alkaline spills of black liquor and bleach plant filtrate were collected and returned to the weak black liquor evaporators. Acidic spills and excess acidic filtrate were collected and returned to the washing system at the brownstock decker. Spills in the recausticizing area were collected and returned to the mud washer.

Many operating problems occurred in the bleach plant[14,16]. Bleaching chemical consumption increased significantly in the DC stage and in the first chlorine dioxide stage due to accumulation of dissolved organic matter in the system. Foaming was a problem at the interface between alkaline and acidic systems, and defoamer addition was necessary. Periodically, deposits related to defoamer use occurred. Deposits of pitch and calcium-based scales were continuing problems. They were particularly severe for hardwood operations,requiring greater purge from the system and a lower percentage of bleach plant effluent recovery. Pitch deposits resulted in some periods of off-quality pulp[16]. Stabilizing water flow through the countercurrent washing systems in the bleach plant was difficult without adding fresh water to seal tanks and increasing the volume of effluent. Storage of filtrate was inadequate to allow coordination between the filtrate producing department (bleach plant) and the filtrate consuming departments (unbleached fiber line, salt recovery plant, and lime kiln scrubber).

Operating problems also arose in the recovery system[15,16]. The most serious was extremely rapid corrosion of the recovery boiler superheater that required temporary discontinuation of bleach plant effluent recovery only 17 months after startup. After repairs, the superheater was reconfigured to lower the final steam temperature before resuming bleach plant effluent recovery. Many problems occurred with the salt recovery process, resulting from scale formation on the evaporator heat transfer surfaces and changing particle size in the first stage of crystallization. Numerous corrosion problems in the salt recovery plant led to significant downtime. Use of bleach plant effluent in the lime kiln scrubber introduced organic matter into the recausticizing system. This caused increased carryover of suspended solids from the white liquor clarifiers and difficulties in lime processing.

Many operating problems associated with overall performance also occurred[16]. BOD and suspended solids were then the only regulated pollutants from the mill. Bleach plant effluent recovery (to the extent achieved) had little impact on BOD discharge performance for the entire mill. Due to the operating problems in bleaching and recovery described above, the percent recovery of bleached plant effluent was low. The values averaged 50% for softwood and 20%–30% for hardwood. Some outright stoppages of bleach plant effluent recovery occurred. Because of the larger recovery of sodium than chloride on a molar basis from bleach plant effluent, the sulfidity of the kraft recovery cycle decreased with time. To restore the proper balance, purchased sodium hydroxide used for bleach plant effluent neutralization was decreased. The system for managing spills and upsets was ineffective due to insufficient capacity, inadequate monitoring facilities, and insufficient catch-up capacity in the systems that would receive the returned spilled material. Energy and bleaching chemical costs were higher than in conventional systems existing then. These performance deficiencies were not resolved during the long, incremental development of this closed cycle system. Bleach plant effluent recovery and salt recovery operations therefore ceased in 1985.

Some design deficiencies existed in the fiber line equipment and process that exacerbated attempts to achieve bleach plant effluent recovery. The continuous digester did not operate smoothly, especially in the first few years. This gave very high variability in unbleached kappa number. Brownstock washing was insufficient to remove black liquor adequately from the pulp before bleaching. This aggravated bleaching chemical consumption. Black liquor evaporator capacity was sufficient for normal processing but insufficient for spill makeup. The detailed mechanical design of the bleach plant washers and seal tanks caused many problems that took several years to resolve.

Despite these difficulties, this path-breaking installation did achieve some successes. Bleached plant water usage of only 16 m^3/t was possible during stable operation. Chlorine dioxide substitution of 50%–70% was practical and produced high-quality pulp. Recycle of 65% of the material dissolved in bleach plant effluent occurred on a sustained basis[14]. The salt recovery process operated successfully and produced salt of sufficient quality to recycle to chlorine dioxide generation.

2.3 Lessons and projected requirements

Many important lessons were learned from this first attempt to eliminate effluent from bleached kraft pulp mills. System design must consider the chemistry of all components of wood and their transformation products. This is an extremely complex system of individual compounds, macromolecules, and process conditions. Equipment and material design must also consider changing process conditions imposed by closed cycle processing. Sensors and process control systems must adequately monitor and manage water flows and chemical concentration to avoid excess concentration of undesirable components. System procedures must consider differences in operating strategies for closed cycle systems vs. traditional designs. Operating personnel must receive this information in appropriate training programs. The concerns about chemistry and engineering expressed above apply not only to steady-state circumstances but also to non-steady-state occurrences such as upsets, startups, shutdowns, catch-up, etc. Many potential problems relate to unanticipated, uncontrolled, interfacial phenomena such as formation of pitch, scale, and foam.

A lesson from the first closed cycle mill is the need to adopt a long-range view of the context in which the business operates. This includes the economy, regulatory requirements, and markets. In this case, economic circumstances changed dramatically after the design because of the sharp rise in energy prices due to the oil embargo of the early 1970s. Much impetus for development was to achieve the zero discharge requirement that the U.S. Congress said it would impose by 1984. In the 1970s, a strong incentive existed to invest in the development of a large and complex process such as a closed cycle mill. By the end of the 1970s, this projected requirement of zero discharge for all industrial enterprises was abandoned. Considerable motivation for solving the technical problems of the closed cycle mill therefore disappeared.

Many significant technological and engineering advances have occurred since the design of this first application in 1974. Continuous and batch digesters are much more sophisticated and have better control. Extended delignification technology provides unbleached pulp of lower kappa number with greater uniformity and cleanliness. Oxygen delignification has changed from a new technology in 1974 to an established, effective technology in the 1990s. Many developments have occurred in high-performance washing equipment. Sensors and computers that were extremely limited in 1974 now offer broad and sophisticated capability for monitoring and controlling processes.

Since 1974, a vast increase in knowledge of process chemistry and in application of this knowledge to sophisticated process simulation has occurred. Coupled with increasingly effective engineering design tools, this facilitates the design of closed cycle systems in modern mills.

An interesting perspective on requirements for further development in 1982[1] described six problems requiring resolution:

- Increased bleaching chemical consumption
- Calcium accumulation in acidic bleaching stages
- Pitch deposition
- Volatile organic compounds
- Water management and control
- Economical and effective chloride control.

Further discussion in this chapter shows that many of these problems have been solved in the intervening 15 years, but some still require solution.

3 System aspects

3.1 Basic technologies

Some technological components of bleached kraft pulp mills must already be in place when attempting to take the final step in mill closure, i.e., recovery of bleaching filtrate. These basic technologies in the fiber line, recovery cycle, and effluent treatment will provide a mill system that produces the following:

- Effluent of low BOD, low COD, and low environmental impact
- Pulp of high quality and uniformity
- Operation that is cost effective, efficient, and stable.

All these are important conditions to meet before taking the final steps to complete, or nearly complete, mill closure by a high degree of effluent recycle. Effluent treatment is a basic technology because spill containment systems, suspended solids removal systems, and biological treatment systems are important components of effective mill operation even for closed cycle systems.

In the fiber line, basic technology begins with wood preparation. The effluent from wood handling such as storm water run-off from wood storage areas and effluent from debarking must be minimal. They require treatment so that they do not have significant impact on the environment. Debarking, chipping, and chip thickness screening are important in closed systems to provide a feed stock to the fiber line that is uniform and clean. This minimizes upsets and deviations in quality that require aggressive treatment to remedy. Extended delignification technology can be important to closing a bleached kraft mill to provide a lower kappa number than has been traditional, i.e., lower than 30 with softwoods. The technology must also provide a low reject rate and steady operation that are possible with modern extended delignification systems. Considerable debate has existed about the ideal kappa number at the digester. Extended delignification systems can produce softwood pulps of kappa no. 12. As noted elsewhere in this book, increased yield loss at low kappa numbers favors a softwood kappa number of 23–25.

Brownstock washing is extremely important for closing systems so that dissolved organic matter is effectively removed and fed to the recovery system. This also avoids operating difficulties in subsequent stages. Washer efficiency and reliability are important. Efficient and reliable pulp screening is also important for maintaining high pulp quality without needing aggressive bleaching to remove otherwise screenable particles.

Oxygen delignification is essential for producing bleached kraft pulp in highly closed systems. Oxygen delignification systems with good delignification efficiency and high selectivity will permit removal of over 50% of the lignin remaining after cooking. With good post-oxygen washing, dissolved organic matter will be returned to the black liquor system. Bleach plants have traditionally used very large volumes of water, not only for washing but also for miscellaneous uses such as washer doctors and seal water for pumps, agitators, and mixers. Before considering bleach plant effluent recovery, one must decrease water used in bleaching through countercurrent washing and implementation of practices to minimize entry of miscellaneous water.

Many bleaching sequences are possible in closed systems, both elemental chlorine free (ECF) and totally chlorine free (TCF) bleaching sequences.

Several essential basic technologies must be in place in the chemical recovery cycle to implement mill closure. Liquor losses from all parts of the recovery cycle must be low. This requires efficient, well maintained process equipment and effective operating practices. To avoid liquor losses, methods must exist to deal with upset conditions, spills, and leaks. Black liquor evaporator sizing, design, and operation must not compromise condensate quality. Condensate requires segregation, and the fractions rich in volatile organics require stripping to remove those organics. Evaporator capacity must be sufficient to accept the extra volume from spills and upsets.

Table 1. National statistics on foundation technologies and environmental performance for 1994[21].

		Mean values (values for 50% cumulative production)			
	Lowest Achieved	**Canada**	**USA**	**Finland**	**Sweden**
Effluent Flow, m^3/adt	30–40	90–100	90–100	90–100	90–100
BOD5, kg/adt	<0.5	2	2	2	5
COD, kg/adt	<10	25	–	30	40
AOX, kg/adt	<0.1	0.7	1	0.3	0.25
				Nordic countries	
Kappa number of softwood pulp to bleaching	10	30	30	15	
Percentage of production with kappa number <20 (after pulping and/or oxygen delignification)	–	25	35	90	

The ultimate definition of a closed system is zero effluent. When considering the development of closed system technology, we must appreciate the present state of kraft mill technology. Table 1 provides some statistics on basic technologies and environmental performance from 1994[17]. Note that the industry is changing very rapidly concerning this issue. Data for following years will therefore show significant change. The values in Table 1 are the mean values for 50% cumulative production. For 50% of the pulp produced, values will exceed this number. Values will be lower than this number for 50% of the pulp produced. The mean flow of effluent is 90–100 m^3/adt, and the lowest achieved in this report was 30–40 m^3/adt. High values were over 200 m^3/adt. Clearly, many basic technologies to decrease water use would be necessary before contemplating any recovery of bleach plant effluent. Extended delignification and oxygen delignification have wide use in the Nordic countries, and the mean kappa number of softwood to the bleach plant is 15. Almost 90% of production has a kappa number lower than 20. For a closed system in Canada and the United States, a significant capital expenditure may be necessary to install extended delignification, oxygen delignification, or both. Note that biological treatment of effluents, an important basic technology for minimum impact mills, is universal in Canada, Finland, and the United States. The practice is not universal in Sweden.

Despite this lack of universality and the absence of closed systems to date, the values of discharge of BOD_5, COD, and AOX are significantly lower than they have been in previous years due to the installation of biological treatment systems in Finland and North America especially, the replacement of chlorine with other bleaching chemicals, and the tightening of mill process systems to minimize pollutant discharge. In 1989, a team of aquatic biologists concluded "that treated bleached kraft effluents from well managed and operated North American mills show little or no adverse impact on receiving water"[18]. Elimination of chlorine from bleaching had an added benefit. "By switching to chlorine dioxide as the principal bleaching agent, we have reduced the environmental effects of chlorinated compounds to the point of insignificance"[19]. Implementation of basic technology including biological treatment to minimize pollutant discharge may provide acceptable environmental performance in all but exceptional circumstances. Particularly for older mills, one must recognize that with each incremental step in closing a mill system the costs increase dramatically and the benefits diminish.

3.2 Water, chemical, and energy balances

Water balance

As discussed above, average use of water in 1994 was 90–100 m^3/adt, with best practice being about 30–40 m^3/adt[17]. Decreasing water use requires judicious design and practice in each department.

Many motivations exist to decrease water use and effluent volume. The cost of supplying fresh water varies from U.S.$ 30 to U.S.$ 100 per 1 000 m^3 [(20)]. Other justifications are limited water supply, avoidance of water or effluent treatment expansion, energy savings, decreased fiber losses, decreased chemical losses, and improved effluent treatment operation. Water reduction projects will therefore often justify themselves. The cost escalates as mill effluent flow decreases. The water reduction cost to achieve zero effluent flow for an older mill previously using 70 m^3/adt was estimated to be about U.S.$ 200 million[20]. Decreased water volume discharge also leads to decreased COD discharge. A strong correlation exists between water consumption and COD discharge in Swedish mills. Data from 1994 show about 50 kg COD/adt at 100 m^3/adt water consumption and 20 kg COD/ adt when the water consumption decreases by half[21].

Figures 4 and 5 provide data for fresh water use and the amount of effluent for each department in a bleached kraft pulp mill[22]. The data show values for conventional design with a total effluent of 40 m^3/adt of pulp and the values for a mill started up in early 1996 that discharges only 12 m^3/adt.

Other authors have described similar proposals to achieve low water balances[5,23]. Table 2 provides an overview of water to and from the mill, and Table 3 gives vapor discharges by department.

Effective washing and water reuse are essential in decreasing effluent volume in bleaching through countercurrent washing and by partial recovery of bleach plant filtrate [24,25].

Table 2. Overall water balance[27].

	(m^3/adt)
Water to mill	
From river	41.1
In wood	1.8
In chemicals	0.2
Total	**43.1**
Water from mill	
To effluent	35.4
To atmosphere	7.7
Total	**43.1**

Table 3. Vapour discharges[27].

	(m^3/adt)
Process vapour discharges	
Recovery boiler	0.7
Lime kiln	0.4
Power boiler	0.4
Process vents	1.5
Process cooling tower	3.8
Effluent cooling	0.9
Total	**7.7**
Combustion vapour discharges	
Recovery boiler	1.4
Lime kiln	0.2
Power boiler	0.5
Total	**2.1**

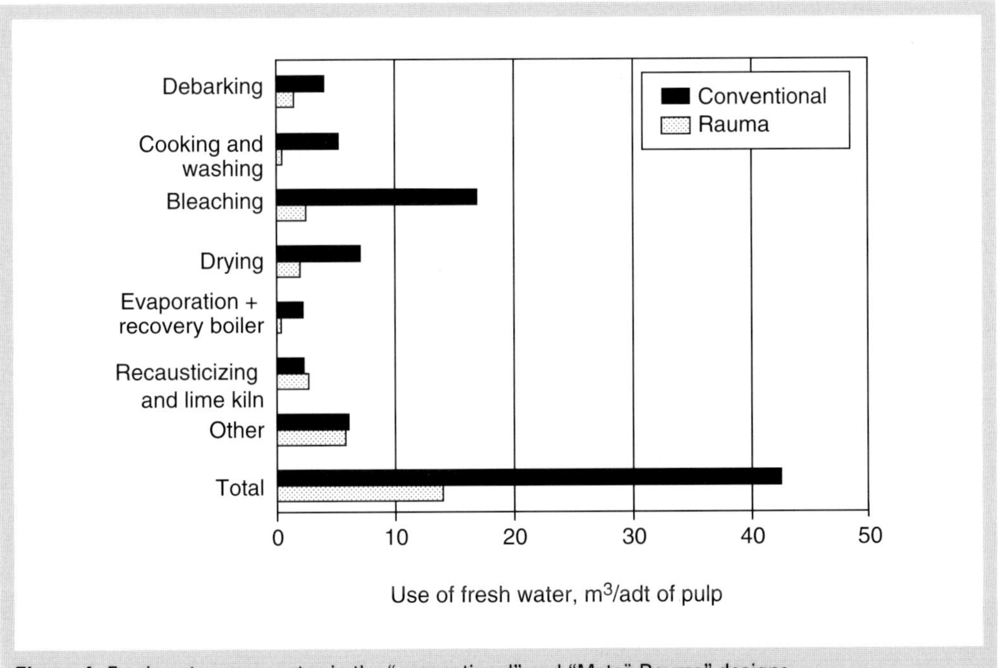

Figure 4. Fresh water usage rates in the "conventional" and "Metsä Rauma" designs.

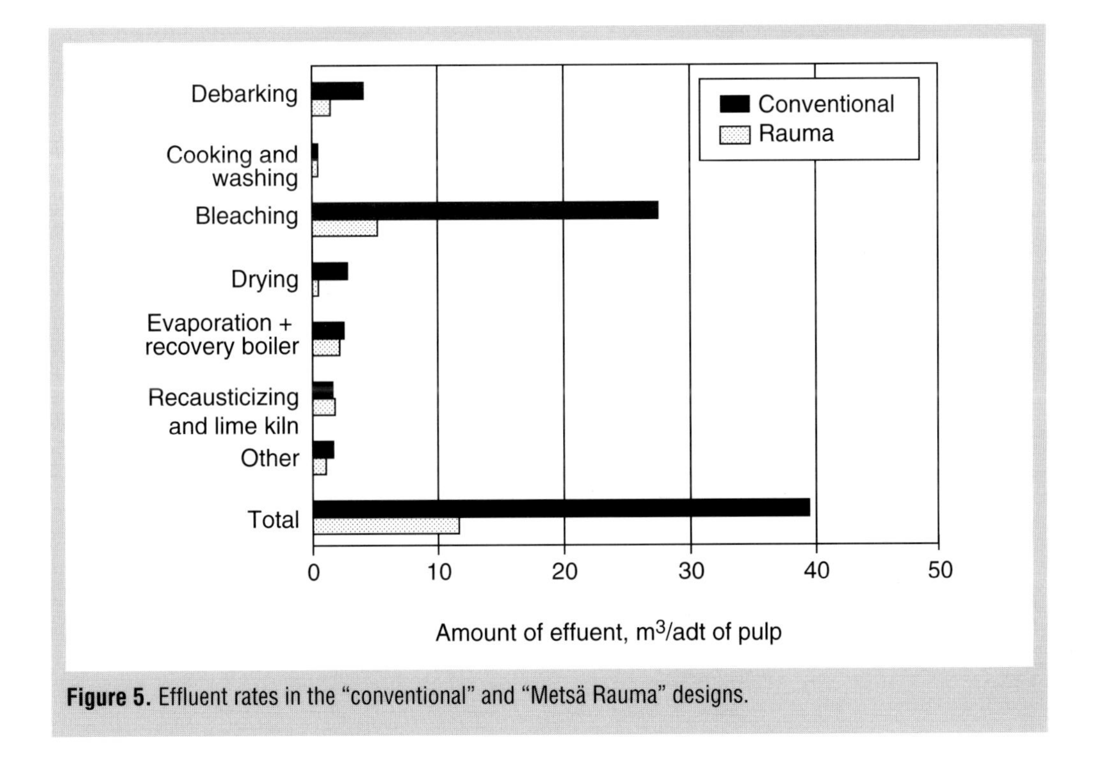

Figure 5. Effluent rates in the "conventional" and "Metsä Rauma" designs.

Often, the first step in bleach plant filtrate recovery is to reuse alkaline filtrate in brownstock washing. Numerous examples are available in the case histories below and in other proposals[26,27]. Alkaline filtrate is more compatible with brownstock washing filtrate than acidic filtrate and is low in chloride ion. A limit exists in the degree to which alkaline filtrate can be recovered due to volume rate limitations in brownstock washing. Another limit exists in the dissolved organic concentration that the system can tolerate without adversely influencing bleaching.

Acidic filtrate from chlorine dioxide, ozone, or chelation stages is more troublesome to introduce into brownstock washing because it first requires neutralization. It also contains dissolved calcium that will be insoluble under the alkaline conditions of brownstock washing and will form deposits. Acidic filtrate can displace liquor in pulp entering the bleach plant to decrease the flow of effluent from the bleach plant. Providing a way to remove dissolved calcium, magnesium, and other cations from the closed loop of recycled acidic bleaching filtrate is essential. One approach is to use some acidic effluent in recausticizing as first practised in the application described above.

Chemical balance

Table 4. An example of sodium and sulfur balances for an "open" mill and a "closed" TCF mill[32].

	kg/adt			
	"Open"		"Closed"	
	Na	S	Na	S
INPUT				
Wood	0	0.2	0	0.2
Bleaching chemicals	0	0.5	14[1]	7[2]
Make-up and scrubber alkali	9	0	0	0
Tall oil plant	0	2.8	0	2.8
Total	**9**	**3.5**	**14**	**10**
OUTPUT				
Pulp	4.5	1.0	1.5	0.5
Rejects, sludges, accidental losses	4.5	1.5	4.5	1.5
Sulfur gases	-	1.0	-	1.0
Purge of recovery boiler ash	-	-	8	5.5[3]
Sulfur purge	-	-	-	2.0
Total	**9.0**	**3.5**	**14**	**10**

Notes: 1. 25 kg NaOH
2. 20 kg H_2SO_4
3. 25 kg Na_2SO_4

Changing the balance of sodium and sulfur input vs. output controls the sulfidity of the pulping chemical recovery cycle. This is also true of “closed” bleach plant effluent recovery systems with some additional complications[28–30]. Table 4 shows an example of inputs and outputs[28]. In open systems, unavoidable input of sulfur is mainly from the use of sulfuric acid in the tall oil plant and a secondary input of magnesium sulfate as a protector in oxygen bleaching. In most mills, additional sulfur makeup would come from the chlorine dioxide generator by–product; sodium sesquisulfate, sodium sulfate, or spent acid. Sulfur may be added as salt cake, i.e., sodium sulfate or other makeup chemicals. The output of the sulfur is in the pulp as entrained liquor, with liquor losses from spills and leakage, and with loss of sulfur-containing gases.

Input of sodium in the open system is mainly through makeup chemicals such as generator byproducts, purchased saltcake, or sodium hydroxide. Mills with scrubbers where sodium hydroxide solution is necessary as a scrubbing medium are exceptions. Losses of sodium are from liquor entrained with pulp or from other causes.

In closed systems, sulfur input may change significantly due to use of sulfuric acid in pulp bleaching to create acidic conditions. In the example given for TCF bleaching in Table 4, 20 kg of sulfuric acid are necessary to create acidic conditions for an ozone stage and chelation. Recovery of bleach plant effluent also brings a significant amount of sodium required to make extraction or peroxide stages alkaline. In this case, it is 25 kg of NaOH/t of pulp. It is assumed that all alkali required for oxygen delignification comes from oxidized white liquor.

Output from the closed cycle system in Table 4 has significant losses of sodium and sulfur from pulp, rejects, etc.[28]. It is surprising that the authors would assume the system to be closed. The input of sodium and sulfur from bleach plant effluent recovery must balance with output. The main point of removal in most closed system proposals is a purge of recovery boiler ash. The principal constituent of this is sodium sulfate. In the example of Table 4, 5.5 kg of sulfur will require purging. This would amount to 25 kg of sodium sulfate dust that is very water soluble. Some locations might discharge this as a solution. In some cases, this will require disposal as a solid waste. In a tighter system than described in Table 4, the purge required might be as high as 14 kg of Na and 9 kg of S. This is approximately 40 kg of recovery boiler dust/t of pulp. Note that the example in Table 4 calls for a separate purge of sulfur from the system. The authors propose achieving this purge by heat treatment of black liquor[28].

Any chlorine dioxide in bleaching stages from which filtrate is recovered will appear in filtrate as chloride ion or chlorate ion. These ions must match with sodium ion in the pulping chemical recovery cycle. They will increase the sodium input required or decrease the sodium necessary for purging sulfur. The balances of sodium, sulfur, and chlorine in the recovery system were troublesome in the first closed cycle mill.

Table 5. The effects of pulping and bleaching options on softwood kraft mill parameters[35].

	ECF			TCF	
Case	**A**	**B**	**C**	**D**	**E**
Cooking					
Kappa number	25	25	15	25	15
Yield, % on wood	47	47	43	47	43
Oxygen delignification					
Kappa number	–	12	8	10	8
Yield, % on pulp	–	94.7	97.5	92.6	97.5
Yield, % on wood	–	44.5	41.9	43.5	41.9
DEoDD bleaching					
Yield, % on pulp	92.6	95.5	97.4		
Yield, % on wood	43.5	42.5	40.8		
ZEoZQP bleaching					
Yield, % on pulp	–	–	–	95.0	96.1
Yield, % on wood	–	–	–	41.3	40.3
Wood, t/adt	2.07	2.12	2.21	2.18	2.23
Recovery bolids					
Dry, t/adt	1.52	1.63	1.76	1.81	1.91
Organic, t/adt	1.06	1.13	1.21	1.24	1.29
Bleach effluent					
Volume, m^3/adt	18.4	6.2	5.9	5.8	5.8
Organic solids, kg/adt	72.7	45.2	26.0	59.6	44.0
White liquor[1]					
EA, kg/adt	414	453	532	515	578
Electrical energy[2] **gross, kWh/adt**	620	650	660	830	835

1 Part must be sulfide-free
2 Gross energy demand at mill site

Other aspects of chemical balance in closed systems are the influence of pulping and bleaching systems on the load of white liquor demanded from recausticizing and the load of organic material sent to the recovery boiler. Table 5 gives an analysis of the effects of changing kappa number from the digester and the effect of ECF vs. TCF bleaching on a partially closed system[31]. As the system is progressively closed, this example shows that organic material sent to the recovery boiler increases. When the kappa number in pulping decreases, more organic material goes to the recovery boiler. In a recovery-limited mill, the increased load per ton of pulp may constrain pulp production. The table also shows the increased white liquor required for this alternative.

Energy balance

In closed systems, dissolved organic matter that would otherwise be discharged will be recovered and directed to the recovery boiler for burning. This will increase the heat load on the recovery boiler and steam generation. Depending on the system, this might be 5%–10%[5,30–32]. In closed system designs, warm process water is cooled by evaporative cooling and recycled to the system. This increases the electrical energy required and also increases the heat loss through evaporation. Energy requirements between open systems and closed systems are very similar in most other respects.

3.3 Nonprocess elements

In closed cycle systems, minor components that unavoidably enter the process with the wood, chemicals, and water may accumulate in the system and cause adverse effects in different parts of the mill. These can include corrosion, formation of scale and deposits, and other operational problems. Proper management of these undesirable nonprocess elements – including minimizing input, monitoring concentration, and facilitating removal – is therefore essential for implementation of mill and bleach plant closure[33–35].

Bleach plant filtrate recovery has two main problems:

- Accumulation of the nonprocess elements that are soluble in alkaline conditions in the unbleached pulp part of the mill. These elements are potassium, chloride, aluminum, and silicon.
- Accumulation of the nonprocess elements that dissolve in the acidic conditions of certain bleaching stages and reprecipitate under alkaline conditions. Elements of particular concern are calcium and manganese.

The following text presents an overview of the behavior of the main nonprocess elements. The discussion does not include some elements that can accumulate in the recovery cycle or lime cycle.

Chlorine and potassium

The main source of potassium is wood[36]. Chlorine enters the system as chloride ion with wood and as a contaminant in chemicals and process water. Recovery of C stage or D stage filtrates is another important source of chloride. Effluents containing high concentrations of chloride ion can be treated separately, such as by evaporation and incineration[33,37].

Chloride and potassium influence the thermal properties of fireside deposits in the kraft recovery boiler. They can potentially accelerate plugging of flue gas passages and increase the rate of corrosion of the superheater tubes[36]. Chloride and potassium are enriched in electrostatic precipitator dust due to their greater volatility. Removal of these ions via ash can occur by discarding the ash or using an evaporative crystallization process described in a later section[38,39]. In a modern kraft mill, the concentration of chloride and potassium in the white liquor should not exceed 0.2 and 0.4 mol/L, respectively[33].

Calcium

Calcium is a nonprocess element in the fiber line. Although wood is the main source of calcium entering the fiber line, calcium can also enter through white liquor when clarity is poor[33–35,40]. Calcium is soluble in acidic conditions, so it will dissolve in acidic C, D, Z, Q, or A stages. Then it precipitates and forms deposits when acidic filtrates are recycled to an alkaline environment. Calcium removal can be accomplished by filtrate purge to recausticizing[14,26], precipitation[33,35], or ion exchange[38,39].

Manganese

Control of manganese is critical when using hydrogen peroxide bleaching because manganese catalyzes the rapid decomposition of peroxide. The maximum desired concentration in kraft pulp is 3 ppm on dry fiber[40]. A metal removal stage is mandatory for efficient P stage bleaching. Manganese removal can be done by acidic washing (ion exchange) and chelation (Q stage)[40–42].

Others

Silicon, aluminum, iron, and barium are other elements of concern in closed cycle bleach plants. Silicon and aluminum enter the system primarily through wood and lime makeup. Scaling in evaporator tubing caused by sodium aluminosilicates can be a major problem[34]. Iron can have adverse effects, especially on oxygen, ozone, and peroxide bleaching, because it catalyzes the decomposition of peroxides[40]. Barium forms unwanted barium sulfate deposits in the bleach plant[40].

3.4 Economics

Many factors require consideration when assessing capital and operating costs for closed vs. open systems. Consider an example of a bleached hardwood kraft pulp mill producing 1500 adt/day for the market. Table 6 gives capital cost estimates by department[43]. For the open case, the capital cost for ECF is identical to the capital cost for TCF. Closing the water loop in a ECF mill requires incremental capital of U.S.$ 71 million. Closing the water loop in a TCF mill requires incremental capital of U.S.$ 55 million. Operating cost estimates for the same case study in Table 7 show ECF open and closed systems as virtually identical at approximately U.S.$ 260/adt. The cost of TCF production is higher by approximately U.S.$ 20 in the open case and by only U.S.$ 11 in a closed case[43]. These estimates are similar to other published estimates. They also conclude that the difference in operating cost of open cycle vs. closed cycle is insignificant, but the operating cost of TCF is higher than the operating cost of ECF[32,44].

Table 6. Capital cost estimates of ECF/TCF mills (1500 adt/d) for market softwood kraft pulp in open/closed configuration (US$ millions)[37].

Mill area	ECF open	ECF closed	TCF open	TCF closed
Site preparation	91	93	92	93
Woodroom	65	66	65	66
Pulping	130	133	130	130
02 delignification	26	28	26	28
Bleaching	46	50	44	48
Pulp Drying	117	121	117	121
Chemicals	29	34	31	30
Evaporators	32	59	32	58
Recovery	79	86	79	83
Recaust	26	28	26	28
Kiln	10	11	10	11
Steam/Power	98	100	98	100
Water/effluent	57	34	57	36
Closed-cycle additions	0	35	0	30
TOTAL	**807**	**878**	**807**	**862**

Table 7. Operating cost estimates of ECF/TCF mills in open/closed configurations (US$/adt)[37].

ITEM	ECF open	ECF closed	TCF open	TCF closed
Fibre	138	138	138	138
Chemicals	47	46	66	57
Energy	5	5	5	5
Manning	38	38	38	38
Maintenance	31	33	31	33
TOTAL	**259**	**260**	**278**	**271**

4 Mill-scale developments

4.1 Champion International Corporation, Canton Mill

As part of corporate strategy to improve environmental performance in their operations, Champion International Corporation has made considerable effort to develop technology to minimize and reuse bleach plant effluents. Their approach for recovering filtrates from kraft pulp bleach plants combines oxygen delignification, 100% chlorine dioxide substitution, oxidative extraction, and a bleach filtrate recycle (BFR) process. This process recovers the filtrates from the first D stage and the EOP stage of an OD(EOP)D bleaching sequence and uses them to wash the pulp in a countercurrent mode through oxygen delignification and into brownstock washing[38,45]. This process avoids many of the difficulties experienced previously at Great Lakes Paper Co. Ltd. in Thunder Bay, Canada, when the Rapson-Reeve closed-cycle concept was implemented[45].

Figure 6 shows the BFR process[45]. Two new sub-processes are essential features of the system:

- A metal removal process (MRP) for removing calcium, magnesium, manganese, and other mineral impurities from the D-stage filtrate.
- A chloride removal process (CRP) for removing chloride and potassium from the recovery boiler precipitator ash.

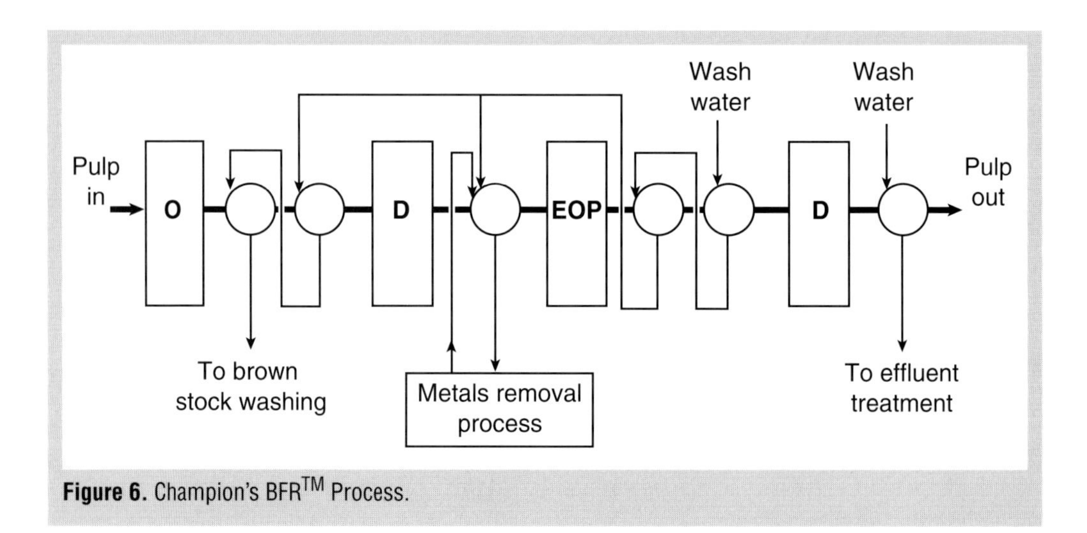

Figure 6. Champion's BFR™ Process.

The MRP has two filtration steps to remove fibers and fine particulates followed by two parallel cation exchange beds, each equipped with an individual regeneration system as Fig. 7 shows[39]. The first stage of filtration uses screens with pore sizes of 50–100 μm. The retained fibers recycle to the D-stage. The second stage of filtration uses three parallel sand filters that

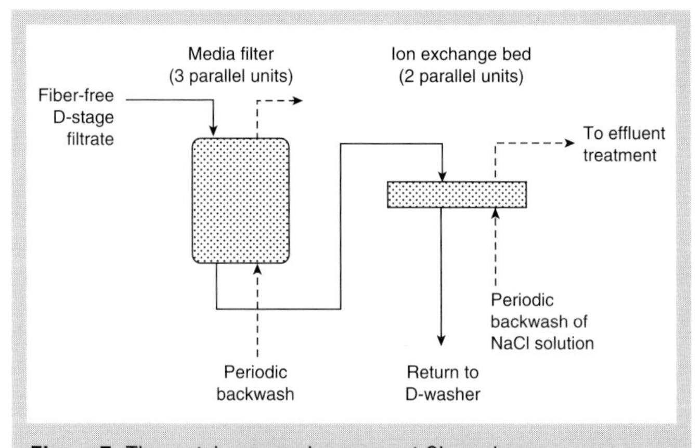

Figure 7. The metals removal process at Champion.

retain materials of more than 1 μm diameter[38,39]. The ion exchange system used in MRP is patented. It differs from conventional ion exchange systems by having a very short cycle time for uptake and regeneration. The resin bed is therefore small. Sodium chloride is the regenerating source.

CRP is an evaporative crystallization process, as Fig. 8 shows[45]. Recovery boiler precipitator ash is first dissolved in hot water to make a 30% dissolved solids solution, which is fed to a crystallizer. Sodium sulfate is the main component of the ash. This crystallizes and is filtered before recycling to the black liquor recovery system. The filtrate that is rich in chloride and potassium is partially recirculated to the crystallizer, and partially discharged to purge these elements from the system[36,38–46]. The process is in use at the Champion Mill in Canton, North Carolina that manufactures 1360 adt/day of bleached board and printing and writing papers. The fiber line has a medium-consistency oxygen delignification stage followed by three stages of post oxygen washing and a D(EOP)D bleach plant[47].

Champion has taken a deliberate, stepwise approach to implementing filtrate recycling. CRP has been a reliable process since its startup in August 1995. Chloride removal efficiency of 95% exceeded the target of 90%. Potassium removal of 47% did not meet the target of 80%[47]. Optimization of the system is continuing. The MRP system has experienced several mechanical problems since its startup in November 1995. Despite these problems and operational challenges, the MRP has proven its ability to remove metals from the D-filtrate[51].

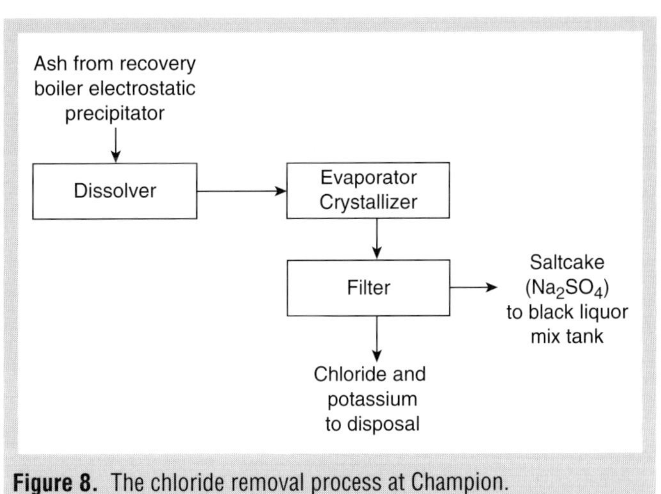

Figure 8. The chloride removal process at Champion.

4.2 Louisiana-Pacific Corporation, Samoa Mill

At the Louisiana-Pacific Corporation pulp mill at Samoa, CA, the in-plant control approach was adopted as an alternative to "end-of-pipe" treatment. Internal process modifications included the use of advanced spill control systems, optimization of the recausticizing area, and improvements to cooking, fiber line, and bleaching areas. These modifications allowed intensive recycling of the bleach plant effluents[48].

The existing CEDED bleach plant was modified to permit use of a TCF sequence, Q(EOP)PPP, to produce 85% brightness softwood pulp. The decision to use a TCF bleaching sequence was based on environmental performance observed during mill trials, increased workplace safety, predictions about the market for TCF pulp, and the greater simplicity of recycling bleaching filtrates without chloride ion[49]. Figure 9 presents a schematic flow sheet of the bleach plant[50].

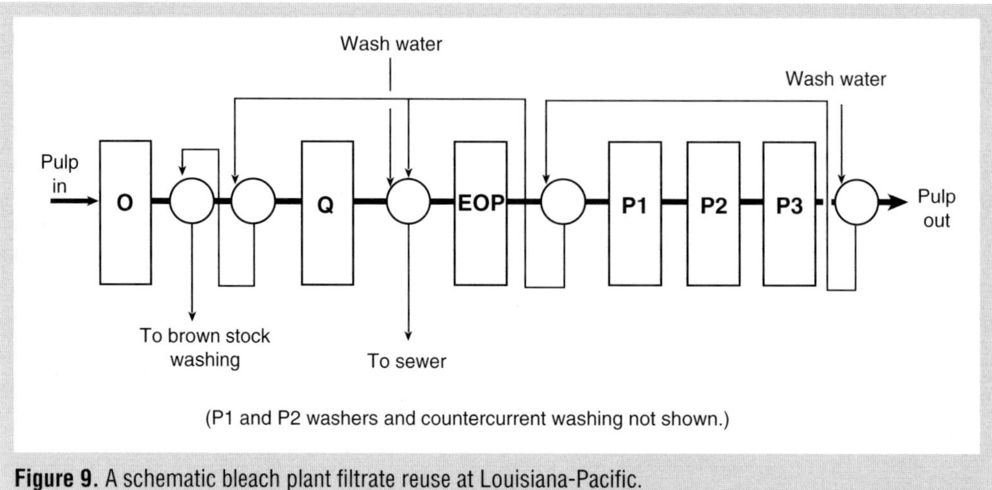

Figure 9. A schematic bleach plant filtrate reuse at Louisiana-Pacific.

Alkaline P-stage filtrates are used countercurrently to wash pulp in the bleach plant and go then to the second post-oxygen washer. Q-stage effluent of 6–7 m^3/adt is not recycled. The major concern that presently prevents recycling the Q-stage filtrate is the potential increase in concentration of metals in the fiber line, especially manganese[51].

Louisiana-Pacific has been progressively improving effluent recycle. As part of the closed cycle project, extensive work has involved monitoring of nonprocess elements (Ca, Mg, Mn, Si, Al, Fe, and Cu), computer simulations, industrial-scale trials, and equipment installation[42,41,52].

A new high-efficiency filter increases removal efficiency of nonprocess elements in green liquor. This allows using some Q-stage filtrate in this area. Mill trials using modified extended cooking have shown a decrease in manganese concentration in the brownstock by 21%, and a 23% decrease in calcium. The company is considering converting the present conventional continuous cooking to Lo-Solids® cooking[42,51].

Improved operations and intensive filtrate recycling have decreased bleaching effluent volume from 47.0 m^3/adt to 6.3 m^3/adt. Mill BOD_5 has decreased from 15 kg/adt to 3 kg/adt. COD has decreased from 54 kg/adt to 9 kg/adt. Using initial energy balances, the recycling of bleaching filtrates produces energy savings of approximately 1.7 MW[48].

4.3 Union Camp Corporation, Franklin Mill

A serious environmental constraint faced by Union Camp Corporation at the Franklin Mill in Virginia was that it could release its treated effluent to the local river only during five months of the year. The river has such low flows and low dissolved oxygen levels during summer that it could not accept the treated effluents from April to October. During this time, the mill retained all effluent in a 2400-acre holding system[49].

Union Camp's strategy to overcome this problem includes recycling the bleach plant filtrates. They selected and installed an OZ(EO)D bleaching sequence, where all of the O and EO-stage filtrates are recycled to brownstock washing, as Fig. 10 shows[53]. The Z-stage filtrate is only partially recycled to avoid calcium scaling. The D-stage filtrate goes to biological treatment[54,55].

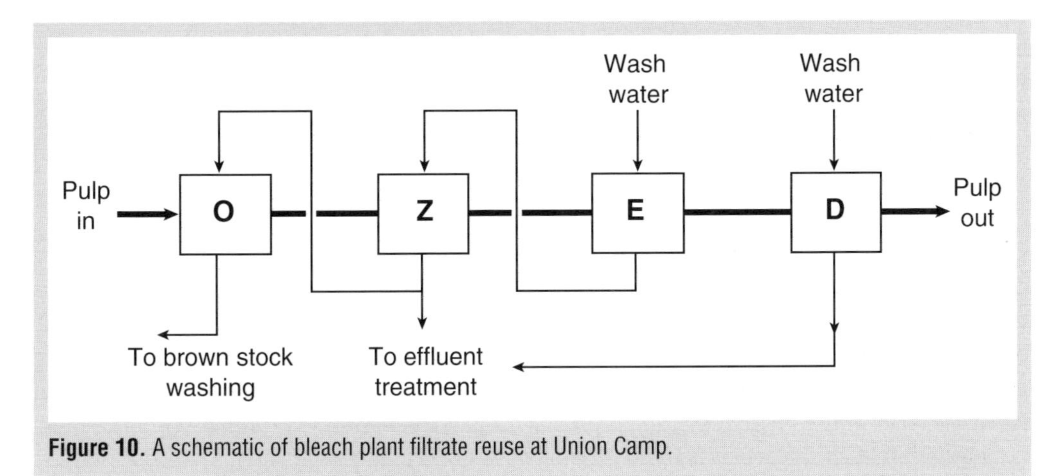

Figure 10. A schematic of bleach plant filtrate reuse at Union Camp.

The total volume of bleach plant effluent discharged is approximately 9.4 m^3/adt. This consists of 7.3 m^3/adt from the D-stage and 2.1 m^3/adt from the Z-stage. Compared with the old CEDED sequence, the following decreases occurred: BOD_5 from 16.3 kg/adt to 4.4 kg/adt, COD from 64.7 kg/adt to 11 kg/adt, color from 185 kg/adt to 3.1 kg/adt, and AOX from 6.5 kg/adt to 0.05 kg/adt. The pulp quality is similar to that from the conventional bleaching sequence[49,55].

4.4 Modo, Husum Mill

Modo's experience in closed-cycle bleaching started in its Domsjö sulfite mill in Sweden. The market demand for TCF pulp and environmental pressure were the driving forces to change the former E(CD)HD bleaching sequence into (EO)P. A unique two-stage sodium-based sulfite pulping process producing pulp of kappa number 9 facilitated recycling of bleaching filtrates that had operated successfully since 1991[56–58]. A new system to remove extractives and nonprocess elements also contributed to the success of bleach plant closure. In the new system, polyethylene oxide (PEO) in solution is added to the filtrate. It reacts with hydrophobic extractives to form a sludge that a flotation unit removes[59].

At the Husum beached kraft mill the primary emphasis was to reduce the effluent discharge by internal control measures as an alternative to end-of-pipe biological treatment. The mill has two production lines: one for hardwood and one for softwood. Bleaching filtrate recovery has been implemented in the hardwood line. The bleach plant has filter washers and a sequence reported to be OQPZP or OQPZD. A small amount of chlorine dioxide(3–5 kg/adt) improves product brightness and lowers cost[57]. During "regular operation", the filtrate from the first two stages of bleaching – assumed to be QP – is recycled to brownstock washing. The filtrate from the last two stages is discharged. About 25% of time, the bleaching filtrates are completely recycled countercurrently. The monthly average bleach plant effluent volume is approximately 5 m^3/adt of pulp[57].

4.5 Munksjö Ab, Aspa Bruk Mill

Munksjö Ab, Aspa Bruk mill in Sweden produces about 140 000 t/year of softwood kraft market pulp. Located on the shores of a lake that is a source of drinking water for approximately 200 000 people, the mill had to decrease its effluent discharge. The strategy for reducing the emissions focused primarily on in-plant control measures as an alternative to biological effluent treatment[60].

Aspa Bruk was the world's first TCF pulp producer in 1990. The bleaching sequence OQPP produced 71% ISO brightness pulp. This bleach plant configuration allowed recycling the P-stage filtrate to the brownstock washing system. The bleaching sequence was then modified to OQPPPP. Trials were performed where Q-stage filtrate was partially used in the recovery area for smelt dissolving and lime mud washing. P-stage filtrate was used for brownstock washing and for Q-stage washing. With further modification to incorporate peracetic acid (PAA), the bleaching sequence OQ(PAA)PPP could produce 90% ISO pulp brightness. This resulted in a new alternative for recycling of the bleaching filtrates, as Fig. 11 shows. The design has a straight countercurrent flow of the filtrates with a bypass of the Q-stage filtrate to smelt dissolving (the white liquor system)[60,61].

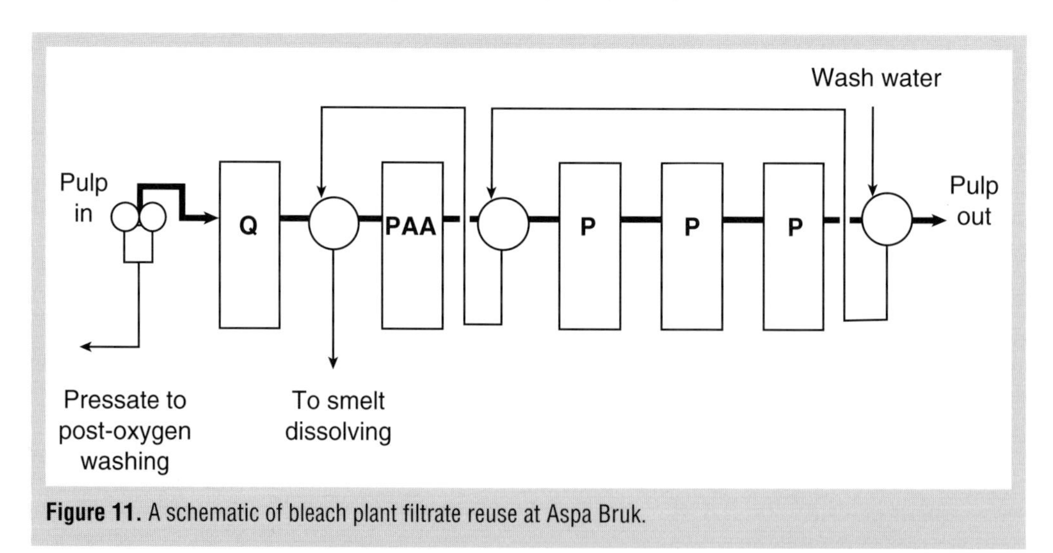

Figure 11. A schematic of bleach plant filtrate reuse at Aspa Bruk.

Other steps have allowed further bleach plant closure. The addition of carbon dioxide to the pulp before the brownstock washing system improved the system runnability and gave better conditions for system closure. A new pressurized peroxide stage should improve the efficiency of the bleach plant. A black liquor impregnation system installed in the digester improved delignification and metal management. The total mill COD discharge was about 30 kg/adt, but this should decrease to 20 kg/adt[60,61].

4.6 Södra Cell, Värö Mill

Södra Cell operates three kraft pulp mills in Sweden and is committed to production of TCF pulp as a corporate strategy in response to market and environmental issues. One mill has an aerated lagoon for effluent treatment, and two other mills have only primary clarifiers. All three mills are moving in steps toward closure.

The Värö mill with an annual capacity of 350 000 adt of fully bleached softwood kraft pulp has had several investments over the years to increase capacity and decrease effluent discharge. Extended cooking, medium-consistency oxygen delignification followed by a pressure diffuser and a wash press, and TCF bleaching were successively installed in this mill. Bleaching now uses a four-stage sequence Q(EOP)Q(PO). Laboratory bleaching tests showed the feasibility of using clean black liquor evaporator condensate for washing in the bleach plant. Using these results, clean condensate partially replaced hot fresh water[62].

Mill trials to recycle the bleach filtrates countercurrently indicated that recycling of filtrate from the first three bleaching stages and reusing them at the post-oxygen washers is possible. No problems with scaling or reduced bleaching efficiency occurred. Reuse of the last PO stage filtrate that carries a high proportion of the COD resulted in a significant decrease in pulp brightness. Different options are under consideration to solve this problem, including biological treatment of this stream. The mill COD discharge in 1996 was about 40 kg/adt[63].

4.7 SCA, Östrand Mill

SCA's Östrand kraft mill (Sweden) is a partially integrated pulp mill. To meet market and environmental demands, the mill has undergone continuous improvement over the last few years. In 1992, they installed isothermal cooking (ITC). In 1994, they began medium-consistency oxygen delignification (replacing a high-consistency system), and in 1995 they started a new TCF bleach plant with effluent recycle. The company is moving toward system closure to achieve a low COD discharge by internal recycling as an alternative to installing biological effluent treatment[64].

Following two-stage oxygen delignification, the bleaching sequence is Q(OP)(ZQ)(PO). Bleach plant filtrates are partially recycled to post-oxygen washing. In attempting to close the mill, severe difficulties with calcium oxalate deposits were encountered, especially around the first Q-stage. In a recent report, the discharged filtrate was about 7 m^3/adt[64], and COD discharge was 15–20 kg/adt. Filtrate flow in the bleach plant is stable. The bleach plant has two spill tanks and a storage tank for filtrate recovery. The pulp produced is clean, and the brightness range is 70%–92% ISO[64].

4.8 Metsä-Rauma, Rauma Mill

Metsä-Rauma has the world's first greenfield TCF pulp mill in Rauma, Finland. It started in 1996, and is designed for very low effluent volume. The drive for system closure was prompted by limited water supply and economical effluent treatment. The choice to produce TCF pulp was based on economics and the desire for complete mill closure[22].

The bleaching sequence is OOZPZP[65]. All alkaline filtrates are recycled to brownstock washing and the recovery system. The Z-stage filtrates go to the biological treatment plant. The volume of acidic effluent is 5 m^3/adt. The total fresh water consumption is only 15 m^3/adt, and the total mill effluent produced is 10–11 m^3/adt. The company projects that total water consumption can decrease further to approximately 5 m^3/adt. Concerns exist that total closure will cause accumulation of nonprocess elements. The mill strategy is to take small steps toward closure and evaluate thoroughly the effects of each change before progressing further[22,65].

5 Relevant technologies under development

Several technologies for treating bleach plant filtrates are under consideration to facilitate the recycling and recovery of the effluents. Chemical precipitation, ion exchange, membrane separation processes, and evaporation followed by combustion are examples of technologies that can be applied individually or combined in closed cycle systems. A brief description of these technologies in treating bleach plant filtrates – and some mill experiences – follow.

Evaporation and combustion

Although evaporation is an energy-consuming process, it is a proven technology with which the pulp and paper industry is quite familiar, especially for black liquor concentration. Figure 12 shows a project for "closing up bleach plants" by evaporation followed by combustion. Mill-scale testing of the project started in 1994[66]. The filtrates are first filtered to remove suspended solids. Then they pass to vapor recompression evaporators. The concentrates are thermally oxidized in a unit designed specially for this purpose. The condensates from the evaporators are subsequently treated in stripping columns, and the resulting clean water is then recycled to the mill[66–68]. Flocculation and electrodialysis have also been considered[69,70].

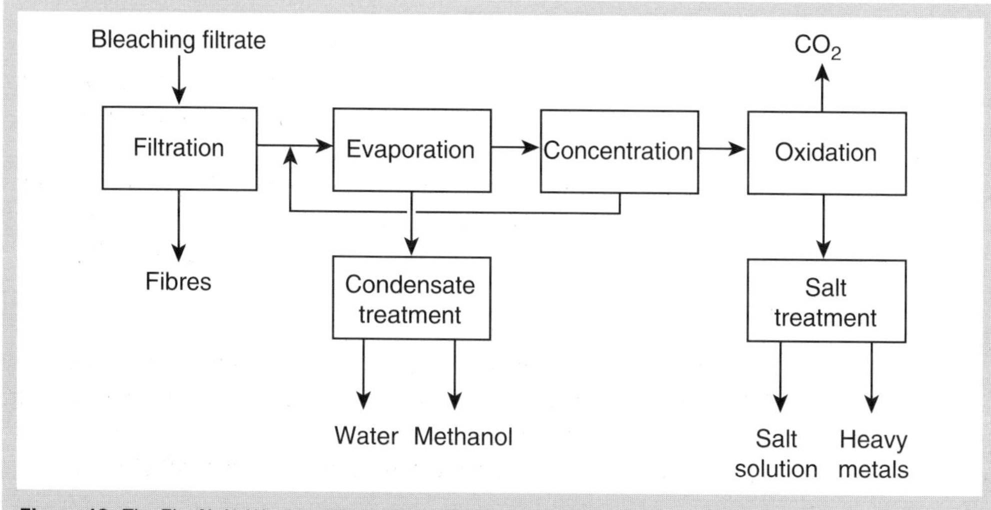

Figure 12. The Eka Nobel/Jaakko Pöyry approach for elimination of ECF bleach plant filtrate.

Another project to eliminate kraft bleach plant effluents uses an approach similar to that used for elimination of effluent from a Canadian BCTMP mill[71]. Another concept for a closed cycle mill also considers evaporation of the bleach plant filtrates in special units[72].

Membrane separation

The use of ultrafiltration membranes for treating bleach plant effluents has been examined to remove color, COD, BOD, and toxicity from alkaline extraction stage filtrates[73–77]. Fouling and flux limitations are two important problems that this technology faces. New membranes and membrane configurations are under development. Research studies have used nanofiltration and electrodialysis to treat bleaching filtrates[78–82].

Ion exchange

Ion exchange removes contaminant cations and anions from water. In the pulp and paper industry, it is routinely used to treat boiler feed water. The MRP described in section 4.1 is a particularly compact ion exchange system[38,39] which can remove undesirable metal cations from acidic bleach plant filtrates[39].

Chemical precipitation

Coagulation and flocculation are traditional means of water treatment often used in the pulp and paper industry. Use of a polyethylene oxide (PEO) polymer in solution can remove extractives and nonprocess elements from bleach plant filtrate. The sludge separated from the filtrate is dewatered and burned[59]. Other types of coagulants, such as biopolymer adsorbents, are under test to remove metal ions in closed cycle bleach plants[83].

Sedimentation, flotation, and filtration

The removal of suspended solids is normally necessary as a pretreatment step before such processes as evaporation, ion exchange, or nanofiltration. This can occur by different methods using sedimentation, flotation, and filtration principles. A new combined flotation and filtration unit is available[84]. The current practice of suspended solids removal from bleach plant effluents includes screening followed by sand filtration[38,39], drum filtration[67], or flotation[59].

6 Minimum impact mill concepts

Several evaluations of closed cycle systems (including elimination of bleach plant effluent) that have occurred in the broader context of sustainable development have led to concepts for the minimum impact mill. Some examples follow:

"The minimum impact mill is one which:

- Maximizes pulp yield and produces high quality products that are easily recyclable and/or safely combustible

- Maximizes the energy potential of the biomass
- Minimizes water consumption
- Minimizes wastes (gaseous, liquid and solid) and disposes of them optimally
- Optimizes capital investment
- Creates sustainable value to shareholders, customers, employees, and local, regional, and national communities[9]."

Another example of articulation of local and global characteristics of a minimum impact mill follows:

"The production of pulp and paper has the opportunity to be a truly ecologically sustainable industrial process. Such an ideal process may be defined as one that:

- Uses naturally renewable raw materials in such a manner that local demand stays in balance with sustainable growth
- Does not require thermal or electrical energy from an outside source, but satisfies its demand by using its own wastes
- Recovers and regenerates any required chemicals from its own waste streams
- Uses makeup chemicals that are abundantly available in nature and which are returned to nature in their original form without causing excessive point loading
- Has waste streams that cause no adverse changes to the local, regional, or global environment[9]."

Other considerations in the characteristics and definitions of a minimum impact mill include the concepts that mills should strive to:

- Minimize resource consumption (water, chemicals, wood, energy)
- Minimize pollution (water, air, noise, solid waste)
- Create a safe, productive, and attractive working environment
- Maximize product quality (brightness, strength, stability)
- Minimize aesthetic disturbances such as plumes of water vapor and noxious odors
- Maximize profits, employment, taxes paid, and returns to shareholders and communities[9].

Another approach to defining the minimum impact mill suggests that a holistic perspective balancing environmental performance and social and economic needs provides the framework for a proper definition of a minimum impact mill[85]. The key elements of this holistic approach are:

- Meeting customer needs for product performance and purity
- Sustainable forestry through ecosystem management
- Clean manufacturing including waste reduction
- Sustainable economics and solid shareholder returns
- Public involvement and open dialogue with regulatory bodies.

Table 8 shows that certain key parameters require monitoring to demonstrate continuous improvement toward a minimum impact mill.

Table 8. Parameters important to demonstrating continuous improvement towards a minimum impact mill[17].

Water	Air	Solid waste	Other
Water usage	Particulates	Solid waste generated	Accidental releases
Bleach plant effluent volume	Total reduced sulfur (TRS)	Solid waste disposition	Non-compliant Events
			SARA 313 releases
Final effluent volume	Methanol Chloroform	Landfill Recycled	Energy use/energy Export
BOD			
COD	Chlorine	Energy	Transportation effects
Suspended solids	Chlorine dioxide	Hazardous waste	
AOX	CO/CO_2	Elimination	Aesthetics
Dioxins and furans	NO_x		Site appearance
Color	SO_2		Odor
Biological tests	VOC		Noise
Nutrients (N and P)	Dioxins and furans		
Heavy metals	Opacity		
Safety	Hazardous air		
	Pollutants (HAPs)		

Another definition of a minimum impact mill provides the following as goals: "to minimize natural resource consumption (wood, water, energy) and minimize the quantity and maximize the quality of releases to air, water and land"[86].

An important concern in considering closed cycle systems for elimination of bleach plant effluent is the environmental impact of process changes. Under what circumstances is it environmentally important or not important to eliminate bleach plant effluent? Conversion of mills to ECF and TCF bleaching virtually eliminates the production of persistent and bioaccumulative chlorinated compounds. Reproductive responses are observed in fish exposed to treated effluent from some bleached pulp mills and also from some unbleached pulp mills. This suggests that efforts to eliminate environmental responses should first focus on diminishing pollutants discharged from the unbleached pulping processes and chemical recovery processes. This applies particularly to eliminating the discharge of spent pulping liquor, evaporator condensates, and other liquors containing wood extractives, and minimizing carryover of spent pulping liquor to the bleach plant[87]. Elimination of spills from pulping and chemical recovery is also essential to environmental improvement. One must recognize that a significant environmental benefit exists in biological treatment of effluents from bleached kraft pulp mills. Compelling reasons – other than normal biological responses – may exist for eliminating bleaching effluent: scarce fresh water, a sensitive ecosystem in the receiving water, and low flow of receiving water.

When considering implementing bleach plant effluent recovery systems in existing mills, some potentially negative effects on mill operation are possible. For existing mills, significant capital costs associated with major rebuilding will be necessary. Increased operating cost is likely. Mill capacity may decrease due to bottlenecks in major equipment such as the recovery boiler. The effects of elimination of bleach plant effluent on solid waste emissions and air emissions require consideration in any evaluation.

Significant risk exists for negative effects on pulp quality in closed systems and for higher operating or capital costs than anticipated. For existing mills, higher costs seem inevitable. Will the higher costs justify the environmental benefits of system closure? In considering closed systems, we must look for processes that are low-cost (low capital cost, low chemical cost, high capacity, and high yield) and produce high-quality pulp (high strength, clean, bright, stable, and uniform). The evolution of closed cycle processes must be linked to environmental improvement[88].

References

1. Reeve, D. W., Pulp Paper Mag. Can. 85(2):T24(1984).

2. Rapson, W. H., Pulp Paper Can. 68(12):T635(1967).

3. Rapson, W. H., in The Bleaching of pulp (R.P Singh, Ed.), 3rd edn., TAPPI PRESS, Atlanta, 1979, Chapt. 16, p. 413.

4. Rapson, W. H. and Reeve, D. W., Tappi 56(9):112(1973).

5. Galloway, L., et al., Technical Report No. 7, National Pulp Mills Research Program, Dickson Australia, 1994, pp. 57,129.

6. Patrick, K., et al., Pulp & Paper 68(3):S1(1994).

7. Towers, M. T. and Wearing, J. T., Technology and Environmental Enhancement in Pulp and Paper Mills,Technology Development and Programs Resource Processing Industries Branch, Industry Canada Workshop Proceedings #3, 1994, p. 61.

8. TAPPI 1996 Minimum Effluent Mills Symposium Proceedings, TAPPI PRESS, Atlanta, (General Ref. only).

9. Axegård, P., et al., TAPPI 1997 Environmental Conference Proceedings, TAPPI PRESS, Atlanta, p. 529.

10. Reeve, D. W. Rapson, W. H., et al., Pulp Paper Mag. Can. 71(13):T274(1970); 74(1):T19(1974); 75(8):T293(1974).

11. Rapson, W. H., Reeve, D.W., et al., Pulp Paper Can. 75(10):T351(1974); 78(6):T137(1977); 77(8):T136(1976); 78(3):T50(1977); 81(2):T43(1980); 81(6):T142(1980); 82(9):T315(1981); and 82(12):T426(1981).

12. Isbister, J. A., Reeve, D. W., and Rapson, W.H., PIMA Magazine 60(80):23(1978).

13. Isbister J. A., DPAT Report No. EPS 4-WP-80-4, Environment Canada, Ottawa, 1980.

14. Pattyson, G., et al., Pulp Paper Can. 82(6):T212(1981).

15. Reeve, D. W., et al., Pulp Paper Can. 84(1):T25(983) and 84(2):T46(1983).

16. Donovan, D. A., TAPPI 1994 Annual Meeting, TAPPI PRESS, Atlanta, p. 159.

17. Mannisto, H., Mannisto, E., and Krogerus, M., TAPPI 1996 Minimum Effluent Mills Symposium, TAPPI PRESS, Atlanta, p. 9.

18. Reeve, D. W., TAPPI 1995 International Environmental Conference Proceedings, TAPPI PRESS, Atlanta, p. 483.

19. *Solomon, K., et al., International Pulp Bleaching Conference, Tecnical Section, CPPA, Montreal, 1994, p. 145.*

20. *Wohlgemuth, G., Mannisto, E., and Mannisto, H., TAPPI 1996 Minimum Effluent Mills Symposium, TAPPI PRESS, Atlanta, p. 141.*

21. *Coméer, I. and Kinell, P., TAPPI 1996 Minimum Effluent Mills Symposium, TAPPI PRESS, Atlanta, p. 159.*

22. *Reilama, I., Envirotech Symposium 96, Finland Trade Centre Consulate General of Finland, Toronto, 1996, Session 3*

23. *Gleadow, P. L. and Hastings, C., Pacific Paper Expo, Vancouver, 1991, p. 130.*

24. *Turner, P. A., et al., in Pulp Bleaching – Principles and Practice (C.W Dence, D.W. Reeve, Ed.), TAPPI PRESS, Atlanta, 1996 Chapt. 3, p. 569.*

25. *Histed, J., McCubbin, N., and Gleadow, P. L., in Pulp Bleaching – Principles and Practice (C.W Dence, D.W. Reeve, Ed.), TAPPI PRESS, Atlanta, 1996, Chapt. 6, p. 647.*

26. *Teder, A., et al., Tappi J. 73(2):113(1990).*

27. *Bryant, P. S., Basta, J., and Johansson, N. G., International Emerging Technologies Conference, Miller Freeman Inc., San Francisco 1997, Session 14–3.*

28 *Warnqvist, B., TAPPI 1996 Minimum Effluent Mills Symposium, TAPPI PRESS, Atlanta, p. 33.*

29. *Oskarsson, I. and Näslund, K., TAPPI 1996 Minimum Effluent Mills Symposium, TAPPI PRESS, Atlanta, p.27.*

30. *Gleadow, P., et al., CPPA Annual Meeting, Tech. sect. CPPA, Montreal, 1996, p. A359.*

31. *Gullichsen, J., International Chemical Recovery Conference, unpublished Keynote speech, Toronto, 1995.*

32. *Mannisto, H., Mannisto, E., and Winter, P., International Environmental Conference, TAPPI PRESS, Atlanta 1994, p. 475.*

33. *Ulmgren, P., TAPPI 1996 Minimum Effluent Mills Symposium, TAPPI PRESS, Atlanta, p. 17.*

34. *Magnusson, H., Mörk, K., and Warnqvist, B., TAPPI 1979 Pulping Conference Proceedings, TAPPI PRESS, Atlanta, p. 77*

35. *Lindberg, H., Engdahl, H., and Puumalinen, R., International Pulp Bleaching Conference, Tech. sect. CPPA, Montreal, 1994, p. 293.*

36. *Shenassa, R., et al., International Chemical Recovery Conference, Tech. sect. CPPA, Montreal, 1995, p. B177.*

37. *Holman, K. L., Golike, G. P., and Carlson, K. R., International Pulp Bleaching Conference, Tech. sect. CPPA, Montreal, 1994, p. 101. (Poster session)*

38. Caron, J. R. and Williams, L. D., TAPPI 1996 Minimum Effluent Mills Symposium, TAPPI PRESS, Atlanta, p. 181.

39. Brown, C. J., Sheedy, M., Paleologou, M., et al., International Emerging Technologies Conference, Miller Freeman Inc., S. Francis, 1997, p. Session 14–2

40. Bryant, P. S., TAPPI 1996 Minimum Effluent Mills Symposium, TAPPI PRESS, Atlanta, p. 95.

41. Bryant, P. S., Robarge, K., and Edwards, L. L., Tappi J. 76(10):148(1993).

42. Jaegel, A., et al., International Emerging Technologies Conference, Miller Freeman Inc., San Francisco, 1997, p. Session 6–1.

43. Parker, G., et al., ABTCP (1974) 27 Congres Anual de Celulose e Papel, Sao Paolo, Brazil, p.167–179.

44. Andersson, O., Jaakko Pöyry Client Magazine, 1:407(1994).

45. Maples, G., Ambady, R., Caron, J. R., et al., Tappi J. 77(11):71(1994).

46. Earl, P. F., Dick, P. D., and Patel, J. –C., 1995 CPPA Pacific Coast Western Branches Technical Conference, CPPA, Montreal.

47. Stratton, S. C. and Ferguson, M., TAPPI 1997 Environmental Conference Proceedings, TAPPI PRESS, Atlanta, p. 423.

48. Jaegel, A. F. and Girard, K. A., International Nonchlorine Bleaching Conference, Miller Freeman Inc. San Francisco, 1995, p. Session 12–3.

49. Miller, S., TAPPI 1997 Bleach Plant Short Course, TAPPI PRESS, Atlanta, Session 2–2.

50. Jaegel, A. F., Jett, S. W., and Spengel, D. B., International Nonchlorine Bleaching Conference, Miller Freeman Inc., San Francisco, 1996, p. Session 11–2.

51. Martin, F., et al., TAPPI 1996 Minimum Effluent Mills Symposium, TAPPI PRESS, Atlanta, p. 167.

52. Brooks, T. R., et al., International Pulp Bleaching Conference, Tech. sect. CPPA Montreal 1994, p. 13.

53. Nutt, W. E., et al., Tappi J. 76(3):115(1993).

54. Joseph, J. C. and White, D. E., TAPPI 1996 Minimum Effluent Mills Symposium, TAPPI PRESS, Atlanta, p. 101.

55. Young, J., Pulp & Paper, 68(9):69(1994).

56. Ahlenius, L., et al., International Pulp Bleaching Conference, Tech. sect. CPPA, Montreal, 1994, p. 195.

57. Ahlenius, L., et al., TAPPI 1996 Minimum Effluent Mills Symposium, TAPPI PRESS, Atlanta, p. 177.

58. Oshinowo, L., MoDo Domsjö bleached sulfite pulp mill, University of Toronto, Sweden – Finland Tour Report, Toronto,1995, p. 12.

59. Rampotas, C., Terelius, H., and Jansson, K., TAPPI 1996 Minimum Effluent Mills Symposium, TAPPI PRESS, Atlanta, p. 317.

60. Fastén, H., Nonchorine Bleaching Conference, Miller Freeman Inc. San Francisco, 1993, p. Session 111–18.

61. Fastén, H., Nonchorine Bleaching Conference, publsher, Miller Freeman Inc. San Francisco, 1996, p. Session 5–1.

62. Moldenius, S., International Emerging Technologies Conference, Miller Freeman Inc., San Francisco, 1997, p. Session 16–3.

63. Larsson, P., Malmström, J., and Igerud, L., 5th International Conference, SPCI, Stockholm, 1996, p. 848.

64. Annergren, G. E., Boman, M. G., and Sandström, P. E., 5th International Conference, SPCI, Stockholm, 1996, p. 832.

65. Anon., Pulp Paper Can. 97(6):16(1996).

66. Myréen, B. and Johansson, H., International Nonchlorine Bleaching Conference, Miller Freeman Inc., San Francisco, 1996, p. Session 12–3.

67. Koistinen, P. R., TAPPI 1996 Minimum Effluent Mills Symposium, TAPPI PRESS, Atlanta, p. 253.

68. Myrén, B., Jaakko Pöyry Client Magazine volume 1,(1995).

69. Johansson, N. G., Clark, F. M., and Fletcher, D. E., International Nonchlorine Bleaching Conference, Miller Freeman Inc., San Francisco, 1995, p. Session 13– 1.

70. Johansson, N. G., Clark, F. M., and Fletcher, D. E., Pulp & Paper 69(6) :71(1995).

71. Evans, T., D., et al., Pulp Paper Can. 96(3):T92(1995).

72. Nykanen, T. and Ryham, R., TAPPI 1994 Minimum Impact Mill Symposium, TAPPI PRESS, Atlanta, p. 111.

73. Muratore, E., et al., Pulp Paper Can. 84(6):T140(1983).

74. Jönsson, A. -S. and Wimmerstedt, R., Desalination 53(9):181(1985).

75. Jönsson, A. -S., Nordic Pulp Paper Res. J.2(1):23(1987).

76. Zaidi, A., Buisson, H., and Sourirajan, S., TAPPI 1991 Environmental Conference Proceedings, TAPPI PRESS, Atlanta, p. 453.

77. Ekengren, Ö., Bjurhem, J.-E., and Filipsson, S., TAPPI 1993 Environmental Conference Proceedings, TAPPI PRESS, Atlanta, p. 403.

78. de Pinho, M. N., et al., TAPPI J. 79(12):117(1996).

79. de Pinho, M. N., et al., TAPPI 1995 Environmental Conference Proceedings, TAPPI PRESS, Atlanta, p. 883.

80. Paleologou, et al., Pulp Paper Can. 95(10):T386(1994).

81. Paleologou, M., Berry, R. M., and Fleming, B. I., J. Pulp Paper Sci.20(2):J39(1994).

82. *Pfromm, P. H., TAPPI 1996 Minimum Effluent Mills Symposium, TAPPI PRESS, Atlanta, p. 291.*

83. *Rorrer, G. L. and Hsien, T.-Y., TAPPI 1996 Minimum Effluent Mills Symposium, TAPPI PRESS, Atlanta, p. 261.*

84. *Guss, D. B., TAPPI 1996 Minimum Effluent Mills Symposium, TAPPI PRESS, Atlanta, p.341.*

85. *Erickson, D., The Changing Chlorine Marketplace: Business, Science and Regulations, Chemical Week conference, New York, 1991 .*

86. *Blum, L., 1997 International Emerging Technologies Conference, Miller Freeman Inc., San Francisco, 1997, p. Session 7–2.*

87. *Kinell, P., Ström, K. -E., and Swan, B., 5th International Conference on New Available Techniques, SPCI, Stockholm, 1996, page 821.*

88. *Reeve, D. W., Second International Conference on Environmental Fate and Effects of Bleached Pulp Mill Effluents, Unpublished Speech, 1994.*

Carl-Anders Lindholm

Appendix

1 Kraft cooking liquors

1.1 White and green liquors

The total, active, and effective alkali of white and green liquor is traditionally determined by titration with indicators such as described in SCAN-N 2:88. In this method, barium chloride is added to a white or green liquor sample to precipitate the carbonate. Effective alkali is determined by titration with acid to the thymolphthalein end point (pH 9.5). Formaldehyde is then added to convert the hydrogen sulfide ions to a strong base and active alkali is determined by continuing the titration to the fenolphthalein end point (pH 8). Finally the titration is continued to the bromophenol blue end point (pH 4) to give the total alkali.

The total, active, and effective alkali can also be determined by potentiometric titration according to SCAN-N 30:85. In this method, a sample of the liquor is titrated with hydrochloric acid of known concentration. The pH value of the reaction mixture and the volume of the hydrochloric acid are recorded continuously. From the recorded data the consumption of acid at the inflection points is determined. The effective, active, and total alkali of the sample are calculated from the amounts of acid needed to reach the three inflection points. The results of this method do not differ significantly from those obtained by the traditional titration method.

The titration methods can be used for normal white and green liquors, but they are not recommended for oxidized white liquors containing significant amounts of polysulfides, alkaline pulping liquors containing significant amounts of sulfite, or black liquor.

SCAN standards are also available for determination of several specific compounds in white and green liquors:

- sulfide, sulfite, and thiosulfate (SCAN-N 3:63)
- chloride (SCAN-N 4:78)
- total sulfur (SCAN-N 5:83)
- sulfate (SCAN-N 6:85)
- sodium and potassium (SCAN-N 29:84)
- sulfide ion (SCAN-N 31:94)
- carbonate (SCAN-N 32:88).

The SCAN standard SCAN-N 34:96 ("Hydrogen sulfide ion concentration") is an extension of SCAN-N 31:94 making it possible to determine the hydrogen sulfide ion concentration in oxidized white liquors.

CPPA Standard J.12 describes methods for analysis of kraft green and white liquors for the following constituents: sodium sulfide, polysulfide, thiosulfate and sulfite, thiosulfate, active alkali, effective alkali, total titratable alkali, sodium hydroxide, sodium carbonate, sodium chloride, total sodium, sulfate, total sulfur and total calcium.

Sodium polysulfide in kraft white liquor can be determined according to TAPPI T 694 om-90. The method is based upon potentiometric determination of sodium sulfide before and after reduction of the polysulfide with sodium amalgam.

TAPPI T 624 cm-85 ("Analysis of soda and sulfate white and green liquors") can be used for accurate analysis of all main components of soda or sulfate white or green liquors and it can also be used as a reference method for establishing quicker or more convenient tests suited for routine control.

TAPPI T 699 om-87 ("Analysis of pulping liquors by suppressed ion chromatography") provides procedures for determination of sulfide, sulfite, sulfate, thiosulfate, chloride, and carbonate in white, green, and black liquors.

Suspended solids in kraft green and white liquor can be determined by TAPPI T 692 om-93, which is based upon filtration through glass fiber discs, with hot water wash to remove the dissolved solids.

1.2 Black liquor

Different methods have been proposed for determination of the dry matter content of black liquor. In the method described in SCAN-N 22:96, a black liquor sample is dried on a glass fiber filter, placed on an aluminum dish, in an oven (or an IR drier) at 105°C for at least 30 min. This standard also provides a procedure for the determination of the fiber content (suspended material) of black liquor. In TAPPI T 650 om-89, black liquor specimens are dried at 105°C for a minimum of 6 h with inert surface extender (sand) and a controlled flow of air to increase the drying rate and eliminate moisture entrapment.

The residual alkali (the hydroxide ion content) in black liquor can be determined according to SCAN-N 33:94. The black liquor is titrated potentiometrically with hydrochloric acid to the first inflection point, which is situated between pH 11.0 and 11.5. Before the titration, sodium carbonate is added to the sample to give a better inflection point and to buffer the titration solution. If the initial pH of the black liquor sample is above 11.0 but below 12.5, a known volume of sodium hydroxide must be added to the sample before titration. The content of residual alkali is calculated from the consumption of the acid at the first inflection point. The hydroxide ion content determined in this way, however, gives an overestimated value due to the acid/base properties of the organic and inorganic substances present in black liquor. In this standard, therefore, the measured value is corrected in order to obtain the true value of the residual alkali content.

The total sulfur content in black liquors can be determined according to SCAN-N 35:96. In this method, a black liquor sample is transferred to a filter paper and burned with oxygen in a Schöninger flask in the presence of hydrogen peroxide. The sulfur is converted into sulfate, which is determined either by potentiometric titration or by ion chromatography.

TAPPI and CPPA standards also provide methods for determining kraft black liquor constituents. TAPPI T 625 cm-85 describes procedures for determining the density or specific gravity, sulfated ash and organic matter, sodium sulfite and sodium thiosulfate, sodium sulfate, total alkali, active alkali and silica, iron, aluminum, sulfur, calcium, magnesium, and sodium in black liquor. CPPA Standard J.15.P presents methods for determining the suspended solids content, total solids, ignition residue, total sodium, total sulfur, sodium sulfide, polysulfide sulfur, thiosulfate plus sulfite, thiosulfate, active alkali, effective alkali, sodium hydroxide, sodium chloride, total calcium, and thermal value.

2 Sulfite cooking liquors

Sulfite cooking liquors can be analyzed for:

- sulfur dioxide by iodimetric titration
- titratable acidity or alkalinity by titration.

The iodimetric titration is based on the following reactions:

$$HSO_3^- + I_2 + H_2O \rightarrow SO_4^{2-} + 2I^- + 3H^+ \quad (1)$$

$$SO_3^{2-} + I_2 + H_2O \rightarrow SO_4^{2-} + 2I^- + 2H^+ \quad (2)$$

I_2 or IO_3^- solutions can be used for the titration. In the latter case, I^- is added to the sample and the following reaction takes place:

$$IO_3^- + 5I^- + 6H^+ \rightarrow 3I_2 + 3H_2O \quad (3)$$

i.e. I_2 is formed in the solution and the titration takes place as described above. Starch is used for indication of the end point of the titration, when the cooking liquor no longer consumes iodide.

2.1 Normal acid bisulfite cooking liquor

Reagents:

Sodium hydroxide (NaOH), 0.1 N

Potassium iodate (KIO_3), 0.1 N

KI solution 50 g/L

Starch solution

Methyl red indicator

Sodium thiosulfate 0.1 N

Procedure

Add 50 mL distilled water to a 250 mL Erlenmeyer flask. Pipet a 2.0 mL specimen of the cooking liquor into the flask, keeping the tip of the pipet below the surface of the liquid. Add 10 mL KI solution and 2 mL starch solution.

Titrate the content of the flask with KIO_3 to a permanent blue color. Avoid shaking the flask until the blue color begins to appear. An approximate value of the KIO_3 consumption is obtained in this way, but some oxidation of sulfite takes place. The titration is therefore repeated so that the specimen is added to a solution that contains almost the whole required amout of KIO_3.

Add 50 mL of distilled water, 10 mL KI-solution and 2 mL starch solution to the emptied and washed Erlenmeyer flask. Add KIO_3 solution to an amount of 2 mL less than the consumption in the first titration. Add 2 mL cooking liquor with a pipet. Titrate with KIO_3 solution to a light blue color. The total KIO_3-consumption is marked a.

Discharge the blue color with not more than one drop of thiosulfate and add 5 drops of methyl red indicator. Titrate with 0.1 N NaOH to a light red color (pH = 7). The NaOH consumption is marked b.

Calculations

Total SO_2, %	$1.60 \times a \times N_1$
Combined SO_2, %	$1.60 \times (a \times N_1 - b \times N_2)$
CaO, %	$1.40 \times (a \times N_1 - b \times N_2)$
MgO, %	$1.01 \times (a \times N_1 - b \times N_2)$
Na_2O, %	$1.55 \times (a \times N_1 - b \times N_2)$
$(NH_4)_2O$, %	$1.30 \times (a \times N_1 - b \times N_2)$

where a is the consumption of N_1 normal KIO_3 solution, mL
b the consumption of N_2 normal NaOH solution, mL

2.2 Other bisulfite cooking liquors

Reagents

Sodium hydroxide (NaOH), 0.1 N

Hydrochloric acid (HCl), 0.1 N

Iodine (I_2), 0.1 N

Starch solution

Procedure

Add 50 mL distilled water to a beaker. Pipet a 2.0 mL specimen of the cooking liquor into the beaker, keeping the tip of the pipet below the surface of the liquid.

Titrate to pH 4.3 with a pH meter (use HCl if the pH of the specimen is above 4.3 and NaOH if the pH is below 4.3. The consumption of NaOH or HCl is marked c. Add 5 mL starch solution and titrate immediately with 0.1 N I_2 solution to a light blue color. The consumption of I_2 solution is marked d.

Calculations

Liquors with pH below 4.3

Total SO_2,%	$1.6 \times d \times N_1$
True free SO_2,%	$3.2 \times c \times N_2$
Combined SO_2,%	$0.8 \times (d \times N_1 - 2c \times N_2)$
Na_2O,%	$0.775 \times (d \times N_1 - 2c \times N_2)$
MgO,%	$0.505 \times (d \times N_1 - 2c \times N_2)$

Liquors with pH above 4.3

Total SO_2,%	$1.6 \times d \times N_1$
Combined SO_2,%	$0.8 \times (d \times N_1 - 2c \times N_2)$
Na_2O,%	$0.775 \times (d \times N_1 - 2c \times N_2)$
MgO,%	$0.505 \times (d \times N_1 - 2c \times N_2)$.

where c is the consumption of N_1 normal NaOH or HCl solution, mL
d the consumption of N_2 normal I_2 solution, mL

2.3 Related methods

Related methods are described in TAPPI T 604 cm-85, "Sulfur dioxide in sulfite cooking liquor" and in CPPA Standard J.13P, "Analysis of sulfite cooking liquors".

2.4 Spent cooking liquor

Spent sulfite cooking liquors can be analyzed for some of the constituents according to the following SCAN standards:

SCAN-N 1:61 "Dry matter in sulfite spent liquor"

SCAN-N 16:66 "Solid matter in sulfite spent liquor"

SCAN-N 17:66 "Nitrogen in sulfite spent liquor"

SCAN-N 18:66 "Phosphorous in sulfite spent liquor"

SCAN-N 19:66 "Sulfate in sulfite spent liquor"

SCAN-N 20:66 "Residual and releasable sulfur dioxide in sulfite spent liquor"

3 Washing losses

On a routine basis, the efficiency of washing is traditionally monitored based on the amount of salt cake (sodium sulfate) remaining with the washed pulp. Water-soluble organic solids are not necessarily extracted in washing to the same extent as sodium. For this reason, it has become common to determine also some properties that give indications of the amount of dissolved organic solids.

3.1 Saltcake losses

Various procedures can be used for determining Na^+ for expressing losses or system efficiency. Often the soda loss is undefined whether it is total soda loss determined by ashing or so-called washable soda determined on a squeezed sample. Because the amount of sorbed (bound) Na^+ may vary significantly between different pulp types, the ratio between washable and total Na^+ losses also varies. In determination of washable sodium losses, the washing procedure, i.e. the amount of washing water also affects the amount of sodium that is extracted.

One proposed method[1] to obtain a correct value for the total sodium content would be to determine the washable content from a liquor sample squeezed from the pulp and then, when the sorption coefficients of the pulp in question are known, calculate the total washing loss.

In the method described in SCAN-C 30:73 ("Sodium content of wet pulp") the pulp is acidified with hydrochloric acid under specified conditions to liberate the bound sodium. The principle is that the pulp sample is diluted to a pulp concentration of about 2% and acidified with hydrochloric acid. The sodium content of a sample of the liquid phase is determined by atomic absorption spectrophotometry or flame photometry. The dry matter content of the pulp is determined gravimetrically and the sodium content is calculated and expressed in kilograms of sodium sulfate per t of o.d. pulp.

Recommended sampling techniques including sample sizes and sampling frequency for use in mills with various pulp-washing systems are described in an appendix to SCAN-C 30:73.

3.2 Chemical oxygen demand (COD)

The standard SCAN-CM 45:91 describes a procedure for determining water-soluble organic material retained in wet kraft pulp. The standard is intended for estimating the efficiency of a washing operation or for measuring the amount of water-soluble organic material carried to a subsequent section of the mill or discharged to the environment. It is designed primarily for wet unbleached pulps produced by the kraft process. The water-soluble organic material is expressed in units of chemical oxygen demand (COD) in oxidation with dichromate. Similar methods for determining the COD are described in ISO 6060 and in CPPA Standard H.3.

The COD value does not unambiguously describe the amount of dissolved organic material, since the composition of the organic material also has some effect. The ratio between COD and dissolved organic material therefore may vary between processes, wood species and the type of filtrate. COD/dissolved organic material ratios of 1.2–1.3 (kg/t:kg/t) have been reported for washed unbleached pine and birch pulps[2].

3.3 Total, fixed, and volatile dissolved solids

More direct values for the amount of dissolved solids in liquor extracted from washed pulp or in washing filtrates can be determined by measuring the total, fixed, and volatile dissolved solids. Various standards describe procedures for such measurements:

- SCAN-W 3:68 "Non-volatile matter in waste waters"
- TAPPI T 656 cm-83 "Measuring, sampling and analyzing waste waters"
- CPPA Standard H.1 "Determination of solids content of pulp and paper mill effluents"

Despite some minor differences, these methods are based on the same principles:

- The total dissolved solids or the dry matter content is the ratio of the mass of the dried residue to the original volume of the sample after drying at 105°C at specified conditions.
- The fixed dissolved solids or the residue on ignition is defined as the ratio of the mass of the ignited dry matter to the original volume of the sample, the dry matter having been ignited at 550–925°C (depending of the standard). The residue on ignition expresses the concentration of inorganic dissolved material in the liquor sample.
- The volatile dissolved solids or the loss on ignition is the difference between the total and fixed dissolved solids (or between the dry matter content and the residue on ignition). The loss on ignition is an estimate of the concentration of dissolved organic material in the liquor sample.

One limitation of these methods is that some organic compounds, such as acetic acid and alcohols are volatilized during drying at 105°C. The methods thus tend to give somewhat low values for the volatile dissolved solids.

According to the standard, the concentrations are given as mg/L. If the liquor sample is extracted from the pulp without dilution, the concentrations can also be expressed in relation to the pulp, i.e., as g/kg o.d. pulp or kg/t pulp. For example, if the liquor sample is extracted from a pulp sample of 10% consistency (1 kg pulp/9 L liquor) and the loss on ignition is 3 200 mg/L one obtains:

$$\frac{3200 mg/L}{1 kg/(9L)} = 28800 mg/kg = 29 g/kg\ pulp = 29 kg/t\ pulp.$$

4 Bleaching liquors

4.1 Concentration of bleaching chemicals

Some standards are available for determining the concentration of bleaching chemicals:

- CPPA Standard J.16P, "Analysis of peroxides"

- CPPA Standard J.22P, "Analysis of chlorine solutions, hypochlorite bleach liquors, and spent bleaching liquors"
- TAPPI T 700 om-93, "Analysis of bleaching liquors by suppressed ion chromatography".

In addition to these methods, the following information strives to present basic methods for determining the concentrations of the chemicals most commonly used for bleaching today.

4.1.1 Chlorine

Principle

The concentration is determined iodimetrically; chlorine oxidizes iodide to iodine in an acid solution:

$$Cl_2 + 2I^- \rightarrow I_2 + 2Cl^- \tag{4}$$

The liberated iodine is titrated with thiosulfate:

$$I_2 + 2S_2O_3^{2-} \rightarrow S_4O_6^{2-} + 2I^- \tag{5}$$

The chlorine content can be calculated from the consumption of thiosulfate.

Reagents

Potassium iodide solution (KI), 10%

Sodium thiosulfate solution ($Na_2S_2O_3$), 0.1 N

Starch solution, 5 g/L

Procedure

Add 15 mL of chlorine water to 10 mL of KI solution in a titration flask, keeping the tip of the pipet below the surface of the liquid. Titrate the liberated iodide immediately with sodium thiosulfate using starch as an indicator. The end point should remain colorless for 30 s.

Calculations

$$c = \frac{E \times n \times V}{a} = \frac{35.453 \times 0.1 \times V}{15} = 0.236 \times V \tag{6}$$

where c is the concentration of chlorine, g/L
E the equivalent weight of chlorine
n the normality of the thiosulfate solution
a the sample titrated, mL
V the thiosulfate consumption, mL.

4.1.2 Hypochlorite

Principle

When a hypochlorite solution is neutralized with acid, the equilibrium moves to the right in the following equations:

$$NaOCl + H^{+} \leftrightarrow HOCl + Na^{+} \tag{7}$$

or

$$Ca(OCl)_2 + 2H^{+} \leftrightarrow 2HOCl + Ca^{2+} \tag{8}$$

When sufficient acid is added to take the pH below 2, the hypochlorous acid quantitatively converts to chlorine:

$$2HOCl + 2H^{+} \leftrightarrow Cl_2 + H_2O \tag{9}$$

The active chlorine of hypochlorite is determined iodimetrically in acid solution where the chlorine oxidizes the iodide into iodine. The liberated iodine is titrated with thiosulfate (Eqs. 4 and 5). The concentration of hypochlorite can be calculated from the consumption of thiosulfate.

Reagents

Sulfuric acid (H_2SO_4), 2 N

Potassium iodide solution (KI), 10%

Sodium thiosulfate solution ($Na_2S_2O_3$), 0.1 N

Starch solution, 5 g/L

Procedure

Add 5 mL of hypochlorite solution to 10 mL of KI solution in a titration flask and acidify the liquid immediately with sulfuric acid (pH < 2). Titrate the liberated iodide immediately with sodium thiosulfate using starch as an indicator. The end point should remain colorless for 30 s.

Calculations

$$c = \frac{E \times n \times V}{a} = \frac{35.453 \times 0.1 \times V}{5} = 0.709 \times V \tag{10}$$

where c is the concentration of hypochlorite, g active Cl/L
E the equivalent weight of chlorine
n the normality of the thiosulfate solution
a the sample titrated, mL
V the thiosulfate consumption, mL.

4.1.3 Chlorine dioxide

The active chlorine of chlorine dioxide solutions can be determined iodimetrically in acid solution. Chlorine dioxide solutions may contain elemental chlorine. A simple iodimeteric titration gives the total concentration of active chlorine in the solution. For separate determination of chlorine and chlorine dioxide, a somewhat different procedure has to be used.

Total active chlorine

Principle
Chlorine dioxide oxidizes iodide to iodine in acid solution:

$$ClO_2 + 5I^- + 4H^+ \rightarrow Cl^- + 2.5I_2 + 2H_2O \tag{11}$$

If the solution contains chlorine, it oxidizes iodide according to Eq. (4).

The liberated iodine is titrated with thiosulfate according to Eq. (5) and the total content of active chlorine can be calculated from the consumption of thiosulfate.

Reagents

Sulfuric acid (H_2SO_4)

Potassium iodide solution (KI), 10%

Sodium thiosulfate solution ($Na_2S_2O_3$), 0.1 N

Starch solution, 5 g/L

Procedure
Add 2 mL of chlorine dioxide solution to 10 mL of KI solution and 20 mL of weak sulfuric acid in a titration flask. Keep the tip of the pipet below the surface of the liquid. Titrate the liberated iodide immediately with the sodium thiosulfate solution using starch as an indicator. The end point should remain colorless for 30 s.

Calculations

$$c = \frac{E \times n \times V}{a} = \frac{35.453 \times 0.1 \times V}{2} = 1.773 \times V \tag{12}$$

where c is the total concentration of active chlorine, g/L
E the equivalent weight of chlorine
n the normality of the thiosulfate solution
a the sample titrated, mL
V the thiosulfate consumption, mL.

Chlorine dioxide and chlorine

Principle

The reactions with iodide depend on the pH of the solution. At pH 8 the following reactions take place:

$$Cl_2 + 2I^- \rightarrow I_2 + 2Cl^- \quad (13)$$

$$ClO_2 + I^- \rightarrow 0.5I_2 + ClO_2^- \quad (14)$$

At pH 2 the reaction is:

$$ClO_2^- + 4I^- + 4H^+ \rightarrow Cl^- + 2I_2 + 2H_2O \quad (15)$$

The liberated iodide is titrated with thiosulfate according to Eq. (5).

Reagents

Sulfuric acid (H_2SO_4), 2 N

Buffer solution, pH 8

Potassium iodide solution (KI), 10%

Sodium thiosulfate solution ($Na_2S_2O_3$), 0.1 N

Starch solution, 5 g/L

Distilled water

Procedure

Add 50 mL of buffer solution and 25 mL of KI solution to a titration flask. Pipet a 25 mL sample of the chlorine dioxide solution into the flask. Titrate the liberated iodide immediately with thiosulfate using starch as indicator. The end point should remain colorless for 30 s. The thiosulfate consumption is marked V_1.

Add 100 mL of distilled water and 10 mL of sulfuric acid. After 5 min repeat the thiosulfate titration. The consumption is marked V_2.

Calculations

The concentration of ClO_2 as active chlorine is calculated from the equation:

$$c_1 = \frac{E \times n \times 1.25V_2}{a} = 0.177 \times V_2 \quad (16)$$

The concentration of chlorine is calculated from the equation:

$$c_2 = \frac{E \times n \times (V_1 - 0.25V_2)}{a} = 0.142 \times (V_1 - 0.25V_2) \quad (17)$$

where c_1 is the concentration chlorine dioxide as active chlorine, g/L
c_2 the concentration of chlorine as active chlorine, g/L
E the equivalent weight of chlorine
n the normality of the thiosulfate solution
a the sample titrated, mL
V_1 the thiosulfate consumption in the first titration, mL
V_2 the thiosulfate consumption in the second titration, mL.

The concentration of chlorine dioxide expressed as g ClO_2/L is 0.3805 x c_1.

4.1.4 SO_2 water

Principle
Iodine oxidizes sulfur dioxide to sulfuric acid:

$$2SO_2 + 2I_2 + 4H_2O \rightarrow 4I^- + 2SO_4^{2-} + 8H^+ \qquad (18)$$

The residual iodine can be titrated with thiosulfate according to Eq. 5.

Reagents

Iodine solution (I_2), 0.1 N

Sodium thiosulfate solution ($Na_2S_2O_3$), 0.1 N

Starch solution, 5 g/L

Distilled water

Procedure
Add 75 mL of 0.1 N iodine solution and 100 mL of distilled water to a titration flask. Add a 5 mL sample of SO_2 water, keeping the tip of the pipet below the surface of the liquid. Titrate the liberated iodide with thiosulfate using starch as an indicator. The end point should remain colorless for 30 s.

Calculations

$$c = \frac{E \times n \times (75.0 - V)}{a} = 0.641 \times (75.0 - V) \qquad (19)$$

where c is the concentration of sulfur dioxide, g/L
E the equivalent weight of sulfur dioxide
n the normality of the thiosulfate solution
a the sample titrated, mL
V the thiosulfate consumption, mL.

4.1.5 Hydrogen peroxide

Principle

Hydrogen peroxide oxidizes iodide to iodine in an acid solution:

$$H_2O_2 + 2I^- + 2H^+ \rightarrow I_2 + 2H_2O \tag{20}$$

Because the reaction is relatively slow, ammonium molybdate is added as a catalyst. The liberated iodide is titrated with thiosulfate according to Eq. 5.

Reagents

Sulfuric acid (H_2SO_4), 2 N

Potassium iodide solution (KI), 10%

Sodium thiosulfate solution ($Na_2S_2O_3$), 0.1 N

Ammonium molybdate solution ($(NH_4)_2MoO_4$), 3%

Starch solution, 5 g/L

Distilled water

Procedure

Add 20 mL of 2 N sulfuric acid to 5 mL of peroxide solution in a titration flask. Then add the following reagents (in this order):10 mL of KI solution, 100 mL of distilled water and 5 drops of ammonium molybdate solution. Titrate the liberated iodide immediately with thiosulfate using starch as an indicator. The end point should remain colorless for 30 s.

Calculations

$$c = \frac{E \times n \times V}{a} = 0.340 \times V \tag{21}$$

where c is the concentration of hydrogen peroxide, g/L
E the equivalent weight of hydrogen peroxide
n the normality of the thiosulfate solution
a the sample titrated, mL
V the thiosulfate consumption, mL.

4.1.6 Ozone

Principle

Ozone can be determined from a gas flow by leading the gas through a gas washing bottle filled with KI solution for a certain time. The iodide liberated by ozone is then titrated with a thiosulfate solution, and the amount of ozone reacted can be calculated from the thiosulfate consumption.

Reagents

Potassium iodide solution (KI), 5%

Hydrochloric acid (HCl) 2 N

Sodium thiosulfate solution ($Na_2S_2O_3$), 0.1 N

Starch solution, 5 g/L

Procedure
Fill a gas washing bottle with about 150 mL of KI solution. Lead the ozone-containing gas through the bottle for a suitable time (2–3 min). Pour the KI solution into a titration flask and acidify it with 10 mL of 2 N hydrochloric acid. Titrate the liberated iodine with thiosulfate using starch as an indicator.

Calculations
Amount of ozone
The quantity of ozone in the gas sample is calculated from the equation:

$$m = n \times V \times 24 \qquad (22)$$

where m is the amount of ozone, mg
n the normality of the thiosulfate solution
V the thiosulfate consumption, mL.

Ozone flow
The ozone flow can be calculated from the amount of ozone and the sampling time:

$$m = \frac{m}{t} \qquad (23)$$

where M is the ozone flow, mg/h
t the sampling time, h.

Ozone concentration
The ozone concentration in the gas mixture can be calculated, if the total gas flow is known:

$$c = \frac{M}{Q} \qquad (24)$$

where c is the ozone concentration, mg/L
Q the gas flow, L/h.

4.1.7 Peracetic acid

Principle

Peracetic acid always contains some hydrogen peroxide that can be titrated with cerium (IV) sulfate in an acid solution in the presence of peracetic acid:

$$H_2O_2 + 2Ce^{4+} \rightarrow 2Ce^{3+} + O_2 + 2H^+ \tag{25}$$

The concentration of hydrogen peroxide can be calculated from the consumption of cerium sulfate.

In the second stage, peracetic acid is determined iodimetrically:

$$CH_2COOOH + 2I^- + 2H^+ \rightarrow CH_2COOOH + 2I_2 + H_2O \tag{26}$$

The liberated iodide is titrated with thiosulfate according to Eq. 5.

Reagents

Cerium ammonium sulfate solution ($Ce(SO_4)_2 \bullet 2(NH_4)_2\,SO_4 \bullet 4\,H_2O$), 0.1 N

Sulfuric acid (H_2SO_4), 5%

Potassium iodide solution (KI), 10%

Sodium thiosulfate solution ($Na_2S_2O_3$), 0.1 N

Ammonium molybdate solution ($(NH_4)_2MoO_4$), 3%

Ferroine solution

Starch solution, 5 g/L

Procedure

Add crushed ice to 150 mL of sulfuric acid so that the temperature of the solution remains below 10°C during the titration. When this temperature is reached, add a 10 mL sample of the peracetic acid solution. Add three drops of the ferroine solution and titrate the solution with cerium ammonium sulfate until the color changes from red to light blue. The consumption of cerium ammonium sulfate is marked V_1.

In the second stage, add 10 mL of KI solution and three drops of $(NH_4)_2MoO_4$ solution and titrate the liberated iodide with thiosulfate. The consumption is marked V_2.

Calculations

The concentration of hydrogen peroxide is calulated from the equation:

$$c_1 = \frac{E_1 \times n_1 \times V_1}{a} = 0.170 \times V_1 \tag{27}$$

and the concentration of peracetic acid from the equation:

$$c_2 = \frac{E_2 \times n_2 \times V_2}{a} = 0.380 \times V_2 \tag{28}$$

where c_1 is the concentration of hydrogen peroxide, g/L
c_2 the concentration of peracteic acid, g/L
E_1 the equivalent weight of hydrogen peroxide
E_2 the equivalent weight of peracetic acid
n_1 the normality of the cerium ammonium sulfate solution
n_2 the normality of the thiosulfate solution
a the sample titrated, mL
V_1 the consumption of cerium ammonium sulfate, mL
V_2 the consumption of thiosulfate, mL.

4.2 Spent bleaching liquors

4.2.1 Residual chemicals

Spent bleaching liquors are often analyzed for the concentration of unreacted, residual bleaching chemicals. Liquor samples for determining the residual chemicals should be extracted from the bleached pulp without dilution or washing. The concentration of residual chemicals is determined essentially in the same manner as described above for fresh chemicals. Because of the lower concentrations, however, larger liquor samples normally have to be used.

The quantity of residual chemicals can be set in relation to the amount of pulp with the aid of the equation:

$$A = c \times \frac{(100 - s)}{10 \times s} \qquad (29)$$

where A is the residual chemical,% on unbleached o.d. pulp
c the chemical concentration in the spent liquor sample, g/L
s pulp consistency, %.

4.2.2 Dissolved solids

Spent bleaching liquors can also be analyzed for the concentration of dissolved solids in the same way as brownstock washing filtrates. The chemical oxygen demand (COD) can be used as a general estimate of the amount of water-soluble organic material. More direct information is attained by determining the total dissolved solids (i.e. dry matter content), the fixed dissolved solids (i.e. the residue on ignition) and the volatile dissolved solids (i.e. the loss on ignition) with some of the methods mentioned in the Section "Washing losses".

In using the COD value as an estimate of the amount of water-soluble organic material, it should be observed that the composition of the material has some effect on the ratio between COD and organic material. The ratio is therefore not the same for material from different bleaching stages. An average ratio of 0.9 between COD and dissolved organic material has been reported for washing filtrates from ECF bleaching[3] and a ratio of 1.3 for TCF bleaching filtrates[4].

References

1. *Männistö, H. Massan pesu, in Puumassan valmistus (N.-E. Virkola, Ed), 2nd. ed. SPIY, TTA, Turku 1983, p. 687 (in Finnish).*

2. *Vehmaa, J., Leaching experiments with mill produced softwood and hardwood kraft pulp, M.Sc. Thesis (in Finnish), Dept. of Forest Products Technology, Helsinki University of Technology, Espoo Finland, 1993.*

3. *Süss, H.U. and Kronis, J.D., The correlation between COD and yield in chemical pulp bleaching, Breaking the Yield Barrier Symposium, TAPPI PRESS, Atlanta, 1998, p.153.*

4. *Lindholm, C.-A., Halinen, E., Henricson, K., ex al, Inverkan av löst organiskt material på MC-ozonblekning, NORDPAP DP2/21, SCAN FORSK-RAPPORT 665, 1996, 111 p.*

Conversion factors

To convert numerical values found in this book in the RECOMMENDED FORM, divide by the indicated number to obtain the values in CUSTOMARY UNITS. This table is an excerpt from TIS 0800-01 "Units of measurement and conversion factors." The complete document containing additional conversion factors and references to appropriate TAPPI Test Methods is available at no charge from TAPPI, Technology Park/Atlanta, P. O. Box 105113, Atlanta GA 30348-5113 (Telephone: +1 770 209-7303, 1-800-332-8686 in the United States, or 1-800-446-9431 in Canada).

Property	To convert values expressed in RECOMMENDED FORM	Divide by	To obtain values expressed In CUSTOMARY UNITS
Area	square centimeters [cm^2]	6.4516	square inches [in^2]
	square meters [m^2]	0.0929030	square feet [ft^2]
	square meters [m^2]	0.8361274	square yards [yd^2]
Breaking length	kilometers [km]	0.001	meters [m]
Density	kilograms per cubic meter [kg/m^3]	16.01846	pounds per cubic foot [lb/ft^3]
	kilograms per cubic meter [kg/m^3]	1000	grams per cubic centimeter [g/cm^3]
Energy	joules [J]	1.35582	foot pounds-force [ft • lbf]
	joules [J]	9.80665	meter kilogams-force [m • kgf]
	millijoules [mJ]	0.0980665	centimeter grams-force [cm • gf]
	kilojoules [kJ]	1.05506	British thermal units, Int. [Btu]
	megajoules [MJ]	2.68452	horsepower hours [hp • h]
	megajoules [MJ]	3.600	kilowatt hours [kW • h or kWh]
	kilojoules [kJ]	4.1868	kilocalories, Int. Table [kcal]
	joules [J]	1	meter newtons [m • N]
Force	newtons [N]	4.44822	pounds-force [lbf]
	newtons [N]	0.278014	ounces-force [ozf]
	newtons [N]	9.80665	kilograms-force [kgf]
	millinewtons [mN]	0.01	dynes [dynes]
Force per unit length	newtons per meter [N/m]	9.80665	grams-force per millimeter [gf/mm]
	kilonewtons per meter [kN/m]	0.1751268	pounds-force per inch [lbf/in]
Length	nanometers [nm]	0.1	angstroms [Å]
	micrometers [µm]	1	microns
	millimeters [mm]	0.0254	mils [mil or 0.001 in]
	millimeters [mm]	25.4	inches [in]
	meters [m]	0.3048	feet [ft]
	kilometers [km]	1.609	miles [mi]
Mass	grams [g]	28.3495	ounces [oz]
	kilograms [kg]	0.453592	pounds [lb]
	metric tons (tonne) [t] (= 1000 kg)	0.907185	tons (= 2000 lb)

Property	To convert values expressed in RECOMMENDED FORM	Divide by	To obtain values expressed In CUSTOMARY UNITS
Power	watts [W] watts [W] kilowatts [kW] watts [W]	1.35582 745.700 0.74570 735.499	foot pounds-force per second [ft • lbf/s] horsepower [hp] = 550 foot pounds-force per second horsepower [hp] metric horsepower
Pressure, stress, force per unit area	kilopascals [kPa] Pascals [Pa] kilopascals [kPa] kilopascals [kPa] kilopascals [kPa] kilopascals [kPa] kilopascals [kPa] megapascals [Mpa] Pascals [Pa] Pascals [Pa] kilopascals [kPa]	6.89477 47.8803 2.98898 0.24884 3.38638 3.37685 0.133322 0.101325 98.0665 1 100	pounds-force per square inch [lbf/in^2 or psi] pounds-force per square foot [lbf/ft^2] feet of water (39.2°F) [ft H_2O] inches of water (60°F) [in H_2O] inches of mercury (32°F) [in Hg] inches of mercury (60°F) [in Hg] millimeters of mercury (0°C) [mm Hg] atmospheres [atm] grams-force per square centimeter [gf/cm^2] newtons per square meter [N/m^2] bars [bar]
Speed	meters per second [m/s] millimeters per second [mm/s]	0.30480 5.080	feet per second [ft/s] feet per minute [ft/min or fpm]
Tear index	millinewton sq. meters per gram [mNm^2/g]	0.0980665	Tear factor computed as: 100 grams-force (gram per square meter) [$100gf/(g/m^2$]
Tensile energy absorption (TEA)	joules per square meter [J/m^2] joules per square meter [J/m^2] joules per square meter [J/m^2]	14.5939 175.127 9.80665	foot pounds-force per square foot [ft • lbf/ft^2] inch pounds-force per square inch [in • lfb/in^2] kilogram-force meters per square meter [kgf • m/m^2]
Torque	newton meters [Nm] newton meters [Nm] newton meters [Nm] micronewton meters [μNm]	0.11298 1.35582 0.0980665 0.1	pound-force inches [lbf • in] pound-force feet [lbf •ft] kilogram-force centimeters [kgf • cm] dyne centimeters [dyn • cm]
Viscosity, dynamic	Pascal seconds [Pas] millipascal seconds [mPas]	0.1 1	poise [P] centipoise [cP]
Viscosity, kinematic	square millimeters per second [mm^2/s]	1	centistokes [cSt]
Volume, fluid	milliliters [mL] liters [L]	29.5735 3.785412	ounces [oz] gallons [gal]
Volume, solid or fluid	cubic centimeters [cm^3] cubic meters [m^3] cubic meters [m^3] cubic millimeters [mm^3] cubic centimeters [cm^3] cubic decimeters [dm^3] cubic meters [m^3]	16.38706 0.0283169 0.764555 1 1 1 0.001	cubic inches [in^3] cubic feet [ft^3] cubic yards [yd^3] microliters [μL] milliliters [mL] liters [L] liters [L]

Index

C

D

M

N

O

P

R

S